Kelvin, Thermodynamics and the Natural World

Kelvin, Thermodynamics and the Natural World

Editors

M.W. Collins
Brunel University London, UK

R.C. Dougal
James Clerk Maxwell Foundation, Edinburgh, UK

C.S. König
Brunel University London, UK

I.S. Ruddock
Strathclyde University, Glasgow, UK

WIT PRESS Southampton, Boston

WIT Series on Design and Nature, Volume 10

Editors:

M.W. Collins
Brunel University London, UK

R.C. Dougal
James Clerk Maxwell Foundation, Edinburgh, UK

C.S. König
Brunel University London, UK

I.S. Ruddock
Strathclyde University, Glasgow, UK

Section opening images taken from a series entitled 'Glasgow' by Marianne Greated, 2014. Indian ink and acrylic on paper

Published by

WIT Press
Ashurst Lodge, Ashurst, Southampton, SO40 7AA, UK
Tel: 44 (0) 238 029 3223; Fax: 44 (0) 238 029 2853
E-Mail: witpress@witpress.com
http://www.witpress.com

For USA, Canada and Mexico

WIT Press International
25 Bridge Street, Billerica, MA 01821, USA
Tel: 978 667 5841; Fax: 978 667 7582
E-Mail: infousa@witpress.com
http://www.witpress.com

British Library Cataloguing-in-Publication Data

A Catalogue record for this book is available
from the British Library

ISBN: 978-1-84564-149-8
eISBN: 978-1-84564-387-4

Library of Congress Catalog Card Number: 2015942775

The texts of the papers in this volume were set individually by the authors or under their supervision.

Printed in Great Britain by Lightning Source UK.

Dedication

Professor Michael W Collins

21/7/1939 – 24/8/2014

It is with deep sadness we are paying tribute to Professor Michael W Collins, Lead Editor of this *Lord Kelvin* book, who passed away on 24 August 2014 at the age of 75.

Michael had an undergraduate degree in Engineering Science from St Peters College, Oxford, followed by a PhD and DSc from City University. There, he went through the academic ranks until becoming a Full Professor. In 2005, he joined Brunel University London as a visiting professor in Mechanical Engineering, but held previously also visiting professorships at the University of Modena & Reggio Emilia and the University of Ferrara in Italy, at London South Bank University and Fudan University, Shanghai, and the Shanghai University of Science & Technology. Besides being a charted engineer and a charted mathematician he was a fellow of eight professional institutions namely the Institution of Mechanical Engineers, Institution of Engineering & Technology, Institution of Chemical Engineers, the Nuclear Institute, the Institution of Mathematics & Its Applications, but also the Institution of Physics & Engineering in Medicine, the Linnean Society and the Royal Society of Arts. His fellowships clearly reflect the breadth of his interests and subject knowledge spanning from colour and the arts, biofluid and thermofluid mechanics, to history and chaos theory. In 1991, he became an Honorary Fellow of the Italian Union of Thermodynamics, *Unione Italiana Termofluidodynamica,* and was awarded the prestigious Knights Cross, Order of Merit, for his services to Polish Sciences. In 1992, he was awarded the Busk Prize from the Royal Aeronautical Society.

Michael published well over 300 papers and book chapters and acted as a guest editor on numerous projects. It is a great pity that Michael did not live long enough to see the completion of his favourite book, particularly as he relentlessly kept perfecting the material, even during periods of bad health, to ensure it being state-of-the-art and of relevance to the modern age. Michael was fascinated by Kelvin's multi-faceted personality as revealed by his life and work. Indeed, there are prominent parallels between Kelvin and Michael, both being dedicated scientists with an immense hunger for knowledge in a wide area of subjects while also being men of faith and spirituality. With WIT Press Michael had a long-standing working collaboration, leading to the publishing of numerous books and Special Issues in the areas of engineering and biomedical technology. He served on the editorial board for the *International Journal of Design & Nature and Ecodynamics* for many years. Michael was an inspiring researcher and gifted networker, but above all he was an extremely kind and gentle person and a true gentleman. He will be much missed as a colleague and collaborator, but above all as a friend.

Carola König & Carlos Brebbia

About the Editors

Richard C. Dougal joined the staff of the Physics Department at Edinburgh University in 1964 after having graduated BSc then PhD there. His research continued on topics in atomic and electron physics. A move to applied optics in 1983 coincided with the early years of the rapid growth of work on the properties and applications of liquid crystals. This accompanied the extension of his mainly classical physics teaching interests to supervision of undergraduate projects in optics. There followed an interest in the history of the early work on colour, which he continued after retiring in 2002, as part of the effort of the James Clerk Maxwell Foundation in Edinburgh, where he has been a Trustee since 2000. He is a Fellow of the Institute of Physics.

Carola S. König is a Research Fellow at the Institute of Environment, Health and Societies and an Associate Lecturer in the College of Engineering, Design and Physical Sciences, both at Brunel University London. She obtained the degree of Dipl-Ing in Mechanical Engineering from the University of Stuttgart, Germany in 1992 and a PhD from Brunel University in 1998. Her expertise is in Computational Fluid Mechanics involving fluid flow with and without heat transfer within a variety of multi-disciplinary environments (e.g. engineering interface with biology, medicine, chemistry, arts). As a Member of the Institution of Mechanical Engineers, she is currently Co-Chair of the Institution's Thermofluids Group and is the Scientific Secretary of the Micro and Nano Flows Conference series. She is also a Fellow of the Institute of the Higher Education Academy.

Ivan S. Ruddock obtained his BSc and PhD degrees from the Queen's University, Belfast and the University of London (Imperial College) respectively. During his postgraduate research on tuneable mode-locked lasers, he produced the shortest duration light pulses observed at that time. He then joined the academic staff of the University of Strathclyde's Department of Physics in 1977 where he conducted research in a wide range of topics including laser physics, single crystal fibres, nonlinear optics, and optical fibre sensors. After early retirement in 2010 as Reader in Physics, he has maintained links with academia via the co-leadership of HOPE (Horizons of Physics Education), a large academic network funded within the European Commission's Lifelong Learning programme. He is a Fellow of the Institute of Physics and the Higher Education Academy and is active in the public understanding of science through writing and speaking on significant historic personalities such as Lord Kelvin.

Contents

Foreword by The Rev. Dr James Collins xvii

Preface: In appreciation of Lord Kelvin xix
D.H. Saxon

Editorial introduction xxv
M.W. Collins, R.C. Dougal, C.S. König & I.S. Ruddock

Personal

Chapter 1: Lord Kelvin **3**
I.S. Ruddock

1 Introduction 3
2 The Thomson family 3
3 Young William 6
4 William Thomson, a Cambridge student 7
5 William Thomson, a professor 9
6 Academia 12
7 The Atlantic telegraph and knighthood 13
8 Lalla Rookh 17
9 Family life 18
10 Glasgow and Cambridge 20
11 Netherhall 22
12 Navigation 23
13 Politics and peerage 25
14 Jubilee 27
15 Retirement 28
16 Industry 29
17 Electricity generation 30
18 Beliefs and controversies 32
19 Health and death 34
20 Honours 35
21 Patents 37
22 Notes 41
References 43

Chapter 2: Kelvin and his world: a cultural overview **45**
M.W. Collins & I.S. Ruddock

1 A complex challenge 45
2 An intellectual colossus 46
3 Kelvin and the industrial world 54
4 The telegraph and Empire 55
5 Kelvin's reputation 60
References 61

Relationships

Chapter 3: James and William Thomson: the creation of thermodynamics **67**
A. Whitaker

1 James Thomson and the Belfast notebooks: the air engine 67
2 The Belfast notebooks: lowering of the freezing point of water 75
3 The Belfast notebooks: reconciling Carnot and Joule 80
4 The denouement 82
5 James Thomson and thermodynamics 88
Acknowledgements 90
References 90

Chapter 4: James Thomson, an engineer and scientist: the path to thermodynamics **93**
A. Whitaker

1 Introduction 93
2 A sketch of the life of James Thomson 96
3 James Thomson's achievements in engineering and science 105
4 James and William: early dilemmas and discussions 113
5 Ideas of heat in the first half of the nineteenth century; the caloric theory 116
6 The theory of Sadi Carnot and its reception by the Thomsons 120
7 The work of Joule, and the response of James and William Thomson 125
Acknowledgements 130
References 130

Chapter 5: Kelvin, Maxwell, Clausius and Tait: the correspondence of James Clerk Maxwell **135**
R.C. Dougal

1 Introduction 135
2 Saturn's rings and the Adams Prize: Maxwell and Kelvin 138
3 Molecular model for a gas: Maxwell and Clausius 138

4 Two laws of thermodynamics: Maxwell, Clausius, Kelvin and Tait 141
5 Conclusion 147
Appendix: Transcript of words on postcard of Fig. 1 148
Bibliography 148
References 149

Chapter 6: Sir William Thomson, Baron Kelvin of Largs and the Institution of Engineering and Technology **153**
S. Hale

1 Introduction 153
2 Lord Kelvin 155
3 The institution after Lord Kelvin 159
Conclusion 160
References 160

The Laws of Thermodynamics

Chapter 7: Engineering Thermodynamics and the Carnot Cycle **165**
M.W. Collins, J.A. Stasiek & J. Mikielewicz

1 Introduction: the Carnot Cycle 166
2 Timelines of characters 170
3 Watt's thermodynamic family tree and place in history 174
4 The p–v diagram: James Watt and the Indicator Diagram 182
5 The Carnot cycle: Carnot's great achievement 185
6 The Carnot cycle: diagrams of Clapeyron and Clausius 190
7 The Carnot cycle for a gas 193
8 From p–v to T–s: Kelvin's absolute scale of temperature 200
9 The 'representational' approach of J Willard Gibbs and the T–s diagram 202
10 The Carnot cycle for steam 205
11 The reversed Carnot cycle: refrigerators and heat pumps 208
12 Contributions to the Second Law of Thermodynamics: Carnot, Kelvin and Clausius 209
13 The perfect heat engine: is thermodynamics Platonic in character? 213
14 Conclusion 215
Acknowledgements 220
References 220

Chapter 8a: The first law of thermodynamics: Kelvin's relationship with Joule **225**
A.K. Wróblewski

1 Introduction 225
2 William Thomson's fascination with Carnot's theory of heat 225
3 Enter James Prescott Joule 227

4 Rudolf Clausius finds the solution 228
5 William Thomson's answer 229
6 Concluding remarks 229
References 229

Chapter 8b: The first law of thermodynamics: The Joule–Mayer Controversy 231
M.W. Collins

1 Introduction 231
2 Joule, Mayer and the First Law: their peer assessment problems 232
3 Joule, Mayer and the First Law: the X-club controversy 234
4 Conclusion: honour satisfied 235
References 235

Chapter 9: The emergence and evolution of the Second Law of Thermodynamics 237
D. Wilkie, R.C. Dougal, M.W. Collins & A. Whitaker

1 Introduction 237
2 Heat and temperature: conceptualised and applied 238
3 Towards the laws of thermodynamics 239
4 The Second Law of Thermodynamics 243
5 Thermodynamic properties and processes 248
6 Statistical thermodynamics 250
7 Later developments: Keenan 253
8 Myron Tribus's thermodynamics: an alternative understanding 255
9 Conclusions 256
References 257

Chapter 10: The teaching of thermodynamics today 261
R. Bartlett & J. Kubie

1 Introduction 261
2 The technological context of Thomson's 1851 paper 261
3 The 'scientific' context of Thomson's 1851 paper 262
4 Enter Thomson 263
5 Death of the Caloric 264
6 The second law of thermodynamics 265
7 Flawed lawmaking 266
8 Returning to Thomson's formulation 267
References 270

Chapter 11: Entropy as thermal charge: an application of bond graphs inspired by Carnot and his cycle **273**
J. Thoma & G. Mocellin

1 Sadi Carnot ... 273
2 Bond graphs ... 274
3 Systemics ... 280
4 Icons for thermofluid machines ... 289
5 The Carnot cycle ... 291
References ... 296

Chapter 12: Thermodynamic entropy and temperature rigorously defined without heuristic use of the concepts of heat and empirical temperature **297**
E. Zanchini & G.P. Beretta

1 Introduction ... 297
2 Aims and structure of the present treatment ... 300
3 Basic definitions ... 302
4 Definition of energy for a closed system ... 310
5 Definition of thermodynamic entropy for a closed system ... 311
6 Fundamental relation, temperature, and Gibbs relation (for a closed system) ... 322
7 Proofs of Clausius and Carathéodory statements of the Second Law and of the Zeroth Law ... 327
8 Conclusions ... 332
References ... 333

Thermodynamics in the wider context of science

Chapter 13: William Thomson (Lord Kelvin) and thermoelectricity **337**
L. Anatychuk

1 Introduction ... 337
2 On the events that preceded Thomson's discoveries in thermoelectricity ... 338
3 Fundamental thermoelectric Thomson relations ... 343
4 Experimental verification of Thomson thermoelectric effect ... 344
5 On the effect of thermodynamic reversibility on Thomson relation ... 345
6 Generalization of Thomson relations pursuant to thermodynamics of irreversible processes ... 346
7 First Thomson relation novel form ... 347
8 On practical applications of Thomson relations ... 350
9 Thomson transversal EMF: anisotropic thermoelements ... 352
References ... 360

Chapter 14: Kelvin and the age of the earth **363**
J.A.T. Halls & D. Wilkie

1 Early estimates ... 363
2 The development of uniformitarianism ... 365
3 Lord Kelvin ... 368
4 Geological time ... 370
5 Growing opposition ... 385
6 Radioactivity as an additional source of heat ... 388
Conclusions ... 389
References ... 390

Postscript

Chapter 15: Kelvin in the twenty-first century **397**
D.H. Saxon

1 'The Kelvin Problem': space-filling foam ... 397
2 Kelvin Waves and the El Nino effect ... 399
3 The Stirling engine ... 401
4 The Stirling engine in outer space ... 402
5 Atmospheric electricity ... 402
6 From the nineteenth to the twenty-first century ... 404
References ... 404

Conclusion

Chapter 16: Honoured by banknotes **407**
I.S. Ruddock, A. Whitaker & M.W. Collins

1 Introduction: the UK, its political history and its banknotes ... 407
2 Kelvin and the University of Glasgow ... 410
3 Queen's University of Belfast ... 416
4 Kelvin and the Giant's Causeway Tramway ... 420
5 King's College, Aberdeen (University of Aberdeen) ... 421
6 Boulton and Watt ... 422
7 Conclusion ... 423
References ... 423

Appendices

Thomson, Brunel and the Atlantic cables of 1865 and 1866 425
I.S. Ruddock

The Model Stirling Engine 435
C.S. König & C.A. Greated

Maxwell's Demon 437
M.W. Collins, R.C. Dougal & C.S. König

Universities of the Heat Engine 443
C.S. König, M.W. Collins & I.S. Ruddock

Honoured by monuments 449
I.S. Ruddock

Lord Kelvin and the Royal Society of Edinburgh 453
C.A. Greated

Foreword

My father invested a good deal of what turned out to be his final years working on this book. I enjoyed (and occasionally endured) numerous conversations with him about Lord Kelvin, his achievements and significance. My father identified with the great man on two levels: Kelvin, although a polymath, was primarily an engineer, an exponent of applied science; Kelvin was also a Christian, a disciple of Jesus Christ. There was an opportunity here then to focus on two things that were of great value to my father, the practical achievements of science and the value of Christian faith.

The achievements of applied science are often not celebrated and remembered in quite the same way as those of more theoretical scientists. Newton, Darwin and Einstein are rightly household names, but there is much less recognition for the likes of Kelvin, those who applied the theories of science for the benefit of the masses. This troubled my father, an engineer himself, and he hoped that this book might amount to a small corrective to this trend. He believed that Kelvin's brilliance was the equal of any of the great theoretical scientists. In particular, my father was much impressed by the laying of the transatlantic cable, perhaps Kelvin's greatest feat, and felt dismayed that Kelvin was not more celebrated for such an attainment.

At an even deeper level, my father was grieved by the persistent perception that 'science' stands in tension with religious faith in general and Christianity, in particular. This widespread perception, more common in my experience among the general population than among scientists, has been fuelled in recent years by the immoderate writings of the so-called 'new atheists'. My father was a moderate person with a bent towards building consensus, so, by character and also by conviction, their writings saddened him even as he recognised and respected the scientific achievements of, for example, Professor Richard Dawkins.

Sadly, immoderate analyses of the relationship between science and religious faith tend to dominate the media and public consciousness. Reactionary believers, who reject wholesale any advances in scientific understanding that they fear may threaten their religious beliefs, play *yin* to militant atheism's *yang*. Neither seems much inclined to recognise or properly engage the *via media* of a thoughtful approach to both faith and science which recognises the value of both in their proper spheres. In editing this volume, my father certainly hoped that the example of a great man of science who was also a man of faith would not be lost on the readers.

My father passed away in August 2014 after a brief and unpleasant struggle with cancer. His wisdom, intelligence, generosity and humility made a great impression on all who knew him. Although his energy gradually deteriorated in the months leading up to his death, his passion for science and his ardent Christian faith sustained his enthusiasm for this life and his hope for the next to the very end. It was a healthy and life-giving combination that I trust will also be discovered in the life of the man honoured in the pages of this book.

A final word of thanks to the publisher, and to Dr Carola König, a colleague and friend of my father who took joint responsibility to get this book completed and published promptly. I am very grateful for your attendance, along with so many colleagues, at my father's funeral and your efforts on behalf of my father. I also remain delighted, although not surprised, by the affection and esteem you hold him in.

The Rev. Dr James Collins
1 December 2014

Preface

In appreciation of Lord Kelvin

D.H. Saxon
Kelvin Professor of Physics, 1990–2008, University of Glasgow, United Kingdom.

Sir William Thomson was ennobled as Baron Kelvin of Largs in 1892, the first scientist to be so honoured. He took the name Kelvin from the river that curls around the foot of the University of Glasgow's spectacular new (1870) campus – the University dates from 1451 but escaped in 1870 from its earlier home after that was blighted by the Industrial Revolution. Largs was the seaside town where he built his mansion.

Born in Belfast in 1824, he was both the youngest (aged 10) and later the oldest (at 75) student in the history of Glasgow University, and in between was, for 53 years, a professor of natural philosophy, making a fortune en route. At his death in 1907, he was the chancellor of the University, and past president of the Royal Societies of both London and Edinburgh. He was the dominant figure in science in the second half of the nineteenth century. His biographer Alexander Russell wrote [1], 'His work lives and will continue to live. To him it has been given to make history which will live so long as intelligent man survives on earth. As the years roll on our indebtedness to him increases.' He is buried in Westminster Abbey next to Isaac Newton, and a nave window there also pays tribute to him as 'Engineer, Natural Philosopher.'

Yet today his achievements are largely unheralded and he is principally remembered for a reactionary approach to the new Physics of the last decade of his life, epitomised by his crisp quotation, 'X-rays will prove to be a hoax'. But did he actually say this, possibly in a weaker form including the word 'probably'? Sources disagree. Thompson says that Kelvin was initially (correctly) sceptical [2], but within weeks was congratulating Roentgen on his discovery and had his own hand X-rayed (Fig. 1).

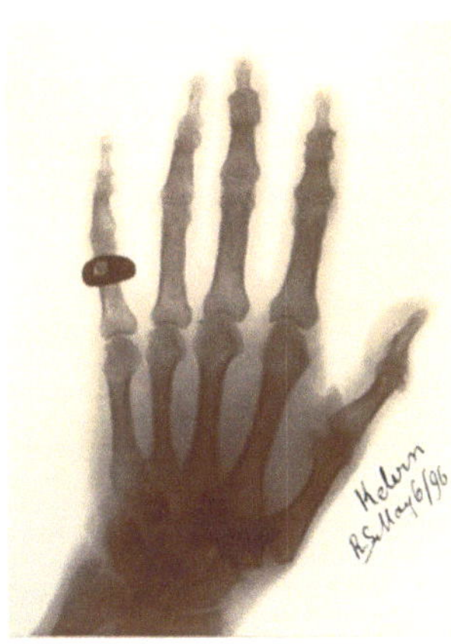

Figure 1: Lord Kelvin's hand X-rayed at the Royal Society.

To appreciate his importance we need to put ourselves back into the frame of 1841, when at the age of 16 he published his first paper. In this he rescued a forgotten paper by Fourier and showed that the well-known instability of Fourier series' descriptions of phenomena close to sharp boundaries did not prevent the use of partial differential equations to study the flow of heat. Thus, the classical physics of continuous media was born. Crucial to this approach, Kelvin took to heart Fourier's message that the behaviour of heat can be described mathematically without knowing what heat is. This was a good thing at the time – before the mechanical equivalent of heat was known (Joule, 1843) and incompatible and incomplete theories of heat, or 'caloric' to use the then current name, were in circulation. He first introduced the word 'thermodynamics' in 1848.

His creativity was by then reaching an astounding peak. The year 1849 saw papers by him on hydrodynamics, electricity, the heating and cooling of buildings by circulating air as well as on the second law of thermodynamics. In 1850, he wrote on steam engines, geometry, electrolysis, the magnetic properties of crystals and regelation and produced a landmark paper on magnetism. This developed the whole field from two contrasting starting points, magnetic poles and current loops, introducing the vector fields $\boldsymbol{B}$ and $\boldsymbol{H}$. He recast physics in terms of energy, being the first to write down the stored energies in an inductor and a capacitor, $\frac{1}{2}LI^2$ and $\frac{1}{2}CV^2$. Electrical capacity, energy, kinetic energy, absolute temperature, simple harmonic motion, magnetic permeability, magnetic susceptibility, stress over strain, bulk modulus, circulation, vorticity and vortex sheet were among the terms he introduced into physics.

The 'Standard Model' of Physics in 1851 faced the same crisis as in 2012 – the problem of impossibilities. A theory of elementary particles without a Higgs Boson, or something very like it, gives rise to impossible predictions at energies within the eventual reach of the CERN Large Hadron Collider. So the Standard Model predicts that an object similar to that discovered in 2012 must exist. The discovery of a new boson at a particular mass removes the impossible predictions and verifies the Standard Model, providing an answer to the question, 'what is mass?'

In 1851, the target question was 'what is energy?' The steam engine had thrown the issue of energy and how to harness it into strong focus. The theory, based on careful measurements, was known to be incomplete as it appeared to allow the construction of limitless energy sources from 'perpetual motion' machines, in striking contrast to reality.

The key discovery that overcame this paradox was made by Kelvin's brother James. It was rather technical in nature: that the temperature at which ice melts falls when external pressure is applied – indeed this is why ice skates work. Suddenly, the contradictions vanished and the first and second laws of thermodynamics could be written down, the absolute scale of temperature defined and the absolute zero (the [unattainable] minimum temperature) determined. Physics was rewritten in terms of energy. Terms such as 'kinetic' and 'potential' energy were introduced by Kelvin and Tait who co-authored the first textbook on Physics.

The second law of thermodynamics can be stated in various ways that turn out to be logically equivalent statements. Kelvin's (1851) formulation is 'It is impossible, by means of an inanimate agency, to derive mechanical effect from any portion of matter by cooling it below the temperature of the coldest of the surrounding objects.' Starkly stated, it has stood the test of time and the efforts of many would-be inventors. Indeed, it has been argued that everything we know in science may be wrong, except the first and second laws of thermodynamics which must be right. In the simplest layman's paraphrase, these laws state (1) you cannot get something for nothing (2) you cannot even break even. These are pillars of experience indeed.

In 1858, the first transatlantic telegraph cable was laid – an epic undertaking with huge practical difficulties, Kelvin was on board as part of the team. No one knew how deep the ocean was. Attempts to measure this always resulted in the breakage of the cable reel. Kelvin solved this. Steamships were needed to follow a straight path. Only by 1866 was a single ship (Brunel's Great Eastern) able to carry the weight of cable needed.

The 1858 cable did not last very long. The insulation could not withstand the high voltages used in an attempt to detect the signal. A parliamentary enquiry vindicated Kelvin's more systematic approach and put him in charge of the 1866 attempt. Along the way, he invented first the mirror galvanometer and then the inkjet printer (the 'siphon recorder') as the receiving and recording mechanism requiring the least signal power. (The only moving part is the ink.)

The achievement of the transatlantic cable shrank the world more than anything before or since. It has the same logical structure as email – digitally encoded, packet switched, and seeking the least crowded route. Never again could a battle be fought (as in New Orleans in 1812) weeks after the peace treaty had been signed but neither General had been informed. The information age had arrived.

It earned Kelvin his knighthood and set him in a path to riches and invention after invention. Perhaps, his most famous was the compass for iron ships, needed to give a true reading of magnetic North in spite of the permanent magnetic moment of the ship and the additional magnetic moment induced in the hull by its orientation in the earth's field. At the same time, he worked to refine the accuracy of electrical units of measurement, ultimately chairing the committee that named the amp, volt, ohm, etc. as we know them today. He pioneered electric light, and in 1881 made his home in Glasgow the first house in the world to be fully lit by electricity, using 106 lamps. International students flocked to work in his laboratory – Gerhard Phillips came from Eindhoven to learn about electric light and Aikitu Tanakadate, the father of Japanese geosciences (important in a country with earthquakes) who was the first person to use Roman letters for Japanese and the first person in Japan to own a typewriter. This arose from the practical need to encode the Japanese language to send it by telegraph.

Famously, Kelvin was the first to apply mathematics to the question of the ages of the Earth and of the Sun, looking at all known energy sources and how long they could sustain the Sun's heat output and how long the Earth's mountains could survive against erosion. The most promising energy source for the sun was gravitational shrinkage. This is ruled out since it is recorded that Alexander the Great saw a solar eclipse on the day he crossed the river Oxus. This set an upper limit to the Sun's size at that date. No one in the nineteenth century knew about either radioactivity in the Earth's core, which drives plate tectonics, or nuclear fusion which powers the sun. As a result, his ages were hugely underestimated, causing uproar amongst evolutionists. His methods were right: the quantitative approach was both new and correct, but the data were incomplete.

To learn something of his character let him speak for himself, via a few quotations attributed to him. Some are ones practising scientists of today would immediately take to heart:

> When you are face to face with a difficulty, you are up against a discovery.
> The more you understand what is wrong with a figure, the more valuable that figure becomes.
> To measure is to know.
> If you cannot measure it, you cannot improve it. [3]
> When you can measure what you are speaking about, and express it in numbers, you know something about it; but when you cannot measure it, when you cannot express it in numbers, your knowledge is of a meagre and unsatisfactory kind. [4]

Other quotations are less twenty-first century:

> I am never content until I have constructed a mechanical model of the subject I am studying. If I succeed in making one, I understand, otherwise I do not. [5]

And finally a challenging pair:

> There is nothing new to be discovered in physics now. All that remains is more and more precise measurement. [6]
> One word characterises the most strenuous of the efforts for the advancement of science that I have made perseveringly during fifty-five years. That word is failure. I know no more of electric and magnetic forces or of the relation between aether, electricity and ponderable matter, or of chemical affinity than I knew and tried to teach to my students of natural philosophy fifty years ago in my first session as professor. [7]

This last quotation, made at his fiftieth anniversary celebrations as professor, would have shocked his audience. Yet it harks back to his beginnings, in the joy he felt in learning that you can describe how heat flows in mathematics without ever knowing what heat is. This is the triumph and tragedy of classical physics. It is brilliant phenomenology, but falls short of explaining how the structure of atoms forces the behaviour of the material. Kelvin's era was closing. It would be a task for others to elucidate the new phenomena – the electron, X-rays, radioactivity, the photoelectric effect and relativity that came crowding into his last decade. We should honour him for what he achieved, and for his yearning for what remained to be achieved. He felt himself to be like Isaac Newton in old age, playing with the odd attractive pebble on a beach while an ocean of truth lay undiscovered before him, and he felt the frustration of this.

His life was characterised by boundless energy that would lead to 661 scientific papers and to 75 patents. He introduced experimental laboratory training as part of the degree course. A modern comparator could be Nobel laureate Richard Feynman. Both were brilliant mathematical physicists and problem solvers. Both made major contributions to many areas of physics, had a wide interest in other areas and were inspirational teachers, and both were key to the success of a government enquiry into a failure. If we are to look for one thing by which to remember him, scientists might pick the absolute scale of temperature as his crowning achievement: members of the public might opt for the telegraph cable across the ocean. Russell's eulogy quoted above is not inappropriate for the scale of his achievements.

We allude in Chapter 17 to a handful of Kelvin-originated topics of immediate relevance in the twenty-first century: space-filling foam and its realisation at the Beijing Olympics 2008 Water cube, the use of Stirling engines to generate electricity from energy sources other than fossil fuels, and the Kelvin wave phenomenon that provides the conduit for the El Nino effect, plus the possibility of predicting earthquakes.

Three selected postage stamps attest to his world-wide impact: from Guinea-Bissau – a vital link in the telegraph network, from Japan – a portrait of Kelvin's eminent visitor Tanakadate, illustrating the Japanese typewriter, and from Serbia – Kelvin's tidal predictor, that remained in use in many countries for almost a century (Figs 2–4).

Kelvin, by being featured in Fig. 2, is the only contributor to the Atlantic Cable project honoured personally in postage stamps .The Cable Ship Monarch is also shown, the fourth of its name. It was built in 1946, was the largest cable ship of its day, firstly being used to repair the cable networks neglected during World War II. The 1866 cable project ship, *Great Eastern* of Brunel, has featured on a number of stamps between 1955 and 1989 [8].

The Meiji Government of the Japanese Emperor Mutsuhito (meiji tenno, *'enlightened sovereign', 1852–1912*) modernised all aspects of the Japanese state, including the import of Western science and technology. Tanakadate, one of the most eminent scientists involved in this, worked with Kelvin in Glasgow 1888–1890, and was so deeply affected that back in Japan he was nicknamed Lord Kelvin by his students. The stamp in Fig. 3 shows the typewriter – vital to encoding Japanese for telegraphy – and illustrates his role as the founder of earth sciences in Japan.

Figure 2: Guinea Bissau, 1983, postal contribution to the *World Year of Communications.*

Figure 3: Japan, 2002, commemorative stamp in the *Cultural Leaders* Series for Tanakadate Aikitsu (1856–1952). (Reproduced with permission of Akira Matuura, Ninohe City Civic Centre Museum & Japan Post Co. Ltd.)

Figure 4: Serbia: 2007, part of three-stamp series commemorating centenaries of international scientists: Kelvin, Dmitry Mendeleev (1834–1907, Russian chemist, periodic table) and Giuseppe Occhialini (1907–1993, Italian physicist, cosmic rays, space research).

The fine portrait of Kelvin in Fig. 4 has as background a tide-predicting machine. This is a 10-component harmonic analyser used for calculating the tides for nearly a century. This particular machine was conceived by Kelvin in 1872–1873, designed by him and collaborators, and is now in the Science Museum in London [9,10].

REFERENCES

This preface is based on material which first appeared in physicsworld.com on 17 December 2007 at: http://physicsworld.com/cws/article/indepth/2007/dec/17/in-praise-of-lord-kelvin

[1] Russell, A., *Lord Kelvin, His Life and Work*, Jack: London, 1912.
[2] Thompson, S., The *Life of William Thomson, Baron Kelvin of Largs* (2 vols), Macmillan: London, 1910.
[3] The four quotations listed above appear in a number of variations. It is not easy to single out one as the first occurrence.
[4] Lord Kelvin, Lectures on electrical units of measurement. *Popular Lectures*, **1**, p. 73, 1887.
[5] Lord Kelvin, Baltimore lectures on molecular dynamics and the wave theory of light, Cambridge University Press: London, 1904.
[6] Disputed – reportedly at the British Association for the advancement of science meeting in 1900. No direct citations.
[7] Gray, A., Lord Kelvin's Jubilee, *Nature*, **54**, pp. 173–181, 1896.
[8] Acknowledgement: assistance from Bill Glover, contributor to *History of the Atlantic Cable and Undersea Communications*, at www.atlantic-cable.com.
[9] Thomson, W., The tide gauge, tidal harmonic analyser and tide predictor, *Institution of Civil Engineers*, **65,** pp. 2–25, 1889.
[10] Acknowledgement: assistance from Emilija Marinkov, Embassy of the Republic of Serbia, London.

Editorial introduction

M.W. Collins[1], R.C. Dougal[2], C.S. König[1] & I.S. Ruddock[3]
[1]*College of Engineering, Design & Physical Sciences, Brunel University London, United Kingdom.*
[2]*James Clerk Maxwell Foundation, Edinburgh, United Kingdom.*
[3]*Department of Physics, University of Strathclyde, Glasgow, United Kingdom.*

William Thomson, Lord Kelvin of Largs (1824–1907) was a prodigiously multi-talented figure in the nineteenth century world of science and engineering. As a teenager, he displayed a combination of the highest mathematical ability with an exceptional level of work, producing a stream of refereed publications by his early 20s. In addition, his brilliant inventiveness resulted in over 60 patented physics-based devices and he excelled as an entrepreneurial engineer, playing a leading part in the great Atlantic cable venture. His lifelong work at Glasgow University included being a pioneer of university engineering. He was an iconic figure in an age of iconic figures. His contemporaries included the British Faraday, Maxwell, Joule and Stokes, together with the Germans Clausius and Helmholtz, and he followed in the footsteps of the Frenchmen Fourier, Regnault and the unique Carnot. He was first among equals, exercising leadership for British science and engineering and achieving Presidency of both the Royal Society (of London) and the Royal Society of Edinburgh. His personal characteristics were transparent and definite: he had a robust Christian faith and worldview and contributed substantially to the Irish aspect of British politics. Finally, his national status was second to none: he became a peer and Privy Councillor, was an inaugural member of the Order of Merit and was buried in Westminster Abbey close to Isaac Newton.

Kelvin lived and worked in the heady British world of high Empire and dominant international influence. His submarine cable success substantially strengthened the effectiveness of Imperial communication and management. Of the highest import in world history was the industrial revolution and Kelvin actively contributed to the two latter stages of its wider definition. Thirdly, it was a time of common Christian faith, which was an integral part of the thinking of Kelvin and his energy peers. This latter contrasted strongly with the agnosticism of Charles Darwin, T.S. Huxley and the X-club, throwing light on the nineteenth century conflict between the British energy and biology specialists.

Kelvin's science especially focused on thermodynamics. In addition to *energy,* he introduced the concept and definition of the *absolute temperature scale.* His Second Law statement, based on heat engines, contrasted with Clausius's focus on the transfer of heat. With Joule, he discovered the Joule–Thomson effect, demonstrating the fall in temperature in adiabatic expansion of a gas. He fully appreciated the probabilistic nature of the Second Law, and he named and brought to birth Maxwell's Demon.

From very early in his career, he was interested in the thermal processes of the sun and earth and correctly posited finite ages for them. Unfortunately, his accuracy of calculation and lack of knowledge of radioactivity led him to make lifelong quantitative estimates of these ages that were too low. The ongoing disagreement between the energy specialists and the biologists, mainly stemming from this, developed into a struggle for general scientific authority, which was never resolved. Much more positively, Kelvin extended his thermodynamics to give a worldview of nature, the cosmos and their history. He correctly identified the source of cosmic energy to lie in gravitational fields, and Prigogine's nonlinear thermodynamics and dissipative structures of today resonate with Kelvin's promulgation of a universal tendency to dissipation of energy.

A key issue in any study of Kelvin is that of his reputation. The supreme national respect he was accorded during his lifetime, has fallen to only a moderate level a century later. This strongly contrasts with the current reputations of his colleague, the physicist James Clerk Maxwell, and with the discoverer of natural selection, Charles Darwin. A number of reasons account for this inconsistency and they are discussed in the same spirit as in the latest multi-author volume on Kelvin stemming from the centenary celebrations of his death in 2007. Here, we stress the following: the lack of his being recognised as a superb engineer as well as a mathematician and physicist, the especial example being the Atlantic cable project, the cultural values prized so highly by the Victorians being interpreted ambiguously until quite recently, and in the overshadowing of his sound cosmological insights by his being wrong-footed over the question of the age of the earth.

Finally, the above analyses are re-interpreted in terms of current thinking. *Each aspect is now of considerable significance or interest*, whether by the way of political and scientific action or by lively debate. Kelvin was proud to have introduced the term *energy* into the scientific vocabulary. The present-day energy crisis and global warming are proof that its significance has endured. It also suggests how the old tensions between biology and thermodynamics have been replaced with a new understanding of the essential place of thermodynamics within biology and, more widely, of the close relationship between biology and engineering. A twenty-first century Kelvin would be involved in all these considerations. They also enable a fresh assessment to be made of the debt of appreciation that we owe Kelvin today.

Personal

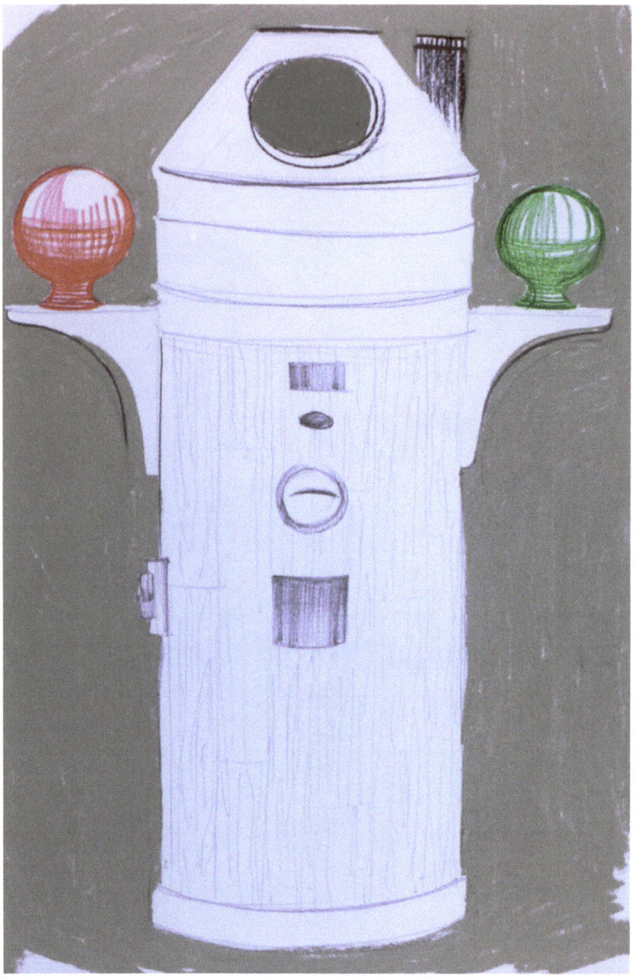

Compass binnacle. (Marianne Greated 2014, pastel and pencil on paper)

Chapter 1

Lord Kelvin

I.S. Ruddock
Department of Physics, University of Strathclyde, United Kingdom.

1 Introduction

In December 1910, film fans in Belfast found a new venue had been added to their list of choices – the Kelvin Picture Palace. Its name was neither based on any of the musical or mythological nouns or adjectives beloved of theatrical impresarios nor obviously named after a Belfast location or personality. The new cinema had been converted from two large terrace houses on College Square East and was to continue in use for more than 60 years under a range of identities. One of the houses, No. 17 College Square East, was the home of James and Margaret Thomson and their large family, and was where their second son, William, was born on 26th June 1824. William Thomson, now better known as Lord Kelvin from his elevation to the peerage in 1892, rose from these rather non-descript beginnings to the pinnacle of the scientific establishment and became a national figure in engineering, commerce and politics. Two years after the opening of the Kelvin Picture Palace, one of the first permanent monuments to his memory was erected, a statue in Belfast's Botanic Gardens, which now helps to perpetuate the title in the city of his birth despite him having spent most of his adult life as a professor of natural philosophy in the University of Glasgow.

2 The Thomson family

To fully understand some aspects of William Thomson's personality and the background to his eventual career, it is essential to look at his origins, in particular, the role and influence of his father, James Thomson. Notwithstanding his undisputed talent, as a young man William Thomson benefitted by his father being in precisely the right place at the right time to help promote his case. Seldom has this been so apparent, but with James Thomson's death coming soon after William Thomson's appointment to the post in Glasgow, he was free to establish his own reputation. His eventual success completely vindicated the faith that the University of Glasgow had in him when he was elected to the chair at such a young age.

The Thomson family into which James Thomson was born on 13 November 1786 had a small farm in the heart of County Down at Annaghmore near Spa, Ballynahinch. Although Spa is now one of the many commuter villages near Belfast, for more than two hundred years it was a popular destination for tourists in search of a cure for rheumatism at its sulphur well. Near the road junction around which the principal buildings were located, there was the church and The Regency with its baths and treatment rooms and where refreshments were available. In its grounds were two small pump houses where water could be sampled. The only other notable feature in Spa was a maze, traces of which can still be seen along with one of the pump houses. The Thomson ancestors had come to County Down from Lowland Scotland to escape the persecution that was associated with the suppression of the Covenanters in the seventeen century. One hundred years later, the late eighteenth century was also a turbulent period in the history of the British Isles. The success of the French Revolution had spread the ideas of liberty and civil rights far beyond France. In Scotland, Thomas Muir of Huntershill, near Glasgow, was a leading campaigner for parliamentary reform, while in Ireland, the United Irishmen, Presbyterians and Roman Catholics, were also demanding parliamentary reform and the removal of the Crown from Ireland. Their campaign culminated in a number of skirmishes including one at Ballynahinch in 1798, which James Thomson witnessed as a 12-year-old boy.

James Thomson was one of a family of three sons and three daughters. Like many neighbouring farms in this locality, the Thomsons grew flax for the linen industry, but James's future did not lie in agriculture. Having displayed the ability to educate himself beyond what his father had taught him at home, he attended a small school run by their Presbyterian minister and was eventually promoted to assistant teacher. Inspired by his mentor, James Thomson planned to follow in his footsteps but at that time there was no college in the north of Ireland. It was usual for prospective students of the ministry, law or medicine to study in Dublin or Glasgow, and it was the latter that James chose in 1810 as a Presbyterian. The session at Glasgow College lasted for 6 months from November to May, but during the other 6 months, he supplemented his income by continuing teaching in Spa. James Thomson and fellow students made their crossings of the North Channel by whatever means was available ranging from a fast customs cutter to a very slow lime smack. On one occasion, they were rewarded by the sight of Henry Bell's Comet on the Clyde between Glasgow and Greenock.

Immediately after graduating with his M.A. in 1814, James was back in Ulster having secured the post of the teacher of mathematics in the school department of the newly opened Belfast Academical Institution (Fig. 1), taking classes in arithmetic, geography and book-keeping. This establishment was a combination of school and college, and is still open as a boy's grammar

Figure 1: The Royal Belfast Academical Institution of 1814. (Photograph by I.S. Ruddock.)

school, the Royal Belfast Academical Institution, having received a Royal warrant from William IV in 1831. With his childhood home at Spa being too far from Belfast to commute from, he lived opposite the college in a new house, No. 17 College Square East, Figs 2 and 3. He also owned its neighbour, No. 18, but rented it to the McAdam family[a] [1]. Here, he settled down to a career of teaching and writing textbooks, with 1 on arithmetic running to 72 editions and still on sale in the early twentieth century. Within a year, he was appointed as a professor of mathematics in the college department that supplied further education in Belfast until the opening in 1845 of Queen's College of the Queen's University of Ireland, now the Queen's University of Belfast.

Figure 2: College Square East, Belfast, in 1886 facing the Royal Belfast Academical Institution. The house occupied by the Thomson family was in the lighter shaded part of the terrace on the right. (Photograph by R.J. French.)

Figure 3: Nos 17 and 18, College Square East, Belfast, as the Kelvin Picture Palace in 1911. James Thomson owned both houses, but lived in the left half, No. 17, and rented No. 18 to the McAdam family. (Photographer unknown.)

3 Young William

In 1817, James Thomson married Margaret Gardner, the daughter of a Glasgow merchant. They had met in Belfast the previous year at the home of Dr William Drennan, a leading United Irishman and a founder of the Belfast Academical Institution [2] while she was visiting a cousin. Their family of six – Elizabeth, Anna, James, William, John, Margaret and Robert – was educated by James Thomson at home. Elizabeth and Anna both married and had children, two of whom were involved with William's subsequent industrial activities, James became the distinguished engineer whose life and career are described elsewhere in this volume, John died of fever while studying medicine and Robert emigrated to Australia where he entered business and settled. The happy marriage lasted for only 13 years. Mrs Thomson died in 1830 at 40 years of age leaving James Thomson with the six young children, two of them infants. Margaret Thomson is buried in Belfast's historic Clifton Street cemetery along with her daughter Margaret who died the following year aged 4. Their graves are near some leading United Irishmen including the aforementioned Drennan [3]. Now for the second time in his life, an opportunity presented itself, on this occasion in the form of the University of Glasgow's chair of mathematics which became vacant in 1832. His application was successful and he was duly elected to the position later in the same year. He could now make a fresh start with his family in a city that was already familiar to him and where his wife's family were prominent citizens.

As in Belfast, James and William and their brothers and sisters were mainly educated at home – this time in Professors' Court beside Glasgow College on its old site on High Street. Then in October 1834, the two eldest boys entered the university at the age of just 12 and 10 years, respectively. They also attended classes in the nearby Glasgow Mechanics' Institution, a body which seceded in 1823 from Anderson's Institute of 1796, a forerunner of the University of Strathclyde [4]. Although there were already boys of 14 attending classes as was the case in the Belfast Academical Institution, James and William were both exceptionally young even for that time. In the history of the University of Glasgow, William holds the record for the youngest undergraduate. Attending a university at this age was not a daunting experience for the two boys. They lived next door and had regularly sat in their father's lectures and sometimes had to be restrained from answering the questions he would throw out to the students. They would also try and repeat at home some of the practical demonstrations they had witnessed being given by the other professors. Both their careers as students at Glasgow were illustrious. Their names appeared regularly in the prize lists for natural philosophy, astronomy, mathematics and even logic and Greek.

During a family holiday in the summer of 1839, William became fluent in French when the four brothers stayed behind in Paris for 2 months while James Thomson took his elder daughters to tour Switzerland and Germany. He put this newly acquired skill to good use by reading Laplace's Mécanique Céleste in preparation for 'On the Figure of the Earth', his prize essay which won a Glasgow University medal in 1840. The family's informal language studies continued the following year when they spent the summer in Frankfurt and Baden. This time, William's holiday reading was *Fourier's Théorie Analytique de la Chaleur* but he had to smuggle it with him in his trunk. His father had forbidden all work and William (Fig. 4) had to slip down to the cellar of their lodgings to read a few pages from time to time. Inspired by it, he wrote his first paper 'On Fourier's Expansions of Functions in Trigonometrical Series' during his last year at Glasgow and had it published in the *Cambridge Mathematical Journal*, under the signature 'P.Q.R.'. The paper was a defence of the details of the Fourier series against apparent discrepancies found by Philip Kelland.[b]

Figure 4: William Thomson in 1840 aged 16 drawn by his sister Elizabeth King. (Illustration from A.G. King, *Kelvin the Man*, Hodder and Stoughton Limited: London, 1925 [20].)

At the end of his sixth year in April 1841, William left the university but without a degree. He had sat and passed the qualifying examinations but decided not to formally graduate. His plans for the next stage in his career might have been prejudiced if he already had.

4 William Thomson: Cambridge student

On 6 April 1841, William Thomson matriculated as an undergraduate in Peterhouse College, University of Cambridge. He had still not reached his 17th birthday. Other notable Peterhouse students from Scotland in those days included Peter Guthrie Tait[c] (Fig. 5) and for a short time, James Clerk Maxwell. Due to its reputation for mathematical training, the college was an appropriate choice of James Thomson for his son's education. Associated with the college was William Hopkins, a graduate of 1829 and later President of the British Association, and the most famous tutor as measured by the success of his students in the annual Wrangler competition.

Life for William at Cambridge was a balanced mixture of academic work, attending social events, learning to play the cornopean – a brass instrument related to the cornet – and the physical exertions of swimming, walking, running and rowing. He was a founder member and president of the Cambridge University Music Society, played in its orchestra and maintained a love of music for the rest of his life. Letters exchanged with his father in Glasgow frequently included bills or requests for financial support. Some were for books that his father suggested he could do without in the meantime, but one that caused considerable difficulties was his shared purchase of a boat. Although his father had to be persuaded to contribute to its cost, his generosity was eventually repaid by William's winning of the Colquhoun silver sculls in November 1843; his daily 2 hours rowing with like-minded students who formed 'the Fleet' having clearly been a worthwhile investment. Although he also rowed in the Peterhouse boat in the annual head of the river event, he and his friends tended to steer clear of the less academic students who made up the rest of the crew.

Figure 5: Peterhouse College undergraduates in the 1850s. Peter Guthrie Tait is first on the left. (Photograph from C.G. Knott, *Life and Scientific Work of Peter Guthrie Tait*, Cambridge University Press: Cambridge, 1910.)

Near the end of his first academic year in the spring of 1842, William started work on two more manuscripts for the *Cambridge Mathematical Journal*. These were to be the first of more than 600 papers, letters and comments he would eventually publish in the following 65 years. They were completed during his summer holidays with his family in the west of Scotland and were closely linked to his holiday reading of *Fourier's Théorie Analytique de la Chaleur* 2 years previously as both concerned heat conduction. 'On the linear motion of heat' presented the solution of a differential equation describing heat in an infinite solid, whereas 'On the uniform motion of heat in homogeneous solid bodies and its connection with the mathematical theory of electricity' pursued the analogy between isothermals in heat conduction in solid bodies and the equipotential surfaces associated with charged bodies. Before the second paper was published, William felt obliged to add references to two mathematicians whose related work he was unaware of, but years later he discovered that he had unknowingly provided alternative solutions to theorems of Gauss and Green. The return to Cambridge in October 1842 marked the start of serious study for William. From then until the final examinations in 1845, his mathematical training was supervised by Hopkins. The aim was to become first in the mathematics Tripos and be acclaimed Senior Wrangler. This competition was effectively discontinued in 1909 when the graduation list was first published alphabetically and not in rank order, but in its day being Senior Wrangler was recognised as the mark of someone who would achieve greatness. The champion was even profiled in *The Times*, but to score the highest marks required stamina. Years of practice in solving problems resulted in the ability to write out familiar solutions as quickly as possible under pressure.

Towards the end of 1842, events started to unfold that would establish William's future career. In December, James Thomson first mentioned in a letter to his son that the current holder of the chair of natural philosophy in Glasgow, Dr William Meikleham, was ill and a successor might soon be needed. By the following spring, he had formed the view that despite youth, his own son might be the most suitable candidate. Fortunately for the relatively immature William, the incapacitated Meikleham stayed in his post with his duties performed by other members of staff, while James supplied a running commentary on Meikleham's health.

In early January 1845, William's cohort sat their Tripos examinations. To the surprise and disappointment of friends and family he was runner-up to Stephen Parkinson, a St John's student. Nevertheless, a couple of weeks later, William had his revenge when he beat Parkinson in the Smith's Prize competition, which was deemed to be a greater test of understanding and application of mathematics than the Wranglership. Also, being Second Wrangler, Thomson was in good company. Among other such 'failures' can be listed James Clerk Maxwell and J.J. Thomson, while William Bragg was Third Wrangler and Bertrand Russell, seventh.

Immediately after this round of examinations, William went to Paris to gain some experience of experimental work in Henri Regnault's[d] laboratory at the College de France and then engaged on investigations into the density of gases. During the 4 months he spent in Paris, William also sought out some of the most influential French scientists including Jean-Baptiste Biot, Joseph Liouville, Jacques Charles François Sturm, Claude Pouillet and Léon Foucault. Soon after returning to Cambridge as a mathematics tutor and coach, he was elected in June 1845 to a college fellowship. Worth £200 per annum and including accommodation, this was a valuable position but one he had to eventually resign in 1852 on his marriage; following revision of the college's statutes in 1860, he was re-elected as a Fellow in 1872. William could have spent the rest of his working life at Cambridge like Hopkins. He was 21 years of age and had already published about 15 papers.

5 William Thomson: professor

Dr Meikleham eventually died in May 1846 and his young successor, William Thomson was elected as the professor of natural philosophy in the following September, Fig. 6, left. In the intervening months, his father had continued to press on him the opportunity that was within his grasp and the need to obtain the right type of testimonial. Although he witnessed his son's initial success, he did not live to see much of his subsequent career in Glasgow nor the national and international reputation he would eventually make for himself. James Thomson (Fig. 6, right) died in January 1849 of the cholera epidemic, which struck Glasgow that winter and he was buried behind Glasgow cathedral in the Thomson family plot in the city's necropolis.

Glasgow College, and the 'department', such as it was, would have been unrecognisable to a 21st century academic. Most of the buildings on Glasgow's High Street dated from the seventeenth century and formed just two quadrangles containing class rooms, one for each subject, Fig. 7; it did not move to its present location in the west end of the city until 1870. The professor was effectively the department and was expected to do all the teaching. It would be a long time before the modern type of university department fully evolved with professors, readers, senior lecturers, lecturers, research fellows and assistants, postgraduate and undergraduate students, technicians and secretaries. But by the time he did eventually retire from the faculty of the university of Glasgow University in 1899, Thomson had gone some way in creating it.

Thomson was expected to meet his classes for 2 hours per day, 5 days per week for a session of 6 months with the rest of the time free for his own investigations. But despite having

Figure 6: Left: William Thomson aged 22 in 1846, the year he became professor in Glasgow. (Photograph from A. Gray, *Lord Kelvin: An Account of His Scientific Life and Work*, Dent: London, 1908.) Right: James Thomson in 1848 aged 62 drawn by his daughter Elizabeth King. (Illustration from A.G. King, *Kelvin the Man*, Hodder and Stoughton Limited: London, 1925 [20].)

Figure 7: Glasgow College in the mid-nineteenth century. Left: inner quadrangle looking east; Glasgow's High Street is through the arch beyond the outer quadrangle (Photograph from S.P. Thompson, *Life of Lord Kelvins*, Macmillan and Co. Limited: London, 1910 [7].) Right: northwest corner of the inner quadrangle; the natural philosophy rooms are in the northern range. (Photograph from D. Murray, *Lord Kelvin as Professor in the Old College of Glasgow*, MacLehose, Jackson and Co.: Glasgow, 1924.)

to devote a considerable amount of time to teaching, he would not have been bothered by department and faculty meetings, examination boards, resit examinations, appraisal schemes, research assessment exercises, teaching quality audits, personal reviews, personal development planning, risk assessments and the other myriad activities that currently occupy an academic's life. Natural philosophy was then purely a Faculty of Arts subject on the same footing as logic and moral philosophy since the Faculty of Science did not come into being until 1893. The arts curriculum was intended only for cultural purposes without any intention of providing training for commercial reward [5].

Students from this period have described [6] how Thomson, dispensing with his academic gown, spoke rapidly but clearly and distinctly, took great pains to be accurate and precise in his language, and was eager, alert and animated in his manner. There was no syllabus for the guidance of the professor, and even if there was, it was unlikely that Thomson would have followed it. He taught natural philosophy without notes, making it clear that he was not simply dictating a textbook on the subject, nor preparing his students for the regurgitation of facts. Instead, he engaged students in a scientific dialogue, working out problems on the blackboard as though he also was not sure where the trail would lead. For their part, the students rarely lost interest or caused a disturbance. To preserve a record of his teaching, he had the bright idea of offering a prize each year for the best set of notes taken in his class, which in turn established the syllabus that was to be taught the following year and so on. He would give students a complete set of notes beforehand with the instruction that they were to study it and be prepared to be examined on it; a style of teaching now sometimes referred to as student-centred learning.

When Thomson was appointed to the professorship, there was also no such thing as an undergraduate teaching laboratory, let alone research facilities. For most of his time in the Old College, he had the services of a private assistant, Mr Donald McFarlane, a former student and teacher, and a laboratory attendant, Mr Tatlock. There was an apparatus room containing equipment for use in lecture demonstrations, and it was heavily used because Thomson believed in this form of teaching. In some ways, he was a pioneer of the modern approach in which computer simulations are used to dynamically illustrate theory in place of sterile two-dimensional diagrams. As a creative scientist, he also needed apparatus and manpower to obtain data to corroborate his theoretical work on mechanics, heat, fluids and electricity. Manpower was no problem as many students in the natural philosophy class were only too happy to volunteer, but what about obtaining hardware as this was long before the establishment of research councils and their grants? A compromise was reached with the college in which the faculty supplied a vacated wine cellar and some funds which Thomson supplemented from his own resources. Thus, one of the very first university research laboratories was established, which is known to the student assistants in the mid-nineteenth century as the Physical Laboratory. Thomson would explain what the experiment was and how it was to be carried out to McFarlane, who then distributed the tasks to the group of students. Thomson would pop in from time to time to see the state of progress. This was easy to do as he lived less than 100 yards away in No. 2 Professors' Court, at the southeast corner of the Duke Street and High Street crossing, on the eastern side of the city centre.

The first lecture each day would be opened by Thomson saying the Third Collect of the Morning Service of the Church of England, followed sometimes by a prayer. The attendance was then taken by the busy McFarlane, who had already prepared the demonstration apparatus and laid out enough pieces of chalk for the following hour's work. On two successive days, Thomson once reached for a fresh piece of chalk only to discover that the usually conscientious McFarlane had forgotten to have it ready. Thomson ordered him to have 100 pieces ready for the next day, all of which were to be counted. The faithful assistant was determined that he would not let his master down again. He found a hundred sticks, and carefully and neatly set them out every couple of inches on the 16-feet long window ledge of the lecture theatre. At the next lecture,

Thomson carefully counted the chalk and finding exactly 100, congratulated the beaming McFarlane to the applause of the audience.

The highlight of the lecture programme was always the demonstration of conservation of momentum using a ballistic pendulum. McFarlane would hand Thomson a big-game hunting rifle loaded with a one ounce bullet. With his ears plugged with cotton wool, Thomson would then aim at the lead bob of the pendulum and fire. The deflection would be measured by McFarlane and written on the blackboard for substitution into the equation to give the speed of the bullet. Meanwhile, the students in the front row would collectively breathe a sigh of relief that once again their professor's aim was good and that the bullet had not missed the bob and ricocheted off its cast iron casing.

Tests, apart from the final examinations, were mainly oral and usually conducted along with exercises during the second session of the day. A few students were selected in turn, but when a question was not answered, it was then passed to another student and so on until a satisfactory response was obtained. Selection of the unlucky candidates was done by using a box of name cards distributed in three compartments. Those still to be tested were in the Purgatory section, those who had successfully answered their questions went to Heaven, while the failures were sent to Hell to be tested again once Purgatory had been exhausted. As in university classes of any era, candidates would do anything to avoid being quizzed in public – sometimes they pretended to be absent or would surreptitiously attempt to move their cards from Purgatory to Heaven only to be thwarted by the ever vigilant McFarlane. Homework was also set but, by the 1880s when the natural philosophy class had increased to over 200 students, McFarlane was also tasked with marking assignments in addition to his existing duties in the laboratory and lecture theatre.

6 Academia

During the 60 years that William Thomson was associated with the University of Glasgow, he contributed to and developed and consolidated much of what is now known as classical physics. His revolutionary ideas on thermodynamics and its applications to understanding work and energy and the debate on the age of the Earth are discussed elsewhere in this volume. His entrepreneurship and industrial activities relating to the electric telegraph and navigational aids are related later in this chapter; but there was much more – mechanics, electricity and magnetism, electrical units, hydrodynamics, optics and, to a lesser extent, the structure of matter. Some idea of the scale of his interests can be gleaned from the titles of his publications [7]. Notwithstanding that many of them are in the form of letters and comments instead of the modern style of research paper, this was a prodigious output considering that he was the sole author in most cases.

Thomson enjoyed some productive collaborations that lasted for many years. While that with James Joule[e] is perhaps the best known, the longest lasting was with Peter Guthrie Tait. During the early 1860s, just after Tait had been elected to the chair of natural philosophy in Edinburgh, they started on 'Treatise on Natural Philosophy' later known as Thomson and Tait or T and T′. The intention was that the whole spectrum of physics would be covered in four volumes and be the first proper comprehensive textbook of the subject. As it transpired, only the first volume on dynamics was completed, but nevertheless it was a success from first publication in 1867 and eventually ran to multiple editions. Used by generations of physics students into the twentieth century, it helped define physics as an academic discipline.

7 The Atlantic telegraph and knighthood

By the middle of the nineteenth century, the electric telegraph had become a practical means of communication. Progress from the bench to commercial exploitation was mainly due to the work of Charles Wheatstone,[f] William Cooke and Alexander Bain[g] in the United Kingdom and the artist and entrepreneur Samuel Morse in the United States. As with many technologies involving different components and approaches and competing patents, there was also a forgotten cast of personalities stretching back to the end of the eighteenth century. Plans for international communication developed rapidly in the 10 years following the adoption of Wheatstone and Cooke's needle telegraph by first the London and North Western Railway in 1837 and then the Great Western Railway in 1839. In 1850, the first experimental cable was laid across the English Channel followed in 1851 by the first working cables between the United Kingdom and France, and in 1852 and 1853 between the United Kingdom and Ireland and the Netherlands, respectively. Once mainland Europe had been reached, progress in wiring up convenient stretches of the British Empire could start in earnest using overland routes. While these endeavours were underway, some entrepreneurs were thinking about the feasibility of reaching North America. With Ireland already connected to Great Britain, and Newfoundland to Cape Breton and thence to New York, all that needed was the intervening Atlantic cable, but compared to constructing and laying under the North Sea or the English Channel, the new challenges were orders of magnitude greater.

In the 1850s, the behaviour of electrical pulses in transmission lines was not well understood. Some experimental evidence existed to show that in long cables, temporal dispersion would greatly restrict the speed at which signalling could be accomplished. This is analogous to the broadening suffered by ultrashort light pulses in optical fibres, which determines the bit rate of modern telecommunication networks. William Thomson was engaged on the problem and showed that the broadening depended on the capacitance and resistance, both of which are proportional to the length of the cable, thus giving rise to what was known at the time as the 'law of squares'. The issue of insulation remained, but this would be compounded in an Atlantic cable by the greater voltage required to ensure a detectable signal was received at the other end. Even the conductivity of the copper varied widely depending on the supplier while debate raged among the proponents of the cable on the optimum diameter of the core to increase the signalling speed. Finally, there was also the practical hurdle to be overcome of finding a ship large enough to transport the enormous volume and weight of thousands of miles of cable and then feed it over the stern without it snapping in a rough sea.

By 1856, the Atlantic Telegraph Company had been formed with Thomson as a director representing Scottish shareholders. If anybody had been designed with ultimate failure as its aim, then this was it. The main participants included Cyrus Field, a New York businessman as a general manager, Sir Charles Bright, an electrical engineer and later politician as an engineer-in-chief and the person who was to become Thomson's nemesis in this enterprise, E.O. Wildman Whitehouse.[h] Following a career as a medical practitioner, Whitehouse had become interested in telegraphy and was now appointed as an 'electrician' to oversee the laying of the cable. He claimed to have disproved the law of squares, but Thomson disputed his experimental method, specifically the choice of battery and detectors used.

With almost indecent haste, the first complete cable was delivered in July 1857. It came in 1100 sections each of 2 miles in length, but had been manufactured by two different companies who unfortunately added the armour sheath with opposite twists. The short lengths were assembled into eight segments of 300 miles each for ease of loading on board ship. Later the same month, the first attempt to lay the cable began with the British and United States governments each lending a warship – HMS Agamemnon and USS Niagara. The plan was for the Niagara to start laying from Valentia Island, Ireland, and then for the Agamemnon to take over and lay the second half to Newfoundland. Whitehouse excused himself from the voyage on account of illness and was replaced at the last moment by Thomson. Not unsurprisingly, after Niagara had been successfully laid over 300 miles, mishandling of the cable by the crew caused it to break and be lost in deep water.

With an additional 700 miles of cable and an improved paying-out apparatus, the company set about the task afresh the following June. Thomson was once more aboard the Agamemnon in place of Whitehouse to supervise the testing of the cable as it was being submerged. However, instead of using Whitehouse's apparatus, he had wisely brought his own more sensitive mirror galvanometer. This time the plan was for the two ships to start from mid-ocean and lay the cable between them as they sailed apart but again the expedition ended in failure. After the second break, the two ships failed to find each other at their rendezvous and had to return once more to Ireland. The directors of the company were now almost on the point abandoning the whole idea of an Atlantic cable, but with Thomson's encouragement it was decided to have at least one more attempt later in the summer. The ships eventually reached their destinations on 5th August, exactly 1 year after the commencement of the Atlantic cable project in 1857 and landed their respective ends of the cable. Both crews had spent eight nerve-racking days after the cable was first slowly lowered into the sea and its continuity checked every 10 minutes by sending a signal back and forth as they sailed apart.

The sense of euphoria after the completion of the first complete cable was short lived. For the first few weeks, there was a dispute between the directors of the Atlantic Telegraph Company in London and Whitehouse in Valentia concerning the possible existence of a fault, the apparatus used at the two ends and the overall performance of the cable. Because of the low sensitivity of the Whitehouse detector at the Newfoundland end, he had been compelled to increase the voltage of the pulses being sent from Valentia, which eventually contributed to the failure of the cable within a few weeks. The Newfoundland clerks did not have the same need to resort to high voltages because Whitehouse sometimes used Thomson's mirror galvanometer as the detector in Valentia, and then passed on the messages as if they had been received by his own equipment. He was sacked on 17th August after he disobeyed his directors by lifting the cable at the mouth of Valentia harbour to inspect it but only succeeding in damaging it.

The target of trans-Atlantic communication was revived at the start of the next decade. Even in 1858 when the cable failed, its design and quality were suspect. The variable performance could have been due to inadequate or damaged insulation as much as Whitehouse's rough handling, but then the science of pulse propagation in cables was not understood as it was not really necessary for short overland links. The 1859 Board of Trade enquiry concluded that an Atlantic cable was feasible but confirmed that the earlier attempts failed because of poor design and manufacture. Following a systematic investigation by a committee established by The Atlantic Telegraph Company which included among its membership Thomson, Wheatstone and Joseph Whitworth,[i] the cable had been completely redesigned by the time of the next attempt in 1865. With increased tensile strength, improved insulation and a copper core of larger cross-sectional area, the new version also required much lower voltages than were being used in 1858. These improvements were, however, at the expense of its weight which had now doubled.

Figure 8: The cable fleet including Great Eastern off Berehaven, County Cork, July 1865. (Metropolitan Museum of Art, New York: Robert Charles Dudley collection.)

The previous attempts had required two of the largest warships then available but now the Atlantic Telegraph Company wanted to lay the cable in a single run. Fortunately, SS Great Eastern, the world's largest ship that had only been launched in 1858, was available having been laid up in a damaged state after its last passenger voyage. Due to its large size, it was a commercial failure but was about to enjoy a successful second career as a cable ship. Now owned by the Telegraph Construction and Maintenance Company, Great Eastern was chartered by the Atlantic Telegraph Company and loaded with the necessary thousands of miles of cable off Sheerness on the Thames. With the cable stored under water in three tanks and Thomson on-board once more, the ship set sail for Ireland on 15 July 1865. There the cable was joined to a shore length, and a week later Great Eastern and its accompanying ships sailed westwards to Newfoundland, Fig. 8. Progress was good despite the cable having to be retrieved on two occasions for repair due to damaged insulation, but at only a few hundred miles from the destination it snapped. The crew successfully grappled for the cable at a depth of almost 3 miles but the grappling rope broke three times as the cable was being lifted, and with insufficient length on board for a fourth attempt, the expedition had to be abandoned.

Although this 1865 attempt had also failed, the various elements showed that they were individually suited to the task. Apart from the cable itself, these included the ship, the paying-out machinery, the continuous testing procedures, the grappling gear and the mooring of buoys to mark a lost cable. One year later in June 1866, Great Eastern again left the Thames to lay a cable but this time it also carried the balance of the 1865 cable. After successfully reaching Trinity Bay, Newfoundland, on 27th July, the fleet returned to look for the previous year's broken cable, Fig. 9. Having witnessed the problems experienced in 1865 during grappling, Thomson's plan was that the cable should be lifted simultaneously by three ships to avoid exceeding its breaking strain and that of the grappling ropes. By 31st August, the end was eventually secured aboard the Great Eastern, and with hundreds of miles of new cable spliced to it, the second complete Atlantic cable was landed in Trinity Bay on 8th September. The quality of the two cables was cleverly demonstrated by the ease with which low-voltage pulses could be detected after passing through almost 4000 miles when they were connected in series for test purposes. The 1865 and 1866 cable laying voyages are described further a *Story in Pictures* in the Appendix.

Figure 9: An artist's impression of the broken 1865 cable being retrieved in 1866 by a grappling hook from Great Eastern. (Illustration from Illustrated London News.)

Within weeks of Thomson returning to the British Isles on Great Eastern, he was offered a knighthood. To the general public, he was the hero of the Atlantic Telegraph and his name was synonymous with its success. He had been involved in solving the practical issues surrounding the cable's bandwidth and insulation, the detection of the small currents it transmitted and the engineering challenges connected with laying cable from a tossing ship and even its retrieval from the ocean floor. As a result, this was the single most important event that transformed him into the public figure he remained for the next 40 years until his death in 1907.

Thomson also benefitted financially from the telegraph. As well as being a much sought after consultant on almost every subsequent submarine cable project as the continents were being connected together, he developed numerous items of instrumentation. He was associated with 11 electric telegraph patents, mostly in the 20 years from 1854 during this period of greatest progress. His most significant was the invention in 1867 of the siphon recorder [8], an improved method of making a permanent record of received messages, Fig. 10. In this device, a continuous line was printed on a moving ribbon of paper by ink transported from a reservoir by siphon and then expelled from a nozzle by electrostatic repulsion. With the siphon attached to a moving coil between the poles of a magnet, as in a mirror galvanometer, the line moved to the left or right depending on the fluctuations of the message's current. With the printing now being effectively frictionless, high-speed transmission using prepared punched tape was now possible. The increase in cable capacity meant that there was extensive world-wide adoption of the siphon recorder in the ensuing decades with an annual royalty of the order of £1000 per company for the duration of the patent.

Figure 10: An example of a message received using Thomson's siphon recorder. (Illustration from Thompson [7].)

8 Lalla Rookh

The steady income from his telegraph apparatus meant that Sir William Thomson was, in the parlance of the twenty-first century, 'seriously rich' and like other prosperous Victorian gentlemen, he needed something to spend it on. The house would come later, but a boat would suffice for now. When he was a student at Cambridge, he had demonstrated a fondness for messing about on the river, but with his recent experiences in cable laying, he had experienced long voyages in various vessels in all sorts of weather conditions. In the summer of 1870 soon after the death of his first wife, he eventually settled on Lalla Rookh (Fig. 11), a two-masted sailing yacht of 126 t, which had been lying in Cowes. It was of such a size that a skipper and crew were necessary and had enough space for six or seven guests to be taken on extended cruises around the British Isles, the Mediterranean or the West African coast.

The first proper cruise took place the following year after some winter maintenance on the Clyde. In the absence of his wife, Sir William had to first seek the advice of Mrs Peter Guthrie Tait as to what soft furnishings were required to make it habitable. Beforehand in March and April, he discussed with her at length the relative merits of cotton and linen and specified the number and quality of table napkins, tablecloths, towels and sheets. Their quality could not be compromised as a yacht of this scale would eventually provide the opportunity to socialise with royalty and aristocracy. Lalla Rookh was frequently sailed to Cowes for the annual regatta and could be found at Inveraray in Loch Fyne when Thomson was a guest of the Duke of Argyll or in Bangor Bay (County Down) when visiting the Marquis of Dufferin at his Clandeboye estate.

His usual pattern was to spend as much time as possible on Lalla Rookh in the 6 months between the university's spring and autumn terms. In contrast with the pressures experienced by a twenty-first century academic, he was able to use the long vacation for whatever purposes he saw fit. Also, although he was an active researcher, he did not need to be based in the university to supervise postgraduate students or undergraduate summer projects.

Lalla Rookh was used in different ways. Thomson not only loved entertaining members of his extended family on board but also invited his scientific friends for cruises where problems in physics and mathematics could be discussed. Among these guests were Tait, Alexander Crum Brown,[j] Hermann von Helmholtz,[k] Sir Charles Siemens[l] and Sir George Darwin.[m] The yacht also became a convenient laboratory when the marine compass or deep-sea sounder was under development.

Figure 11: Lalla Rookh, its crew and guests. (Photographer unknown.)

Despite the vessel having a professional skipper, there was always the risk that there would be conflict between him and the owner. Not surprisingly in the light of his later interests, Thomson had a reputation for being a good navigator and had faith in the 'flying' soundings he could take with his new sounder. On at least one occasion the skipper, Captain Flarty, was heard to say 'You will not rest until you have your boat at the bottom'. This was not a completely empty comment because Lalla Rookh was involved in a potentially serious incident in July 1882 when it ran aground on rocks off Scalpay in the Inner Hebrides.

9 Family life

When William Thomson was preparing to apply for the Glasgow position in the summer of 1846, he discovered he had a potentially strong rival – Archibald Smith. A native of Glasgow, Smith (1813–1872) was Senior Wrangler and Smith's Prizeman at Cambridge in 1836 and co-founder of the *Cambridge Mathematical Journal*. Despite this glittering start, he moved from academia to law and practiced as a property lawyer in Lincoln's Inn for the rest of his working life. He continued to contribute to mathematics and physics, focusing mainly on terrestrial magnetism and the effect of iron hulls on the deviation of marine compasses. His efforts were rewarded by election to the Royal Societies of Edinburgh and London in 1837 and 1856, respectively, and the award of the latter's Royal Medal in 1865 [9, 10]. Smith eventually decided not to let his name go forward for election, but as the Thomson and Smith families were already acquainted in Glasgow, it was natural that William would come to know Archibald's sister, Sabina [11]. By the spring of 1852, he had proposed three times to her but was continually refused, partly on the advice of her brother, even though he had not been humiliated in anyway by William's success.

In July of 1852 at the age of 28, his engagement was suddenly announced to a family friend and second cousin, Margaret Crum (Fig. 12, left), the 21-year daughter of Walter Crum, his father's first cousin. Crum was a Glasgow cotton merchant from Thornliebank in the south of the city and during 1848–1865 was the president of Anderson's University, the name by which Anderson's Institution, now the University of Strathclyde, was known during 1828–1886 [12]. But this step did not completely mark the end of Sabina Smith's involvement in Thomson's life. When he was in Belfast attending the September 1852 meeting of the British Association for the Advancement of Science, she travelled to Ireland to give him the opportunity of proposing for a fourth time. Thomson decided that three unsuccessful offers were enough, a decision that still preyed on Sabina Smith almost 40 years later. He and Margaret were married later the same month but unfortunately his new bride soon developed prolonged ill health of an unspecified nature.

Mrs Thomson's fragility did impact at times on Thomson's freedom to accept engagements, but she was also talented in literary matters and assisted him in drafting manuscripts as well as being a published poet in her own right [13]. They frequently travelled in the summer to continental Europe for mountain and spa holidays. One such occasion at Bad Kreuznach provided the opportunity for Thomson to visit and make the personal acquaintance of von Helmholtz who he had earlier invited to attend the 1855 British Association meeting in Glasgow. Von Helmholtz was in Königsberg, East Prussia, now Kaliningrad in the Russian exclave of Kaliningrad Oblast. Throughout their 17 years of marriage, Thomson was a dedicated nurse until Margaret's premature death in June 1870 as Lady Thomson, which she had become in 1866.

For the next few years, Thomson concentrated on his academic and industrial activities, but could not have foreseen that they would be the catalyst for him meeting the woman who would be his second wife and remain his partner until his death in 1907. In the summer of 1873, he was on board the cable-ship Hooper carrying the first 2500 miles of cable for laying between Lisbon and Brazil. The shore-end was first landed in Lisbon on 20th June and then the ship sailed westwards to Madeira. But as a fault had been found in one of the 500 mile lengths soon after leaving Plymouth, it was decided to repair it in Funchal Bay before attempting to lay any more cable. During the 16-day enforced break, Thomson was entertained by Charles Blandy, a leading wine producer and father of three daughters. Communication was quickly established between the Hooper and Blandy and his daughter Frances Anna (Fig. 12, right) by signalling lamp.

The hope of romance blossomed sufficiently during these 2 weeks for Thomson to decide to return at the earliest opportunity. The following year he arrived back in Funchal Bay again on cabling business but this time independently in Lalla Rookh, just in advance of the Hooper.

Figure 12: Lord Kelvin's wives. Left: Margaret Thomson, née Crum in 1858. Right: Lady Kelvin, Frances Anna Blandy, with Lord Kelvin in 1906. (Photographs from Thompson.)

Mooring at mid-day on Sunday 10 May 1874, he proposed as soon as possible after the Blandy family emerged from the morning church service. After spending the next 2 months in Madeira, Thomson and Francis were married there on 24 June; he was 50 and she was 36. In contrast with his early experience of the first Lady Thomson, he was excited by her ability as a sailor as he confided in a letter to von Helmholtz written on the eve of the wedding. During their 23 years together, Francis, or Lady Kelvin from 1892, travelled most summers with her husband to mainland Europe and to North America on his 1884 and 1902 visits. In September 1907, after returning home from one of these holidays in Aix-les-Bain, she suffered a stroke with some associated loss of speech. She made a partial recovery during the next months but eventually died in Largs in 1916, outliving her husband by 9 years who had died in December 1907.

10 Glasgow and Cambridge

Midway through Sir William Thomson's professorial career, the University of Glasgow moved from its cramped medieval location to the present campus in the west of the city on Gilmorehill. It was the fourth oldest university in the English-speaking world having been founded in 1451 when King James II of Scotland persuaded Pope Nicholas V to grant a bull authorising Bishop William Turnbull of Glasgow to set up a university. For its first 9 years it was based at Glasgow Cathedral, but then for more than 400 years it occupied the collection of Renaissance-style buildings on High Street where Thomson had established his laboratory in 1846. Unfortunately, in addition to Thomson's own expansionist plans, there was competition for space in this part of Glasgow. While the city was rapidly expanding, the burgeoning railway companies also needed large tracts of land to build yards and depots to deliver and collect goods. As early as 1845, the Caledonian Railway had offered £30,000 for the old college with an additional £70,000 to assist with its relocation. Then, in 1863, the City of Glasgow Union Railway made a similar offer which was accepted, but more fundraising and government support was necessary before work could begin in 1867. The chosen location was a green field site enclosed by a large loop of the River Kelvin.

The new building and its spire were designed by Sir George Gilbert Scott and his son John Oldrid Scott as a veritable cathedral of learning that still dominates the Glasgow skyline when viewed from most of the city, Fig. 13. Although the demolition of the ancient High Street buildings is regarded by some as an act of architectural vandalism, Glasgow gained a fine addition to its collection of Victorian commercial and industrial buildings. But not everything on High Street was lost. Pearce Lodge and the Lion and Unicorn Staircase were moved stone by stone from the old site to the new and reconstructed as the gate lodge near the north end of Kelvin Way. The last degree congregation in the old college was held on 1 May 1870 and then the Gilmorehill campus opened later the same year in time for the 1870–1871 session. Thomson had helped to raise funds, but by the time he had moved into No. 11 in the new Professors' Court (Fig. 14), Margaret Thomson's long battle with ill health was ended when she died in June 1870. With Thomson in mourning, the job of relocating the natural philosophy apparatus fell to the faithful McFarlane and his other assistant and Thomson's nephew, James Bottomley. There is further description of the new buildings in 'Chapter 17: Conclusion: honoured by banknotes'.

The rebuilding period was an unsettling time for Thomson in other ways. In November 1869, he had received an informal invitation to become a praelector or lecturer at his old college in Cambridge. This move was linked to the findings of the Royal Commission on Scientific Education and the proposed establishment of a Physical Laboratory at Cambridge, later to become

Figure 13: The new University of Glasgow buildings at Gilmorehill. (Illustration from Thompson [7].)

Figure 14: Left: Lord Kelvin's house at No. 11 Professors' Square, University of Glasgow. Right: the plaque erected by the Institute of Physics on No. 11 Professors' Square, University of Glasgow. (Photographs by I.S. Ruddock.)

the Cavendish Laboratory part-funded by the Duke of Devonshire. Although Thomson turned down the approach on the grounds of his wife's illness, he also asked for full particulars of the post and an indication of when a decision had to be made. By December 1870, 6 months after her death, Peterhouse's Master now wrote inviting him to apply for the newly created Professorship of Experimental Physics. Since Thomson's own circumstances had changed in the intervening year with the opening of the new campus, he decided to decline on the grounds of the improved facilities, the convenience of Glasgow as an engineering base and his unwillingness to start afresh. This was not the end of his involvement in Cambridge's search for a candidate. At the request of George Gabriel Stokes,[n] he approached his old friend von Helmholtz to see if he was interested, but von Helmholtz had only recently been appointed as the director of the new Institute of Physics in Berlin and was also not going to apply. Eventually, an equally eminent candidate was found during the following year when James Clerk Maxwell was elected in March 1871 and held the chair until his premature death in 1879.

11 Netherhall

In 1874, soon after his marriage to Frances Blandy, Sir William Thomson bought land near Largs and started designing and then acting as clerk of works for the construction of the house that later became known as Netherhall. In keeping with his status as a public figure and helped by his now substantial private income, this was a large three-storey detached gentleman's residence in the popular Scottish baronial style, Fig. 15. Little expense was spared as the total cost came to an enormous, for that time, £12,000. Netherhall is still standing in more or less unchanged external condition, although no longer in the countryside on the edge of Largs.

The property has now been converted into apartments having been a Church of Scotland conference centre for many years, but its previous owner is still remembered in the name of one of the driveways. It was also possibly the second house in the British Isles to be lit by electricity following Cragside, Northumberland, where an engineer and an industrialist Lord Armstrong[o] had installed an early hydroelectric scheme. Thomson's Glasgow home at No. 11 Professors' Square was also an early conversion to the new form of energy, and along with Netherhall used dynamos driven by gas engines. Although his second marriage was childless like his first, Netherhall was not a sterile and empty building. When it was not filled with nephews and nieces, it was used for business meetings, dinner parties and visits by many of the famous scientists and engineers active in the latter half of the nineteenth century, sometimes in connection with cruises on Lalla Rookh.

Figure 15: Netherhall near Largs in Ayrshire. (Illustration from Thompson [7].)

12 Navigation

The aim of marine navigation is to know one's position on the surface of the Earth, i.e. the latitude and longitude. Latitude is determined via by the measurement of the sun's elevation at local mid-day. The modern sextant was invented in the early eighteenth century for such a purpose, but for hundreds of years before this, the measurement was made by variants of the astrolabe or backstaff. Longitude proved the more difficult parameter to obtain as it required the time difference between local noon and that at Greenwich to be known at the rate of 15° of longitude per hour. This could only be achieved by the ship carrying with it the time existing at its previous port. Although the basis of the solution to the problem was known for a long time, the inherent difficulties in constructing a practical marine chronometer were not finally solved until the 1770s. John Harrison managed to construct a clock, which was accurate to a fraction of a second per day by incorporating thermal compensation for the balance spring and improved bearings. Due to expense and limited production, universal adoption of the chronometer by the world's merchant fleet did not take place until the nineteenth century. As a result, longitude was still being measured by the method of lunar distance where the angle between the moon and a selected celestial object only occurs at a specific Greenwich time given by the nautical almanac while the local time and hence the time difference are obtained by observing the elevation of the sun.

Knowledge of one's latitude and longitude is not the only requirement in practical navigation. To sail from one port to another, a compass bearing must be decided upon and followed. This is basically the technique known as dead-reckoning, where a compass and log is used to track the passage of a ship on a chart. However, assuming that the various navigational steps have been accurately performed within the precision of the available charts, caution must still be exercised in approaching even a familiar coast as the depth of the water under the ship needs to be known and landmarks identified. The problems facing seafarers in the nineteenth century were the same as in previous centuries except that advances in ship construction and performance exacerbated some of them; for example, compasses that were perfectly adequate for wooden sailing ships could not be relied upon in iron and steel ships, and with the increased speeds of steam ships, sounding using a knotted line was impractical.

From the 1850s, Sir William Thomson had become an increasingly experienced sailor having been on numerous cable-laying expeditions and of course since 1870 as the owner of Lalla Rookh. But the sea had also caused him to experience personal tragedy. In 1875, one of his nephews, David Thomson King, son of his sister Elizabeth, was lost when a cable ship ran aground on Ushant, an island off the Brittany coast on passage to Brazil. His involvement with the marine compass which eventually became an important academic and commercial part of his life was first sparked by an invitation to contribute an article to a new general interest magazine. Investigative work on the marine compass had already been performed almost thirty years earlier, most notably in the United Kingdom by Sir George Airy, the Astronomer Royal. The basic principles of the relationships among the stability, vibration period and moment of inertia of the compass card and the magnetic moment and size of the needle were known, but not fully applied to its design and construction. Techniques were also in place for compensation of the fixed magnetism of the ship and any temporary perturbations caused by its direction by placing soft iron spheres on either side of the compass card. Thomson's main contribution was the analysis of the effect that needle size and ultimately compass card size had on its performance at sea. The Admiralty's Standard Compass of 1842, with its four needles attached to a card of 7½ inches in diameter and with a total mass of more than 100 g, had replaced the fairly primitive compasses used until then with their single long needles of up to 15 inches in length. Thomson advocated even smaller needles of only a couple of inches in length such that compensation could be achieved by the iron globes placed nearby instead of requiring hundreds of tons as had been estimated for long needles. In 1889, his compass was adopted by the Admiralty and consisted of eight such needles suspended by silk threads from a card of 10 inches in diameter with a total mass of less than 11 g. With its period of vibration being double that of the previous compass and the friction at the pivot being much reduced due to the decrease in weight, the compass's overall stability in rough seas or in a warship during gunfire was greatly improved, Fig. 16.

The other navigational aid whose development Thomson contributed to and benefitted from financially was the deep-sea sounder. The aim was to replace the simple sounder comprising a weight on the end of a rope, which could only be used when the ship was more or less stationary. Instead of simply measuring the length of the rope between the surface and the seabed, Thomson's sounder effectively measured the water pressure by the calibrated movement of a piston in a cylinder containing air. Since the depth was deduced from the pressure, this meant that the line between the ship and the seabed did not have to be vertical and would in practice be some distance astern of the ship if the sounder was thrown overboard while underway. Deep-sea sounders of this design were produced first by James White's business in Glasgow in the late 1880s, and then by successor companies until the middle of the twentieth century, and were still being used by the Royal Navy in the 1950s before being superseded by electronic devices. Kelvin's industrial interests are described further in Section 16.

As well as the compass and deep-sea sounder, he was involved during the last quarter of the nineteenth century in a campaign for lighthouses to have characteristic flashes instead of being fixed or simply flashing [14], the derivation of the navigational corrections that arise from the assumption that the Earth is a perfect sphere and tide prediction. While the first two were more

Figure 16: Lord Kelvin with a compass binnacle in 1902. (Photograph from Thompson [7].)

or less academic activities, tide prediction led once more to the construction and manufacture of instruments from which an income would accrue. There were two distinct aspects to this exercise. First, there was the need to continually monitor the rise and fall of the tide at the port of interest,[p] and second, by analysing the historical record, the tide at this location could be predicted for any time in the future. It was already known from the work of Laplace and Airy [15] that tidal movement could be represented by a series of harmonic functions with different phases and amplitudes but its de-convolution demanded extensive tedious arithmetic. By adapting the disc–globe–cylinder integrator recently developed by his brother, James, Thomson devised his Tidal Harmonic Analyser in 1876. This was a mechanical analogue computer capable of solving linear differential equations and integrations, and with five sets of disc–globe–cylinders it could find the constants of the first and second orders. The final step was then to plot the predicted tide patterns using these values as the settings of a Tide Predictor.

13 Politics and peerage

Among Thomson's other extra-mural activities was an interest in politics. In 1869, he was invited to stand as the Liberal candidate for the Glasgow and Aberdeen Universities' parliamentary seat. This had only been created in the previous year, but the holder, James Moncreiff who was also Glasgow University's Rector, was appointed for a fourth time as Lord Advocate for Scotland and resigned. On account of the likely conflict between attendance at Westminster and his other commitments, Thomson decided not to accept. In the 1880s, Home Rule for Ireland became a major issue which would eventually bring down the government of the day. Although Thomson had spent most of his life, and indeed all of his career in Glasgow, he still considered himself an Ulsterman, and as such had an interest in the future of his place of birth and those who lived there and shared his political, religious and social views. Not surprisingly then, it was during this dramatic period in British history that Thomson found himself embroiled again in west of Scotland Liberal party politics.

The fundamental situation in Ireland was the same in the nineteenth century as it is now. In simple terms, the majority protestant population in the northeastern part, the descendants of English and Scottish settlers preferred to remain within the United Kingdom while the Roman Catholic indigenous population in the rest of Ireland aspired to self-government and eventual independence. In the 1885 General Election, the Liberal party led by the Prime Minister, William Ewart Gladstone, won the largest number of seats but was left without an overall majority. With the balance of power in the House of Commons between the Liberals and Conservatives held by Charles Stewart Parnell's Irish Parliamentary Party, Gladstone then committed himself to Irish nationalism and introduced the First Home Rule Bill. In early 1886, opposition among his party led to what was first thought to be a temporary split by a group who became known as Liberal Unionists. Included among these pro-unionists were Liberals with large estates in Ireland who feared their confiscation in an independent Ireland.

The First Home Rule Bill or Government of Ireland Bill was defeated in the Commons in June 1886 by a revolt of 63 such Liberal MPs. With Gladstone having little option other than to go to the country again, Parliament was dissolved and another General Election was set for July. As a public figure and an Ulsterman, Thomson effectively became a leader of the Liberal Unionists in Scotland and campaigned vigorously for unionist candidates. These included opposition Conservative candidates such as Col. John Malcolm in the Argyllshire constituency, who with the support of the Liberal Unionists defeated the Crofters' Party incumbent. With 77 Liberal

Unionist candidates being elected in Great Britain, Lord Salisbury's Conservative Party won the most seats but, like Gladstone the previous year, relied for his parliamentary majority on another party, in this case the Liberal Unionists.

Thomson's peerage was announced in Queen Victoria's New Year's Honours list of 1892 on the recommendation of Salisbury. With him being the first scientist to be honoured in this way, his elevation was warmly received as he would be able to contribute to this aspect of the work of the House of Lords. But there was also the legacy of his role in Salisbury winning the 1886 General Election. He would not have become a leading figure in the west of Scotland politics without first being a Glasgow-based scientist, engineer and businessman of international renown, but would his scientific achievements alone have been sufficient for him to become the first scientific peer more than 60 years before the creation of modern life peerages made this more common? Smith and Wise [16] in their 1989 biography of Kelvin discussed the link between his political campaigning and the reward of the peerage, but this was not a uniquely revisionist view. In his 1910 biography of Kelvin [7], Sylvanus Thompson[q] pointed out that The Daily News in January 1892, when commenting on Kelvin's reputation as a man of science, also reminded its readers of his claims on the current government.

The newly ennobled peer took his seat in the Lords as Baron Kelvin of Largs on 25th February having been introduced by Lord Rayleigh who had succeeded to his own baronetcy in 1873. Thomson's title neatly combined the location where he worked – the River Kelvin flows through Glasgow's West End and past the University – and lived – Netherhall. His coat of arms also reflected his career with the shield supported on the left by a student in academic dress and holding a voltmeter and on the right by a sailor holding a sounding line, with 'Honesty without fear' as the motto, Fig. 17.

Having established a London residence in Eaton Place, Belgravia (Fig. 18), Kelvin could now apply his scientific and engineering expertise to the work of parliament. Within 2 months, he

Figure 17: Left: Baron Kelvin of Largs' coat of arms. Right: Lord and Lady Kelvin dressed for the coronation of Edward VII in 1901. (Illustration and photograph from Thompson [7].)

Figure 18: Lord Kelvin's London home at 15 Eaton Place, Belgravia, with the blue plaque erected by English Heritage in 1996. (Photographs by I.S. Ruddock.)

was a member of a committee considering private bills promoting underground electric railways in London. During the remaining 15 years of his life, he also campaigned for the flashpoint of paraffin to be raised, spoke against the imposition of death duties on private collections of art and scientific instruments and the inclusion of penalty clauses in mail-steamer contracts on the grounds of safety, and was chairman of a Royal Commission enquiring into arsenic poisoning in beer. He also enjoyed being present when the 1893 Irish Government Bill or Second Home Rule was defeated by the Lords although being passed by the Commons, unlike its 1896 predecessor.

14 Jubilee

In 1896, the University of Glasgow celebrated the golden jubilee of Lord Kelvin's appointment to the chair of natural philosophy [17]. Delegates from every university and scientific body in the British Isles, together with many from the Empire and foreign locations came to Glasgow to take part in the 3-day event. The programme opened on Monday 15th June with a conversazione at which the guests were joined by staff and students from the university, representatives of local government and prominent citizens of Glasgow and the west of Scotland. The telegraph companies set up siphon recorders to receive numerous congratulatory telegrams from, among others, the Viceroy of India and the Postmaster-General of New Zealand. To demonstrate the state of the art of telegraphy, one message was sent from the university to Kelvin via Newfoundland, New York, Chicago, San Francisco, Los Angeles, Florida, Washington and New York and Newfoundland again taking a total of 7½ minutes; his reply of thanks, sent by the same route, however, only took 4 minutes. The next day's meeting focused on the presentation of more than 90 greeting addresses from around the world by the visiting delegates and commenced with that from the Prince of Wales. Kelvin received the degree of LL.D and then taking the chair, conferred 14 more honorary degrees. Among the recipients were some of Kelvin's contemporaries including Gabriel Lippmann,[r] Woldemar Voigt[s] and Christian Christiansen.[t] The day was concluded by a grand banquet in the city's St Andrew's Hall at which there were more addresses, toasts and replies, including Kelvin's own reminiscences of his early days in the old college on Glasgow's High Street. On the third and final day, the University of Glasgow's Senate, presaging the pattern of the modern scientific conference, and no doubt to the relief of some of those present, organised a cruise in the Firth of Clyde for about 250 of the 2500 participants in the Jubilee.

15 Retirement

Soon after his 75th birthday in June 1899, Lord Kelvin notified the University of his intention to retire later in the same year, Fig. 19. This took effect from 1st October when his colleague Andrew Gray succeeded him to the chair of natural philosophy. However, the retirement was short-lived; Kelvin immediately registered as a research student thus being in his lifetime both the youngest and oldest student in the history of the university. That he did not slow down towards the end of his career can be judged from the range of public duties undertaken. He spoke in the House of Lords, acted as chairman of public meetings, gave scientific talks, visited engineering establishments, opened new buildings, unveiled monuments, while at the same time continuing as an academic and industrialist.

Taking 1897 as an example, on 13th January he gave in Glasgow an evening public lecture entitled 'Molecular Dynamics of Crystals', on 1st May he dined and spoke in London at the Royal Academy's banquet, on 8th May he acted as the chairman of an event in Peterhouse College, Cambridge, at which a presentation was made to the Master, on 21st May he gave a Friday Night Discourse at the Royal Institution, London, on 'Contact Electricity of Metals', on 28th May he spoke in the House of Lords against death duties for private collections of

Figure 19: Lord Kelvin's last lecture in 1899 at the age of 75. (Photograph from Thompson [7].)

works of art or scientific instrumentations, on 2nd June he spoke to the Victoria Institute[u] on the age of the Earth, on 26th June he was on board the White Star liner Teutonic to witness Queen Victoria's Diamond Jubilee Naval Review, on 28th June he opened the Shoreditch power station, on 1st July he was invited as a peer to the corresponding Military Review at Aldershot, the home of the British Army, on 2nd July he participated in the presentation of an endowment to University College, London, for the establishment of a laboratory and chair in electrical engineering in memory of Sir John Pender, on 15th July he was a member of a deputation from the Royal Society to Windsor to present a congratulatory address to Queen Victoria, on 29th July he visited the training ship H.M.S. Worcester and presented it with a compass, on 4th August he visited Greenock to open a factory on behalf of British Aluminium, on 7th August he sailed with Lady Kelvin to North America to attend the Toronto meeting of the British Association for the Advancement of Science, travelling as far as Vancouver in the west and Philadelphia in the south, visiting industry and academia, returning home on 16th October, on 29th October he was in Cambridge for a meeting of the Fellows and in early November entertained Joseph Chamberlain, the Rector of the University of Glasgow, at Netherhall.

16 Industry

On 21 March 2003, the world's largest ocean liner, RMS Queen Mary 2, was launched in Saint-Nazaire. Although it is one of the more recent additions to the United Kingdom shipping register, it is equipped with navigational equipment designed and manufactured by Kelvin-Hughes, one of the world's longest established instrument makers. The association between Kelvin and Hughes was created in 1947 when Kelvin, Bottomley & Baird Ltd, Glasgow merged with Henry Hughes Ltd, London. Kelvin, Bottomley & Baird was the final independent manifestation of the company that arose from Lord Kelvin's industrial collaboration and enterprise in the late nineteenth century. It had its origins in the firm of James White, Optician and Philosophical Instrument Maker in Glasgow founded in 1850 to supply and maintain apparatus in the University of Glasgow. Within a few years, White was producing electrical instruments such as electrometers and electrical balances to Thomson's designs. The successful laying of the Atlantic cable in 1866 and the invention of the siphon recorder and deep sea sounder greatly strengthened the link. Their collaboration is best known for the marine compass which has been in continuous production since 1876.

After White's death in 1884, Thomson continued to maintain his interest in the business and raised most of the capital needed to construct and equip new workshops in Glasgow's Cambridge Street. In 1900, Lord Kelvin became a director of a newly formed company, Kelvin & James White Ltd, which acquired White's business. His nephew, James Thomson Bottomley, also joined the firm at about the same time and eventually became chairman in 1913 of Kelvin, Bottomley & Baird Ltd as it was by then known to reflect its new partner, Alfred Baird, a compass adjuster with the firm since 1884. On the basis of being the supplier of compasses and deep-sea sounders to the Royal Navy and much of the British merchant fleet as well as manufacturing electrical apparatus, the company expanded from Glasgow in 1934 when an additional instrument factory was opened in Basingstoke. The subsequent link with Hughes was prompted by the destruction of the latter's London offices in the Blitz of 1941. Henry Hughes & Sons Ltd was an ideal partner having already developed an echo sounder in 1923. The two companies fully merged in 1947 and became a subsidiary of Smith's Industries Ltd, which since 1935 had had a controlling interest in Hughes. In the 1950s, the company's origins was acknowledged by naming its research vessel Lalla Rookh II. The connection with Smiths lasted until 2007 when it became independent again following a demerger, but it continues to manufacture and sell all aspects of marine equipment from paper charts to military and civilian maritime radar.

17 Electricity generation

Lord Kelvin's direct and indirect involvement in submarine telegraphy has already been described, but he was also engaged as a consultant in the emerging electrical power industry. He had been building a scientific reputation in both the basic theory of electricity and magnetism and practical aspects such as instruments to measure voltage and current. Linked to the development of meters and balances, there was a parallel contribution to defining the units of electrical measurement by membership of the British Association's and Board of Trade's Electrical Standards Committees. In the 1880s, Kelvin advised the promoters of the Giant's Causeway, Portrush and Bush Valley Tramway and Railway Co. Ltd on the north Antrim coast. This pioneering system was one of the world's first electric railways and the first to use hydro-electric power [18], Fig. 20.

Two further examples of the early exploitation of this means of generating electricity serve to illustrate his enthusiasm for it as a source of power. First, when he opened the British Aluminium factory in Greenock in 1897, he was reported as saying:

> ...That magnificent piece of work of the Aluminium Company was the beginning of something that would yet transform the whole social economy of countries such as the Highlands, where water abounded. He looked forward to the time when the Highlands would be re-peopled also with industrious artisans doing the work which that utilization of the water would provide for them.

Figure 20: A tram of the Giant's Causeway, Portrush & Bush Valley Railway and Tramway Co. Ltd at Dunluce Castle on the North Antrim coast before 1899 when the lineside rail for current collection was abandoned in favour of an overhead cable. See also Chapter 17. (Photograph by R.J. Welch.)

But along with this prescient view of the future of hydroelectricity in Scotland, he added:

> It is only at a distance that the sentimental question, 'What is to become of the beautiful Falls of Foyers (Loch Ness)?' was asked ... when the time came that every drop of water that now fell over the Falls of Foyers was used for the benefit of mankind, no wise man, no man who thought of the good of the people, would regret that the power in the waterfall was developed for the benefit of mankind.

Second, in 1890 he chaired the Niagara Commission on behalf of the New York Cataract Company to assess detailed proposals to construct a power station at the Falls and during his 1897 visit to Canada, which immediately followed his visit to British Aluminium, he inspected the area and in a press statement reiterated the previous comments by again predicting the expansion of hydroelectricity:

> ...I look forward to a revival of life and prosperity in the Highlands of Scotland, and to the present crofters being succeeded by a happy industrial population occupied largely in manufactories rendered possible by the utilization of all the water power of the country. It seems to me a happy thought that the poor people of the country will be industrious artisans, rather than be mere guides to tourists.

He betrayed again his lack of sentimentality in the possible environmental impact of such installations by adding;

> I look forward to the time when the whole water from Lake Erie will find its way to the lower level of Lake Ontario through machinery, doing more good for the world than even that great benefit which we now possess in contemplation of the splendid scene which we have before us in the waterfall of Niagara. ... I do not hope that our children's children will ever see the Niagara cataract.

This comment followed his talk to the British Association in Canada. On that occasion, he appeared to be adopting what was for the time a fairly radical concern for the environment, his focus being primarily the threat to the country's energy reserves and practical problems connected with combustion:

> Great Britain's coal was more than could be burned by all the oxygen in the air immediately over the British Isles. In regard to the effect of sunlight in storing energy and fuel, the present rate of sunshine was equivalent to the production of two tons of vegetation per square metre, per thousand years; an estimate agreeing very closely with the growth of German forests and of English hayfields. We might, as coalfields become exhausted, have to think of growing hay for fuel, as more economical than raising coal. But if we burn up our fuel supplies so fast the oxygen of the air may become exhausted, and that exhaustion might come about in four or five hundred years.

Despite Kelvin's eminence as an electrical engineer, he was an advocate along with Thomas Edison of direct current supply in the face of what would now be regarded as overwhelming evidence for the advantages of alternating current for distribution, conversion and in powering electrical machines. Nevertheless, he was eventually convinced by the Westinghouse Electric Company's demonstration of three-phase alternating current transmission using Nikola Tesla's apparatus at the 1893 Chicago World's Fair, and in 1896 allegedly declared magnanimously 'Tesla has contributed more to electrical science than any man up to his time'. [19].

18 Beliefs and controversies

Although Lord Kelvin had been brought up in an Ulster Presbyterian home, he usually attended the Anglian style Scottish Episcopal Church in Glasgow or Largs, probably due to his time in Cambridge, and only occasionally the (Presbyterian) Church of Scotland. When he was seeking appointment to the Glasgow chair, he first had to satisfy a religious test. This legacy of the era of the Covenanters involved declaring oneself a member of the Church of Scotland by signing the Westminster Confession of Faith, the fundamental doctrinal statement of the Church of Scotland that is still subscribed to by its clergy and Elders. He happily agreed to it although his father was strongly opposed to the existence of such tests during his own time as a Glasgow professor.

Mention has already been made of Kelvin's habit of beginning his morning lectures with an act of devotion from the Church of England prayer book. This reverence also extended to his private life as was the case in many other families in the nineteenth century. Agnes Gardener King, his niece and the daughter of his sister Elizabeth, recalled an occasion in 1888 when he was visiting her home in London on a day that was too wet to go to church [20]:

> Bibles were brought out, old version and new, and he began turning up passages, discussing the genuineness of Old Testament history, comparing dates about the prophecies and consulting Biblical dictionaries in relation to the debated points, all with the liveliest interest.

Kelvin's private religious views became public knowledge and the subject of comment when he contributed to the debate on the age of the Earth and the origin of life. His involvement in the former arose from attempts in the 1860s to estimate it by solely thermal arguments. Using data on the change in temperature with depth below the Earth's surface, he calculated the time taken for a sphere of molten rock to cool to its present temperature and arrived at a value of approximately 100 million years. Similar calculations concerning the age of the Sun and its source of energy also produced erroneously small values. With the benefit of the hindsight that twenty-first century physics offers, it is clear that his mistake was to ignore the heating that nuclear activity provides. Although the calculated age was viewed by Victorian evolutionary biologists and geologists as too short for the observed deposition and erosion of rocks to have occurred and for life to have developed, it was a commendable result for the period. Despite being frequently ridiculed over a century later, at the time many people still held to the Genesis story of creation including an Earth whose age was no more than a few thousand years. Notwithstanding his clear belief in an ancient universe, Kelvin's opinion of the origin of life was more conventional. Again quoting Agnes Gardner King:

> In the evening mother read aloud Darwin's confession of faith, or rather his confession of want of faith. Uncle William said of his views about the absence of evidence of design, that he considered such views utterly unscientific. He expressed himself very

> vehemently on the subject, and said that our own power of discussing and speculating about Atheism and Materialism was enough of itself to disprove such a theory. With regard to Evolution, he said that it could not in the least degree explain the great mystery of nature and creation; if all things originated in a single germ, then that germ contained in it all the marvels of creation, physical, intellectual and spiritual to be afterwards developed. It is absolutely impossible that atoms of dead matter should come together so as to make life.

While these specific comments were for the benefit of his close family, he made no secret of his belief in some kind of creation event. Take, for example, what happened in 1903. *The Times* of 2nd May reported on a lecture at University College, London, under the headline 'Lord Kelvin on Science and Religion' [21]. Kelvin had attended 'Present day rationalism: an examination of Darwinism' given by the Rev. George Henslow[V] as the first of five public lectures on 'Christian Apologetics' organised by the College's Christian Association. The event might have warranted little more than a mention had it not been for his controversial comments when proposing the vote of thanks. Although the content of the lecture was not recorded, the vote of thanks was reproduced by *The Times* almost verbatim. In it, Kelvin said he wished to make a personal statement including:

> (He was) in sympathy with Henslow in the fundamentals of his lecture but could not say that with regard to the origin of life, science neither affirmed nor denied creative power. Science positively affirmed Creative Power … Modern biologists were coming once more to a firm acceptance of something and there was a vital principle … There was nothing between absolute scientific belief in creative power and the acceptance of the theory of fortuitous concourse of atoms. Was there anything so absurd as to believe that a number of atoms by falling together of their own accord could make a crystal, a sprig of moss, a microbe, a living animal?

Whether this would have been end of the matter is unknown because Kelvin then wrote to *The Times* on the day it appeared. Although not doubting the accuracy of his reported statement, he said he wished to delete the reference to a crystal because time had prevented him from explaining the difference between its structure and that of an animal or plant. He added,

> … while fortuitous concourse of atoms is not an inappropriate description of the formation of a crystal, it is utterly absurd in respect to the coming into existence, or the growth, or the continuation of the molecular combination presented in the bodies of living things. Here scientific thought is compelled to accept the idea of Creative Power.

The letter drew an immediate response from Sir William Thiselton-Dyer, Director of the Royal Botanical Gardens, Kew. His father-in-law, Sir Joseph Hooker and Henslow's brother-in-law, had been the previous director. While acknowledging Kelvin's eminence, he took exception to him commenting on this subject:

> In the domain of physics, to the exploration of which Lord Kelvin has devoted an honoured lifetime, he would be a bold man who would cross swords with him. But for dogmatic utterance on biological questions there is no reason to suppose that he is better equipped than any person of average intelligence.

Thiselton-Dyer was also of the opinion that Kelvin had displayed inconsistency in his views on scientific thought:

> In the (attitude of scientific enquiry) to inorganic nature, he claims for scientific investigation the utmost freedom; in (organic nature), scientific thought is compelled to accept the idea of Creative Power … Lord Kelvin, in effect, wipes out by a stroke of the pen the whole position won for us by Darwin.

Thisleton-Dyer's letter set the tone for much of the ensuing correspondence. Like most arguments it had simple beginnings but developed during 3 weeks into a heated exchange focusing on Kelvin's views on a 'Creative Power' in nature, his right to comment on biology, and the apparent promotion by University College, London, of theological propaganda. Sir John Burdon-Sanderson, Oxford's Regius Professor of Medicine and a past president of the British Association rebuked Thisleton-Dyer by defending Kelvin's right to comment on biology while also disagreeing with him. The exchange between Thiselton-Dyer and Burdon-Sanderson prompted some light relief from an anonymous correspondent. 'Que scais-je' suggested that if he was in Lord Kelvin's place, he would rather be criticised by the former than be defended by the latter whose three-paragraph letter was no different than a nursery sandwich for delivering medicine – 'sugar:aloes:sugar'. In addition to other contributions from William (W.H.) Mallock, a novelist and philosopher, and Sir Ray Lankester, a zoologist and natural history director of the British Museum, Karl Pearson, the leading biological statistician of the day, accused University College of abandoning its non-sectarian tradition by apparently promoting this series of lectures. As well as being illustrative of Kelvin's religious views, the episode also demonstrates another, perhaps unknown, aspect of his personality. Before giving the vote of thanks, he must have been aware of the interest his participation would generate. His letter to *The Times* was probably deliberately provocative and perhaps even tongue in cheek. He chose to take no part in the subsequent discussion and remained silent even when there were contributions from some of the era's leading scientists. These actions suggest that behind that familiar and austere facade, an element of mischief was at work at this late stage in his life.

19 Health and death

Despite not retiring from his chair in Glasgow until after his 75th birthday and then living for another 8 years, Lord Kelvin was no stranger to ill health. He had first to contend with the infirmity of his first wife during most of their marriage from 1852 until 1870, but during this period his own health started to impact on his life and career. The first problem occurred in late 1860 when he suffered a serious fracture of his left hip while indulging in curling at Largs. Due to differences in opinion among the attending physicians from Largs, Glasgow and Edinburgh as to the diagnosis and treatment, the ensuing delay resulted in his leg being permanently shorter once it had healed by the spring of 1861. Unfortunately, as well as being lame for the remaining 46 years of his life, he was occasionally incapacitated by recurring pain from the old injury. However, it was during the period of convalescence that he started writing in the first of his famous green backed notebooks of which he would eventually fill over 100 volumes. To compound his mobility problem, from 1895 he was also afflicted by severe neuralgia caused by his trigeminal or fifth cranial nerve or as he referred to it, his 'No. 5 demon'. One of the worst bouts took place in the winter of 1904–1905 soon after his installation in November as the Chancellor of the University of Glasgow. Agreeing to undergo surgery in March 1905, the state of his health was of sufficient public concern that King Edward VII twice telegraphed to enquire after his progress, but its success was limited as he again complained of 'a very wicked two days and nights' attack' in July.

In 1907 in his 83rd year Kelvin was as active as ever. In January, he was invited to stand as the President of the Institution of Electrical Engineers for the third time. At first he declined, but later agreed on the understanding that his duties would be lightened where possible with the formal election taking place in May. Sylvanus Thompson described meeting him in March [22] and noted how struck he was by his feebleness, lameness and deafness. Kelvin also described to Thompson the trouble he had been experiencing with his sight, in particular the almost total loss of vision in the central region of the retina in his left eye. By the time of the British Association meeting in Leicester in the summer, he had rallied sufficiently to engage in vigorous debate. In mid-August, he and Lady Kelvin then travelled to Aix-les-Bain for their annual month's holiday with the intention of returning home in time to open new laboratories in Queen's College, Belfast, on 20th September.

Soon after arriving home in Largs on 14th September, it was Lady Kelvin's health that gave cause for concern when she collapsed with the stroke and suffered loss of speech and mobility. Lord Kelvin decided not to go to Belfast and sent in his place his nephew James Thomson of Newcastle-upon-Tyne, son of his brother James Thomson. Although Lady Kelvin's condition improved slightly during the next month, he spent most of this time at home. Other missed engagements included the opening of the new session of the Institution of Electrical Engineers at which he would have given his inaugural presidential address. In late November, he was confined to bed with a chill but this developed into a fever. His health continued to worsen during December until he slipped away in Netherhall during the night of the 17th with his wife at his bedside.

Lord Kelvin's funeral took place in Westminster Abbey on Monday 23 December 1907. The congregation that assembled was a glittering cross-section of the scientific and political establishment of Edwardian Britain. King Edward VII was represented by the Duke of Argyll and the Prince of Wales, later George V, by Sir Arthur Bigge, his private secretary. Academia was represented by the Royal Societies of London and Edinburgh, the Universities of Glasgow and Cambridge and many other British and foreign seats of learning and learned societies. Local politicians present included the Lords Provost of Glasgow and Edinburgh and the Lord Mayor of London. He was buried near Sir Isaac Newton and his grave was marked by a slab engraved with the simple legend 'William Thomson, Lord Kelvin, 1824–1907'.

Some of the major monuments to Kelvin are described in *Honoured by Monuments* in the Appendix.

20 Honours

Lord Kelvin was showered with honours of every description during his working life and later during the period of his supposed retirement. The highlights for an Irish or British scientist were much the same in Queen Victoria's reign as they would be now in her great-great-grand-daughter's and included Fellow and President of the Royal Society (1851 and 1890–1894, respectively), knighthood (1866), peerage (1892), Order of Merit (1902), Privy Councillor (1902) and Chancellor of the University of Glasgow (1904). An abbreviated list of honours and awards is given below under the categories of 'universities', 'national and municipal' and 'learned societies and academies'.

The following is adapted from *The Life of Lord Kelvin*, S.P. Thompson, Macmillan (1910).

20.1 Universities

Twenty-one honorary degrees from universities in the United Kingdom, Canada, the United States of America, Hungary, Italy, Germany, Norway and Serbia.

Chancellor of the University of Glasgow, 1904.

20.2 National and municipal

Knight Bachelor, 1866.

Freedom of the City of Glasgow, 1866.

Commander of the Imperial Order of the Rose (Brazil), 1873.

Legion of Honour (France), Commander, 1881.

Legion of Honour (France), Grand Officer, 1884.

Knight of the Prussian Order Pour le Mérite, 1884.

Commander of the Order of Leopold (Belgium), 1890.

Freedom of the City of London, Honorary Liveryman of the Clothworkers' Company, 1891.

Barony of Largs in the County of Ayr, 1892.

Deputy Lieutenant of the County of the City of Glasgow, 1894.

Knight Grand Cross of the Royal Victorian Order, 1896.

Honorary Colonel of the Electrical Engineers (Volunteers), 1897.

Master of the Clothworkers' Company, 1900.

Order of the First Class of the Sacred Treasure of Japan, 1901.

Order of Merit, 1902.

Privy Councillor of Great Britain, 1902.

20.3 Learned societies and academies

Lord Kelvin was a member, fellow or president of more than 80 national academies and learned and cultural societies in the United Kingdom, the United States of America, Canada, Sweden, Germany, Hungary, France, the Netherlands, Italy, Austria, Switzerland, Denmark, Russia, Belgium, Australia, New Zealand, Finland, Norway and Mexico including the following.

Royal Society of Edinburgh:
Fellow, 1847; Keith Medallist, 1864; Gunning Victoria Jubilee Prize, 1887; President, 1873–1878, 1886–1890, 1895–1907.

Royal Society of London:
Fellow, 1851; Royal Medallist, 1856; Copley Medallist, 1883; President, 1890–1895.

Mathematical Society of London:
Member, 1867; President, 1898–1899.

Royal Astronomical Society:
Fellow, 1868.

Institution of Electrical Engineers:
Foundation Member and Vice President ('Society of Telegraph Engineers'), 1871; Honorary Member, 1899; President, 1874, 1889, 1907.

Institution of Civil Engineers:
Member, 1874; Honorary Member, 1889.

Physical Society of London:
Member, 1875; President, 1880–1881.

Royal Irish Academy:
Honorary Member, 1878.

Royal Society of Arts:
Albert Medallist, 1879; Member, 1880.

21 Patents

Lord Kelvin's name was associated with over 60 patents. The majority were in his name only, except for those from 1900 which were mainly filed by Kelvin and James White Ltd. Apart from a few on fluids and mechanical devices, their subject matter reflected his commercial interests starting from involvement in the Atlantic cable in the 1850s: the electric telegraph, the marine compass and sounder, and instruments to measure electric current, voltage and electricity consumption.

Adapted from 'Life of Lord Kelvin', S.P. Thompson, Macmillan and Co.: London, 1910.

Year	No.	
1854	2547	Thomson (W.), Rankine (W.J.M'Q.) and Thomson (J.). Improvements in electrical conductors for telegraphic communication.
1858.	329	Improvements in testing and working electric telegraphs. (Disclaimer added in 1871.)
	437	Improvements in apparatus for applying and measuring resistance to the motion of rotating wheels, shafts or other rotating bodies.
1860	2047	Thomson (W.) and Jenkin (F.). Improvements in the means of telegraphic communication.
1865	1784	Thomson (W.) and Varley (C.F.). Improvements in electric telegraphs.
1867	2147	Improvements in receiving or recording instruments for electric telegraphs.
1870	3069	Improvements in electric telegraph transmitting, receiving, and recording instruments, and in clocks.
1871	252	Improvements in transmitting, receiving and recording instruments for electric telegraphs.
	810	Improvements in clocks and apparatus for giving uniform motion.
1873	2086	Thomson (Sir W.) and Jenkin (F.). Improvements in telegraphic apparatus.
1876	1095	Improvements in telegraphic apparatus.
	1339	Improvements in the mariner's compass and in the means of ascertaining and correcting its errors. (Disclaimer and memorandum of alteration, 1877.)
	3452	Improvements in apparatus for navigational deep-sea soundings.
	4876	Improvements in the mariner's compass and in appliances for ascertaining and correcting its errors.
1879	679	Improvements in the mariner's compass and in appliances for correcting its errors.
1880	781	Improvements in navigational sounding apparatus.
1881	3032	Improvements in regulating electric currents, and in the apparatus or means employed therein.

	5668	Improvements in dynamo-electric machinery, and apparatus connected therewith.
1883	2028	Improvements in apparatus and processes for generating, regulating, and measuring electric currents.
	4617	Apparatus for generating, regulating, measuring, recording, and integrating electric currents.
	5675	Improved navigational sounding apparatus.
	5676	Improvements in the mariner's compass and in the means for ascertaining and correcting its errors.
1884	4655	New or improved suspensions for electrical incandescent lamps.
	5335	Thomson (Sir W.) and Ferranti (S.Z. de). Improvements in dynamo-electric machinery.
	6410	Improvements in breaking electric contact to prevent over-heating by imperfect contact.
	10,530	Thomson (Sir W.) and Bottomley (J.T). Safety fuses for electric circuits.
	11,106	Improvements in apparatus for measuring electric currents.
1885	12,240	An improved navigational sounding machine.
1886	9016	Improved apparatus for measuring the efficiency of an electric circuit (amended, 4 October, 1897).
1888	18,035	Electrostatic apparatus for measuring potentials.
	18,035a	An improved ampere–gauge and connections.
	18,035b	Improved apparatus for continually measuring potentials or currents.
1889	4923	Improvements of compass cards for use on hoard ship (amended, 1 July 1890).
	5471	Improvements in valves for water, steam, or other liquids or gases.
	5769	Apparatus for measuring and recording electric currents.
1890	8959	Improvements in the mariner's compass.
1891	1004	An improved indicator for electric potentials.

	3864	Improvements in valves for water, steam, or other liquids or gases.
	18,436	Improved apparatus for measuring and recording electric currents.
1892	10,230	An improved electric condenser.
1893	2198	Improvements in balances.
	2199	An instrument for measuring electric currents.
	5733	Improved arrangement for reading the deflections of electric instruments.
	24,471	Improvements in electric supply meters.
	24,841	Improvements in the mariner's compass.
	24,979	Improvements in instruments for measuring and recording electric pressures and currents.
1894	5034	Improvements in instruments for measuring electric currents.
1895	22,661	Improvements in apparatus for indicating and recording electric supply.
	24,868	Improvements in recording instruments for telegraphic and other purposes.
1896	4207	Improvements in instruments for measuring electric currents.
1897	18,438	Improved coil for electric instruments.
	25,178	Improvements in navigational sounding machines.
1898	18,522	An improved electrolytic apparatus for the production of alkali.
	21,716	Improvements in electric measuring instruments.
1900	3937	Apparatus for indicating and recording electric pressure and current.
1902	17,267	Improved electrostatic high-pressure voltmeter.
	22,030*	Improvements in and connected with navigational sounding machines.
	22,031	Improvements in the mariner's compass.
	22,032*	Improvements in the mariner's compass.
	28,588*	Improved magnetic brake and damper of vibrations.

1903	24,526*	Improved sounding machine.
1904	22,695	Improvements in the mariner's compass.
1905	4897	Kelvin (Lord), Blakiston (R.), Hope (W.), and Richards (G.B.). Improvements in liquid meters.
	20,813*	Improvements in and connected with navigational sounding machines, specially for flying soundings.
	20,813a*	Improvements in and connected with navigational sounding machines, especially for flying soundings.
	26,759*	An improved apparatus for adjusting the compass at sea without sights.
	26,759a*	An improved apparatus for adjusting the compass at sea without sights.
1906	26,132*	An improvement in connection with navigational sounding machines, especially for flying soundings.
1907	12,783*	A combined level and shadow pin for use at sea in connection with the mariner's compass.
	23,570*	An improved compass suspension for use on board ship.

N.B. Those patents marked with an asterisk in the above list were taken in the name of Lord Kelvin and the firm Kelvin & James White, Ltd.

22 Notes

[a] Robert Shipboy McAdam (1808–1895), a Presbyterian, was an Irish scholar and founder of the Ulster Gaelic Society as well as being an inventor and owner of a Belfast foundry.

[b] Philip Kelland (1808–1879) was born in Somerset and graduated in 1834 from the University of Cambridge as Senior Wrangler and Smith's Prizeman. He was appointed as a professor of mathematics in the University of Edinburgh in 1838.

[c] Peter Guthrie Tait (1831–1901) from Dalkeith, near Edinburgh, graduated in 1852 from the University of Cambridge as Senior Wrangler and Smith's Prizeman. He was a professor of natural philosophy in the University of Edinburgh (1860–1901) and was a long-term friend and collaborator of William Thomson and co-author with him of Treatise on Natural Philosophy (1867).

[d] Henri Victor Regnault (1810–1878) was a French chemist and physicist best known for his careful measurements of the thermal properties of gases. In 1840, he was appointed as the chair of chemistry of the École Polytechnique, and, in 1841, as a professor of natural philosophy in the Collège de France, both in the Paris environs. The Regnault crater on the Moon is named after him.

[e] James Prescott Joule (1818–1889) was born into a brewing family in Salford, Lancashire. He is chiefly associated with the First Law of Thermodynamics, the mechanical equivalent of heat and the Joule–Thomson effect.

[f] Sir Charles Wheatstone (1802–1875) was an English scientist and inventor who contributed widely to optics, acoustics and electricity, and, with Sir William Fothergill Cooke (1806–1879) developed the electric telegraph into a practical system of communication.

[g] Caithness born Alexander Bain (1811–1877) was a pioneer of electric horology and the printing electric telegraph. He had a long running feud with Wheatstone over priority concerning the electric clock, while his attempts to build a telegraph empire in the USA were eventually crushed by Samuel Morse over alleged patent infringement. He is best known as the inventor of the facsimile machine in which he demonstrated the concept of the raster line scan for the first time and electrical transmission of images.

[h] Edward Orange Wildman Whitehouse (1816–1890), an English surgeon, was recruited by Cyrus Field as chief electrician of the transatlantic telegraph cable for the Atlantic Telegraph Company.

[i] Sir Joseph Whitworth (1803–1887) was an English mechanical engineer and inventor who developed machine tools, munitions and a steel-making process. He is best remembered for devising the British Standard Whitworth screw thread in 1841.

[j] Alexander Crum Brown (1838–1922) was a physician and professor of chemistry in the University of Edinburgh (1869–1908) and made numerous contributions to the developing subject of organic chemistry.

[k] Hermann von Helmholtz (1821–1894) was a German physician and physicist who is best known for his work in physiology (ophthalmology and perception of sound) and physics (energy and electromagnetism).

[l] Carl Wihelm (Sir Charles) Siemens (1823–1883) was a German born engineer and inventor, and brother of Ernest Werner von Siemens, founder of the Siemens engineering company.

[m] Sir George Darwin (1845–1912), Charles Darwin's fifth child was a lawyer, astronomer and mathematician.

[n] Sir George Gabriel Stokes (1819–1903) was an Irish mathematician and physicist who made important contributions to fluid mechanics, optics and spectroscopy. A near contemporary of Kelvin at Cambridge, he was Senior Wrangler and Smith's Prizeman in 1841.

[o] Sir William Armstrong (1810–1900), first Baron Armstrong, was a Tyneside engineer and industrialist and one of the founders of the Armstrong Whitworth engineering company.

[p] At the time of writing, 130 years' worth of data from the Belfast Harbour Commissioners' Kelvin Tide Gauge are being analysed in a joint project between the School of Geography, Archaeology and Palaeoecology, Queen's University of Belfast and the Northern Ireland Environment Agency for evidence of long-term sea-level trends.

[q] Silvanus Thompson (1851–1916) was a physicist and author of textbooks and the definitive biography of Lord Kelvin.

[r] Jonas Ferdinand Gabriel Lippmann (1845–1921) was a French physicist and Nobel laureate best known for inventing a method of colour photography based on optical interference.

[s] Woldemar Voigt (1850–1919) was a German physicist who contributed to crystal physics, thermodynamics and electro-optics.

[t] Christian Christiansen (1843–1917) was a Danish physicist who worked in spectroscopy and optics and invented the Christiansen optical filter based on dispersion in a suspension of a transparent dielectric in a liquid of slightly different refractive index.

[u] The Victoria Institute was founded in 1865 in response to the publication of 'On the Origin of Species' and now known as Faith and Thought, it is a forum for those wishing to reconcile Christian faith and science.

[v] The Rev. George Henslow (1835–1925) was the younger son of John Stevens Henslow (1796–1861), an English botanist and geologist and mentor to Charles Darwin. George Henslow was a school teacher and later lecturer in botany in London's St Bartholomew's Medical School.

References

The primary source is 'Life of Lord Kelvin', S.P. Thompson, Macmillan and Co.: London, 1910.

[1] Moore, A.S., *Old Belfast*, Carter Publications: Belfast, pp. 69–70, 1951.
[2] Smith, C. & Wise, N.W., *Energy and Empire*, Cambridge University Press: Cambridge, p. 12, 1989.
[3] Merrick, A.C.Y. (comp.) and Clarke, R.S.J. (ed.), *Old Belfast families and the New Burying Ground*, Ulster Historical Foundation: Belfast, 1991.
[4] Muir, J., *John Anderson and the College He Founded*, John Smith & Son Limited: Glasgow, p. 119, 1950.
[5] Thomson, R.Y., (ed.), *A Faculty for Science – A Unified Diversity*, The University of Glasgow: Glasgow, 1993.
[6] Murray, D., *Lord Kelvin as Professor in the Old College of Glasgow*, MacLehose, Jackson and Co.: Glasgow, 1924.
[7] Thompson, S.P., *Life of Lord Kelvin, Baron Kelvin of Largs*, MacMillan and Co. Limited: London, 1910.
[8] Thompson, S.P., *Life of Lord Kelvin, Baron Kelvin of Largs*, MacMillan and Co. Limited: London, pp. 570–575, 1910.
[9] Obituary of Archibald Smith, *Proc. Roy. Soc.*, **22**, pp. i–xxiv, 1873.
[10] National Dictionary of Biography, vol. 53, Smith, Elder and Co.: London, pp. 16–17, 1898.
[11] Smith, C. & Wise, N.W., *Energy and Empire*, Cambridge University Press: Cambridge, pp. 140–146, 1989.
[12] Muir, J., *John Anderson and the College He Founded*, John Smith & Son Limited: Glasgow, p. 115, 1950.

[13] Thomson, M., *Verses by M T*, privately printed by MacLehose and MacDougall: Glasgow, 1866 and 1874.
[14] Thomson, W., On lighthouse characteristics. *Paper read at the Naval and Marine Exhibition*, Glasgow, 11 February, 1881. Reprinted in *Popular Lectures and Addresses*, vol. **iii**, pp. 389–421, 1881.
[15] Thompson, S.P., *Life of Lord Kelvin, Baron Kelvin of Largs*, MacMillan and Co. Limited: London, p. 730, 1910.
[16] Smith, C. & Wise, N.W., *Energy and Empire*, Cambridge University Press: Cambridge, pp. 779–811, 1989.
[17] *Celebration of Jubilee of Lord Kelvin 15th, 16th, and 17th June, 1896*, Glasgow University Press: Glasgow, 1899.
[18] Hidden, A.E. & Latimer, C.J., *Science and Technology: Belfast and its region*, Queen's University of Belfast: Belfast, 1987.
[19] O'Neill, J.J., *Prodigal Genius*, I. Wasburn: New York, 1944.
[20] King, A.G., *Kelvin the Man*, Hodder and Stoughton Limited: London, 1925.
[21] Ruddock, I.S., Lord Kelvin, science and religion. *Physics World*, **17(2)**, p. 56, 2004.
[22] Thompson, S.P., *Life of Lord Kelvin, Baron Kelvin of Largs*, MacMillan and Co. Limited: London, p. 1197, 1910.

Chapter 2

Kelvin and his world: a cultural overview

M.W. Collins[1] & I.S. Ruddock[2]
[1]*College of Engineering, Design and Physical Sciences, Brunel University London, United Kingdom.*
[2]*Department of Physics, University of Strathclyde, United Kingdom.*

Abstract

Any study relating to William Thomson, Lord Kelvin, is a challenging task. In this introductory overview, we address four themes: Kelvin himself, his faith, his politics and his relationship with the industrial revolution and the industrial world. The important question of the change in Kelvin's reputation from the beginning to the end of the twentieth century is addressed in parallel.

1 A complex challenge

This volume concerns the towering figure of Victorian science and engineering, William Thomson, Lord Kelvin. We see a teenage prodigy who by virtue of his sheer native ability, personal energy and influential connections was awarded a key Glasgow professorship at the age of 22. In turn, this gave him a very long working lifespan which he exploited to the full. He attained national status in the Britain of high empire, and his achievements had world-wide significance in science and engineering.

In consequence, a study of virtually any aspect of Kelvin is demanding, in terms of scale and complexity, as shown in this volume's initial chapter *Lord Kelvin* by Ivan Ruddock. Crosbie Smith and Norton Wise's biography [1], for instance, is not only over 900 pages long, but includes 23 pages of bibliography and almost 2000 footnotes. When the other publications of Smith, addressing the cultural aspects, are included [2, 3], a grand total of some 1670 pages is involved, while the first substantial biography was Silvanus Thompson's 1910 epic two-volume *Life of Lord Kelvin* with 1297 pages [4]. But to understand Kelvin we have to go further. He was an inheritor of the first stage of the Industrial Revolution (Chapter 5) being born only 5 years after the death of James Watt. The three-volume biography of Watt by Richard Hills [5–7] similarly comprises over 1000 pages of text. Then, there is the intellectual galaxy of Kelvin's British and international peers, and the overall scientific context of the age in which they worked. Combining these aspects is the vision of Donald Cardwell [7, 8] and Bruce Hunt [8, 9]; the subtitles of [7] and [8] given in the References list also underline this vision. Cardwell and Hunt's relatively modest page counts cannot obscure their long and deep devotion: '…I have long been interested in the relationships, historical and contemporary, between science and technology, or the industrial arts', for Cardwell [8, p. ix], and 'This book has had a long gestation … the leave I needed to complete this book', for Hunt [9, p. vii].

What greatly increases the task is addressing the dialogue between the biological and energy experts. This requires the consideration of a parallel enterprise to that of Smith & Wise for Kelvin. Adrian Desmond's biographies of *Darwin* [10] and *T.S Huxley* [11, 12] add a further 1700 pages of information to be absorbed! Only by doing this, however, can issues such as the X-Club be assimilated in the round. (The X-club was a dining club of nine prominent scientists, including Huxley, established in 1864 in the wake of the publication of the *Origin of Species*. Although ostensibly created as a social and scientific network, the membership espoused causes such as naturalism, liberal Anglicanism and the professionalisation of science, and wielded considerable influence within the nineteenth century scientific establishment for almost 30 years.)

The publishing of material does not stop. For Kelvin, there was David Lindley's *Degrees Kelvin* [13] in 2004, and then a multi-authored book, simply entitled *Kelvin* [14] that stemmed from the 2007 Centenary commemoration events of his death held at Queen's University of Belfast on 7th December.

The sub-title of [2] – *A Cultural History of Energy Physics in Victorian Britain* – describes Crosbie Smith's approach to the development of thermodynamics and energy in nineteenth century Britain. Kelvin features throughout, from the first line of page 1 to the last paragraph of page 324, the latter via his doctrine of 'energy dissipation'. Smith explains his approach in a closely argued and information rich Introduction. We will highlight the, *for us*, crucial argument. Smith points out that recent studies of Charles Darwin represent the battle of the X-club of Huxley *et al.* as being principally with Anglican Cambridge, i.e. a combination of establishment science and theology to the Church of England. 'I argue' says Smith 'that the battle lines need to be redrawn to include as a third group our North British scientists of energy, a group not assimilable to Cambridge Anglicanism even though alliances between Scotland and Cambridge were commonplace' [2, p. 7]. In our own book, which focuses on Kelvin himself, we follow Smith's approach, re-expressing it somewhat as between thermodynamics and biology. Mainstream biology now fully accepts that the Laws of Thermodynamics apply just as much to the living as to the engineered world *and seeks to apply them*. Also, since a main motive for this volume is to examine why Kelvin's reputation is only moderate, and to seek to aid in its recovery, we show how enduring is Kelvin's thermodynamic worldview. Specifically, how his principal of universal dissipation anticipated the rationale of Ilya Prigogine's Dissipative Structures, now starting to appear in medicine. In 2], Smith continually addresses the underlying theological background of the energy specialists, and in *Energy and Empire* [1], Smith and Wise include the interaction between Kelvin's activities and the high period of the British Empire.

2 An intellectual colossus (Crowther [15, p. 291])

2.1 His mathematics, physics and engineering

He deserves to be called a mathematician, as well as a physicist and even an engineer.

Denis Weaire [16, p. 57]

... we use the term 'practical' ... it suggests the all-pervasive spirit of engineering and industry that we elaborate in his life and work.

Smith and Wise, introducing their biography [1, p. xxi]

At the age of thirty-five, after a dozen years of almost unparalleled fertility in pure scientific research, he became one of the best and most famous engineers of his time.

***After the first Atlantic cable attempt*: James G. Crowther [15, p. 274]**

Such, then, were the first fruits of mathematical physics ... the era of the second industrial revolution had truly dawned.

***After the successful fourth Atlantic cable attempt and his knighthood*: Smith and Wise [1 p. 683]**

Thomson had virtually an entire career as an engineer, as well as one of a physicist ... one of the great engineers of an age of engineering.

David Wilson, comparing Kelvin with his great friend G.G. Stokes [17, p. 12]

The above quotes identify Kelvin as a mathematician, physicist and engineer. Firstly, his mathematical ability was of the highest level. Expected to graduate as Senior Wrangler (best mathematics undergraduate) at Cambridge, he came second, although he reversed the order by winning the Smith's Prize [1, pp. 80–82]. Despite 'only a fraction of his published output' being 'predominantly mathematical' the total was 'so vast' that the mathematics was 'more than a lifetime's work' [14, pp. 55–57]. His innate ability and extreme energy combined to lead to Weaire's claiming him as a mathematician.

We turn to physics, the world in which Kelvin's mathematical genius was applied. His interdisciplinary character was there from the start. The influence and contributions of his elder brother James Thomson ('little-known engineers ... James Thomson', dust jacket text [1]) are often unjustifiably forgotten (and addressed by Chapters 3 and 4 in this volume by Andrew Whitaker). In introducing James's portrait at 16 years of age, Smith and Wise describe the zeal of two *teenagers*: 'James's engineering enthusiasms from an early age provided enduring inspiration for his brother's physics' [1, p. 52]. Kelvin's understanding and application of physics was little short of phenomenal with over 600 papers and 60 patents [16, p. 55]. Smith and Wise refer to him as a mathematical physicist ([1, dust jacket text, p. xxii] and Chapter 6, for example), indeed as Kelvin did himself [1, p. 682] but that hardly does justice to his brilliant inventiveness.

The combination of mathematics and training in experimental methods won Kelvin his Glasgow chair. His university subject area of natural philosophy comprised '... the two instruments of investigation, experiment and mathematics' [1, p. 122]. His university laboratory became an 'empire' in the old University of Glasgow building (p. 132) and thence to a 'vastly superior' laboratory (p. 135) when the University moved in 1870. It was a further four years before the Cavendish Laboratory at Cambridge was set up, 'part stimulated.by ... Glasgow' [15, p. 238].

Returning to his inventiveness, the most publicly spectacular demonstration of it concerned those 'numerous telegraphic patents' [1, p. 667] for the great Atlantic submarine cable enterprise. This required four attempts with increasing input from the plain William Thomson. It was here that his double expertise in physics-oriented mathematics and invention fused into true engineering and resulted in his knighthood [1, p. 682; 15, p. 254].

Smith and Wise take his mathematics and physics further into engineering values as follows: 'for the engineer, economics and natural philosophy were not separate subjects' [1, p. 666] and consistently after the cable success: 'Sir William regarded the question of economy as paramount' [1, p. 684].

All in all, then, Kelvin had added brilliance in engineering to that he had already shown in mathematics and physics.

2.2 An iconic figure in scientific history

It was at such venues that the once-local names of Faraday, Kelvin, Joule and Watt entered into a 'universal' language of science.

Ben Marsden and Crosbie Smith [3, p. 219]

This communication from the young professor, not yet twenty-three years old, to Faraday astounds and stupefies the imagination.

Crowther, on Kelvin's letter of 11 June 1847 [15, p. 251]

Constructing knowledge in private: the Joule Thomson debates (1847)

Smith, Section Heading [2, p. 78]

Maxwell and Thomson agreed not only on the purpose of the demon, but in large degree on its nature.

Smith and Wise, on 'Maxwell's demon' [1, p. 623]

Fifty-eight years of friendship: Kelvin and Stokes.

Alastair Wood, Chapter Title [14, p. 64]

Helmholtz's theory and his own meteoric theory might complement one another.

Smith and Wise, on Kelvin's 1854 address to the British Association [1, p. 526]

To mathematical physics, and especially to Fourier, Sir William Thomson owed his strength in commercial telegraphy.

Smith and Wise, after the success of the Atlantic cable [1, p. 682]

The scientific world in which Kelvin worked was replete with iconic names in the historical development of science. He interacted with all of the following scientists listed below, almost all of whom are named in *The Complete Family Encyclopedia* [18]. For our purposes, this reference is important in so far as it gives a guide to *assessed reputation with the general public*, a significant issue in this introduction.

(a) In the United Kingdom:
Michael Faraday (1791–1867) (37 lines)
'English chemist and physicist', 'three laws of electromagnetic induction and two laws of electrolysis, all proposed originally by Michael Faraday'.
James Clerk Maxwell (1831–1879) (17 lines) 'contributed to every branch of physical science'.
James Prescott Joule (1818–1889) (27 lines) 'British physicist whose work … led to … the first law of thermodynamics'.

George Gabriel Stokes (1819–1903) (14 lines) 'Irish physicist … in 1840s studied the viscosity of fluids'.

(b) In Germany:
Rudolph Julius Emanuel Clausius (1822–1888) (9 lines).
'One of the founders of the science of thermodynamics' *Hermann Ludwig Ferdinand von Helmholtz* (1821–1894) (14 lines) 'German physicist … formulated the law of conservation of energy and did important work in thermodynamics'.

(c) In France:
Jean Baptiste Joseph Fourier (1768–1830) (10 lines) 'French applied mathematician whose formulation of heat flow 1807 … '
Victor Regnault (1810–1878) (not referenced in [18]) 'If Fourier stood as the heroic master of mathematical analysis in the theory of heat, Regnault stood as an equal master of experimental measurements' [1, p. 128].
Nicolas Leonard Sadi Carnot (1796–1832) (15 lines)
'French scientist … who founded thermodynamics'.

Information about Kelvin's interactions with Joule and Maxwell is given in Chapters 8 and 9 of this volume by Andrzej Wroblewski and David Wilkie *et al.*, respectively.

In conclusion, we return to the question of public impact and reputation. Firstly, in this volume's Chapter 7, Section 14.1, a wider comparison exercise has been undertaken which includes the biologists. There it is noted that biology has a much greater public impact than thermodynamics. Secondly, in his Foreword to the 2008 *Kelvin* volume [14], Sir Brian Pippard says that Kelvin's researches hardly get a mention 'in a present-day physics students lecture course'. Regarding Stokes, too, 'his mentor and lifelong friend Sir George Stokes has fallen into similar obscurity' [14]. For two of the Editors (MWC and CSK) writing as engineers, the Navier–Stokes equations are the foundation of Computational Fluid Dynamics, the joint names being encountered perpetually.

2.3 First among equals

…. 1896 saw Glasgow University celebrate the 50th Jubilee of its most distinguished professor … certificates of membership of the 80 or so learned societies across the world which had elected Thomson to their ranks ….

Mark McCartney [14, p. 21]

… the Royal Society … in 1924 to celebrate the centenary of Kelvin's birth … from Japan stated: The National Research Council of Japan specially desires to recall with gratitude … his eminent disciples as organisers. Dyer, Gray, Ayrton, Perry, Ewing and Knott, personally recommended by Kelvin himself, came to the Far East.

Colin Latimer [14, pp. 213–214].

Kelvin was first among equals in the world of British science and engineering, notwithstanding the competition of the biologists. A national level of scientific attainment in the UK is recognised by Fellowship of the Royal Society (of London) (RS), which dates back to 1645. There is a Scottish equivalent, the Royal Society of Edinburgh (RSE), which originated in 1783. Kelvin was a Fellow of both. He was elected to the RS in June 1851, at the same time as his friend Stokes and antagonist Huxley. More significant than Fellowship, and unique at any one time, is the elected Presidency. The doubly unique Kelvin served both as President of the RS from 1890 to 1895 and as President of the RSE no less than three times totalling 21 years (1873–1878, 1886–1890 and 1895–1907). This is a self-evident manifestation of the highest regard that Kelvin enjoyed among his peers; moreover, that regard spanned more than three decades. Regarding the RS, there was a problem with the biologists. T.S. Huxley, besides being a radical for *science*, was a reactionary when it came to *technology* and *engineering*. 'He lashed the techno-flunkeys … the 'Engineers, Chemical traders … who find it helps them to appear to the public as if they were men of science'…He feared the veneration switching to science's products … its 'froth and scum', its engines and telegraphs, as he angrily put it to Disraeli.' [12, pp. 249–250]. Consistent with this, Huxley did not want the RS 'exploited by enterprising commercial gents who make their profit out of the application of science'. So he was persuaded to stand for President 'because this would keep out Kelvin' ([14, p. 281], quoting from [19]). In the event, Huxley had to resign early on health grounds, Stokes followed him and then Kelvin followed Stokes.

The position of President of the RS carried considerable national influence which Huxley used as much as he was able: 'It also gave Huxley the ear of Prime Ministers and Privy Counsellors' [12, p. 149]. Subsequently, no doubt Huxley felt justified when Kelvin did indeed quote his President's position in commercial advertising, bringing upon himself considerable criticism. Finally, turning to Glasgow University, 5 years after retiring from his Chair [1, p. 115] in 1904 [14, p. 13] Kelvin was elected as the chancellor. Including his years as undergraduate and retirement 'research student', Kelvin was a member of the University for 67 out of his 83 years lifetime. Moreover, given the extreme age of the university (founded in 1451), this represented an amazing one-seventh of its entire existence.

2.4 Kelvin's faith and worldview: Huxley and Darwin's agnosticism

Kelvin, who was a practising Christian … In his view, a creative power was needed for animate – but not inanimate – objects.

Ivan Ruddock, on the 1903 debate in The Times on the origin of life [20].

… the cosmic evolution of the earth as an abode fitted for life seemed to have been the result of benevolent and wise design by a God who had chosen well …

Kelvin's 'reply to Hooker's criticism of natural theology … 1868', Smith and Wise [1, p. 641].

The debate between the Thomson brothers contributed to a major discourse among Scottish natural philosophers. During the 1860s and early 1870s, Maxwell, Jenkin, Tait, and Balfour Stewart (1828–1887) all participated. Their special concern with thermodynamics and molecular physics in relation to free will was at first largely a response to their German friend Helmholtz.

Smith and Wise [1, p. 617].

We have the sober scientific certainty that heavens and earth shall 'wax old as doth a garment' [Psalm 102: 26] … the theological framework was an unambiguous expression of his own deepest convictions.

Smith and Wise, quoting Kelvin & Tait in Energy, 1862 [1, p. 535].

He portrayed agnosticism not as a rival 'creed', but as a method of inquiry. The sciences ... 'are neither Christian, nor Un-Christian, but are Extra-Christian', in a word, 'unsectarian'.

Desmond, on Huxley speaking to the Young Men's Christian Association [11, p. 374].

The mystery of the beginning of all things is insoluble by us; and I for one must be content to remain an Agnostic.

Charles Darwin, autobiographical material on Religious Belief [21, p. 54].

... a man undoubtedly can be 'an ardent Theist & an evolutionist', look at Charles Kingsley and Asa Gray. For himself, he had 'never been an atheist ...'

Desmond and Moore, on Darwin's views late in life [10, p. 636]

In current scientific and engineering publications, matters of personal belief are conspicuous by their absence. After all, the reliability of both predicted and experimental scientific data should be reproducible and independent of the views of the researcher. However, in scientific history and certainly here, *it is virtually obligatory* to address such belief systems. They formed an essential part of Kelvin and colleagues on the one hand, paralleled by their positive lack in the worldview of the biologists on the other. This fundamental disagreement became a matter of intense interest in British public and scientific life in the late nineteenth century

Kelvin had a clear Christian faith, in common with his North British energy peers. Smith describes it thus: 'The scientists of energy belonged to a post-evangelical culture whereby human beings, in their moral and material actions, had an obligation to aspire to the perfections of nature and of Christ' (Chapter 2, Note 15 [2, p. 317]). The conflict with Tyndall and the X-club biologists is also represented in precise and careful terms by Smith as between: '*scientific naturalism* (united by a stand against Christian doctrine) and *the science of energy* (promoted as a natural philosophy in harmony with, though not subservient to, Christian belief)' (ours in italics [2, p. 172]). The three aspects of *design, free will* and *finite duration and universal dissipation of the cosmos* are highlighted by the relevant quotes above. Here, the significance is that Kelvin's faith was strongly based on an intelligent interpretation of the Bible: the above aspects of his basic convictions became part of a thermodynamics worldview.

In absolute contrast to the explicit Christian faith of Kelvin and his energy peers was the *agnosticism* promulgated by Huxley and his fellow biologists. In fact, Huxley actually *invented the word*. The current dictionary meaning of *agnostic* is virtually identical to Huxley's: 'a person who believes that we know nothing of things beyond material phenomena – that a Creator, creative cause, and unseen world are unknown or unknowable things [coined by T.S. Huxley in 1869 from Greek *agnostos*, unknown, unknowable]' (Chambers Concise Dictionary [22, p. 18].

Finally, there was Darwin himself. The first quote shows that agnosticism was also his position, but his writings betray a certain reluctance to obliterate the traces of being a 'Theist' [21 pp. 53–54]. In fact, he was 'Never an Atheist' (also Title of Chapter 41 [10]). In contrast to Kelvin's robust health, Darwin seemed to be perpetually in poor health. No reader can remain untouched by the graphic record by Desmond and Moore of the intensity of his suffering when he died. Darwin's uncertainty, too, was reflected on that day: 'Oh God', he cried helplessly, 'Oh Lord God' [10, p. 662].

2.5 Kelvin's politics: Irish Home Rule

Already privately converted to Home Rule, Gladstone took over by February 1886.

Smith and Wise [1, p. 803]

... his (Kelvin's) opposition, a continuation of ... the liberal and political outlook of his father.

Smith and Wise [1, p. 803]

The X had regrouped ... Tyndall wanted them to draw up 'a scientific declaration' for the Union.

Desmond [12, p. 168]

Kelvin's politics are a significant issue in two respects. They were certainly a contributory, possibly the principal, factor in the award of his peerage. Of equal importance, given our interest in the cultural background, is his reasoning for being against Irish Home Rule. The story is told in some detail by Smith and Wise in Chapter 23: Baron Kelvin of Largs [1].

The complex political structure of the United Kingdom and its history are given in this volume's chapter Conclusion: Honoured by Banknotes. Over the whole of Kelvin's lifetime, Ireland was in political union with Britain (1801–1921) sending Members of Parliament to Westminster. However, the political unity hid a consistently troubled history involving much violence. This resurfaced in the latter half of the nineteenth century, with the disestablishment of the Irish Church and land reforms occurring around 1870. After 1885, it dominated British politics. In the general election of that year, the Irish MPs held the balance of power. The Liberal Prime Minister Gladstone [1, p. 803] led a Government pledged to grant Home Rule. This irrevocably fractured an already divided Liberal party and the Home Rule Bill of 1886 was defeated. The election of 1886 produced a new government led by the Conservative Salisbury. It was a strong coalition including Liberal Unionists, meaning that Irish Home Rule was a dead political issue. (History would cruelly vindicate both Gladstone and Salisbury within 10 years of Kelvin's death – Home Rule was successfully achieved via armed rebellion but it was accompanied by partition, with Northern Ireland remaining in Union with Britain).

Within this story, Kelvin who had been a 'faithful supporter' [1, p. 803] of the Liberals, left them at the time of the split and campaigned heavily for the Liberal Unionists in the 1886 election. Smith and Wise are in no doubt – Salisbury offered Kelvin a peerage essentially because of this support [1, p. 807]. However, although they dismiss the idea (current at the time in the press and subsequently by biographers) of the peerage being awarded primarily for his national scientific prominence, the latest Kelvin volume [12] repeats this view. '...his peerage...was in considerable part for his support of the Liberal Unionists in Lord Salisbury's coalition government, but it was generally assumed that it was for his work in science and engineering and, as such, it was extremely widely welcomed' (Whitaker [14], p. 279).

Kelvin's reasons for supporting Unionism are illuminating (explained in detail in [1, pp. 802–806) and resonate with our theme of cultural background. On the one hand, 'Britain and the British Empire were now the best guarantee of liberty for all the citizens of Ireland': Kelvin 'had in mind the British notion of equality before the law'. On the other hand industry and manufacturing (in the wake of the industrial revolution) had brought 'unprecedented economic growth' to Belfast, widening the gap between the north and south of the island. Kelvin wanted to see 'the growing economic prosperity of the province – founded on industry, technology and ingenuity'.

Unionism was one issue that united Kelvin and the biologists. 'Huxley thought the destruction of the Union 'a cowardly wickedness' … Tyndall wanted the X to draw up 'a scientific declaration' for the Union' [12, pp. 167–168]. Interestingly, though, Huxley 'scuppered this' being aware of the Separatist groundswell among the younger Fellows of the Royal Society.

2.6 Kelvin's national status

When his peerage was awarded … it was extremely widely welcomed.

Andrew Whitaker [14, p. 279]

Buried beside Newton in Westminster Abbey.

McCartney [14, p. 13]

The national status of Kelvin was high in the extreme, with his *peerage*, his *Order of Merit* award, appointment to the *Privy Council*, and *burial in Westminster Abbey.*

In January 1892, he became the 'first scientist' to receive a peerage [3, p. 218]. In fact, the honours list citation quoted 'for most valued service to science and progress in this country' [17, p. 84]. In 1887, his lifelong friend Stokes had been elected to Parliament as an Independent for the Cambridge University seat, and made a baronet (essentially a hereditary knighthood [14, p. 79]) in 1889 . Kelvin viewed the situation as giving them parallel voices for science, and he himself spoke in the House of Lords on a number of occasions [14, p. 84].

Twenty-first century Britain has two highly restricted and classless national honours, *Order of Merit* and *Companion of Honour*: both are termed as 'orders of chivalry'. The Order of Merit was founded by King Edward VII in 1902 and restricted to 24 members, with the Companion of Honour founded by George V in 1916 and restricted to 65 members [18, pp. 205, 616]. They are without title, but with designatory letters OM and CH, respectively. *Kelvin became an inaugural member of the OM in 1902.*

The *Privy Council* originated with the Norman kings, became the English governing body under the Tudors and was displaced by the Cabinet (the existing British system) in 1688. Cabinet members were and are automatic members of the now formal body [18, p. 674]. To be made a Privy Councillor (PC) is both an honour (with the title Right Honourable) and a position of national influence. Kelvin was made one in 1902.

This links with the biologists, because Huxley, too, was made a PC but earlier than Kelvin in 1892. Huxley was 'dead against' decorations, but Kelvin, by then a Lord, 'assured him that it was something he could accept' and Huxley's appointment made him 'the first Crown adviser on science' [11, pp. 210–211]. What is equally telling, however, is that scientific disagreement with the biologists was not personal with Kelvin.

Finally, Kelvin received the ultimate double accolade of burial in Westminster Abbey and that beside Isaac Newton [14, p. 13]. Again, this can be compared with the biologists: in this case

Darwin, who substantially predeceased Kelvin in 1882. The story of Darwin is detailed by Desmond and Moore in Chapter 44 [10] and what is helpful for the reader is the explanation of how it was done and the criteria involved [10, pp. 664–668). As with Kelvin later, Newton was the gold standard of comparison: 'Was he not the greatest Englishman since Newton … Newton's honour had to be Darwin's' [10, p. 668].

3 Kelvin and the industrial world

3.1 The timescale of the industrial revolution: Kelvin's participation

…. Lord Kelvin, the most famous mathematical physicist of the nineteenth century Britain … … William Thomson identified himself as much with the shipyards and engineering works of his adopted city of Glasgow as with the democratic education offered in its university

Smith and Wise, dust jacket text, Energy & Empire [1]

Timescale estimates of the traditional understanding of the Industrial Revolution (I.R.) are given in Table 1. Toynbee's original dating, given in 1884, of 1760-1840 has been followed until relatively recently. Crump's very recent 'Brief History' extends the start to the early 1700s. Toynbee's *finishing date* is also important for us, as it implies that Kelvin was simply *an inheritor* of that Revolution.

These days a more reasoned interpretation of the Industrial Revolution gives it three stages: *The First, The Spread,* and *The Second,* as summarised in Table 2 using Koeller's analysis [25]. Smith and Wise also discuss this chronology [1, pp. 652–653].

Kelvin's working lifetime, also given in Table 2, therefore covers over half of the second and the whole of the third stages of the redefined I.R. This is really significant: it means that Kelvin is no longer just an inheritor, but a *participant in the wider Industrial Revolution of 1700–1900.* In the interests of Kelvin's reputation we expand the point.

Table 1: The Industrial Revolution according to Toynbee's traditional timescale.

Ref.	Comment	Timescale
[23]	Arnold Toynbee's original estimate (1884)	1760–1840
For example, [18, 22]	Range from encyclopaedia	1750–1850
[24]	Thomas Crump's 'Brief History … .' (2010)	c. 1700–1850

Table 2: Kelvin and the industrial revolution after Koeller [25].

Stage of the I.R.	Subject area	Timescale
First I.R.	Textiles and steam	1712–1830
Spread of the I.R.	Steamships, steel, oil	1830–1875
Second I.R.	Electricity and chemicals	1875–1905
Second I.R.	After Smith and Wise [1, p. 652]	c. 1865–1900
Kelvin's working lifetime at G.U.	Appointment to Chair	1846
	Retirement from Chair	1899
	Active until death	1907

Firstly, Kelvin now should be recognised as an *increasingly important figure in the Spreading Stage of the Industrial Revolution.* Koeller's first event in this stage is the inception in 1840 by Samuel Cunard of the transatlantic steamship service [25]. The date resonates with Kelvin's 1840s activities related to 'the magnetism of iron ships and its effect on compasses' [1]. This continued for the decades up to 1880 by when his patented instruments had become the standard of the commercial fleet of the Empire. The 1840–1880 timescale agrees with Kubicek's 'mid-Victorian period' in which 'Kelvin developed telegraphic, directional and sounding devices of great importance ... on ocean routes' [26]. Secondly, Kelvin should be seen as a *principal British participant in the Second Industrial Revolution's primary subject area of electricity.* This affiliation includes the Atlantic cable, but is much wider; indeed, Smith and Wise devote three chapters of [1] (some 150 pages) to Kelvin's achievements.

4 The telegraph and Empire

4.1 The 1866 Atlantic cable and Kelvin

When the ... cable finally reached the American coast on 27 July 1866, having been successfully unrolled and dropped along the ocean floor by Isambard Kingdom Brunel's mighty Great Eastern, it was plainly the dawn of a new era.

Niall Ferguson [27, pp. 168]

... the spread of the telegraph network meant that details from even the most far-flung battlefields could reach Britain in twenty-four hours, the time it took ... in June 1896

Lawrence James [28, p. 211]

... the one triumphantly unifying factor was cable construction: 'the nerves of the empire' by the 1890s girdled the world with 121,000 miles of telegraph wires. The web of submarine cables transformed the conduct of trade and diplomacy, as well as colonial administration.

Ronald Hyam [29, p. 19]

The British Empire has been described as a web of submarine cables, and they so excited Kipling that he wrote a poem about them.

Hyam [30, p. 212]

... communications technology, especially the telegraph and undersea cables, that made London (and Britain) the information hub of the world.

John Darwin [31, p. 112]

Lord Kelvin, more self-effacing but perhaps an even more significant figure (than Sir Robert Murchison), developed telegraphic, directional, and sounding devices of great importance to the maintenance of Britain's hegemony on ocean routes.

Robert Kubicek [26, p. 256]

Table 3 addresses the following question: to what extent does credit for this accrue to Kelvin? It summarises whether the texts of a wide range of recent references on the British Empire give

Table 3: References to cables, telegraphs and Kelvin in recent publications on the British Empire.

Ref.	Author	Q. 1	Q. 2	Q. 3	Q.4
[28]	Lawrence James	Yes	Yes	0	0
[32]	Piers Brendon	Yes	Yes	0	0
[33]	Kwasi Kwarteng	0	0	0	0
[34]	Denis Judd	Yes	0	0	0
[35]	Richard Gott	0	0	0	0
[27]	Niall Ferguson	Yes	Yes	Yes	0
[36]	Philippa Levine	0	0	0	0
[29]	Ronald Hyam	Yes	0	0	0
[30]	Ronald Hyam	Yes	Yes	0	0
[37]	Sarah Stockwell	Yes	0	0	0
[38]	John Darwin	Yes	Yes	0	0
[31]	John Darwin	Yes	Yes	0	0
[39]	Frank McDonough	Yes	Yes	Yes	0
[26]	Robert Kubicek	Yes	Yes	Yes	Yes
[40]	Trevor Lloyd (concise general history)	Yes	0	0	0
[41]	Martin Wiener (judicial system of empire)	Yes	0	0	0
[42]	Schreuder and Ward	Yes	Yes	0	0

explicit information about cables, telegraphs and Kelvin. The four test questions are: (1) Are cables and telegraphs mentioned?; (2) Is the cable/telegraph network discussed?; (3) Is the 1866 laying of the Atlantic cable mentioned? and (4) Is Kelvin mentioned?

The six quotes at beginning of this section, show that the successful laying of the Atlantic cable in 1866 was a climactic event in the integrated global history of transoceanic communication and travel. And yet, Table 3 reveals *only one explicit mention* of Kelvin. Even then, the quote, given in full from Kubicek [26], does not unambiguously connect the Atlantic cable with him. Whereas at the time in Britain, Kelvin's achievement was fully appreciated in the awarding of his knighthood, the appreciation of him today is a completely different matter. Why is it that Kelvin's name has been almost air-brushed from these histories? This question is now addressed.

4.2 The British Empire

4.2.1 Marxist interpretation of the Empire

Since the Atlantic and subsequent cables greatly helped communications within the British Empire, any negatively critical history will be unlikely to treat Kelvin sympathetically. Until comparatively recently, the historical appreciation of the empire has been subject to a Marxist approach. Marxism as a political creed may be summarised as an extended set of keywords: 'capitalist society … the exploitation of the proletariat by the bourgeoisie' … 'communism will be achieved when the class struggle results in the overthrow of capitalism by the dictatorship of the proletariat …' ([22, p. 684], item on *Marxism*). The eminent British Marxist Eric Hobsbawm (1917–2012) whose Obituary was published recently (October 2, 2012 [43]), 'was widely considered to be one of the greatest British historians of his age'. As a Communist Party member, 'he wielded enormous influence during the 1960s and 1970s' but still 'persisted' in his apologetics for 'totalitarian communism' in the 1990s, when that system had collapsed and even its proponents found it to be 'an unmitigated catastrophe'. The Obituary explains both his brilliance and controversialism in considerable detail, the former reflected in his being made Companion of Honour in 1998. His book *The Age of Empire 1875–1914* [44] expresses this Marxist approach. Hobsbawm's Obituary coincided within a few days, with a Book Review published by the same newspaper (30 September 2012 [45]) on John Darwin's *Unfinished Empire* (our reference [38]). Andrew Roberts concludes his review thus: 'Darwin's book might at long last herald the victory of the post-Marxist phase of imperial historiography, and not a moment too soon'.

4.2.2 Negative assessments of the Empire

Despite Roberts's confidence, however, consistently negative assessments of the Empire are still being made. A significant example is Piers Brendon's *The Decline and Fall of the British Empire* [32], published in 2007. (The title harks back to the iconic 18th century work of Edward Gibbon, *The History of the Decline and Fall of the Roman Empire*, 1776–1788 [46] being a current well-illustrated abridged version). Brendon's rationale is that it was 'a fundamentally weak empire … It enshrined a belief in freedom that would fatally undermine its authority. It left a contested legacy: at best a sporting spirit, a legal code and a near-universal language; at worst: failed states and internecine strife' (dust jacket text). Despite its length (xxii + 786 pages), 'the development of the dominions … . is only a sketch' (p. xxi) and the clear accent is, like Gibbon, on the negative aspects. ('Dominions' refer to the self-governed major territories of Australia, Canada, New Zealand and South Africa). The reviewer Robert McCrumb is enthusiastic and, again, negative in tone: 'If the First World War reduced the Empire to a near-bankrupt charade, the Second World War, more genuinely global, finished it off completely' [47]. Kwasi Kwarteng's *Ghosts of Empire* ([33], 2011) follows Brendon's line more sharply. Its focus again excludes 'the white dominions'. For Kwarteng, 'The British Empire was not merely undemocratic; it was anti-democratic. By comparison, the United States's avowed values could not be further removed from those of the British Empire' [33, p. 7]. He concentrates on residual chronic post-imperial problems such as Kashmir. Even more damning is Richard Gott's *British Empire, Resistance, Repression and Revolt* [35, 2011]. For him the violence exerted in the Empire was comparable to 'the exploits of Genghis Khan or Attila the Hun 'rather than Alexander the Great' [35, p. 3]. Richard Drayton's review of [35] ([48], 7 December 2011) founds Gott's approach lacks balance: 'nor does he explore how the economic and technological bases of British power changed between 1750 and 1850 … the Empire was made by more than violence'.

We conclude as follows: from any negative assessment of the Empire, whether Marxist or otherwise, *the reputation of Kelvin suffers by neglect.*

4.2.3 Balanced assessments of the Empire

The other authoritative surveys referenced here give more balanced interpretations, and, more importantly, *make a virtue of that balance.* Trevor Lloyd, in his concise *Empire* ([40], 2001) introduces it thus '…advocates of empire pointing to the tranquillity … enemies of empire pointing to the wars of conquests … .the important thing is to explain it' [40, p. x]. Similarly, Martin Weiner's in-depth analysis of imperial justice (Brendon's above positive factor of a 'legal code') *An Empire on Trial* [41, 2009] highlights the 'fallacy of thinking of the Empire in terms of an 'imperial project' whether one of good or evil. Rather, the Empire was a site of many simultaneous 'projects', worthy of both celebration and indictment, often at the same time' [41, p. 231]. Finally, there is the twin enterprise of John Darwin: *The Empire Project* ([31], 2009) and *Unfinished Empire* ([38], 2012). These latter two comprise a total of some 1300 pages, that is, on a par with the core material on Kelvin and on Darwin/Huxley. So we read '...the place of the political history of the dominions has been all but ignored by two generations of imperial historiography ... Revision is long overdue.' [31, pp. 15–16]. Also, Darwin refers to the Marxist interpretation: 'This depiction of Empire as (more than anything else) a system of racial oppression was the sharpest prong of a much wider attack ... There is no need to take up a dogmatic position on the truth or untruth of these various claims ...' [38, pp. 3–6]. Finally, at about the same time (2010) Ronald Hyam, 'the leading British authority on the end of Empire' [30, p. xii], published his 'swansong' [29, p. xiii], *Understanding the British Empire* [29]. 'The book cannot be a comprehensive guide ... It is not an advice manual, more of an academic confection. Nor with one exception ... ..does it engage much with other historians' theories' (p. xiii).

But do these 'balanced' assessments help Kelvin's reputation? Broadly speaking, from Table 3, they do not.

4.2.4 A positive assessment of the Empire

Nothing, in short, had prepared him for Sydney ... 'It is a most magnificent testimony to the power of the British nation'.

Charles Darwin on seeing Sydney, Australia, 1836 (Desmond and Moore [10, p. 177])

I was constantly struck by its (the Empire's) ... ubiquitous creativity ... the glorious skyline of Sydney ...

Niall Ferguson on visiting Sydney in 2002 [27, p. xxii]

Of the histories of the British Empire referenced, hardly any are explicitly positive. Niall Ferguson's *Empire* is an exception. Above, we have chosen a less generally featured aspect, that of its enduring built environment. (Also, Hyam's [29] cover illustration is an 1880 view of 'The Great Indian Peninsular Railway, Victoria Terminus, Bombay'.) Brendon gives a quite complimentary description of 1810's Sydney as a: 'dignified metropolis ... full of handsome residences and smart shops' [32, pp. 71–73]. While Brendon does not update this, Sydney astonished Darwin only 15 years later, and 170 years after that was able to more than inspire the twenty-first century Ferguson. Much more significant is Ferguson's general conclusion: 'In truth, the imperial legacy has shaped the modern world so profoundly that we almost take it for granted' [34, p. 365]. Further, Ferguson then goes on to give a more positive exposition of the questions of liberty, the rule of law and peace and violence, the concerns of the 'negative' histories above.

Does Ferguson's treatment have any helpful consequence for Kelvin? It does!

4.3 Light breaks through

'Before the coming of the telegraph........'

Hyam [30, p. 1]

'The crucial development from the point of view of imperial rule was the construction of durable under-sea cables … When the cable … finally reached the American coast on 27th July 1866 … .it was plainly the dawn of a new era. That the cable ran from Ireland to Newfoundland made it clear which power was most likely to dominate the age of the telegraph'.

Ferguson [27, p. 168]

It is clear from Table 3 that while the cable/telegraph system is recognised as a crucial element of the British Empire, the key Atlantic facilitation of that system remains an untold story. Hyam's six-word quote above from the very first page of his exposition takes for granted the system's importance, while his quotes in Section 4.1 expand on this. But it is *Niall Ferguson alone* who tells the Atlantic cable story, and moreover, uses *'Passing the Cable on Board the Great Eastern'*, 1866, as the lead of his second illustration set. In this present volume, the whole achievement constitutes the premier *Story in Pictures* in the Appendix.

4.4 Kelvin and Brunel

4.4.1 The magnificent achievement: Kelvin, Brunel and seamanship

Isambard Kingdom Brunel (1806–1859), the builder of Britain's Great Western Railway, was an (or possibly *the*) iconic Victorian engineer. In 2002, he was voted as the second Greatest Briton in the BBC's national poll behind Sir Winston Churchill, and together with Churchill, appeared as a major character in the opening ceremony of London's 2012 Olympic Games. Brunel's engineering glory is still conspicuously evident in twenty-first century Britain in his railways and bridges. That glory impacted national life from the start, immediately leaping the cultural divide. In the General Introduction to Volume 6 of *The Cambridge Cultural History of Britain*, covering the years 1785–1851, Boris Ford writes: 'Turner's wonderfully evocative painting of the Great Western Railway offers a glimpse of the energy and bustle of the new age and of the misty grandeur of the neo-classical world it was superseding'. Brunel's inventive genius also included three 'firsts' in shipping: the mammoth *Great Eastern*, launched in 1858 shortly before his death was the world's largest ship, by the end of the nineteenth century, it was still the largest and was used in 1866 as the cable carrying ship [27, p. 168].

Therefore, the core technology for the Atlantic cable laying was a combination of Kelvin's design of undersea cable, with the ability of *Great Eastern* to carry the entire cable and be navigated for the actual laying. In subsequent publicity, *Great Eastern* became very prominent (for example, front cover of [3]), but (while Kelvin's name is unfortunately omitted by Ferguson), the illustration mentioned above comprises both the cable and *Great Eastern*.

It was a huge triumph of seamanship and British engineering, partnered by US/UK organisation and funding. Daniel Gooch, Brunel's closest colleague and the Chairman of the British Cable Company, was on board Great Eastern at the time. In his diary entry for 26–27 July 1866 [49, pp. 141–146], he witnessed that 'all seemed mad with joy', the 'old hands' from a previously unsuccessful attempt 'danced round' the cable end, cheering at the top of their voices'! A service of thanksgiving was held 'in the little wooden church in the village' and soon after, a reporting cable was sent to Britain, resulting in a congratulatory cable being sent by Queen Victoria to the US President. Daniel Gooch, eventually to receive a Baronetcy, had the *Times* leader (cabled of course to *Great Eastern)* posted on its deck. On reading it, borrowing its words 'one of the sailors said "I say, Bill, we be benefactors of the human race". "Yes" says Bill, "we be", and strutted along … two inches higher!'

While Brunel, sadly, had died some years before, in essence he was represented by his friend Gooch. *For Kelvin, it was a working voyage and he was deservedly present at this momentous event.* The further spectacular achievement of the recovery of the previous attempt's broken cable and successful linking is also told by Gooch, as is the immediate financial income from the (now) two cables. More details are given in the *Story in Pictures.*

4.4.2 *Kelvin loses out, Brunel is remembered*

We have shown that in histories of the British Empire, the cable/telegraph system was recognised to be a crucial element. Moreover, the 1866 success enters world history as a primary part of the Second Stage of the Industrial Revolution. It is particularly unfortunate, therefore, that only one British Empire history referenced here addresses that event, and even that single instance credits only Brunel, not Kelvin. The pattern of neglect is repeated in [39] which gives (pp. 115–117) an extensive chronology of the Empire from 1815 to 1914 but 1866 simply does not appear.

As a postscript to this, in Adrian Desmond's biography of Huxley [11], the cable-laying exploit *does* feature ('… Brunel's gigantic *Great Eastern* paid out the enormous tonnage of telegraphic cable', p. 351). However, the engineering achievement is incorporated by Desmond into 'the revolution' led by 'human antiquities more ancient than the Bible's, of evolution more stirring than Genesis' [11, p. 351]. Again, incidentally, Brunel is credited, Kelvin neglected.

4.4.3 *Kelvin and Brunel are credited*

However, there is an additional Industrial Revolution issue. Via their contribution to shipping, Kelvin and Brunel additionally participated in the 'spread of the Industrial Revolution'. In this case, however, both have been named.

> *For Kelvin*: 'Lord Kelvin … ..developed telegraphic, directional, and sounding devices of great importance to the maintenance of Britain's hegemony on ocean routes', Robert Kubicek [26, p. 256]
>
> *For Brunel:* 'Yet the Great Britain, built by Isambard Kingdom Brunel in 1843, provided an irresistible model of strength and speed … Leviathans constructed on the principles of the Great Britain secured the nation's rule over the waves', Brendon [32, p. 146].

This latter is a welcome corrective to previous comments about the negative aspects of Brendon's study.

5 Kelvin's reputation

An ongoing theme is the question of the tepid reputation of Kelvin today, although there are signs this is changing. Wherever appropriate this has been addressed. Far from questioning his diversification as Crowther [15] and others have done, we conclude that his engineering and thermodynamic achievements on which this book focuses deserve to be highly honoured. What, then, are the problems? We have found the following:

Firstly, in the historical development of the British Empire, the successful laying of the Atlantic cable was a landmark event for which Kelvin received his knighthood. But our analysis shows that with few exceptions Kelvin's name is unjustifiably absent.

Similarly, it is comparatively only recently that the Industrial Revolution has been interpreted (by Koeller) as three stages. Kelvin made substantial contributions to the latter *two* stages. One aspect was that the cable laying was significant in international terms.

Finally, the common faith of Kelvin and his energy colleagues helped him to develop a thermodynamic cosmic worldview, consistent with today's thinking. Crosbie Smith has addressed this aspect comprehensively.

We live in a world dominated by energy considerations, including security of supply, living standards for the third world or carbon-reduced emissions to name perhaps the most important. Energy was Kelvin's word, and he needs to speak to us today.

The final message of this overview is a straw in the wind. Reputations need to enter popular culture and Marcus Chown's *We need to Talk about Kelvin* [50], subtitled *What everyday things tell us about the universe,* is written for a popular readership. Sure, Kelvin's under-prediction of the sun's age is there (pp. 86–88), but joint with Helmholtz, and as part of a constructive story that puts it as one stage in a scientific progression. What is more significant is his unqualified tribute to Kelvin – one of the greatest scientists of the nineteenth century, responsible for the temperature scale used by all scientists today and for the laying of the first transatlantic telegraph cable. In praise of Kelvin, Chown pretty much trumps all the other recent publications in our reference list. Hopefully, it is a sign of things to come: it's about time!

References

[1] Smith C. & Wise, M.N., *Energy & Empire,* Cambridge University Press: Cambridge, 1989.
[2] Smith, C., *The Science of Energy. A Cultural History of Energy Physics in Victorian Britain,* The Athlone Press: London, 1998.
[3] Marsden, B. & Smith, C., *Engineering Empires. A Cultural History of Technology in Nineteenth-Century Britain,* Palgrave, Macmillan: Basingstoke, 2007.
[4] Thompson, S.P., *Life of Lord Kelvin,* Macmillan and Co.: London, 1910.
[5] Hills, R.L., *James Watt. Volume 1: His Time in Scotland, 1736–1774,* Landmark Publishing: Ashbourne, 2002.
[6] Hills, R.L., *James Watt. Volume 2: The Years of Toil, 1775–1785,* Landmark Publishing: Ashbourne, 2005.
[7] Hills, R.L., *James Watt. Volume 3: Triumph through Adversity, 1785–1819,* Landmark Publishing: Ashbourne, 2006.
[8] Cardwell, D.S.L., *From Watt to Clausius. The Rise of Thermodynamics in the Early Industrial Age,* Heinemann: London, 1971.
[9] Hunt, B.J., *Pursuing Power and Light. Technology and Physics from James Watt to Albert Einstein,* Johns Hopkins University Press: Baltimore, 2010.
[10] Desmond, A. & Moore, J.R., *Darwin,* Penguin: London, 1992.

[11] Desmond, A., *Huxley: The Devil's Disciple*, Michael Joseph: London, 1994.
[12] Desmond, A. *Huxley: Evolution's High Priest*, Michael Joseph: London, 1997.
[13] Lindley, D., *Degrees Kelvin*, Joseph Henry Press: Washington, 2004.
[14] Flood, R., McCartney, M. & Whitaker, A. (eds), *Kelvin*, Oxford University Press: Oxford, 2008.
[15] Crowther, J.G., Part I. William Thomson 1824–1907, in *British Scientists of the Nineteenth Century*, Volume 2, Pelican Books, Penguin: Harmondsworth, 1941.
[16] Weaire, D., William Thomson (Lord Kelvin) 1824–1907, Chapter 8, *Creators of Mathematics: the Irish Connection*, Ed Ken Houston, University College Dublin Press: Dublin, 2000.
[17] Wilson, D.B., *Kelvin and Stokes*, Adam Hilger, IOP Publishing Ltd: Bristol, 1987.
[18] *The Complete Family Encyclopedia*, Helicon Publishing for Fraser Stewart Book Wholesale Ltd.: London, 1992.
[19] Hall, M. B., *All Scientists Now: the Royal Society in the Nineteenth Century*, Cambridge University Press, Cambridge, 1984.
[20] Ruddock, I., Lord Kelvin, science and religion, *Physics World*, **17**, p. 56, February 2004.
[21] Darwin, C., *Autobiographies*, M. Neve & S. Messenger (eds), Penguin Classics: London, 2002.
[22] *Chambers Concise Dictionary*, New Edition 1991, W & R Chambers: Edinburgh, 1991.
[23] Toynbee, A., *The Industrial Revolution* (original publication 1884), Beacon Press: Boston, MA, 1964.
[24] Crump, T., *A Brief History of How the Industrial Revolution Changed the World*, Robinson: London, 2010.
[25] Koeller, D.W., The industrial revolution 1700–1900, available at http://www.thenagain.info/WebChron/WestEurope/IndRev.html [accessed 18 December 2014].
[26] Kubicek, R., British expansion, empire and technological change (Chapter 12), *The Nineteenth Century*, Volume III, The Oxford History of the British Empire, A. Porter (ed.), Oxford University Press: Oxford, 1999.
[27] Ferguson, N., *Empire. How Britain Made the Modern World*, Penguin: London, 2004.
[28] James, L., *The Rise and Fall of the British Empire*, 2nd edn., Abacus: London, 1998.
[29] Hyam, R., *Understanding the British Empire*, Cambridge University Press: Cambridge, 2010.
[30] Hyam, R., *Britain's Imperial Century, 1815–1914. A study of Empire and Expansion*, 3rd edn., Cambridge Imperial and Post-Colonial Studies, Palgrave, Macmillan: Basingstoke and New York, 2002.
[31] Darwin, J., *The Empire Project, The Rise and Fall of the British World-System 1830–1970*, Cambridge University Press: Cambridge, 2009.
[32] Brendon, P., *The Decline and Fall of the British Empire 1781–1997*, Alfred A. Knopf: New York, 2008.
[33] Kwarteng, K., *Ghosts of Empire. Britain's Legacies in the Modern World*, Bloomsbury: London, 2011.
[34] Judd, D., *Empire. The British Imperial Experience from 1765 to the present*, Taurus: London, 1996.
[35] Gott, R., *Britain's Empire. Resistance, Repression and Revolt*, Verso: London, 2011.
[36] Levine, P., *The British Empire. Sunrise to Sunset*, Pearson, Longman: Harlow, 2007.
[37] Stockwell, S., (ed.), *The British Empire. Themes and PeRSpectives*, Blackwell Publishing: Malden, MA, 2008.
[38] Darwin, J., *Unfinished Empire, The Global Expansion of Britain*, Allen Lane, Penguin: London, 2012.

[39] McDonough, F., *The British Empire 1815–1914*, Access to History Series, Hodder Education, Hachette, UK: London, 1994.

[40] Lloyd, T., *Empire: The History of the British Empire*, Hambledon and London: London and New York, 2001.

[41] Wiener, M.J., *An Empire on Trial, Race, Murder, and Justice Under British Rule, 1870–1935*, Cambridge University Press: Cambridge, 2009.

[42] Schreuder, D.M. & Ward, S. (eds.), *Australia's Empire*, The Oxford History of the British Empire, Companion Series, Oxford University Press: Oxford, 2008.

[43] Obituary for Eric Hobsbawm, historian whose dedication to Marxism fuelled a career that was brilliant and controversial. *Daily Telegraph: London*, 2nd October 2012.

[44] Hobsbawm, E., *The Age of Empire 1875–1914*, Abacus: London, UK, 1994.

[45] Roberts, A., Book of the Week. Andrew Roberts welcomes an account of Empire that puts paid to the Marxist model, *Seven, Sunday Telegraph: London*, 30th September 2012.

[46] Williams, R. (ed.), *Gibbon's Decline and Fall of the Roman Empire, Abridged and Illustrated*, Grange Books: Rochester, UK, 2000.

[47] McCrum, R., Review: The Decline and Fall of the British Empire by Piers Brendon. *Piers Brendon's The Decline and Fall of the British Empire Is a Triumph of Scholarship and Narrative*, The Observer: London, 21st October 2007.

[48] Drayton, R., Review: Britain's Empire: Resistance, Repression and Revolt by Richard Gott. *The Violence at the Heart of Colonialism Is Exposed in Richard Gott's History*, The Guardian: London, 7th December 2011.

[49] Gooch, D., *The Diaries of Sir Daniel Gooch*, first published in 1892. With Introduction by Sir Theodore Martin, Nonsuch: Stroud, 2006.

[50] Chown, M., *We Need to Talk About Kelvin*, Faber and Faber: London, 2009.

Relationships

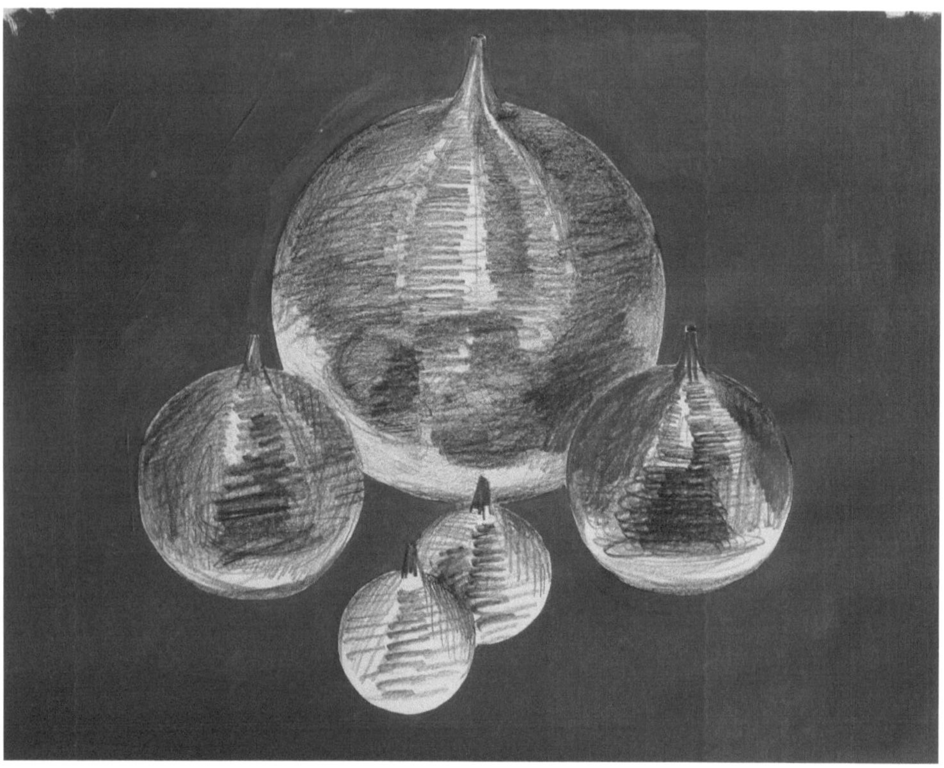

Helmholtz resonators. (Marianne Greated 2014, gouache and pencil on paper)

Chapter 3

James and William Thomson: the creation of thermodynamics

A. Whitaker
Department of Physics, Queen's University Belfast, United Kingdom.

Abstract

In his Belfast notebooks, James Thomson demonstrated deep thinking about the nature of heat and work, and his analysis of the lowering of the freezing point of water under pressure was a crucial step towards a complete theory. Throughout this period, James and William Thomson worked together, but when this theory emerged as thermodynamics, it was credited to William, and to Rudolf Clausius and MacQuorn Rankine. James's claim to be added to the list of founders of the theory is discussed.

1 James Thomson and the Belfast notebooks: the air engine

Enthused by many aspects of the Carnot theory, in April 1848 James had started what we shall call the Belfast notebooks [1]. These consist of three fairly substantial exercise books, crammed with a considerable variety of investigations into aspects of Carnot's theorem, with the overall title *Motive Power of Heat*. It will be remembered that, at this period of his life, James had returned to Glasgow because of ill-health, but was working on a variety of projects. On the front cover of Book A is a note that James attached in September 1863, which says: 'This book contains, at page 58, under date May 1848, my original manuscript investigation of the Amount of the Lowering of the freezing Point of Water by Pressure'.

On the inside cover of Book A, James has written for reference the basic laws of gases, essentially the laws for the ideal gas, another reminder that in this period these were still far from standard. Indeed, the correction between the Centigrade scale and the air thermometer scale is originally given as 267°, later replaced by 273° as more accurate results became available.

Book 1 starts with a full account of the contents of the Clapeyron paper, titled *Air or Gas Engine Imagined by Carnot: Preparatory Investigation*. James starts by giving what he calls an important formula combining the laws of Mariotte and Gay Lussac, which James writes as $PV = P_0V_0(1 + Et)$, again essentially the ideal gas law. He then works through the Carnot theory in considerable mathematical detail. He is working within the caloric theory and makes a point of arguing explicitly that the heat given out to the cold reservoir is equal to that gained from the hot reservoir. James's isothermal processes are actually taken to be infinitely close together, and so are his adiabatics, as was standard for both Carnot and Clapeyron. Thus, James says that each pair 'must be parallel, being portions of curves lying infinitely near one another'.

James then calculates the total work gained as $\frac{dV}{V}(P_0V_0ET)$, and says: 'Now this work has been obtained by the passage from t to $t–T$ of a quantity of heat which we have denoted by $\mathrm{d}Q$ and is the quantity absorbed or evolved by the gas when with a constant temperature t or $t–T$ it alters its volume by $\mathrm{d}V'$. The other quantities, including E are as in the equation above. A little analysis leads him to the result that the work gained from H units of caloric is given by $\frac{1}{C}HT$ where $\frac{1}{C}=\frac{EP_0V_0}{V\frac{dQ}{dV}}$. C, he says, is a 'remarkable quantity, to be determined experimentally, constant for all gases but varying with the temperature'.

In this way, James has made the Carnot theory very much his own, and, of course, he is extremely interested in it as a fundamental set of ideas of tremendous importance. However, the notebooks make it absolutely clear that, as an engineer, and indeed as an engineer who was particularly interested in the question of efficiency and waste at both the theoretical and the practical levels, and as an engineer who must have been extremely keen to make some money and pay his way at home, he also clearly regarded it as a technical challenge. Carnot had sketched in a very general way the means of maximising the amount of work obtained from a given amount of fuel; James decided to design an engine at least broadly along these lines. Much of the content of his notebooks consists of detailed descriptions of the engine as it evolved, together with comprehensive diagrams of practically draughtsman standard. It was to be an air-engine.

The significance of the choice of air-engine is as follows. In his paper based on the work of Carnot [2], Clapeyron had stressed that, for maximum efficiency of any engine, the work produced should correspond to maximum possible fall of temperature of the vapour. This implied that the lowest temperature reached by the vapour in the condenser should be as near as possible to that of the atmosphere. More importantly, though, it implied that its highest temperature should be as high as possible, and that the fact that the temperature of the boiler was between 1000°C and 2000°C below that of the furnace corresponded to a major waste of heat.

It would be dangerous to increase the boiler temperature of the steam engine because of the high pressures that would be produced and the corresponding risk of explosion. It was, therefore, clear, then, that an air-engine might in principle be much more efficient than a steam engine because it could utilise a much greater fall of temperature. These ideas had been mentioned by Carnot, and in his 1849 paper, Kelvin [3] also had a short section titled 'Comparison of the relative advantages of the air-engine and the steam-engine'. He commented that: 'A "perfect air-engine" would be a much more valuable instrument than a 'perfect steam-engine', but admitted that: 'Neither steam-engines nor air-engines, however, are nearly perfect; and we do not know in which of the two actual kinds of machine the nearest approach to perfection may be actually attained'.

Kelvin went on to mention that: 'The beautiful engine of Mr Stirling of Galston may be considered as an excellent beginning for the air-engine', though obviously capable of much further improvement. Rev. Robert Stirling (Fig. 1), who was a Presbyterian minister from the town of Galston in Ayrshire, 20 miles from Glasgow, had taken out his first patent for an air-engine as early as 1816, and only 2 years later an engine that he had built was used for pumping water

Figure 1: Robert Stirling.

Figure 2: John Ericsson.

in a quarry. The engine incorporated an 'economizer' or 'regenerator' designed to make use of heat that would otherwise be wasted. This idea was independently invented by John Ericsson (Fig. 2) in the 1830s [4, p. 296], and a range of Ericsson engines, very similar to Stirling engines have been designed and built. Robert Stirling and his brother James, who was a civil engineer, produced further patents in 1827 and 1840, and an engine of 45 horsepower was used in James's factory in Dundee [5, p. 48]. The reason for using this engine was to avoid the disastrous consequences of an explosion of a steam-engine, all too common at that period. However, the Stirling engine itself failed three times in 4 years and unfortunately had to be replaced with a steam engine.

In the late 1820s, the Stirling brothers had presented a model Stirling engine to the Natural Philosophy Department at Glasgow University, and in 1847, this engine was discovered by William Thomson in a store crammed with old and unwanted equipment, clogged with dust and oil. William immediately set about having it cleaned and put into a working condition.

The principle of the Stirling engine (Fig. 3) is very simple. It works with a fixed mass of gas and can use any source of heat. Today's engines may use, for example, solar power, waste heat or nuclear energy; a table-top demonstration version uses a candle or the heat from a coffee cup. The engine has a sealed cylinder with one part hot and one part cold and the working gas is moved from the hot side to the cold side. When the gas is on the hot side, it expands and pushes up on a piston; when it is on the cold side, it contracts and pulls down the piston. The job of the displacer in Fig. 3 is to move the air from the hot side to the cold side as the engine runs. The regenerator is a temporary heat store – a lump of matter arranged so that the working fluid

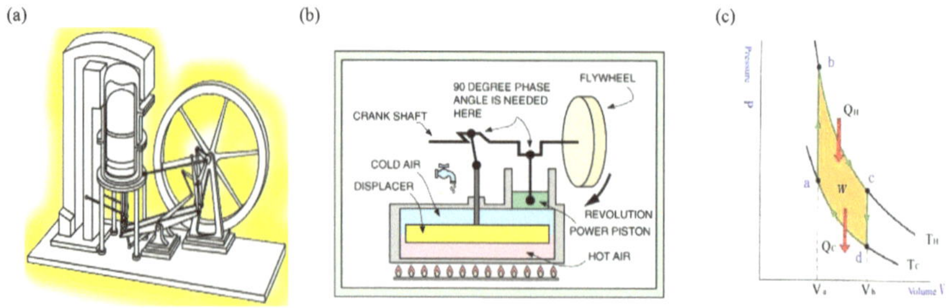

Figure 3: (a) Robert Stirling's engine of 1816, (b) a simple diagram of a modern Stirling engine and (c) a PV-diagram for a Stirling engine.

Figure 4: Robert Napier, 'The father of Clyde shipbuilding'.

passes through it first in one direction and then in the other; it enables heat to be retained in the system that would otherwise be lost to the environment. (It may be mentioned that some Stirling engines today have an extra piston rather than the displacer; this makes them effectively equivalent to the Ericsson engine.)

We shall return to Robert Stirling and his engine shortly. For the moment we may say that by 1847, Lewis Gordon was attempting to set up a substantial company for the manufacture of the Stirling engine. Smith says that: 'Over the next few years, Joule, Rankine and [Robert] Napier (Fig. 4) would all take up the challenge set by Carnot's theory to find a workable air-engine which could bring fame and fortune to its designer [5, p. 95]'. James Thomson also took up that challenge.

On 24 April 1848, he wrote that

> I have just found out a new air engine which leaves nothing farther to be wished for in the *theory* of the action of air, gas or vapour, in developing motive power from a fall of heat from the highest temperature which a vessel (without rubbing parts) will bear to as low a temperature as we have at command … I have been led to the new air engine from having seen a fundamental fault theoretical and practical in Stirling's air engine, and from having observed that Ericsson's Air Engine is subject to the same objection, as well as having found the fundamental difference between the theory of Carnot's imaginary engine and these two real ones.

The following day, James wrote that:

> Until this week I had for a long time (more than a year) seen various faults in Stirling's Air Engine … But I could not see *in every* respect the mode by which the motive power was obtained from the fall, or rather I failed in all attempts to make it out to be an apparatus, not perfect in operation indeed, but still an apparatus for putting in practice the ideal action required in Carnot's imaginary engine. It was only yesterday, however, that I found that the principle of Stirling's and Ericsson's is fundamentally different from Carnot's. I then was able to alter this principle in such a way as to produce a *new imaginary engine* quite distinct from Carnot's, but perfect as well as it; one namely which would utilise *all* the fall of heat supplied at one temperature and removed at another. Then this new ideal engine has the character that, with the materials we possess, an apparatus can be constructed which will practically almost quite fulfil all the conditions required by the theory of the ideal engine. I have often tried to adopt an apparatus to the four processes of Carnot's but have failed.

James went to considerable effort to understand, establish and explain the differences between his own engine, and the proposed one of Carnot on the one hand, and those of Stirling and Ericsson on the other. In part, this may have been for his own understanding, but it is quite obvious that he hoped to patent his proposed engine, and for that purpose he would have to demonstrate the differences between his own engine, and the published one of Carnot, and the patented ones of Stirling and Ericsson.

It will be noted that the Stirling cycle follows Carnot in using two isothermal processes, but replaces the adiabatic processes of Carnot by isochoric processes – processes at constant volume. In the latter, heat does, of course, flow under a difference of temperature in conflict with the Carnot criterion; the regenerator was included to minimise not only heat loss but also this dissipation. In theory, the Stirling cycle was still capable of providing an efficiency very little reduced from that of Carnot, but in practice materials were not available to allow the high temperatures required to take full advantage of the fall in temperature of the caloric, and limitations of design meant that efficiencies were very low.

James considered that the Stirling engine was better than that of Ericsson; his criticism of both engines focused around the fraction given by volume of the cylinder through which the displacer moved relative to that of the entire volume of gas used. If the fraction is large, then it is not possible to build up a significant increase of pressure, because only a small part of the volume of gas will become hot, and there will also be large heat losses from the heat passing round the displacer to the air around it. This loss could be reduced if the strokes were quick enough, but then there would be insufficient time for the heat to pass from the air to the water. On the other hand, if the fraction is small enough to avoid these problems, then the work done at each stroke will be very small. James's summing-up of the Stirling engine was that:

> In almost every respect in Stirling's air engine it turns out that if we make the best adaptations for any one condition we make the worst for some other or others and what is of great importance when we take a middle course we violate all the conditions *greatly*.

In contrast, James considered that: 'In mine the best adaptations in various respects can be made at the same time'.

James also stressed that his engine was different from that of Carnot. On 25 October 1848, he included a slip of paper titled: 'Paper showing that the engines imagined by Carnot and used by him for assisting in his investigations as regarding the Motive Power of Hear are *different in principle from mine*'. He drew a detailed diagram comparing the two schemes, with his own engine in red and Carnot's in black. James, in fact, replaces Carnot's adiabatics by isobaric processes – processes at constant pressure. He comments that:

> If Carnot were to use an infinite number of infinitely small engines to perform the work represented [by James's figure], still every one of these engines will be *different in principle from mine* ... So I do not at present admit that my engine is a practical apparatus for working out Carnot's principles when taken to embrace even their widest range.

James then stressed the most important difference between his scheme and that of Carnot:

> Carnot reduces the temperature by expanding while I reduce it by abstracting caloric in such a way that it may be restored ... If in any of Carnot's investigations he goes on the supposition that the lines for II and IV [which for Carnot are adiabatics] may be of any form whatever, then these investigations will comprehend my own.

In James's engine, the main change from Stirling was that the displacer was made to follow the motion of the piston; in his description, rather than the term 'displacer'; in fact, he uses the term 'plunger. In later versions of his design, the piston has become conical, and he uses the work 'cap'. He writes that:

> The fundamental principle then which I adopt is to retain the *pressure constant* during the stroke of the plunger by permitting a *proper* change of volume at that period, and to obtain the necessary change of pressure on the piston not by heating and cooling of the air due to the motion of the plunger, but by the operation of Mariotte's Law, the cylinder being made so that the plunger can follow the piston and this expel all the air from the cylinder to the hot part there.
> *process 1* Take in heat by raising plunger. Volume to more than double, but raise plunger along with piston so as to keep all air in hot place. Pressure follows Mariotte's Law.
> *process 2* Let down plunger and allow volume to diminish by descent of piston [cap] pressure remaining constant but the whole air being transformed from hot to cold.

> *process 3* Push down piston [cap] pressure following Mariotte's Law heat being given out at the lower temperature.
> *process 4* Raise plunger the piston rising at the same time (but more slowly) to retain the pressure constant. The process ends when plunger by overtaking piston has put all the air from cold to hot.

'By keeping pressure constant and making the plunger conical', he writes, 'the action of the regenerator is rendered theoretically perfect and in practice almost perfect'.

Actually in May 1848, James met Robert Stirling in person. James warned Stirling that he was working on similar ideas himself, so asked him not to report anything that might be regarded as confidential. Stirling did say that he was making improvements to his engine but was quite relaxed in his conversation, remarking that anything he might say was already in the public domain. James, though, took care to have witnesses: his father, William and William's classroom assistant, Robert Mansell. Fortunately or unfortunately, though, James was to realise that he had little to gain from Stirling's cooperation, but equally little to fear from his competition. James wrote in his notebook that: 'I found that, as I had previously thought, he does not understand his own engine; not knowing at all the way in which the heat is expended in generating work'.

For a start, Stirling did not appear to appreciate the significance of the changes of temperature produced by changes in pressure. In today's theory, these would be explained by the fact that work would have been done either on or by the gas, but at the time they would have been described as a change between latent and sensible caloric. More particularly, Stirling had convinced himself that though heat might be lost at some stages of his cycle, an average might be struck, so that through the complete cycle, and if friction and similar effects might be ignored, all the heat might be retained.

James pointed out that there must be *some* loss of heat, or some flow of heat between furnace and condenser, as otherwise he would be in possession of a perpetual motion machine. Even this did not worry Robert Stirling unduly; surely, he said, if friction is ignored, perpetual motion is not out of the question. James replied that that is the case, but such machines cannot be perpetual sources of power! It may be noted that with our current understanding of conservation of energy, anyone holding Stirling's belief that a certain quantity of heat could provide essentially an infinite amount of work would certainly be regarded as invincibly ignorant. Such a belief was not so ridiculous for believers in caloric, for whom, in the production of work, heat was not transformed but merely changed from a state of 'concentration' to one of 'diffusion'. It was, perhaps, by no means out of the question that the regenerator could reverse this change in the caloric, so that it could be used to produce more work.

The exchange must have convinced James that he and his brother were probably in possession of far more understanding of the theoretical and experimental facts than all but a few people in the world, and they realised that it is was certainly very worthwhile continuing their discussions and James continuing his attempts to design and build a useful air-engine.

James's design work continued for several years. Sections of the book were marked as written at the Causeway Head and at Cupar, Fife, so the work must have continued through family holidays. At Blair Logie, he had written a section on 'Calculations and Data to the Power of the Engine'. Finally, on 7 March 1850, he wrote: 'I have just finished the drawing of the air engine began in 1848'.

Certainly, the work must have benefitted James immensely in terms of his understanding of the science involved, and in experience in engineering design. It certainly helped him in the important work to be described in the next section. Sadly, though, from the point of view of actually building the engine, and possibly making a great deal of money, all the effort seems to have been abortive.

Smith [5, pp. 157–165] tells the immediate story of the air-engine project in detail. In 1852, John Ericsson was in the course of powering a ship of nearly 2000 t, *The Caloric Ship Ericsson* (Fig. 5) using an enormous Ericsson engine, which had perhaps the largest piston ever built. When he heard about this ship, Rankine approached the industrialist, Napier, and together they planned a strategy involving basic engineering science to be carried on by Rankine, with engine construction, market know-how and capital to be provided by Napier. The use of isochoric processes backed up with the regenerator, rather than adiabatic processes, meant that the volume of the engine could be small, as required for operation on a ship; this should have been a massive advantage over Ercisson's own huge engine, and also much less heat would be wasted than on Stirling's scheme. By 1853, all seemed set fair; contracts had been exchanged both for a non-functioning model, to be built by James White, the Glasgow instrument-maker, who worked so closely with William Thomson for many years, and for an experimental working air engine.

It was not to be. By 1854, White had not delivered the model, and the experimental engine did not yet work. *The Ericsson* had a sea trial in which the engine ran well, but it seemed to be underpowered. Then some time after the trial, the ship was hit by a tornado coming out of a cloudless sky, and it sank. It was raised, but only for its air-engine to be removed and replaced by a steam engine. For at least the foreseeable future, it was the end for air-engines, and James Thomson's stood very little chance even of gaining attention.

Over the subsequent years, Stirling engines have had a chequered history. Through the first half of the nineteenth century, as steam power became safer and cheaper, they found it difficult to compete. However, in the second half of the century, they were often used when a reliable source of low-to-medium power was required, such as raising water or providing air for church

Figure 5: The Caloric Ship Ericsson.

organs. They were inefficient, but, unlike a steam engine, they could be operated by anybody who could light and control a fire. Then during most of the twentieth century, they were unable to compete with electric motors and small internal combustion engines, but they have recently experienced a revival in interest because of increased fuel prices and environmental concerns. Stirling engines are compatible with all fuel sources, including renewables, and applications today include space-based astronautics, and generation of electricity from solar energy, biomass and domestic refuse. It should be mentioned that the Stirling engines of today are by no means always air-engines; a variety of different gases are used.

2 The Belfast notebooks: lowering of the freezing point of water

The credit for the important discovery of the lowering of the freezing point of water under pressure is normally allocated to James Thomson [6] for the theoretical part and William for the experimental part. Since the issue is open to some debate, we shall leave that till later in this section and concentrate for the moment on the physics. The issue emerged during discussion between the brothers on the means of operation of the Stirling engine, but in fact it was a direct application of the Carnot paper itself. In fact, both Carnot and Clapeyron, as well as dealing with the general gas case we have discussed so far, dealt with the case of phase change, but it was specifically William Thomson who argued as follows [4, pp. 296–298; 6].

Let us consider a Stirling engine, with the usual lower hot reservoir and upper cold reservoir, but to which no heat is applied; we also assume that there are no losses due to friction. Essentially the engine would work opposite to the usual way; the hot reservoir would cool and the cold reservoir would heat up until their temperatures were equal; the engine would be turned but without expenditure of work. If we wished to continue this process, to cause a situation in which the upper reservoir is hotter than the lower one, we would need to provide work; we would at this point essentially be taking advantage of the reversibility of the engine by working it in reverse. If, however, we keep both reservoirs at the melting point of ice, we may keep a stream of water running across the upper part of the engine and immerse the lower part in a basin of water at the same temperature. By turning the engine forwards, heat would constantly be taken from the lower region and deposited in the stream above, so the water in the basin would be converted gradually to ice, but no work would be required.

We may express the situation in terms coming more directly from Carnot's work. We may have a Carnot cycle with both isothermals at the melting point of water. Heat will be systematically moved from lower to upper reservoir, but because there is no difference in temperature between the reservoirs, no work will be required.

At this point, though, one or both of the brothers (to be discussed later) saw the following objection to the scheme. In fact, since ice is less dense than water, water expands when it freezes, and clearly this expansion could be used to produce useful work. Thus, in operating the engine as above, there would be a net gain of work and possibly a perpetual motion machine could be constructed.

The only answer to this dilemma that the brothers could come up with was that, in fact, the temperature in the two isothermal processes could not be the same. It will be appreciated that if the freezing water does external work, this must come from work done to turn the engine. This necessitates the pressure of the gas being higher when it is expanding than when it is being compressed. For net work to be performed by the person turning the engine, and hence by the engine itself in order to freeze the water, the Carnot principle tells us that the temperature must have been lower during the expansion at higher pressure than during the contraction at lower pressure. In other words pressure must lower the freezing point of water. One may obtain the same result directly from a Carnot cycle; to produce net work, a Carnot cycle must be between two different temperatures.

It is certain that it was James who then performed the necessary calculations in the Belfast notebooks in May 1848. In the model he used, as usual he followed four processes, using a cylinder that contains ice at 0°C. In the first, the cylinder is put into contact with an infinitely large lake at 0°C, and the motion of the piston has the effect of melting enough ice in the cylinder to produce a cubic foot of water. In this process, heat is gained from the lake. In the second process, the cylinder is removed from the lake, and the piston is pushed down a little further until the pressure is increased by a quantity p. During the process the temperature decreases; at any moment it is equal to that corresponding to the instantaneous pressure; at the end of the process, it is $-t$°C.

In the third process, the cylinder is in contact with a second infinitely large lake at a temperature of $-t$°C. During this process, the pressure and temperature remain constant, and the water increases its volume by freezing. Then, in the fourth process, the cylinder is removed from the lake and the piston is moved back to the position that it occupied at the beginning of the first process.

The work produced in the process could be calculated in a straightforward way in terms of p, but using figures from experiments of Regnault, James could also calculate the work produced from knowledge of the fall in temperature t°C. He produced the formula $t = 1/138\ n$, where t is the lowering of the freezing point in degrees centigrade and n is the pressure above atmospheric pressure in atmospheres. For an excess pressure of 1 atmosphere, clearly the lowering of the freezing point is around 0.0073°C, an extremely small quantity to attempt to measure.

There then followed the publication by James of the famous paper in the *Transactions of the Royal Society of Edinburgh* [6]. This paper had been communicated to the Society on 2 January 1849. The calculations in the paper follow those in the notebooks quite closely, though with very minor changes in data and results, and James also provided a full explanation of the ideas involved. Having demonstrated the origin of the effect in general terms, but before providing the calculations, James commented that:

> The fact of the lowering of the freezing point being thus demonstrated, it becomes desirable, in the next place, to find what is the freezing point of water for any given pressure. The most obvious way to determine this would be by direct experiment with freezing water. I have not, however, made any attempt to do so in this way. The variation to be appreciated is extremely small, so small in fact as to afford sufficient reason for its existence never having been observed by any experimenter. Even to detect its existence, much more to arrive at its exact amount by direct experiment, would require very delicate apparatus which would not be easily planned out or procured.

It is for this reason, he contented himself with the calculations.

This makes even more memorable the fact that by the end of 1849, William Thomson had constructed the necessary apparatus and performed the experiment with great success. William, it may be remarked, had only been in position as a professor of natural philosophy (Physics) for around 3 years, previously having been mostly regarded as a mathematician with very little practical experience. Robert Mansell constructed an ether thermometer for him [5, p. 95], which was, William assured his class at Glasgow, 'the most delicate that ever there was made, there being 71 divisions [and between 2 and 3 inches] in a single degree of Fahrenheit' [4, p. 282]. At first, not surprisingly, there were great problems – it was difficult to compare the thermometer with any other. However, before the end of the year, he was triumphant. He told his class:

> A pressure of 18 atmospheres was applied and the temperature was lowered by $17\frac{1}{2}$ of the divisions, 0.246 was the lowering according to experiment; 0.2295 according to theory ... Under a pressure of 8.1 atmospheres the temperature was lowered $7\frac{1}{2}$ divisions. 0.106 is the lowering by experiment; 0.109 by theory. This is within 1/3000 of a Fahrenheit degree [4, p. 282].

William mentioned James's calculations in his paper on Carnot [3]. His own experimental results were read to the Royal Society of Edinburgh in January 1850 and published in 1851 [7].

Clearly, the work of the Thomson brothers had been exceptionally important and successful. As well as providing theoretical and experimental knowledge of great significance in its own right, it was, or at the very least appeared to be, a triumphant confirmation of the truth of the Carnot argument. Certainly, it would now be difficult to deny that at least the main thrust of Carnot's argument had been validated. Still, though, the brothers felt that at least some parts of Joule's argument had also to be included in a final synthesis, but for the moment they were unable to see their way through to a final resolution.

Let us, though, briefly return to the question of the division of credit between the brothers. It is quite clear, of course, that the credit for the experimental work goes to William, that for the calculations to James. It has been usual to give the credit for the very idea that there is a reduction in the freezing point to James also. This is natural in any case, as the fact was announced in his paper together with the calculations.

To study the situation in more detail, we note that, at the start of this paper, he made it clear that William had given the argument for freezing without the necessity to provide work, as published earlier [8]. He then remarked that: 'This at first appeared to me to involve an impossibility ...', proceeding to the argument already given in this paper that, because of the expansion of water on freezing, it seemed that work could actually be produced by the process. He then says that: '[To] avoid the absurdity of supposing that mechanical work could be got out of nothing, it occurred to me further that is necessary farther to conclude that *the freezing point becomes lower as the pressure to which the water is subjected is increased* [italics in original]'.

This seems to establish that the idea itself is due to James. However, Smith and Wise [4, p. 298] have drawn attention to a letter William wrote to Forbes in December 1848 in which he appeared to claim credit for himself. He wrote that: 'In conversing with my brother James about this proposition, it struck me that we may prove as a consequence that the freezing point of water under heavy pressure must be lower than 32'. Smith and Wise suggest three explanations for the seeming contradiction: '(i) that William allowed James the credit in the published paper as a form of recompense for his elder brother's other disappointments in health and career and as an acknowledgement of his guidance in the subject; or (ii) that William had simply given an erroneous account to Forbes; or (iii) that the idea was so much the product of both minds that a separation of their contributions was impossible'.

William could indeed have been generous, but James was a man of the greatest integrity, and it seems unlikely that he would have explicitly claimed individual credit that he knew he was not due. It seems likely that (iii) is nearest the truth, that in the course of their discussions the idea reached each of their minds independently. William was then delegated to do the experiments, James to do the detailed calculations and it naturally followed that James would go on to write the paper giving the account of the idea. It is thus suggested that both brothers were telling nothing but the truth – as, indeed, one would expect.

James may thus have to share the credit for this important discovery with his brother, but it is clear that, over the entire course of their work, his contribution was immense. Another contribution was to follow. As was common in these days, his paper [6] was reprinted in the *Cambridge and Dublin Mathematical Journal* in November 1850, but for this republication he introduced a small but important change.

In the first version, he stated, broadly following Carnot and Clapeyron, that the third process may be described as follows: 'Continue the motion till all the heat has been given out to the second lake at $-t°$, which was taken in during [Process 1] from the first lake at 0°'. This follows the usual description of the Carnot process in terms of a caloric theory: all the heat gained by the system from the hot reservoir in the first isothermal process is deposited in the cold reservoir in the second isothermal process.

As a footnote in this original version, he says that:

> This step, as well as the corresponding ones in Carnot's investigations, it must be observed, involves difficult questions, which cannot as yet be satisfactorily answered, regarding the possibility of the absolute formation or destruction of heat as an equivalent for the destruction or formation of other agencies, such as mechanical work; but in taking it, I go on the almost universally adopted supposition of the perfect conservation of heat.

However, by the time the paper was reprinted, he had come to an important decision. In his description of the third process, he inserted the words:

> Continue the motion till so much heat has been given out to the second lake at $-t^o$, as that if the whole mass contained in the cylinder were allowed to return to its original volume without any introduction or abstraction of heat, it would assume its original temperature and pressure. This, if Carnot's principles be admitted, as they are supposed to be throughout the present investigation, is the same as to say, – Continue the motion … [continuing as in the first version]

Thus, James avoids the assumption of conservation of heat by continuing the third process until we reach a point where the fourth process will take us back to the original state of the system.

This change, which is spelled out [9, p. 201] in his Collected Papers, perhaps seems a minor point to modern readers, but it is interesting that, in his Obituary of James [9, p. xcv], Bottomley drew particular attention to this change, saying that: '[F]or the first time, Carnot's principle was stated, and Carnot's cycle described, in words carefully chosen, so as not to involve the assumption of the material theory of heat, or rather, as Thomson himself puts it, the supposition of the 'perfect conservation of heat"'. In their Biographical Sketch in the same volume [9, p. xxxv], Larmor and James Thomson Junior repeat essentially the same words.

That James's change was looked on as extremely important is seen from the fact that, when William prepared his Carnot paper [3] for publication in his collected papers (MPP), he inserted a note of 5 November 1881 in his description of the third process: 'The specification of this operation ... is the only item in which Carnot's temporary and provisional assumption of the materiality of heat has effect. To exclude this hypothesis, Prof. James Thomson gave the following corrected specification for the third operation [continuing in analogy at James's words above ...]' [3 MPP, pp. 122–123, also 127–128].

Truesdell's opinion [10, p. 86] may be noted, that, with this change James was actually correcting not Carnot but William, but, from any point of view, we see yet again how highly William regarded James's sustained contributions to thermodynamics. However, we conclude this section with another application of Carnot's ideas, this time due to William alone.

Through the latter part of the eighteenth century and the early part of the nineteenth, a vast amount of work was performed on thermometry, with particular attention being paid to the heating of gases [11]. By mid-nineteenth century, it was fair to conclude that the gas thermometer was the most satisfactory, because the behaviour of gases when heated fell into a number of distinct classes, depending, as we would recognised today, on the atomicity of the particular gas. In retrospect again, we would be clear that the behaviour of ideal gases was particularly simple, and could, in principle, be used as the basis of an absolute scale of temperature. In practice, though, no gas would be ideal, different gases would behave differently when heated, and so the gas thermometer was extremely useful but could not be an absolute one.

By 1848, when William Thomson had assimilated the ideas of Carnot, he realised that it would be possible to use a Carnot cycle to define an absolute temperature scale. At this stage, of course, William was working within a caloric theory, so his account of his ideas [12] was incorrect in detail in modern terms, but the basic idea certainly worked. He wrote that:

> The characteristic property of the scale which I now propose is, that all degrees have the same value; that is, that a unit of heat descending from a body A at the temperature $T°$ of this scale, to a body B at the temperature $(T{-}1)°$, would give out the same mechanical effect, whatever be the number T. This may justly be termed an absolute temperature scale, since its characteristic is quite independent of the physical properties of any specific substance.

The argument was indeed perfectly satisfactory, but an awkward feature is that the scale does not approximate to any gas scale. We are not here implying that gas scales should be correct; the point of the argument is that they are not correct and should be compared to an absolute scale. But we would like them to be nearly correct. This may be achieved with a different definition of an absolute scale, as later discussed by William in his magisterial account of the new theory of thermodynamics, written between 1851 and 1855, and collected together in MPP [12].

Here (MPP, p. 235) he wrote that: 'The temperatures of two bodies are proportional to the quantities of heat, respectively, taken in and given out in localities at one temperature and at the other, respectively, by a material system subjected to a complete cycle of perfectly reversible thermodynamic operations'. To explain the idea simply, in modern thermodynamics we assume absolute temperatures, T_h and T_c for the temperatures of the hot and cold reservoirs of a Carnot engine, and deduce that the quantities of heat taken in at the hot reservoir and deposited in the cold reservoir, Q_h and Q_c, respectively, are related by $Q_c/Q_h = T_c/T_h$.

What William Thomson does is essentially the opposite process to this. He measures Q_h and Q_c, and then defines T_h and T_c by using the equation $T_c/T_h = Q_c/Q_h$. With one suitably chosen fixed point – today we would use the triple point of water, an absolute scale has been set up, independent of the properties of any substance. However, the scale is identical to the (hypothetical) ideal gas thermometer, so it will be close to any thermometer using a real gas.

To return to 1848, it was clear at that point how useful the Carnot cycle could be, but still Joule's ideas could not be ignored. In the next section, we shall explore a short section of the Belfast notebooks where James made what may in retrospect seem a rather desperate attempt to reconcile the two without giving up the idea of caloric.

3 The Belfast notebooks: reconciling Carnot and Joule

On 20 May 1848, James sketched in his notebook an extraordinary attempt to reconcile Carnot and Joule by an analogy with mechanics. [This section of his notes has already been included in the paper of Smith [13].] James presented two columns, one containing mechanical terms, the other terms related to the relationship between heat, temperature and work. As well as momentum, the mechanical side stressed what we would now call kinetic energy, though that term had not yet been coined, and James still called it (half) *vis viva*.

James stressed that momentum, or mass multiplied by velocity, which he also called quantity of motion, is not lost in impact [in a collision], but some *vis viva*, in general, is lost [except in an elastic collision]. 'During the passage of a given quantity of motion from a velocity V to a velocity v', James said, 'work is given out'. 'Work', he spelled out, 'is equivalent to force multiplied by space, equivalent of ½ *vis viva*, equivalent to ½ mass of body multiplied by velocity^2'. So: 'By impact or mutual frictions: *vis viva* is lost, but quantity of motion is not lost. ½ *vis viva* is locked up in *velocity*'.

He then established his analogy as follows. The equivalent of the conserved quantity of motion, or momentum, is the conserved quantity of caloric. James says that the quantity of motion in a mass or set of masses is equal to the momentum that would be given out in stopping 'so as to be at rest with reference to a thing regarded as stationary'. By analogy, the quantity of caloric in a set of masses is 'the caloric given out in descending to 0°C'.

James explicitly stated that the analogy of mass is capacity [which is what we today would call thermal capacity], and the equivalent of velocity is temperature. Thus, just as the quantity of motion is mass multiplied by velocity, the quantity of caloric is [or is proportional to] capacity multiplied by temperature.

To any believer in a caloric theory, that is to say, anybody who believes that one may talk of the heat of a set of masses, these equations are little more than obvious. However, to a modern reader no less than to James Joule, the analogy between caloric and momentum must seem dangerous in the extreme; caloric or heat, we and Joule will say, should be equivalent to energy, and might better be the analogy of *vis viva* than of momentum.

However, James was able to attain some formal consistency in his next step. We saw earlier, for example in Fig. 2, that on a caloric theory, the motive power or work produced is proportional not to temperature but to its square. Therefore, James wrote that motive power or work is equivalent to half the capacity of the body multiplied by the square of its temperature, directly analogous to the mechanical quantity we had earlier, half the mass of the body multiplied by square of its velocity.

James has said that the motive power on the mechanical side of the analogy is known as *vis viva*. By analogy, he gave the name of the motive power on what we may call the thermodynamic side as *vis calida*. Just as in the passage of a given quantity of motion from one velocity to another, work is given out, so, in the passage of a given quantity of heat from a temperature T to a temperature t, work is also given out, and in the two cases, the work is equivalent to $\frac{1}{2}$ *vis viva* and $\frac{1}{2}$ *vis* calida. Just as *vis* viva may be lost but quantity of motion or momentum is not lost, so 'by conduction or radiation: *vis calida* is lost, but quantity of heat is not lost'. Just as $\frac{1}{2}$ *vis viva* is work locked up in *velocity*, so '$\frac{1}{2}$ *vis calida* is work locked up in temperature'.

At this point in the argument, James was ready to claim that it does more than just establish an analogy: he has shown, he suggests, that *vis viva* and *vis calida* are mutually convertible. Both are equivalent to work, or we may say, may be transformed into work. He considered that his argument was an authoritative statement of the position of William and himself, and continued by stating, based on this argument, 'the question at issue between Joule and ourselves'. The question posed is 'Is a certain *quantity of heat* equivalent to a certain *vis viva* the two being mutually convertible?' This was certainly Joule's position – work could be transformed to heat, and *vice versa*.

For James, though, since *vis viva* and *vis calida* were equivalent, Joule's position was that a certain quantity of heat was equivalent to and convertible to a certain *vis calida*. But for James this implied a contradiction. He had defined the quantity of heat as the product of capacity and temperature, but *vis calida* as $\frac{1}{2}$ the product of capacity and the *square* of the temperature. Clearly, it was completely impossible that the two could be equivalent, as, ignoring the factor of $\frac{1}{2}$, James wrote 'quantity of heat = *vis calida*/temperature'. This was an argument, he said, that 'a certain *vis calida*/ temperature [should be] equivalent and convertible to *vis calida*, which appears to be nonsense', as he rightly says. He added in brackets: 'Is my reasoning good?'

James immediately turned to a consideration of Davy's experiment in which had he melted two pieces of ice by their mutual friction, which suggested that work or *vis calida* actually produced a quantity of heat, and thus very much supported Joule against the position reached in James's notes; as he said '[It] would be incompatible with my view of the subject'. He asked himself in brackets: 'Is the experiment to be trusted?'

Ironically, within a few years, when William had fully accepted a dynamical theory of heat as an important component of his theoretical construction of thermodynamics, he was to acknowledge Davy's work in particular, even more than that of Rumford, as that which had established the theory.

As to James's own analysis, one may delight in its boldness and its inventiveness. From a modern perspective, it has the difficulty that seems inevitable in a caloric theory: the work obtained is proportional not to the temperature difference through which the heat has dropped, but to its square. It will be remembered that, in Williams' view, the 'blunder' in Joule's presentation at the BAAS was to disagree with this rule, and exactly the same factor is involved in James's difference with Joule in these notes. In retrospect, the cure for the problems faced by the brothers was to dispense neither with Carnot nor with Joule, but to bring them together and to expel caloric.

4 The denouement

By the spring of 1850, the Thomson brothers and Joule had been circling the problem of heat and work for a number of years. The Thomsons had shown themselves willing to question the caloric theory, with its conservation of heat and its upholding of heat as a state function, but they were seemingly unable to move effectively beyond it, and thus were not able to discuss waste or dissipation in a useful way. Joule had moved beyond caloric in his full acceptance of what we would today call conservation of energy, but was still unable to understand fully what happened to energy that seemed to have been wasted. It was in this spring that Joule wrote to William [4, p. 316] saying that:

> [A]s your brother said, there ought to be some connecting link between the results I have arrived at and those deduced from Carnot's theory. Perhaps you will succeed before long in discovering it. For my own part it quite baffles me.

However, this honour was not to go to William but to Rudolf Clausius (Fig. 6), then only 28 years old. Clifford Truesdell was not one to shower compliments on the creators of thermodynamics. He describes William's paper on Carnot [3] as 'enthusiastic, rambling, and sometimes vague [10, p. 168]' and his 1851 paper [14] as 'verbose and rambling [10, p. 224]'. Of Clausius, he asserts that: 'Few mathematical physicists have shown so little sense of the right mathematics for the job [10, p. 206]'. So we must take Truesdell very seriously when he [10, p. 204] says that:

> There is no doubt that Clausius with this paper created classical thermodynamics. Compared with his work here, all preceding except Carnot's is of small moment. Clausius exhibits here the quality of a great discoverer: to retain from his predecessors major and minor – in this case, from Laplace, Poisson, Carnot, Mayer, Holtzmann, Helmholtz, and Kelvin – what is sound while frankly discarding the rest, *to unite disparate theories*, and by *one simple if drastic change* to construct a complete theory that is new yet firmly based upon previous partial successes.

Figure 6: Rudolf Clausius.

Figure 7: Macquorn Rankine.

Clausius's paper (included in Ref. [2]), published in April 1850, was titled 'On the motive power and on the laws which are deducible from it for the theory of heat', and the 'simple if drastic change' was to reject the conservation of heat, and to replace it by the equivalence of work and heat.

Clausius gave great credit to William Thomson for his analysis [3] of the state of the subject at the end of the 1840s, and his explanation of the problems that he discussed. Indeed, Clausius admitted that his knowledge of the work of Carnot was only through the accounts given by Clapeyron and by William Thomson. Clausius made the following comments on Carnot's work:

> [Carnot] says expressly that no heat is lost in the process, but that the *quantity of heat* remains unchanged, and adds: 'This fact is not doubted; it was assumed at first without investigation, and then established in many cases by calorimetric measurements. To deny it would overthrow the whole theory of heat, of which it is the foundation'. I am not aware, however, that it is has been sufficiently proved by experiments that no loss of heat occurs when work is done; it may, perhaps, on the contrary, be asserted with more correctness that even if such a loss has not been proved directly, it has yet been shown by other facts to be not only admissible, but highly probable.

To this end, Clausius discussed the work of Joule, in which, he said: 'Heat is produced in several different ways by the application of mechanical work'. These experiments, he said, 'have almost certainly proved ... the law that the quantity of heat developed is proportional to the work expended in the operation'. Various experiments, he said, had suggested that heat was not a substance, but consisted in the motion of 'the least parts of bodies', so Clausius argued that such a motion may be transformed to work, the loss of *vis viva* being proportional to the work accomplished.

Having supported Joule, Clausius then gave an eloquent account of William's difficulties in reconciling his work with that of Carnot:

> [Thomson] speaks of the obstacles which lie in the way of the unrestricted assumption of Carnot's theory, calling special attention to the researches of Joule, and also raises a fundamental objection which may be made against it. Though it may be true in the case of the production of work, when the working body has returned to the same condition as at first, that heat passes from a warmer to a colder body, yet on the other hand it is not generally true that whenever heat is transferred work is done. Heat can be transferred by simple conduction, and in all such cases, if the mere transfer of heat were the true equivalent of work, there would be a loss of working power in Nature which is hardly conceivable. Nevertheless, he concludes that in the present state of the science, the principle adopted by Carnot is still to be taken as the most probable basis for an investigation of the motive power of heat, saying: 'If we abandon this principle, we meet with innumerable other difficulties – insuperable without further experimental investigation – and an entire reconstruction of the theory of heat from its foundations'.

However, Clausius then suggested that we should not be daunted by these difficulties, but rather familiarise ourselves with the consequences of the idea that heat is a motion. Indeed, he actually did not feel the difficulties were as serious as William Thomson had suggested. Carnot's theory did not have to be discarded, a step which Clausius said, in agreement with William, we would find difficult to take, since it has been amply verified by experiment. Clausius's new method, he claimed, was not in contradiction to the essential principle of Carnot, only to the subsidiary statement that no heat is lost. Clausius then summed up his new theory, very much the basis of today's thermodynamics, as follows:

> It may well be the case that at the same time a certain quantity of heat is consumed and another quantity transferred from a hotter to a colder body, and both quantities of heat stand in a definite relation to the work that is done.

Clausius also introduced two other central components of thermodynamics. The first, which he stated explicitly, was, in the terms we have been using, that heat was not a state variable:

> It is common to speak of the *total heat* of bodies, especially of gases and vapours, by which term is understood the sum of the free and latent heat, and to assume that this is a quantity dependent only on the actual conditions of the body considered, so that, if all other physical properties, its temperature, its density, etc., are known, the total heat contained in it is totally determined. This assumption, however, is no longer admissible if our principle is adopted.

Clausius gave detailed examples of how the concept of total heat broke down. If, for example, a gas moved from one physical state to another along different routes on a *PV*-diagram, then in general different amount of work would be done on or by the gas along the different routes.

It follows then also that different amounts of heat will have been lost by or given to the gas during the two processes, so the idea that the gas 'contains' specific amounts of heat at the beginning and at the end of the process is untenable. (Kelvin had actually recently made similar remarks [4, pp. 312–314] but had not quite known how to follow through his ideas.) Of 'latent heat', Clausius wrote that: '[It] is not merely, as its name implies, *concealed* from our perception, but it is *nowhere present*; it is *consumed* during the changes in doing work'.

Clausius showed, though, that there was an important new state function. When heat is provided to a system, he said that, in his terminology at the time, two types of work may be done. The first is what we today would just call 'work', though Clausius called it 'external work'; this may be a gas expanding or a liquid being vapourised and having to push back the external pressure to make room for itself. This clearly depends on the external pressure, so 'work' for us, 'external work' for Clausius, is, like heat, not a state function. However, what Clausius called 'internal work' related to overcoming the mutual attraction of the particles, and so was independent of pressure and was a state function. Today it would be called 'internal energy', and with our increased knowledge of the kinetic theory of gases, we would wish to make it clearer than Clausius that it included what we would call kinetic energy as well as potential energy. For an ideal gas, of course, the potential energy is zero, and so internal energy is solely kinetic energy.

Ever since Clausius, the quantity has been given the symbol U. Clausius said explicitly that it is a function of volume and temperature, 'and of being therefore fully determined by the initial and final conditions of the gas, between which the transformation has taken place'. His paper, in fact, included the equation:

$$dQ = dU + AR\frac{a+t}{v}dv$$

where A is the heat equivalent of work, R is the gas constant, $a = 273$ and t is the temperature in degree centigrade. This is our present first law of thermodynamics for the case of an ideal gas. (Here, U is clearly to be considered in units of heat.) Clausius's adaptation of Carnot's work essentially also provided the second law.

If we should say that Clausius had laid lasting foundations for classical thermodynamics, we must admit that there was still much for Macquorn Rankine (Fig. 7) and William Thomson to do. In fact, even before Clausius's work had been published, Rankine, on the basis of rather elaborate and perhaps unconvincing models involving the motion of atmospheric particles forming vortices around atomic nuclei, had produced detailed results quite analogous to many of those of Clausius. William Thomson was to have a stimulating debate with Rankine through 1850 and into 1851 [5, pp. 102–106].

Thomson knew of Clausius's theory as early as August 1950, and indeed discussed it, at least in general terms, with Rankine from that time. However, he was to write, in an early draft of his important paper of 1851 [4, p. 324]:

> The same conclusion has been arrived at by Clausius, to whom the merit of having first enunciated and demonstrated it is due. It is with no wish to claim priority that the author of the present paper states that *more than a year ago* he had gone through all the fundamental investigations depending on it which are at present laid before the Royal Society.

By February 1851, it is clear that he felt he had reconciled Carnot and Joule in his own mind, and was persuaded by Joule to lose no time in writing an account of his new theory. Thus were written the first three parts of what would eventually become the great paper of 159 pages eventually put together and published in Thomson's collected papers [14]. (The title used in the collected papers is the title of these first three parts.) The fourth and fifth parts followed in 1851 and the sixth, which dealt with thermoelectric currents, an area where Thomson made exceedingly important discoveries, in particular, the so-called Thomson effect, in 1854. The seventh, on thermoelastic and thermomagnetic properties of matter, was mostly written in 1855, though major additions were made for the collected papers. Lastly, appendices in the version included in the collected papers had originally been published in 1851, 1853 and 1854.

Here, we may concentrate on just a few notions from the beginning of this paper. It should incidentally be made clear that, like his 1849 paper [3], he regarded it as less a statement of a new discovery, more a general coherent account of the significance of the new theory, and a full working out of the conclusions that might be drawn from it. Thus, he acknowledged the contributions of Rankine and Clausius at the outset. He then spelled out what he called two propositions. The first, which he attributed to Joule, was that whenever heat is produced from thermal sources, or lost in thermal effects, equal amounts of heat are put out of existence or generated. He, thus, completely abandoned the caloric theory, accepting Joule's ideas in their entirety, and becoming perhaps the main advocate from that moment of what he called the *dynamical theory* of heat. In this role, he was to celebrate the memory of the work of Davy and Rumford half a century before; why, he asked, had scientists largely ignored their proofs of the dynamical nature of heat? (Thomas Kuhn [15] made the fair rejoinder that Kelvin himself had been using the caloric theory only a year earlier!)

The second proposition, attributed by William to Carnot and Clausius, states that the most efficient engine acting between particular temperatures of source and refrigerator is a reversible one. William expressed this proposition in his own terms a little later in the paper: 'It is impossible, by means of inanimate material agency, to derive mechanical effect from any portion of matter by cooling it below the temperature of the coldest of the surrounding objects'. This became his famous statement of the second law of thermodynamics.

So far he had broadly repeated the concepts of Clausius, and to an extent, of Rankine. But the difference between them was that Clausius's interests were rather narrow, being largely restricted to the theory of heat engines [5]; it might be said that his paper was an admittedly brilliant but technical solution to a technical problem. Rankine too was a comparative newcomer to the topic. William, in contrast, had spent the best part of a decade worrying practically incessantly about the conceptual and cosmological significance of the problems he had been considering. With the solutions now understood in principle, he had much to say about their implications. He began in the remainder of this paper, and continued for the rest of his life.

First, he became the apostle for the new idea of energy. Until that time, physics had been constructed around the Newtonian idea of force, which was immensely useful in mechanics but not so useful elsewhere in physics. With the understanding that heat, light, sound, electricity and magnetism could all be expressed in terms of energy, with the full appreciation of what we now call kinetic and potential energy, and paying regard to the idea of transformations between the various types of energy, it became clear, at least to William, that all the various areas in physics could very fruitfully be discussed in the new paradigm of energy. The treatise he and Tait wrote together [16] is a celebration of this new belief, and today their evangelism has been so successful that it is almost regarded as tautological to describe physics as the science of energy.

Even more significant was William's long concern, together with James, with the question of waste. What for Clausius was little more than a logical explanation of what happened when heat could have produced work but had failed to do so – extra heat was deposited in the cold reservoir, for William became the solution to his central conceptual problems, and the key to his new worldview. This heat was not lost in the material world, thus satisfying William's demand that only God could create or destroy. Nevertheless, this energy is 'lost to man irrecoverably'. Thus, William's worldview was one of dissipation and irreversibility, with an arrow of time leading to the so-called heat death, where everything is at the same temperature, and any interesting features in the universe have been lost. It was a most beautiful solution to the worries that William and James had shared over many years, and this major conceptual development played a deservedly large part in building up William's towering reputation through the second half of the nineteenth century [17].

We may only sketch the subsequent history of thermodynamics. For many years after 1851, Clausius showed little interest in the idea of energy, but in the 1860s he turned his thoughts to such general matters, and then, around 1865, introduced the idea of 'entropy'. Entropy remains constant in a reversible process but increases in an irreversible one; indeed, such is the content of one of the most useful statements of the second law of thermodynamics.

However, it was at about this time that Tait, who had taken on the self-appointed task of William Thomson's bulldog, apparently decided that William's initial support of Clausius had been premature and unwise. It is true that William himself felt that Clausius's own proofs of some of the most fundamental portions of thermodynamics had been unsatisfactory, in particular being restricted to the case of ideal gases; he felt that his own proofs were the first satisfactory and general ones. Tait [18] took the process much further, arguing consistently against Clausius's priority for thermodynamics, and also the suitability and correctness of his methods [5, pp. 255–260]. At least to an extent Maxwell appeared to support Tait, and Clausius had to resort to complain of Maxwell's *Theory of Heat* [19] that:

> [In Maxwell's book] my writings are left quite unmentioned; and my name occurs only once, when it is said I introduced the word *entropy*; but it is added that the theory of entropy had already been given by W. Thomson. Hence anyone who derives his knowledge of the matter solely from this book must conclude that I have contributed nothing to the development of the mechanical theory of heat.

It seems to have been Josiah Willard Gibbs (Fig. 8), the American physicist, who was the first to endorse fully Clausius's achievements. Gibbs himself, of course, was to contribute massively to foundational studies in thermodynamics [5, pp. 260–261].

The other member of the triumvirate credited with the founding of thermodynamics, Rankine, was active on both technical and theoretical fronts in the years after 1851 [5, pp. 150–166]. We have already seen his efforts to develop an air-engine, and he was also involved in the production of a marine compound engine, attempting to use the new thermodynamics to produce the most efficient and compact heat engine for long-distance navigation [5, p. 5]. On the

Figure 8: Josiah Willard Gibbs.

theoretical side, he wrote extensively on the structure of thermodynamics, though always from an engineering perspective. Rankine is credited with bringing the actual word *thermodynamics* into general use in the subject in 1859, though William Thomson had introduced it in a more restricted context a few years earlier. Rankine also established the terminology of the *first* and *second laws* of thermodynamics and was responsible for the term *kinetic* energy. In 1855, all this work gained Rankine the prestigious Chair of Engineering at Glasgow in succession to Lewis Gordon, and, of course, it was Rankine's early death in 1872 that paved the way for James Thomson's own return to Glasgow. Mention of James brings us back to the central topic of this paper and encourages us to ask – in the final emergence of modern thermodynamics, why was James's name absent?

5 James Thomson and thermodynamics

From the beginning of their interest in the study of work and heat right up to 1849, it is quite clear that James and William worked as a partnership. This applied to the discussion of their early dilemmas, though their reception of the work of Carnot, and later of that of Joule, and their later approach to these theories and their attempts to weld them into a single unified theory. It also applied to the discovery of the lowering of the freezing point of water under pressure.

Indeed it may be remarked that not only was there a partnership but, if either of them should be described as the senior partner, it was certainly James. He had perhaps the greater desire and ability to reach the most fundamental aspects of any set of ideas, and also his practical engineering experience, and particularly his intense interest in the means of minimising waste and maximising efficiency in any process, made his interest in, and his appreciation of, what was to become the subject of thermodynamics extremely strong.

For example, Smith [13] remarks that: '[M]uch of the early interest in Carnot and Clapeyron lay with James, and it appears to have been his enthusiasm which inspired William to go beyond a mere passive awareness of Clapeyron's memoir'; and that: "James as the 'theoretical engineer' had rather more specialised and concentrated interests relating to the motive power of heat"than William, whose interests 'were much wider in the field of natural philosophy and mathematics generally'. It may also be remarked that, at least during the latter part of the 1840s, as a newly appointed professor, and even before then as an aspirant to the position, William was especially busy, while James was at home with much more free time. He was, of course, recovering from illness, but, as we have seen, after presumably a period during which work was not easy, he actually was able to work hard and extremely constructively.

Yet from 1850, it appears that William was working effectively on his own. The initial parts of his paper *On the Dynamical Theory of Heat*, which were completed in 1851, do not mention James. Since his claim to be regarded as one of the founders of thermodynamics rests mainly on that paper, it follows that James's contribution has been scarcely known, and again it should be mentioned that Crosbie Smith's work in drawing attention to his contribution has been of the utmost importance.

Why did this separation of the interests of the brothers occur? It will be remembered that it was at this period that, following the death of their father, James left Glasgow in a desire to avoid any breakdown of family harmony over his decision to become a Unitarian. He only returned to his family, and then to Belfast not Glasgow, at the request of Anna in 1851. It is certainly the case that William stayed with James in London for over a month in the summer of 1849, and of course at this time the discussions on the lowering of the freezing point and its experimental prediction were ongoing. However, in the subsequent 2 years or so, it may be that not only were the brothers not living in the same house, but communication by post was somewhat curtailed. Ironically, these years, so important in the sudden flowering of thermodynamics, were perhaps the one time during their lives that the brothers may not have been in very close contact either in person or by mail. It is still difficult to believe William would not have wished to discuss the work of Rankine and Clausius with James, but there does not seem to be evidence that this actually happened.

When William wrote the paper, it will be remembered that, ostensibly at least, he regarded it as a means not of providing new results, but rather of expressing the results of others in an informed and informative form. Thus, it might be argued that, since, again it must be said ostensibly, he was not attempting to claim credit for himself for the work reported, by refraining from mentioning James, he was not denying James any credit either.

Yet, this argument seems unconvincing. William must have been aware that there was much that was new in his paper. As well as the fact that he considered the mathematical proofs much more satisfactory than those of Clausius, there were, in particular, the arguments on waste, dissipation and the future of the Universe, which went well beyond anything Clausius had provided or indeed was interested in. And much of the basis of this work was the discussions of the two brothers over the years up to 1849. It is undoubtedly a simplification, but perhaps not a gross one, to suggest that many of William's most important contributions to thermodynamics were based on the answers provided by Clausius to the profound questions discussed by William and James through the 1840s. James could really have hoped for more credit.

So where does our argument leave James's position in the history of science? It is clear that his contribution to his brother's important publications was crucial. Whether that actually entitles him to be regarded as one of the founders of thermodynamics must be a matter of opinion, depending on the extent to which one is prepared to take into account intellectual support and guidance, as distinct from crucial publication. In more general terms, though, it is clear that he should be recognised as a major figure in the history of science and technology in his own right, with a wide range of discoveries and achievements, not merely as worthy of attention only as the brother of Lord Kelvin.

In a more local context, he may be regarded as an important contributor to the nineteenth-century Irish tradition in mathematics and science, or, to put this more precisely, the tradition built up between the northeast of Ireland and the southwest of Scotland [20]. What Crosbie Smith said of Tait [21] may amply be said of James Thomson: '[H]e formed an integral part of a North British scientific intelligentsia, whose members often circulated around academic posts and institutions in Ulster and in Scotland but whose influential networks of science and engineering extended far beyond these shores'.

Acknowledgements

I thank Deirdre Wildy, Senior Subject Librarian at Queen's University Belfast, for permission to use and quote from the material in the Special Collections in the Library, and Diarmuid Kennedy, Special Collections Librarian and his staff for their courtesy and help in obtaining material from the James Thomson archive. I also thank Joan Whitaker, Physics Subject Librarian, and the other staff of the Science Library at Queen's University Belfast, for considerable support in obtaining books and papers.

I also like to acknowledge Mark McCartney for many interesting and useful conversations on the topics of James and William Thomson and other aspects of nineteenth century physics.

Finally, I am grateful to Michael Collins for suggesting that I should write this paper.

References

[1] Thomson, J., Notebook A 14 (B), *Thomson Papers*, Queen's University Belfast, 1848.

[2] Carnot, S., *Reflections on the Motive Power of Fire, and on Machines fitted to Develop That Power* (translated and edited by R.H. Thurston, 1890), Dover: New York, 1960.

[3] Thomson, W., An account of Carnot's theory of the motive power of heat, with numerical results deduced from Regnault's experiments on steam. *Transactions of the Royal Society of Edinburgh*, **16**, pp. 541–74, 1849; also in Thomson, W., *Mathematical and Physical Papers* (MPP), Cambridge University Press: Cambridge, pp. 113–55, 1882.

[4] Smith, C. & Wise, M.N., *Energy and Empire: A Biographical Study of Lord Kelvin*, Cambridge University Press: Cambridge, 1989.

[5] Smith, C., *The Science of Energy: The Construction of Energy Physics in the Nineteenth Century*, Athlone: London, 1991.

[6] Thomson, J., Theoretical considerations of the effect of pressure in lowering the freezing point of water. *Transactions of the Royal Society of Edinburgh*, **16**, pp. 575–580, 1849; reprinted with some slight but important alterations in Thomson, J., *Cambridge and Dublin Mathematical Journal*, **5**, pp. 248–253, 1850.

[7] Thomson, W., The effect of pressure in lowering the freezing point of water, experimentally demonstrated. *Proceedings of the Royal Society of Edinburgh*, **2**, pp. 267–271, 1851; also in MPP, pp. 165–168.

[8] Thomson, W., Notice of Stirling's air engine. *Proceedings of the Philosophical Society of Glasgow*, **2**, pp. 169–170, 1847.

[9] Thomson, J., *Collected Papers in Physics and Engineering*, Cambridge University Press: Cambridge, 1912 (eds. J. Larmor & J. Thomson).

[10] Truesdell, C., *The Tragicomical History of Thermodynamics 1822–1854*, Springer-Verlag: New York, 1980.

[11] Fox, R., *The Caloric Theory of Gases: From Lavoisier to Regnault*, Clarendon: Oxford, 1971.

[12] Thomson, W., On an absolute thermometric scale founded on Carnot's theory of the motive power of heat, and calculated from Regnault's observations. *Philosophical Magazine*, **33**, pp. 313–317, 1848; also in MPP, pp. 100–106.

[13] Smith, C.W., William Thomson and the creation of thermodynamics: 1840–1855. *Archive for History of Exact Sciences*, **16**, pp. 231–288, 1977.

[14] Thomson, W., On the dynamical theory of heat, with numerical results deduced from Mr Joule's equivalent of a thermal unit, and Mr Regnault's observations on steam. MPP, pp. 174–332.

[15] Kuhn, T., Energy conservation as an example of simultaneous discovery. *Critical Problems in Modern Science*, ed. M. Clagett, University of Madison Press: Wisconsin, pp. 321–356, 1959; also in Kuhn, T., *The Essential Tension*, University of Chicago Press: Chicago, pp. 66–104, 1977.

[16] Thomson, W. & Tait, P.G., *Treatise on Natural Philosophy*, Cambridge University Press: Cambridge, 1883.

[17] Whitaker, A., Kelvin – the legacy. *Kelvin: Life, Labours and Legacy*, eds. R. Flood, M. McCartney & A. Whitaker, Oxford University Press: Oxford, pp. 278–306, 343–345, 2008.

[18] Tait, P.G., *Sketch of Thermodynamics*, Douglas: Edinburgh, 1877.

[19] Maxwell, J.C., *Theory of Heat*, Longmans, Green: London, 1871.

[20] Whitaker, A., Is Kelvin representative of an Irish tradition? *Kelvin and Ireland*, eds. R. Flood, M. McCartney & A. Whitaker, Institute of Physics Conference Series: Bristol, volume 158, conference 1, 2009.

[21] Smith, C. & Tait, P.G., Queen's College and Ulster-Scots natural philosophy. *Queen's Thinkers: Essays on the Intellectual Heritage of a University*, eds. A. Jackson & D.N. Livingstone, Blackstaff: Belfast, 2008.

Chapter 4

James Thomson, an engineer and scientist: the path to thermodynamics

A. Whitaker
Department of Physics, Queen's University Belfast, United Kingdom.

Abstract

James Thomson, brother of William Thomson who later became Lord Kelvin, was also a highly creative practical engineer but also an extremely deep thinker in many areas of science. After early difficulties in his career due to ill-health, he was a professor of civil engineering at Queens University Belfast from 1853 to 1873, and then at the University of Glasgow until 1889, and he also became a Fellow of the Royal Society. Through the 1840s, the brothers worked together to analyse the logical problems of the then current theories of heat and work, and also together they studied the seemingly contrasting ideas of Sadi Carnot and James Joule, in the desire to combine them to create a new successful theory.

1 Introduction

By any criterion but one, James Thomson (1822–1892) (Fig. 1) was indisputably one of the very most able and important engineers and scientists of the nineteenth century. As a creative engineer, he was responsible for a large number of interesting and useful technical inventions. Much of his engineering work related to a wide range of studies of fluid motion, and his interest in this topic extended from engineering into the more strictly scientific field, while other important scientific achievements included work on the relations between the gaseous, liquid and solid states of matter; on elasticity and the strength of materials and on various aspects of freezing. In the latter category was his famous paper on the lowering of the freezing point of water by pressure [1], which was exceptionally important, not just as a major discovery in its own right, but also because of its central role in the establishment of the formal structure of thermodynamics.

His fertile imagination extended well beyond engineering and physics to what might best be described as geophysics, and he studied such problems as the flow of rivers, glaciation, and the prismatic structure of the Giant's Causeway, in all cases providing novel approaches, and clear and convincing analysis.

In career terms, he spent 20 years as a professor of Civil Engineering in Belfast, his birthplace, and a further 16 in Glasgow where he occupied what had been the first chair of engineering in the world, and was still one of the very most prestigious. His teaching was recognised as being devoted and inspiring. He became a Fellow of the Royal Society and was chosen to deliver its important Bakerian Lecture in the year of his death, and he received honorary degrees from the Queen's University, the University of Glasgow and the University of Dublin.

Outside academia, he worked as a successful consulting civil engineer, occupying a number of important civic posts. At a time when Belfast was expanding rapidly from a fairly small town to a large industrial city, with all the social problems that the process involved, James was an engineer to the Water Commissioners, and as such was responsible for the crucial task of providing fresh water, and for the removal of sewage; he became a well-known authority on such matters He was also influential in the provision of public parks.

(The most notable account of James Thomson's life and work is the collaboration between the famous physicist Joseph Larmor, and James's son, also called James; this account is included in his Collected Papers [2]. More recent shorter articles are by Bernard Crossland and John Moore [3], and by Peter Bowler [4]. Apart from a discussion of Ref. [1], none of these accounts provide much information on Thomson's contributions to thermodynamics, the main topic of the present paper, for which, see later.)

By any criterion bar one, then, James was a highly successful man. Unfortunately, the criterion by which he has most often been judged is by a comparison of his achievements with those of his younger brother, William Thomson, later Lord Kelvin (1824–1907) (Fig. 2) [5–7]. Kelvin was, of course, not only a scientist and engineer of spectacular achievement, but also extremely important in the development of scientific education as the founder of the first undergraduate laboratory. His technical work made him extremely well known and indeed rich, mostly as a result of his work on the first Atlantic telegraph cable, and after that on a range of important inventions. He worked closely with government on committees and inquiries, and, partly at least as a result of his political activities, became Lord Kelvin. He was to be buried in Westminster Abbey. It would have been impossible for James to compete with Kelvin on Kelvin's own ground!

But James had his own considerable strengths, many of which were in areas where William was comparatively weak. While this will emerge during the rest of this paper, some preliminary

Figure 1: James Thomson.

Figure 2: William Thomson, Lord Kelvin.

remarks may be made at this point. With two brothers, one, William, nominally a scientist, the other, James, nominally an engineer, one might expect that it would be the engineer who would be less wedded to the study of fundamental principle, but with the Thomsons it was, if anything, the other way round. Larmor writes of James as 'the philosopher, who plagued his pragmatical brother' [6, pp. 282–283]. As an early example, when the brothers made toy boats in their boyhood, James was not satisfied until a toy boat he might be making was perfect in every detail, while William was happy as soon as his floated at all [2, pp. xvii–xviii]!

Similarly in later life, before James would publish a paper or make public an invention, he would have considered every detail long and carefully. Thus, his theories stood the task of investigation, and his engineering inventions worked. The less patient William, on the other hand, often threw off ideas at white heat, and, though they may often have been brilliant, they frequently required substantial correction or amendment.

Smith and Wise [6, p. 282] contrast James' 'determined single-mindedness, manifested in his compulsive involvement with engineering problems' with 'William's impulsive attacks on a variety of loosely related problems'. They also quote from a letter of Hermann Helmholtz to his wife as follows: 'James is a level-headed fellow, full of good ideas, but cares for nothing but engineering, and talks about it ceaselessly all day and all night, so that nothing else can be got in when he is present. It is really comic to see how the two brothers talk at one another, and neither listens, and each holds forth about quite different matters. But the engineer is the most stubborn, and generally gets through with his subject.'

Other interesting comments come from John Nichol. Nichol was the son of John Pringle Nichol, who was a professor of astronomy at The University of Glasgow when James Thomson's father, yet another James whom we shall call James Senior, was a professor of mathematics. John Pringle Nichol was particularly responsible for William becoming a devotee of the work of Joseph Fourier, who was to be an important influence on him right the way through his career. John Nichol was friendly with the Thomson children when they were young, and later he served with James Thomson when John himself was a professor of English Language and Literature at Glasgow from 1862 to 1889, so his opinions are of great interest.

John Nichol [2, p. xxxviii] wrote: 'Of the sons I liked James best. He was crotchety and apt to be sulky with those who would not enter into his crotchets; here his faults end. He was steadfast, independent, straightforward, quiet, unobtrusive ... His passion was engineering; he was always on the eve of something that was going to revolutionize trade. He used to show me lots of models, and often he would walk out to try his boats on the streams, as a chemist goes to make an experiment that will test the worth or worthlessness of years of toil ... I believe that some of the inventions were excellent, but there was always some practical obstacle which prevented their bringing to the inventor either the fame or the fortune they merited. James was an idealist in his way'.

We have now met some of the facets of James' mind, intelligence and character. In the following two sections, we shall look briefly at his life and then give a summary of his general achievements. After that, the rest of the paper is devoted to his contributions to the development of thermodynamics. It is true, of course, that this was the area of pure science for which his brother William has deservedly become most famous. It is usual to regard him with Rudolf Clausius and MacQuorn Rankine as the three founders of the subject. In apparent contrast, apart from

Ref. [1], James Thomson published next to nothing in the area. However, it is very much to the credit of Crosbie Smith [8, 9; see also 6] that he has investigated in detail James's contributions to William's own work, and found them to be extremely important. It would seem that James's determination and clear-sightedness blended well with William's more impulsive and perhaps imaginative qualities. James also gathered considerable experience in practical working with machines and studying their various efficiencies. Smith's publications have stimulated the question that this paper is designed to study, even if it may not actually succeed in answering: Were James Thomson's contributions to the emergence of thermodynamics so great that he deserves to be 'promoted' and regarded as one of its founders?

As has been implied, Thomson's contributions to science and engineering in general, let alone thermodynamics, where most of his main contributions were admittedly unpublished, have been less well known than they should have been. In the Preface to his collected papers [2, p. v], which was published as soon after his death as 1912, it is said that the importance of his scientific activity is well known, but 'chiefly through references to his work by other physicists such as his brother Lord Kelvin, Thomas Andrews, James Clerk Maxwell, Osborne Reynolds [and] J Willard Gibbs'.

Yet one man – the highly perceptive John Perry [10], who had been a brilliant pupil of James Thomson at Belfast, and who in his turn became an exceptional engineer, physicist and educator – was clear-sighted enough to raise precisely the topic of this paper. 'As early as 1847', Perry has written [2, p. li], '[James] applied Carnot's cycle and first principles to the study of the pressure and temperature of melting ice, and he may be said to have really shown to Clausius, Rankine and his brother (Lord Kelvin) the method of attack which has led to the development of thermodynamics.'

2 A sketch of the life of James Thomson

James and William were the third and fourth children of James Thomson Senior, a man perhaps as remarkable as either of his two more famous sons. James and William had two elder sisters, Elizabeth (1818–1896) and Anna (1820–1857), while a younger sister, Margaret, died as an infant. Anna, in particular, was to play a crucial role in James's life. Of their two younger brothers, John (1826–1847) became a medical student, but sadly, while still young, he contacted fever in the course of his duties and died. Robert (1829–1905) entered the insurance profession and emigrated to Australia. Here, he was to suffer mixed fortunes. As actuary to a life insurance company he was renowned for his enthusiasm, but he was dismissed when he was unable to account for considerable sums of money, and he became bankrupt several times in the 1860s. He was subsequently able, though, to rebuild his actuarial career with great success, being renowned as 'a man of genius but erratic working habits' [11].

James Senior (1786–1849) had begun life as a farmhand on a family-owned farm near Ballynahinch, around 15 miles south of Belfast, but, as a man of great mathematical ability and enormous determination, he was able to receive a basic education locally, and then to assist with the teaching, a job he was able to maintain in the summer months while studying for a degree at the University of Glasgow in the winter months. Having obtained his degree in 1812, his natural progression would have been to a ministry in the Presbyterian church, but for an exciting academic opportunity that presented itself.

In 1798, under the United Irishmen, a rebellion had taken place in Ireland, effectively bringing together Catholics and Presbyterians [dissenters], and uniting then broadly against the combination of the British state and members of the established church in Ireland, the Irish Church, which could be regarded as the branch of the Anglican Church in Ireland. At this time, only members of the Irish Church had full political rights. The rebellion had been inspired by the American and French revolutions, and its ideals were those of liberty, equality and freedom from sectarian discrimination. These were very much James Thomson Senior's own beliefs, which he passed on to a very great extent to his children, and he strongly supported at least the principles of the United Irishmen. The rebellion was crushed, and, as is well-known, the tentative unity of Catholics and dissenters was fairly soon broken down, giving way to the strong division between Catholic on the one hand, and Protestant (Presbyterian *and* Anglican) on the other which has largely persisted till the present day.

However, in Belfast, in particular, where the United Irishmen had been mostly Presbyterian rather than Catholic, the ideals of tolerance and non-sectarianism were to remain pronounced for roughly the first third of the nineteenth century. Central to 'liberal Belfast' was the establishment of the Belfast Academical Institution, universally known as 'Inst'. This had school and college departments (the latter being effectively a Scottish university on a small scale), and was avowedly non-sectarian, its secretary stating that: 'The subscribers to the Institution are composed of all religious persuasions'. As well as its liberal perspective, Inst aimed at providing a practical education for its students, and in this and indeed every way James Thomson was ideal for Inst, and Inst was ideal for James Thomson. He became both a teacher of mathematics in the school department (also teaching geography) from 1814, and a professor of mathematics in the college department from 1815, and he immediately became a strong political presence in the administration of Inst. Wherever he was employed, he was to become a leader of men, coupling clear ideas of what should be done, with the political ability actually to get it done in the majority of cases.

As well as his teaching duties, Thomson used his courses as the basis of a series of practical and commercially oriented textbooks, to write which he got up around 4 O'clock in the morning before his day at Inst began. Fortunately, these books sold extremely well and thus made him a lot of money. He certainly felt no need to apologise for his comparative affluence, believing that the labourer was worthy of his hire, another belief that was passed on to his sons. William was clearly the more successful in money making, though John Nichol's comments in the previous section suggest that James was by no means averse to using his engineering knowledge for commercial purposes, though not as successful as his younger brother.

James Senior married Margaret Gardner in 1817, and the family listed above were born over the next 10 years or so; sadly the hard-working but happy and prosperous existence the family enjoyed came to a sudden end when Margaret died in May 1830, and by December 1831 James Thomson Senior had been appointed as the chair of mathematics at the University of Glasgow. It was a new start for the family, in which, particularly in the absence of a mother, the influence of the father on his children was to be immense and of supreme importance.

James Senior himself adapted very quickly to the new situation. The university was at the time something of a hotbed of conservatism, privilege, patronage and nepotism. In particular, religious tests ensured that only members of the (Presbyterian) Church of Scotland could be appointed to chairs. James set himself the task of reform, but this was only achieved slowly, mostly by the death of the reactionary professors and their replacement by those more willing to encompass fresh ideas; the process was to take the remainder of his life and longer, and in fact the religious tests were only to be abolished in 1858.

Due to a misunderstanding, Professor Thomson was initially not nearly as well off in Glasgow as he had expected, and as a result the children were taught at home. Neither James nor William was to regret this, as they felt the free flow of ideas when working directly with their father easily surpassed the routine drill they might have been subject to at school. In fact, they had already attended some of their father's lectures informally before, in 1834, they both matriculated at the university. James was 12 and William 10; it is true that the age of matriculation was lower then, but not normally younger than 14.

Despite their youth, both boys were regular prize winners over the next few years, often finishing first and second in the various lists [5]. But inevitably, it was William, the younger boy who was first, with James second. Larmor writes [2] that 'no cloud of jealousy ever marred the friendship of the pair', as 'the younger brother was even then adored by all the family as a genius', and one may believe him and yet still wonder whether there may have been a dampening effect on James' confidence, which may not have been restored until the early 1850s.

James took a BA in 1839, and an MA in mathematics and natural philosophy in 1840. Then, following a family holiday in Germany, including a walking tour of the Black Forest, obsessed with engineering as he was, he entered the Dublin office of John McNeil, a professional engineer. William studied in Glasgow a further year and could have taken his MA, but did not do so as he was to enter Cambridge University as an undergraduate in 1841, and was not sure whether a holder of an MA was eligible. Despite James's devotion to engineering, it may have occurred to him at the time that William was to have the more comfortable billet, a feeling that would be expressed on at least one occasion in the next few years.

In fact, even before William left for Cambridge, James had returned home after around 3 weeks, as a result of a knee injury he had suffered while walking in the Black Forest that had left him quite lame. As it turned out, he was able to return to the University of Glasgow to study engineering under the new professor, Lewis Gordon for the 1841–1842 session; Gordon was actually the first Professor of Engineering in the world.

Having improved considerably in health, James was able to spend some time in 1842 working under the guidance of John McClean in Walsall. McClean and James had been fellow students in his father's class, and McClean was already advancing in his engineering career that was to bring him to an important position in the early days of railway construction. At the time, he was in charge of a number of canals, and James took the opportunity of visiting recently constructed locks, and measuring and drawing their various components.

Then, early in 1843 James went to the Horseley Iron Works, Tipton, Staffordshire, where he stayed for a few months as a pupil in the drawing office. The first iron steamer to put to sea had been constructed in Horseley, rather strangely [6, p. 288] as it is very far from the sea! James, however, was highly critical of some of the methods the firm adopted; the costing of an extremely substantial contract was delegated to young apprentices, and quantities were estimated by eye rather than measured precisely. Perhaps not surprisingly the firm encountered severe financial difficulties.

James, though, moved in the autumn of 1843 to Millwall in London, where William Fairbairn had set up a very substantial shipbuilding works, a subsidiary of his Manchester engineering firm, which were the first millwrights and among the first engine makers in Britain. James was a premium apprentice, an arrangement for which his father paid a fee of £100 (perhaps the equivalent of £10,000 today). He learned about heavy engineering techniques, studied marine engines and iron ships and made drawings for engines. In fact, his studies of marine engines in Millwall led him to consideration of their fundamental principles, and he had some important and deep discussions with William, which would bear fruit within a few years with the development of thermodynamics.

While at Millwall, James modified the dynamometer of the French engineer, A.J. Morin, and he suggested that this device could be used to measure the work produced by Fairbairn's steam engine. The idea was that the excess work produced by the engine, rather than being wasted, could be sold to neighbours who needed motive power, and James's dynamometer could be used to ascertain exactly how much work they used so they might be charged accordingly [9, p. 40]. Again we see James's intense interest in the efficient use of the power derived from natural resources, and indeed his time in Millwall was useful for him in very many ways. However, he found the atmosphere at Millwall damp and unhealthy, and he suffered from a practically perpetual cold.

The Millwall works were not profitable and by the summer of 1844 they were being run down, prior to being sold, and in the autumn James himself was moved to the Manchester works where he would work more closely with Fairbairn, concentrating on the building of steam-engines. He seems to have become fairly despondent, writing to William that: 'I am not getting settled in any works. Just when I begin to take root I have to leave the place.' Writing again before William's final examinations at Cambridge, he said that: 'I wish my apprenticeship was as nearly done as yours – but when it is done, I fear I shall have no comfortable berth to step into as that which is probably waiting for you' [6, p. 292].

Indeed, James's immediate prospects were far from good. His health became steadily worse. There was some derangement of the action of the heart; his pulse was far too fast, and he felt weak and ill. At the end of the year he had to return to Glasgow, where his Calvinist doctor gave his the traumatic news that he had heart disease and might die at any moment.

One would have imagined that he would have collapsed under this news. If he did, such a collapse was temporary, because in fact rather remarkably in the next few years he was to develop a variety of intellectual pursuits, many of which were to be continuing themes through the rest of his life and will be discussed in more detail in the remaining sections of this paper. They included the recovery of pure water from sea water; the properties of whirling fluids; the horizontal waterwheel or vortex turbine; the problem of design in creation; the expansion of ice; the glacial period and glacial markings; the formation of lakes; the parallel roads of Lochaber; elasticity and the strength of materials and atmospheric circulation. The extreme breadth of his interests is remarkable, from engineering design to more general questions of engineering science, and on to study of problems of general scientific interest.

Part of his interest in most of these topics, and always in the background, was the fundamental question about the relationships between force, work, heat and temperature, and how these relationships were seen in the working of engines and the properties of different substances. An important document is what we shall call his 1848 Belfast notebooks [12], which summarise

James's ideas at that time. The centrepiece of this work may be said to be the important paper on the depression of the freezing point of water by pressure [1].

In 1846, William was appointed to the Chair of Natural Philosophy at Glasgow to the delight of his father and the whole family. There is no doubt that James shared in this delight as he was never jealous of his younger brother, but he would not have been human if he had not felt the contrast between the positions of the brothers, one established in a position of considerable prestige, the other older one an invalid, perhaps unlikely to work again and maybe not expected to live long.

In fact, James's prospects improved markedly in 1848 when a new doctor told him that his health was vastly improved, and there was no reason why he should not regain his strength and move to a normal life. It is not to be supposed that the rather outlandish treatments and draconian diets suggested by the previous doctor had been successful; probably, the preceding prolonged exposure to noisy and dirty machinery, noxious substances and drastically uncomfortable surroundings had left him in a severely unhealthy state, and a considerable amount of time was required for recovery. Though he was keen to return to engineering, he was careful never to place himself in the type of environment that had caused his breakdown in health.

While this improvement of health clearly great news for James, the late 1840s were still not a time of much happiness for him. First in 1847, his younger brother John, who had become a medical student, died of typhus, probably as a result of overworking during an uncontrollable outbreak of fever. Perhaps even worse, in 1849, James Thomson Senior, on whom all the family, but perhaps particularly James, had relied so much, died of cholera.

Even before this second death, James had been passing through a severe moral crisis. He was an extremely devout man, but he had come to the conclusion that he could not accept much that would be consider essential in the Presbyterian faith in which he had been brought up. Though believing in God, he did not believe that Jesus was the Son of God. In fact, he had become in fact a Unitarian, believing in only one person of the godhead, rather than a Trinitarian believing in three (Father, Son and Holy Ghost). Smith [9, p. 310] suggests that it was '[T]he imperfect world of harsh Calvinism' that had driven James to take this step. He had confided his belief to his father, but following the death of the latter, and after a brief visit to his sister Anna, who had married William Bottomley and was living back in Belfast, he determined to move to London where he would perform some collaborative work with Lewis Gordon, hoping to produce a patent for the water-wheel he had designed [6, p. 301].

James had felt that if he was to disclose his views on religion to his family he would cause great distress, and probably a break in relations might be inevitable. So: 'However painful the effort was, and gloomy the way before me, I tore myself away from my friends [2, p. xl].' He then wrote to his brothers and sisters and a very few close friends saying: 'I believe in one God, the Creator, the Preserver, and the Governor of all other things; and I believe that He has given us, while we are in this world, no other revelation of Himself but that contained in his wonderful works, and in the ideas which He has implanted in our minds, or rather which He has adapted our minds to receive. I think we should never neglect to pray to Him…' An interesting aspect of this 'confession' was that he felt no requirement to 'sugar the pill' by giving even lip-service to the idea of Jesus as a unique human being, worthy of the greatest admiration, even if not worship as a member of the godhead. Much as he must have longed for his family at least to appreciate his views, he was too honest to dissemble even in the smallest way.

In fact, he was delighted to receive letters from his family 'indicating such a spirit of true kindness and love on their part as has made me quite happy'. Yet he stayed in London until 1851, and it was only then that he felt able to agree to Anna's plea to come back to Belfast where he could work as a civil engineer. She had done him a great sisterly service, for this was the final turning-point in his life. Having recovered from his illness, with his religious dilemma solved – he was never to question the conclusion he had reached, and his position was accepted by his family – he was able to rebuild his life. For all his past difficulties he was still only 29, and he had before him nearly 40 years of success and happiness until the almost inevitable trials of old age. (Sadly, though, Anna herself was to die in 1857.)

It has been said [2, p. xl] that in Belfast 'conditions were more favourable for his inventions, coal being dearer and less used than in England, and water power more available'. In fact, a very efficient version of the type of turbine invented by Benoît Fourneyron in France in 1827 had been produced by the Macadam Brothers in Belfast with considerable success in the 1840s [13]. (An example of the MacAdam-Fourneyron turbine from the Catteshall paper mill in Godalming, Surrey has been preserved.)

In Belfast, James's engineering practice gradually grew; in particular, in 1853 he became a resident engineer to the Belfast Water Commissioners, a post he held until 1857. In this capacity, he became involved with the measurement of flow of water in rivers and streams, and proposed simple methods that gave excellent results demonstrating 'Dr Thomson's genius as an experimenter' [2, p. xlv]. His expertise was called on by authorities in several other towns and cities, and he also explored many other interesting and significant ideas to do with water flow, which will be described in later sections. In 1855, William wrote to him from Glasgow saying:

> People often ask about you here, and I say that I believe you are getting a good deal of business.... The water-wheels, too, I hear sometimes spoken of, and I hope they are going to get a name yet of practical value. Have they begun to be profitable? I hope they will sooner or later.

As well as a fruitful working environment, in Belfast James found himself among, not just family but friends. He joined a number of cultural societies – the Literary Society, of which his father had been President some 30 years before, the Belfast Social Enquiry Society, the Belfast Natural History Society, and on its foundation a few years later, the Belfast Naturalists' Field Club. Attendance at these meetings stimulated James's interest in the very wide range of academic topics that has already been noted. He was to give a number of papers, including one to the Belfast Social Enquiry Society on the provision of parks in growing cities; this paper was to lead to the provision of the very large Ormeau Park in Belfast. He also met congenial people, in particular Thomas Andrews, Professor of Chemistry and Vice-President of Queen's College Belfast, with whom James was later to carry out exceptionally important work, and Neilson Hancock, Professor of Political Economy, whose sister, Elizabeth, he was to marry in 1853. The marriage was extremely happy and they were to have three children.

As it happened, their honeymoon was interrupted by a request from Queen's (Fig. 3) that James should stand in for Professor John Godwin in the latter's absence from the Chair of Engineering,

Figure 3: Queen's College Belfast in the nineteenth century.

and 4 years later he was appointed to the Chair permanently when Godwin retired. (The formal name of the Chair was Civil Engineering and Surveying.) There can be no doubt that in this position, James had found his métier. He gave up the great majority of his other work, the exceptions being mostly public positions where he worked for the general good, and he dedicated himself to his students and his research. His predecessor Godwin had been very much a railway engineer; he had been actively involved in introducing the railway into Ulster as a resident engineer and manager to the Ulster Railway since 1837. In fact, his college work had been rather a side-line, as his salary from his railway employers was nearly three times the one that he drew from Queen's; his lecturing was restricted to 2 hours on 2 days a week, plus a substantial drawing-class held in his office at the railway company on Saturdays [14, pp. 119, 173]. James took on much heavier teaching duties, partly because the engineering curriculum at Queen's was much extended in 1859. As a teacher, he was concerned and sympathetic, though also a stickler for rectitude.

These qualities come out in comments made in later life by John Perry [2, pp. li–lii]. Perry had entered Queen's without the intention of taking a degree, and, when he decided he wished to do so, it was discovered that some of the requirements were lacking. First, while at Queen's he had not produced enough engineering drawings; Thomson went to the bother of chasing down and studying drawings that Perry had made while employed in a foundry in Belfast, and was able to pronounce them a satisfactory alternative. However, Perry had also registered too low an attendance at classes, and on this ground, Thomson could not be persuaded, despite the efforts of Andrews, to give him the required credit. It was only when a right of the President to exercise leniency in such matters was discovered, or perhaps invented, that Perry was allowed to take the examination. Nevertheless, Perry declared Thomson to be 'one of the kindest of men', though one with 'a very strong sense of duty'.

With a guaranteed salary, Thomson was able to devote his research talent to problems of interest and importance across a wide spectrum of academic and industrial topics, and this work will be sketched in the following section. A particular interest that started when he moved to a new house in 1866 and discovered defects in the drainage was to the whole question of the drainage, ventilation and heating of buildings. Over the succeeding years, many of his suggestions have become standard practice.

James Thomson's years at Queen's were successful and happy, and when in 1872 MacQuorn Rankine, who had become Professor of Civil Engineering at the University of Glasgow, died at the early age of 52, Thomson had little inclination to apply for the post. He felt he was carrying out successful and important research, particularly in his collaboration with Andrews. He himself enjoyed his work very much, and in addition he did not wish to uproot his wife and youngest daughter, neither of whom was physically strong. He may also have recalled that 40 years before when his father had made his own move from Belfast to Glasgow, financial difficulties had occurred; Glasgow was a more expensive city than Belfast, so to maintain the same standard of living, a higher salary would be required.

His brother, though, now Sir William Thomson as a result of his success with the Atlantic cable, was determined that James should join him in Glasgow. He overcame all James's objections; what probably clinched the matter for James was the mere fact that his brilliant younger brother did indeed feel that he was the best man for the post. Next, William organised the campaign for James, just as their father, James Senior had done for William himself 26 years before. In contrast to his father's demand in the previous campaign that William obtained as many testimonials as possible, for his brother's campaign William was content with five, which were all of the highest quality. It is interesting that, for a chair in Engineering, they were all from physical scientists – Maxwell, Andrews, Joule, Helmholtz and Tait (Figs 4–8). Three of these are quoted in the Biographical Sketch [2, pp. lxvi–lxix].

Hermann von Helmholtz wrote that: 'I regard Mr James Thomson as a man of very sound and unusually acute judgement in questions relating to physical, mechanical, and mathematical science, and a very extended amount of knowledge in these same branches. His theoretical prediction of the alteration of the freezing temperature of water by pressure was an original idea of first-rate importance. Also his hydrodynamical considerations, and their practical applications for the construction of water wheels for his jet pumps, for the theory of oceanic and atmospheric

Figure 4: James Clerk Maxwell.

Figure 5: Thomas Andrews [Queen's University Belfast].

Figure 6: James Joule.

Figure 7: Hermann von Helmholtz.

Figure 8: Peter Guthrie Tait.

currents, are of a very original turn of mind, and prove that he is able to follow out his speculations to the very fundament of the question. Besides I know from personal acquaintance his scrutinizing way to penetrate into the very heart of scientific questions, and I should think that such a man ought to be the very best teacher for young engineers'.

P.G. Tait listed much the same set of achievements, adding that: 'I was for some years his colleague in Belfast, and I can thus state of my own knowledge that he is an excellent teacher, and works in thorough harmony with his brother Professors. As an original discoverer in Physics, and especially in the Dynamic Theory of Heat, he holds a very high place indeed… I certainly do not believe that there can be found in Britain any one so well qualified as he is to fill the place of our lamented friend Rankine.'

When James Joule was asked to give a testimonial, he noted that he had already written one for another candidate, as he had no idea that James would be able to leave Belfast. He added that: 'In the event of his seeking the Chair of Glasgow College, it is impossible that there can be any serious competition, inasmuch as [James's] European fame as a discoverer combined with his eminent success as a teacher of Engineering science, are such as to place him altogether without a rival.'

James Thomson was duly appointed to the Chair, and commenced his duties in November 1873. In a letter to William, Thomas Andrews noted that: 'Your brother will indeed be a sad loss to us all here – and to no one more that to myself. I have long been greatly attached to him. He is indeed a rare character, so truthful and gentle and wise withal.'

The marked divisions in the professoriate at Glasgow between conservatives and reformers, which had dominated so much of the efforts of James's father in particular 40 years before, had disappeared; his battles had long been won, and the newly appointed James Thomson was to be as happy and successful in Glasgow as he had been in Belfast. His renown also increased; he became a Fellow of the Royal Society in 1877, and 7 years later he was elected the president of the Institution of Engineers and Shipbuilders in Scotland. Like his brother, he was a convinced opponent of Gladstone's adoption of Home Rule for Ireland as the official policy of the Liberal

It reads:

In commemoration of the scientific
eminence and educational services of
James Thomson
D.Sc., LL.D., F.R.S.
Professor of Engineering at Queen's
College Belfast
1857-1873
and afterwards in the University of
Glasgow
1873-1889
Born in Belfast 1822
Died in Glasgow 1892

Erected by friends of himself
and of his younger brother Lord Kelvin

Figure 9: Plaque erected in the entrance hall of Queen's College Belfast.

Party, and James and William both became ardent members of the breakaway Liberal Unionist Party; this was to lead eventually to William's peerage as Lord Kelvin.

However, in 1889, James was struck with a rapid deterioration of his eyesight. His retina became detached in the middle, so he could barely read and could write only on large sheets of cartridge paper which his wife or one of his daughters would then copy. Fortunately, he never became totally blind, but could only to a certain extent recognise the faces of friends. He resigned his chair, but was actually able to continue to work at his research, in particular at a topic he had long been interested in, that of atmospheric circulation. He wrote a new paper recapitulating much of his work over 30 years or so, and this was given the honour of being selected by the Royal Society as its Bakerian Lecture for 1892. William was then president of the Royal Society, and he read it on James's behalf at the meeting of the Society on 10th March. Sadly, though, even before this paper was published, on 8th May James Thomson died. Within a week, his wife and daughter Bessie also died, all of what was termed 'a severe cold', though it is now suspected of pneumonia. Figure 9 shows a plaque erected in his memory in Queen's College Belfast.

The last few years had not been pleasant, and of course there was the period of great concern earlier in his life, but in both cases James was able to use the time to great profit. In between, there had been a considerable period of happiness and achievement.

As a last note, it may be mentioned that James had sympathies in some areas that could be termed today broadly leftward. As well as his work for healthy and sanitary conditions in housing, and his calls for the setting up of public parks, which have already been mentioned, he wrote an article for the *Northern Whig*, arguing for the nationalisation of public works, in particular railways; this paper was included with his Collected Papers [15].

3 James Thomson's achievements in engineering and science

In this section, we sketch those contributions of James to engineering and science that lie outside the specific area of the creation of thermodynamics, the area which will then occupy the remainder of the paper. It will become clear, though, that much of this work outside the specific area was actually quite closely connected with his work on thermodynamics – for example, his studies of engines of different types, in which he was always concerned with the efficiency [or 'economy'] of operation, his collaboration with Thomas Andrews on the properties of liquids and gases, and his studies of glaciation, which were essentially applications of his important work on the lowering of the freezing point under pressure.

Perhaps, his most important work in engineering was the invention of his vortex turbine. James had always had a strong interest in the utilisation of natural sources of power in rivers and waterfalls, and was particularly impressed by the waterwheels of Holland. He also learned about the French turbine from Gordon's classes in Glasgow; this was a horizontal rotating waterwheel, in which the water enters at the inside of the wheel, flows along the buckets and escapes at the outer circumference. Thus, the centrifugal force fulfils the role played by gravity in a vertical waterwheel.

The idea of the horizontal rather than vertical waterwheel was to satisfy the criteria announced by Lazare Carnot (Fig. 10) in 1803 for maximising the economy of any waterwheel [9, p. 35; 16, p. x]. These are that the water must enter the wheel without 'shock' and must leave it without velocity. Today, we may easily explain the meaning of the criteria in terms of the requirement to make the greatest possible use of the kinetic energy of the incoming water. Kinetic energy will be lost if there is an inelastic initial collision (or 'shock') between the water and the blade, and of course any kinetic energy still possessed by the departing water is wasted. But at the beginning of the nineteenth century, there was at most a rudimentary understanding of these ideas; any genuine appreciation of the idea of energy and its conservation came with the ideas of thermodynamics and the central importance of this quantity throughout physics was particularly due to William Thomson, Lord Kelvin.

Lazare Carnot's criteria are in many ways analogous to, and in some ways maybe the basis of, the celebrated criterion reached two decades later by his son, Sadi Carnot (Fig. 11) [16], and discussed and developed as a centrepiece of thermodynamics by Émile Clapeyron (Fig. 12), the Thomsons and Clausius. Sadi Carnot's criterion is that, for maximisation of the efficiency of a heat engine, one must ensure that heat never flows between sub-systems between which there

Figure 10: Lazare Carnot.

Figure 11: Sadi Carnot.

Figure 12: Emile Clapeyron.

is any difference in temperature greater than an infinitesimal one. Sadi Carnot's work will be extremely important through the remainder of this paper.

In 1846, James made a small model of his own design of horizontal waterwheel, which resembled the French turbine; his first ideas had actually occurred to him as early as 1842. The essential new feature was that pivoted inlet guide vanes were used, which were controlled by an external linkage; this allowed the flow of water into the turbine to be controlled and thus the output power to be varied [3]. To avoid jarring, one of Lazare Carnot's criteria demanded that the water was injected at the same speed as the wheel but at high pressure. As the water moves inwards, the vanes curved from radial to nearly tangential, so the water was effectively always kept as close as possible to a state of equilibrium; the pressure decreased as the water moved inwards, and it left at the centre with an extremely low velocity, again to satisfy Lazere Carnot's criteria [6, pp. 412–413]. The flow of the water was clearly analogous to a natural vortex.

In 1847, while visiting a friend, Walter Crum, who ran a large engineering works, James had been able to have a much larger model constructed and carefully tested. Despite some defects in the manufacture of the model, it still produced around 70% of the total work available in the flow of water, which compared with around 75% obtained from the best overshot wheels. (An overshot wheel is a vertical wheel in which the water passes over the top of the wheel. Thus, the gravitational energy of the water, as well as its original kinetic energy is available for the provision of useful power. So, overshot wheels should be the most effective vertical wheels, though careful design is required; if Lazare Carnot's criteria are not obeyed at least approximately, much, even most, of the initial energy may be wasted in jarring or in the final kinetic energy of the water.)

At this time, James's health was still not good, but by 1850, with his health much improved, he was able to take out a patent, titled 'Hydraulic Machinery and Steam Engines', on what he now called a vortex water-wheel; it would later be called a vortex turbine. He had received much help from William, who, on his travels in France, had discussed James's wheel [17] with Jean-Victor Poncelet, a leading authority on waterwheels and turbines. James came to an agreement with Williamson and Brothers of Kendal, which later became the internationally well-known firm of Gilkes and Gordon, to make his turbine [3], and in 1856 Williamson produced the first turbine to his design. This was known as the Vortex Williamson Number 1, described in the Williamson catalogue as 'designed by Professor James Thomson'. It produced 5 horsepower under a fall of 31 ft with a 9 inch diameter supply [13].

At the South Kensington International Exhibition of 1862, the turbine was awarded a medal for 'originality of design, practical success, and good work', and in the article on Hydro-mechanics in the 1881 edition of the Encyclopaedia Britannica, SG Greenhill and WC Unwin wrote [2, p. xxxiii] that: 'James Thomson's inward flow or vortex turbine (Fig. 13) has been selected as the type of reaction turbines. It is one of the best even in normal conditions of working, and the mode of regulation introduced is decidedly superior to that in most reaction turbines; it might almost be said to be the only mode of regulation which satisfies the conditions of efficient working, and it has been adopted in a modified form in the Leffel turbine, which is now largely used in America.' It may perhaps be surmised that the vortex turbine was at least a partial exception to John Nichol's comment, mentioned in the first section, that for James's inventions there was *always* some practical obstacle preventing fame or fortune.

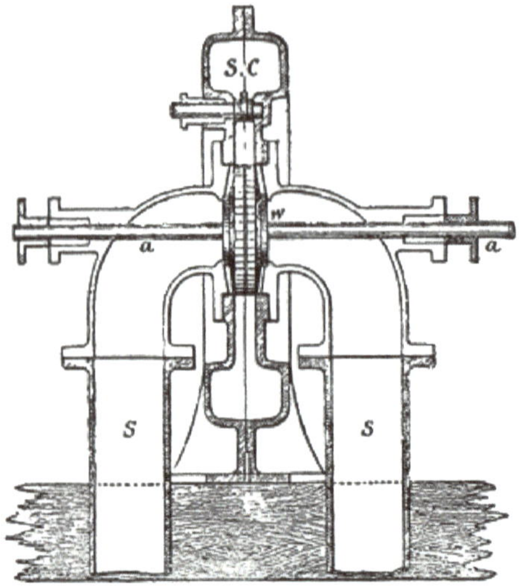

Figure 13: James Thomson's vortex turbine.

The turbine has suction pipes, which permit the turbine to be placed at any height less than 30 feet above the tail-water level. The water enters the turbine by cast-iron supply pipes at a, and is discharged through two suction pipes S. The water on entering the case distributes itself through a rectangular supply chamber SC, from which it finds its way equally to the four guide-blade passages G. In these passages it acquires a velocity about equal to that due to half the fall, and is directed into the wheel at an angle of about 10 or 12 degrees with the tangent to its circumference. The wheel receives the water in equal proportions from each guide-blade passage.' — Source: *The Encyclopaedia Britannica, New Warner Edition* (New York: The Werner Company, 1881)

Many turbines were made under this patent, and they were erected in England, Scotland and Ireland, as well as in many other countries. In fact, the basic design of James's vortex turbine has been exploited right up to the present for large, power generating, water turbines. However, they are usually called Francis Turbines, being named after JB Francis, who designed a similar turbine with fixed guide vanes [3].

James's work on the vortex turbine led to more fundamental studies of vortex motion, and he called his 1852 study of exact hydrodynamic observation and theory the 'vortex of free mobility'. In such a scheme, essentially each particle had the same total energy, however, far it might be from the central axis. This prevented the necessity of work being performed by centrifugal forces when particles moved, which would lower the efficiency of any device. These considerations came 6 years before Helmholtz's own major work on vortex motion [2, p. xlii].

This concept was used when James was involved in the design of extremely large centrifugal pumps. In particular, around 1858, he designed and constructed such a pump with exterior whirlpool for drainage of lands in Jamaica, and also in Demerara in British Guiana, lands which lay generally below the level of high tide. The provision of the whirlpool is to make use of the kinetic energy of the water leaving the pump. James stated [18] that: 'The fluid revolving assumes the condition of a vortex or whirlpool, which has been designated by the writer as the Vortex or Whirlpool of free mobility.'

Another important development was his so-called jet pump, or 'apparatus for drawing up water by the power of a jet'. This was originally designed to drain the pits of his turbines when they required cleaning, but its greater importance was in draining swamps and shallow lakes [19]. The jet pump has no moving parts and only needs a high head of water [3]. In Ref. [18], James reports that a jet pump had recently been erected by William Forster of Ballynure near Clones, and the pump had behaved very successfully in removing the water from wet lands adjacent to his house.

James's involvement with turbines and his position as an engineer to the Belfast Water Commissioners led him to an interest in the measurement of the flow of water in streams and rivers. His great achievement here was the proposal to replace rectangular notches by triangular or V-shaped ones. V-shaped notches are amenable to exact calculation, because the issuing jet is of the same shape; however, much of the notch that is occupied by the stream. James was in fact able to deduce that quantities of water flowing through V-shaped notches are proportional to the $5/2$-power of their linear dimensions [20]. His experiments were performed on open streams with the simplest of equipment, and it is interesting that when, nearly 50 years later, far more sophisticated and expensive experiments on the same problem were carried out at Glasgow, the experimenter reported that, for the gauge notch used by James, his results were remarkably near to the true value. He concluded that: 'It is a proof of Dr Thomson's genius as an experimenter that, with the crude apparatus at his disposal, he was able to arrive at such an accurate result [2, p. xlv].'

James was also interested in the accurate measurement of power, and he constructed a new form of friction brake dynamometer, which he showed to his classes as Glasgow. It consisted of fast and loose pulleys on a shaft, with a rope which was attached to the loose pulley but lapped partially round the fast pulley with weights at each end. The loose pulley adjusted itself to give the required arc of contact, winding or unwinding the cord on the running pulley. Much later William Ayrton was to say that: 'Another plan was to vary the arc of contact. It was curious that every person who had arrived at that idea seemed to regard it as his own. It had been invented over and over again… [I believe] it was due initially to Professor James Thomson [2, pp. xlv–xlvii].'

James's interest in fluid flow was further exhibited in a major study of the flow of water through orifices [21], and he also addressed fundamental questions about the winding of rivers in alluvial plains [22]. The process of river meanders and the formation of oxbow lakes are usually imagined to occur, he says, through the water rushing at the outer bank and washing it away, at the same time causing deposition on the inner bank. James pointed out, however, that this simple explanation could not be true. The hydraulic principle, which he suggested is not generally known, says that a stream flowing into a curve must flow with diminished velocity along the outer bank, and an increased one along the inner bank, this suggesting that it is the inner bank that would wear away the most.

James's solution was that the water turning the bend is subject to a centrifugal force, but a thin layer of the water at the bottom is retarded by friction with the bottom, and so it is subject to a smaller centrifugal force than the great body of the water flowing over it. Consequently, the bottom layer flows across the channel towards the inner bank and rises up between the inner bank and the rapidly flowing water. It protects the inner bank from the rubbing of the rapid water and brings with it sand and detritus from the bottom of the stream, which it deposits along the inner bank. James also said that the same phenomena take place in connection with flow in pipes, with considerable effect on the behaviour of frictional resistance. He was later to set up a simple model with a small artificial river in which were anchored at various depths threads of suitable length.

A last topic in this general area of fluid motion is that of atmospheric circulation. James's first work on this subject was in 1846, when his father asked him to write a paragraph on trade winds for a new edition of his *Geography*. He published his developing views through his life, and, as has been said, his Bakerian lecture [23] of 1892 was based on the ideas he had in his last few years. In the lecture, he prefaced these ideas with a comprehensive and extremely useful historical survey of the topic, discussing the contributions of, among others, Martin Lister, Edmund Halley, George Hadley and John Dalton.

Another range of research topics emerged from James's famous freezing point paper; these were the melting of ice and its importance in the movement of glaciers. These studies brought him into discussion with many of the leading British scientists of the day, glaciology being a special interest of several British scientists, including James Forbes and John Tyndall, while Michael Faraday had been extremely interested in the transition between water and ice. James's first paper, published in 1857 [24], discussed regelation (a term introduced by Tyndall) – the phenomenon that occurs when a fine wire is looped over a block of ice, and a heavy weight is attached to it. The ice will melt under the wire, and the wire may thus move right through the block, though the ice freezes again after the wire has passed, and the block remains solid at the end of the experiment. James called this phenomenon the *plasticity* of ice and showed that it is a fairly direct result of his own discovery of the lowering of the freezing point under pressure.

These ideas are clearly relevant to the movement of glaciers. There were differing opinions on the mechanism involved in this movement; some scientists thought that they slid like solids, others that they flowed like viscous liquids, others still that they crawled by alternate expansion and contraction, or by alternate freezing and melting. Forbes, who argued that ice was viscous, became implacably opposed to Tyndall, who believed in regelation [25]. (This controversy was the second for Forbes in the area of glaciology; earlier he had been an equally virulent opponent of Louis Agassiz.) James Thomson's intervention did little to calm the situation, and, in fact, today it would be accepted that many of the proposed explanations play a role in the overall behaviour. In 1865, no less an authority than Hermann Helmholtz had adopted James's theory, or, as it might be said, his branch of Tyndall's theory.

A related problem investigated by James was the phenomenon of ground or anchor ice, and its effects on navigation in the St Lawrence River. Ground ice is ice that forms from the bottom of a river, in contrast to the usual surface ice we are familiar with on the surface of a lake. James's explanation [26] was that the crystals of ice are frozen from the water at any depth where cold may be introduced by contact, or by loss of energy by radiation; they may also be supplied by snow. These ice crystals may be whirled about in currents or eddies until they come into contact with fixed objects to which they may adhere – rocks, stones, pieces of ice jammed in crevices or ground ice already growing. Such pieces of ice, Thomson says, will adhere together with a continually increasing firmness, even when the surrounding water is above the temperature of freezing.

Rather earlier James had become engrossed by another problem involving glaciers – the so-called Parallel Roads of Lochaber in Scotland. These were horizontal indentations on the upper levels of valleys, and it had been realised that they were the remains of ancient beaches. Charles Darwin had deduced that, at the time of their production, Scotland must have been totally sunk below the ocean, but this, he was later to admit, was his greatest mistake [4]. The concept of an *ice age* had been introduced by Agassiz in the 1840s, and, as we have already seen, many British scientists had become extremely interested in the topic of glaciation.

James had never visited the district, but when he studied the subject in detail from books and maps, he came to the conclusion, universally accepted today, that the 'roads' or beaches had been formed by glacial lakes, created when glaciers dammed the lower reaches of the valleys. When the ice melted, the water drained away. James was able to show that certain locations of glacial barriers would explain the positions of all the terraces. His paper [27] was read to the Royal Society of Edinburgh in March 1848; it was his first paper he had written, and it received an enthusiastic reception, the Secretary inviting him to send further papers. This must have been extremely pleasing for the young man, still regarded as an invalid, and his biographers [2, p. xxxiv] make the interesting comment that: 'Thus encouraged, he turned his mind more and more to pure science rather than to practical engineering.'

We now turn to the topic of the liquefaction of gases. Following his work on the lowering of the freezing point, James was to retain his interest in phase transitions for the remainder of his life, and while he occupied his chair in Belfast, he was fortunate to collaborate between 1861 and about 1872 with Thomas Andrews, an exceedingly gifted practical chemist, whose talents matched those of James in an extremely profitable way [28]. James was fully able to interpret the results of Andrews' brilliant experiments. A number of exceedingly important conclusions were to emerge from their work. First [29] was what is usually simply called *Andrews' Experiment*, which demonstrated the essential difference between a gas and a vapour: below its so-called *critical temperature*, a substance is a vapour and may be liquefied by pressure alone, while above this temperature it is a gas and cannot be liquefied using only pressure. While this is obviously a crucial practical point for the technical task of liquefying gases, the existence of the *critical point* is also exceptionally important for the general understanding of the nature of substances and their different phases.

A second very important result of this collaboration was James Thomson's pointing out that there was a unique value of temperature and pressure where a substance can exist together in gaseous, liquid and solid states. He [30] gave this point in the now famous name of the *triple point*. The third result that James also introduced and discussed informatively [31, 32] was what he called 'a difficulty of making a beginning of [a] change of state', or what would now be called superheating and supercooling.

While the above is sufficient to show the great success of the collaboration between James and Andrews, Rowlinson [28] points out that, but for a variety of factors, even more might have emerged. Andrews was ill in 1872 and again in the late 1870s, and he resigned his chair in 1879, and James Thomson, of course, moved to Glasgow in 1873. After Andrews' death in

1885, P.G. Tait published some of their remaining results posthumously, and it has gradually become clear that Andrews and Thomson had anticipated some of the important results of the Dutch physicists, Johannes van der Waals and Heike Kamerlingh Onnes, both of whom were to be awarded the Nobel Prize for Physics in the new century. Andrews and Thomson did indeed draw diagrams showing what would now be called the *van der Waals loops*; Rowlinson comments that: 'The two ideas [of Andrews and Thomson] were combined in a masterly away in the equation of state that van der Waals put forward in his thesis at Leiden in 1873, and for which he borrowed the title of [Andrews'] Bakerian Lecture.' And many of the results obtained by Andrews in the late 1870s were reproduced and explained by Kamerlingh Onnes in the 1890s.

Another interesting and important piece of work concerned the prismatic structure of basalt as seen at the Giant's Causeway (Fig. 14) in the North of Ireland and elsewhere. James concluded that the older theories could not be correct, and he developed his own theory [33] over a period of years. His account explained the columnar structure as the result of shrinking and cracking during solidification, and said that the cross joints are circular fractures commencing in the centre of the column and moving out, or 'flashing', towards the circumference. He provided a detailed discussion of many more special features of the Causeway, and most of his suggestions are now taken as standard.

Yet another interesting piece of work concerned integrating machines, which had in principle many uses, but their most important one was predicting the tides. Around 1874, William was interested in the possibilities of designing and constructing such machines, and James told him of a machine he himself had designed many years before, probably, in fact, around 1845. It was called a disk–ball–cylinder integrator, and it enabled any fraction of the motion of a revolving disk to be communicated to a cylinder pivoted above it, by the intermediation of a heavy metal ball that rested on the disk and pressed against the cylinder.

Figure 14: The Giant's Causeway, County Antrim in Victorian times.

William immediately realised that it could be used not only to perform simple integrations but also to solve differential equations and to perform harmonic analysis, by which any complicated periodic motion can be broken down into a sum of sinusoidal terms, the last application being central for study of tidal and meteorological phenomena. James published a paper [34] on his invention, and William followed this paper with several on conceivable applications. The so-called Harmonic Analyser for calculating the elements of tides was actually constructed [5, pp. 692–694, 730–731]; the first working model had five disk–ball–cylinders, and later instruments could provide a curve giving the rise and fall of the tide at a particular location for a whole year, the computation taking a mere 24 minutes.

An early topic of interest to James was the distillation of fresh water from seawater on steamships. While this could be achieved very effectively using coal, it must be remembered that space is at a premium on such ships, as the amount of cargo must naturally be maximised. Thus, James suggested two economical methods of obtaining fresh water but was particularly pleased with the one that used the labour of sailors rather than coal, thus saving space for extra money-earning cargo.

Both methods involved using the heat given out in the condensation of the steam to generate more steam. In the first method, this was achieved in a fairly straightforward way using coal. In the second method, which James favoured, the labour of the sailors was used to pump steam from the seawater boiler to a vessel within it, thus increasing the pressure and temperature of the steam. Heat was conducted from this vessel to the boiler, so that more steam would be generated in the latter, while condensation of steam in the vessel created fresh water [9, p. 43]. James had shrewdly noticed that, while on sailing ships the sailors were often busy in adjusting the sails in order to obtain maximum advantage from the wind, on steamships they had little to do except when entering or leaving port or at times of bad weather, so they had plenty of time to provide the pumping and save coal.

We will conclude this section by mentioning briefly a few other tasks and achievements of James at different periods of his life. When he was 16 or 17, he invented a mechanics for feathering the floats of the paddles of paddle steamers [1, p. xcii]; engineers on the Clyde were extremely interested in the idea, but unfortunately it turned out that the same idea had been invented and patented a few months earlier. A few years later, he proposed [1, p. xx] the idea of such a steamer with paddles actuating jointed legs reaching to the bottom of the river, thus being able to propel the steamer upstream against the current. A completely different topic was that of the refraction of light passing through the atmosphere, to which James also made important contributions [1, pp. lxv–lxvi]. Lastly, we may mention that James invented many new scientific terms [1, pp. ciii–civ]; among those which have stood the course of time are 'radian', 'interface', 'ergometer' and 'torque'.

4 James and William: early dilemmas and discussions

Many of the fundamental difficulties that James and William were to discuss over several years, and which were eventually to lead to their work on thermodynamics, emerged much earlier and in very different guises, some involving the ideas of heat and work themselves, some based on more general scientific ideas and some indeed more to do with religion than science.

We will start with the question of the long-term future of the Universe. While it may be said that Newton's mathematics demonstrated that the solar system acted like a clockwork model, Newton himself insisted that God was required to 'operate' as well as to create the Universe [35]. Yet, by the first decade of the nineteenth century, with the famous work of Laplace [36], it appeared to have been proved that the solar system would remain stable eternally. It seemed natural to assume, and it generally was assumed, that the Universe itself would last for ever, that changes would be cyclic in nature, that there was no direction to the development of the Universe, no end-point to which the Universe was heading, no 'arrow of time' [6, pp. 89–92].

Yet, this comforting belief was challenged towards the beginning of the eighteenth century by observations on Encke's comet. This comet was actually discovered by Jean-Louis Pons in 1805, and when he suspected that an apparently new comet of 1818 was a reappearance of his 1805 discovery, it was Johann Encke who was able to analyse the orbit and confirm Pons's belief. Thus, Encke's comet is one of the few to be named after someone other than its discoverer! At that time, all known comets had orbital periods of at least 70 years, for example, that of Halley's comet, the best-known, being 76 years, and their aphelia lay beyond Uranus. In contrast, the period of Encke's comet was as low as 3.3 years, and its aphelion lay within the orbit of Jupiter. Thus, great interest was shown in the comet, particularly when in 1832 Encke was able to demonstrate the surprising result that its period decreased by 212 hours in each revolution.

This apparently showed that there must be a frictional medium in interplanetary space. In turn, this implied that all motion would be retarded and eventually cease, and that the solar system would in time collapse into the sun, and thus it would follow that there was a direction to the behaviour of the Universe, an arrow of time heading towards an end-point of the Universe. Along with all other scientists and mathematicians, James and William worried about these matters. Around 1841, for example, James argued that the tides on earth caused by the motion of the moon around it must lead to retardation either in the motion of the moon itself, or in the rotation of the earth [6, p. 285].

The dilemma had a parallel in a theological debate of the period [6, pp. 92–99], again concerned with a contrast between conservation and decay. On the one hand, it would be generally believed that God had created the Universe, and that man had the power neither to create nor to destroy. This was a clear call for conservation, and in scientific terms it was the *credo* of James Joule and was to become that of William Thomson and his brother. Joule was to write [6, p. 306] of: 'Believing that the power to destroy belongs to the Creator alone, I entirely coincide with Roget and Faraday in the opinion that any theory which, when carried out, demands the annihilation of force [to be interpreted as what would now be called 'energy'] is necessarily erroneous,' and William explicitly agreed.

Yet equally, common experience and the Bible spoke clearly of decay – human beings died, buildings crumbled and fell down, dynasties disappeared. In the draft of one of his first accounts of thermodynamics, William [6, pp. 330–331] was to write: 'Everything in the material world is progressive. The material world could not come back to any previous state without a violation of the laws which have been manifested to man, that is, without a creative act or an act possessing similar power. I believe the tendency in the material world is for motion to become diffused, and that as a whole the reverse of concentration is gradually going on.' He quoted the Biblical lines: 'The earth shall wax old like a garment'; analogously the hymn-writer says: 'Change and decay in all around I see.'

This dichotomy seems to have a considerable effect on William's eventual formulation of thermodynamics. Of course in saying that theological ideas influenced scientific beliefs, one is not suggesting that raw undigested religious ideas were or should have been inserted into scientific theories in an *ad hoc* way. However, nearly all the British men involved in the creation of thermodynamics were strongly religious, and they very naturally developed a worldview that used the fruits of observation, contemplation and experimentation within a religious framework. It turned out that this worldview was extremely helpful in creating the scientific paradigm based on energy that is at the centre of our ideas of physics today.

The third area [6, pp. 285–288] where the brothers had intense discussions, which were again to be extremely fruitful, was in the contrast between different situations where what the brothers were happy to call 'power' was provided to a system. In some cases, a useful 'effect' resulted; in others, it seemed that the power was simply wasted. In a letter of 1862 to William, James referred to a particular discussion they had had around 1842 when James was working at Walsall. Watching the barges go through the locks on one of the local canals, they observed that the purpose of the efforts of the lock-keeper or traveller in raising water was, of course, to provide power to raise the barge, but for much of the time the raised water merely cascaded over the lock gates and it seemed that no useful 'work' was performed. Similarly, water falling from a height could turn a water-wheel and thus provide useful work, or alternatively just fall under gravity with no useful result.

In part, the ideas of the brothers followed those of William Whewell (1794–1866), the important Cambridge mathematician, which had recently been discussed in his book *The Mechanics of Engineering* [37]. Whewell had effectively put together a set of ideas corresponding to much of the strictly mechanical side of our present understanding of conservation of energy. 'Labouring force' or an input of 'work' would be provided to a system. This amount of work could be equated to half the *vis viva* which it could produce; *vis viva* was a time-honoured term, literally 'living force', which was equal to mv^2, and so this was a broad appreciation of the idea of what would today be called kinetic energy. The result of the process could be useful work – a weight raised or a spring extended. Some or all of the work, however, might be 'wasted' in frictional processes.

However, when discussing this 'waste', Whewell was thinking principally of solids, where friction could be explained as permanent wearing or distortion of the solids in contact, for example, overcoming cohesion, or permanently changing the elastic properties. For the Thomson brothers, thinking of water rather than of a solid, it seemed inconceivable that friction could be explained in this way, but they had no alternative set of ideas with which to replace those of Whewell.

The problem was analogous to that concerning the steam-engine that was to give the brothers much cause for thought. Heat leaving the boiler of a steam-engine and eventually reaching the condenser was responsible for the production of external work. Yet equally, one could arrange affairs so that the heat could travel directly from the boiler to the condenser by conduction and no work would be produced. One was bound to ask the question: In the second case, what happened to the work that would have been produced in the first case?

The various dilemmas discussed in this section were only to be resolved in the creation of the subject of thermodynamics, its first and second laws, which included, of course, a new understanding of the nature of heat. It should be pointed out that, in the discussion of the steam engine in the previous paragraphs, it is not to be supposed that the brothers thought in terms of transformation of heat to work. Rather they were working within the 'caloric' theory in which heat was conserved, but its fall from high to low temperature was, where appropriate, associated with the provision of external work. We discuss this topic further in the following section.

5 Ideas of heat in the first half of the nineteenth century; the caloric theory

It is easy to say that, with the coming of thermodynamics, the view of heat held by William Thomson, and indeed the majority of other scientists changed from a theory of 'caloric' to a 'dynamical' theory. However, it is essential to clarify what is implied (and indeed what is *not* implied) by these terms.

Today, the caloric theory is usually taken to be explicitly a theory, widely held between about 1780 and about 1850, in which 'heat' or 'caloric' consisted of particles of a subtle, weightless and highly elastic fluid. The presence of these particles in the pores of the atoms of a solid could be used to explain thermal expansion, while their emission from surfaces explained the well-known Newton's law of cooling. One of the features of caloric theories that made them popular is that there were several other highly developed theories in which physical phenomena were explained in terms of similar subtle elastic fluids; the phenomena so explained included electricity, magnetism and light. Thus, many of the main branches of science were explained in a unified way [38, pp. 14–15].

It is today taken for granted that the caloric theory stood in opposition to what we shall here call a 'kinetic' or 'vibrational' theory in which heat was taken to be the motion of the particles of the particular medium. It is widely held that this type of theory is very much the precursor of, and in fact practically identical to, today's dynamical theory.

It is further usually believed that the caloric theory, like, in fact, the other 'fluid' theories, was naive and, even for its day, old-fashioned and foolish. Textbooks tell us that by the beginning of the nineteenth century the caloric theory had been refuted by the 1898 experiments of Count Rumford, in which repeated boring of a cannon produced a seemingly inexhaustible amount of heat, and those of Humphry Davy around 1799, in which he is supposed to have melted two pieces of ice by rubbing them together. It is implied that the scientific authorities of the time were almost wilfully blind in ignoring these 'refutations' of the caloric theory.

However, Mendoza [16, p. xv] suggests that the current accounts of the caloric theory and its clash with the kinetic theory have been derived fairly directly from those given by William Thomson in his later years. Mendoza suggests that, perhaps wishing to distance himself as much as possible from his earlier beliefs, Thomson tended to emphasis the contrasts between the caloric theory and more modern ones, and to pour scorn on the caloric theory. Thus, he argues, our current perspective is distorted. In this section, we shall attempt to provide a more balanced account.

Incidentally, Smith [9, pp. 170, 186] suggests a rather different motive for Thomson, and in particular for that of his friend P.G. Tait in his *Sketch of Thermodynamics* [39]. By raising the profile of the Rumford and Davy experiments, and thus implying that the dynamical theory of heat had been established as early as the late 1790s, they took away any possibility that the work of Julius Mayer in the 1840s should be awarded any priority. In 1862, Mayer's work had been championed by John Tyndall, the scientific opponent of Thomson and Tait. Thomson and Tait themselves were keen that the priority for the proof that heat and work were interchangeable should be shared between Rumford and Davy for the early suggestions, and James Joule, their own colleague, for his own work in the late 1840s, which was much more authoritative than that of Mayer.

We shall now say a little more about the caloric theory. It is clear that it may handle readily the situation where two bodies at different temperatures come into contact; caloric is believed to repel itself, so some caloric moves from the hotter body where it is more dense to the cooler one where its is less dense. As it does so, the hotter body cools and the cooler body heats up.

However, there are situations which are much more difficult to explain on the simplest ideas of caloric. These would include adiabatic compression or expansion of a gas, in which there is a change of temperature although no heat or caloric may leave or enter the system [38, pp. 39–60]. They also include isothermal expansion or compression, in which heat *does* leave or enter the system but there is no resultant change of temperature.

This apparent anomaly is answered by the statement that the total quantity of caloric, or absolute heat, of a body is composed of two components, the 'free' or 'sensible' or 'perceptible caloric' which affects a thermometer, and the 'latent' or 'combined caloric', which does not [38, pp. 104–156]. The word 'combined' indicates that the caloric was supposed to be combined, perhaps even chemically, with the matter of the gas, but could be 'squeezed out' by compression. The difficult situations of the previous paragraph may then be handled trivially by the transfer of caloric between the two categories. In an adiabatic compression, for example, an amount of caloric is transferred from the latent component to the perceptible component.

Such a technical manoeuvre may be said to typify the behaviour of 'scientific research programmes' in the theory of Imre Lakatos [40]. A band of 'auxiliary' or 'protective hypotheses' may be concocted to avoid any clash between the central theory and experimental results. In the present case, it may be suspected that such manoeuvres will be available for practically any suggested 'anomaly' of the theory. Appropriate movement of caloric from the latent to the perceptible category or vice versa would seem to be able to repel practically any criticism.

Indeed, Fox [38, p. 121] says that:

> With caloric being called upon to explain such a wide diversity of phenomena as radiation, temperature, specific heat, expansion, change of state, and the repulsive forces between gas particles, some degree of complexity was to be expected, but the multiplicity of forms that it was thought caloric could adopt was truly formidable around 1800. The easiest of solutions was, after all, simply to postulate as many sorts of caloric, each of them with its own characteristic properties, as there were phenomena to explain, and although no one went quite as far as this, the current terminology reflects how attractive this type of solution was. Thus we find such terms as combined, latent, radiant, sensible, and free heat (or caloric) all in common use, to say nothing of the quantities absolute

> and specific heat. However it is important to note that in all versions of caloric theory only one fluid was concerned and its properties were merely modified by its state of combination with ordinary matter.

A possible exception to the idea that the theory of caloric could not be refuted might appear to be Rumford's experiment which seemed to suggest that an *infinite* amount of heat must be present in latent form. However, Cardwell [41, p. 102] sums up the response of the supporters of the caloric theory:

> Rumford's challenge to the accepted theory of heat stirred up a hornet's nest of counter-argument and experimental refutations. The consensus would seem to be that while the generation of frictional heat was rather difficult to explain on the caloric theory it was not necessarily impossible to do so, and in any case there were very grave objections to any alternative theory. If heat can be created *de novo*, what happens to the fundamental axiom of the conservation of heat on which the whole science rested? No scientific theory is ever flawless, and the caloric doctrine was no exception. Rumford pointed to a weakness in the doctrine, but that was by no means the same as refuting it.

We will move on to consider how such arguments did finally lose ground in the middle of the century.

First, though, we point out that the direct physical understanding of caloric was not universal. Truesdell [42] follows Fox [38] in saying that: 'The term 'Caloric Theory' when current meant very different things to different schools of thought. Some of these are now strange even to historians of physics; still familiar are some particular models such as subtle fluids or atoms of caloric.' For example, Cardwell [41, p. 143] mentions the belief by some workers in 'caloric of space'. So, certainly we should attempt to follow broad features of caloric theories rather than particular facets, or particular physical models.

It may be said that the most fundamental aspect of the caloric theory was actually that heat was a conserved quantity. One may read the important works of those highly convinced of the caloric theory, such as Sadi Carnot and Clapeyron [16] and the early James Thomson [12], and this is the point that is constantly mentioned, perhaps more often implicitly than explicitly because it was broadly taken for granted. Clifford Truesdell, in his masterly but often savage account of the creation of thermodynamics [42, p. 34], for example, writes that 'For Laplace, heat is never created nor destroyed. We shall refer to this assumption by the traditional term *Caloric Theory of heat*.'

If heat is always conserved, it would seem difficult to interpret those processes which today we describe by saying that heat is converted to work. Before the Carnot cycle, which we shall explain shortly, became well known, the most important example was the steam engine. As Cardwell stresses [41, Chapter 3], around the nineteenth century water power and steam power were the two important and rival sources of power for industry, the former being gradually displaced by the latter. Thus, it was unsurprising that the conceptual analysis taken for granted for water power was assumed to apply also to its rival. In water power, whether it be a waterwheel or a column-of-water engine [41, pp. 85–88, 183–185] in which an engine is driven by the weight or pressure of a column of water, it is water that is the source of the work produced by the system, but, of course, it would not be suggested that water is 'converted' to work! Rather, in the course of its passage through the system, the 'state' of the water is changed and it is that change that is responsible for the production of work.

Similarly, it was taken for granted that it was the passage of heat from high temperature to low that caused the production of work in the stream engine, but it was certainly not supposed that heat was, in any sense, converted to work. One may speak of a 'fall' of heat from 'high intensity' to 'low intensity', or from a state of 'concentration' to one of 'diffusion' [6, p. 243]. Thus, a change in the nature of the heat could be responsible for the production of work, but it was not to be assumed that there was a simple relationship between the amount of heat, its fall in temperature, and the amount or work produced. We shall see how James Thomson in particular produced a detailed analysis along these lines in his Belfast notebooks.

While conservation of heat is certainly the crucial idea in the theory of caloric, it may be expressed in an even more potent way. This is that we may talk of the 'amount of heat' in a system, or say that, in today's terms, heat is a 'state function'. This implies that, for a gas, for example, we may take it round cycles which may be as complicated as we wish, containing, for example, adiabatic or isothermal expansions or compressions, processes at constant volume or constant pressure, and so on. However, we must assume that if the system is returned to the same state at the end of the cycle as it had at the beginning, the total amount of heat lost is the same as the total amount gained, so the amount of heat 'in the system' at the end is the same as that at the beginning.

Carnot [16, p. 19], for example writes that:

> We tacitly assume in our demonstration, that when a body has experienced any changes, and when after a certain number of transformations it returns to precisely its original state, that is, to the state considered in respect to density, to temperature, to mode of aggregation – let us suppose, I say, that this body is found to contain the same quantity of heat that it contained at first, or else that the quantities of heat absorbed or set free in these different transformations are exactly compensated. This fact has never been called into question. It was first admitted without reflection, and verified afterwards in many cases by experiments with the calorimeter. To deny it would be to overthrow the whole theory of heat of which it serves as a basis.

William Thomson was to quote this remark in his 1849 paper in which he explained his dilemmas at that time, dilemmas which would be resolved over the next two years.

Similarly, Truesdell [42, p. 85] writes that: Carnot introduces a *heat function* $H_C(V,\theta)$, the quantity of a heat in a fluid body, to within an additive constant. (There must be an additive constant because, like several other conserved quantities in physics, it is only differences in the function that may be discussed or measured, not absolute values.) Truesdell goes on to say that in his book (or 'tragicomedy' as he calls it), the term *Caloric Theory* will refer to the existence of a heat function, not to broader ideas. It must be said, incidentally, that Truesdell appears to have put his own gloss on what Carnot actually said.

Smith and Wise [6, pp. 315–316] stress that William Thomson could not have thought of heat as a substance; even when he still supported the theory of caloric, as late as the spring of 1850, he had no problem in accepting the conversion of work to heat. Rather they say that he treated heat as a 'state function', implying that no heat could be converted into work in a cyclic process.

Overall then, we shall regard the idea of heat as a state function, or the right to talk of a heat function or *total heat* of a system as the fundamental postulate of the caloric theory. We shall see that, in his crucial paper announcing the new thermodynamics, Clausius [16] spent much effort in showing that it is this idea that much be sacrificed in the new theory.

Earlier we looked at what we called kinetic theories and clearly these theories were in strong contrast with those caloric theories in which heat consists of particles or a fluid. However, they may now be seen to be good examples of our broader classification of caloric theories, based on conservation of heat and heat being a state function. For it is trivial to construct theories in which the *vis viva* or, in today's terms, the total kinetic energy is conserved, and may certainly be regarded as a state function.

Thus, it is clear that while the physical picture presented by those kinetic theories based on experiments such as those of Rumford and Davy are superficially analogous to today's dynamical theory of heat, they may miss the essential point that, in the modern theories, heat is not a state function. In the following sections, we shall follow how ideas of caloric were central to the extremely important work of Sadi Carnot and Clapeyron, and the early work of the Thomson brothers, but did not apply in the new theory of thermodynamics produced in the middle of the nineteenth century.

First though we make one last point. We said that in the caloric theories, the quantity of latent heat was available to explain discrepancies in the caloric bookkeeping. Clearly, this was a somewhat different concept from that of latent heat used in changes of phase today. The latter is defined in our dynamical theory in terms of energy transmitted rather than energy stored. Nevertheless, it may be said that terms such as 'latent heat', 'specific heat' and 'thermal capacity', though hallowed by time, may be somewhat misleading and more appropriate for a caloric type of theory.

6 The theory of Sadi Carnot and its reception by the Thomsons

Sadi Carnot's theory, published in his *Reflections on the Motive Power of Fire* [16], which was centred round his so-called Carnot cycle, was a triumph of the use of fairly abstract and general argument to make crucially important statements, which on the one hand were directly applicable to such practical objects as steam engines, but which on the other could be extended to encompass the widest range of phenomena involving heat and work. The theory was clearly pre-thermodynamic in that it was based on the theory of caloric, and was, in any case, produced in 1824, just over a quarter of a century before the emergence of thermodynamics. Yet, as well as having a direct importance for those designing air-engines and steam engines, such as the young James Thomson, it played a crucial part in the development of thermodynamics itself. With some adaptation, in particular to meet the demise of the caloric theory, it remains a crucial part of any account of the subject today, and may be said to have stimulated, fairly specifically, the production of the all-important second law. (William Thomson, for example, was to ascribe the second law to Carnot and Clausius [6, p, 329].)

It may be mentioned that Carnot, who was born in 1796 the son of Lazare Carnot, studied at the famous École Polytechnique, which was a training school for army engineers [16]. For much of his short life he served as an army officer, but, with the bewilderingly rapid changes in French politics in the period, Sadi's social status oscillated dramatically according to that of his famous father, whose fate was intimately tied in with the rising and plummeting fortune of Napoleon himself. Frustrated by this, he retired from the army on half-pay in 1820, and after much further study wrote his famous treatise. Subsequently, he had a brief spell back in the army, and produced further writing, some aspects of which will be referred to briefly later in this section, but he was to die from cholera in 1832. As Mendoza [16, p. xiv] points out, by a strange coincidence another young French genius, the mathematician, Evariste Galois (1811–1832), had died just a few days before.

Joseph Larmor has described Carnot's work as 'perhaps the most original in physical science'. For a reader of today, it is, at first sight, difficult to appreciate the full significance of Carnot's achievement, but this is only because so many of his insights have become the central components of thermodynamics today, and thus are easy to take for granted.

His first insight was the necessity to consider a *cycle*, in which, for a gas, successive changes of temperature, pressure and volume would take place, but the final state of the system would be identical to its initial state. Clearly, this approach was able to model actual engines working through many cycles. Equally importantly from a theoretical point of view, it focussed attention on the crucial question of what quantities should be regarded as state functions and so would have a value that must clearly be the same at the end of each cycle. We saw in the previous section that these ideas became the best definition of a caloric theory, and so were to become the best description of its eventual failure. Carnot, as we have said, was a believer in the caloric theory, so, of course, for him it is the heat of the system that must have the same value at the beginning and end of each cycle. His terminology [16, p. 8] is that: '[I]n steam-engines the motive-power [or work] is due to a re-establishment of equilibrium in the caloric.'

His cycle, described for a gas, consisted of two isothermal stages, (12) and (34) in Fig. 15, in which the pressure and volume change; the stage (12) corresponds to a higher temperature than (34). In a steam engine, the temperatures are those of the boiler and the condenser, respectively;

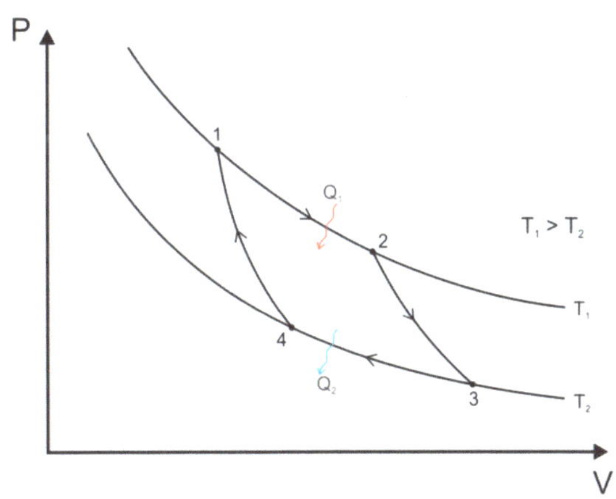

Figure 15: The Carnot cycle.

it should be emphasised that, for all his theoretical and conceptual powers, Carnot was entranced by practical engines. In a more abstract approach, these isothermals may be regarded as 'thermal reservoirs' of effectively infinite extent, so that they may transfer heat to objects with which they are in thermal contact, without their own temperature changing.

The isothermals are joined by two adiabatic stages, (23) and (41), in which no flow of heat takes place between the gas and its surroundings but there are changes in the temperature, pressure and volume of the gas. The gas is expanding and thus doing work in the stages (12) and (23), while in stages (34) and (41), work is done on the gas. However, from the point of view of work, these stages do not cancel each other out; overall, net work is done *by* the gas, and the amount is equal to the volume inside the cycle in Fig. 14. Physically, we may say that this is because the expansion is performed at higher pressures than the contraction; alternatively, we may say that the gas expands at higher temperatures than those at which it contracts.

Heat flows into the gas in the stage (12) and out at (34). Our current theory, of course, tells us that the flow in is greater than that out, the balance being taken up by the work done by the system, all according to conservation of energy. For Carnot, though, all the heat gained in the stage (12) must be lost in stage (34) as heat is conserved. Nevertheless, there are a number of crucial insights obtained by Carnot. First is the point that the very requirement for a cycle of this nature implies that there must be two reservoirs, a cold one as well as a hot one, and that at least some heat must be deposited in the cold one. Once today's dynamical theory is understood, Carnot's argument is trivially translated into our second law of thermodynamics, in which not all heat taken in by the system from the hot reservoir may be transformed into useful work; some must be wasted by being deposited into the cold reservoir. (This is usually called the Kelvin statement of the second law.)

Another central point in today's thermodynamics is that, as mentioned earlier, for maximum efficiency, there must be no heat flow between systems at different temperatures, that is to say no conduction of heat. Carnot expresses this fact in the following way: 'The necessary condition of the maximum is, then, *that in the bodies employed to realize the motive power of heat there should not occur any change of temperature which may not be due to a change in volume*' [italics in original]. Carnot brilliantly achieves this condition by separating the changes of temperature in stages (23) and (41), which as has been said, are achieved by changes of volume and pressure and *not* by flow of heat, from stages (12) and (34), where there *is* flow of heat but no change of temperature. Heat flow is always between systems at the same temperature, or to be precise one may say that an infinitesimal temperature difference must be allowed to cause the actual flow.

Carnot introduces another central feature of today's thermodynamics in his discussion of maximising the effectiveness of such an engine – that of *reversibility*. His own cycle is reversible, and he puts together a simple but brilliant argument to show that no cycle can be more effective than his own. For that let there be a more effective one. Then, it may be used to run a Carnot cycle backwards. But this hypothetical engine, being more efficient than the Carnot cycle, will produce more power than is required to do so, or in other words, a net gain of work will be produced, so that the two machines together are acting as a perpetual motion machine, invariably regarded as impossible in principle. Therefore, our assumption of an engine more effective than that of Carnot must be incorrect.

It may be noted that in the previous paragraph the word 'effectiveness' has been used rather than 'efficiency' as the latter would seem to be less appropriate in a theory based on the ideas of caloric than in our present one, in which one may naturally think of the amount of work done as a fraction of the heat provided. Of course, though, the ideas translate immediately to our present ones when one changes from a caloric theory to a dynamical theory of heat.

Once we have reached the idea of a maximum effectiveness, it is clear that this will be independent of the type of engine being considered. This important fact is stated by Carnot as follows: 'The motive power of heat is independent of the agents to realize it; its quantity is fixed solely by the temperatures of the bodies between which is effected, finally, the transfer of the caloric.'

Carnot's ability to come to more detailed conclusions, like that of his followers until at least the middle of the century, was limited, not just because he was using the caloric theory, but because a detailed understanding of the behaviour of gases was only developing along with the basic ideas of thermodynamics. Carnot, Clapeyron and the Thomsons all include substantial sections in their workbooks and papers struggling with the laws of Mariotte (or Boyle for English people of today!) and Gay-Lussac, and the experimental results of workers such as Dulong and Petit. Most of those involved specialised almost entirely to the case of an ideal gas, but even so the going was difficult, and it was really only with William Thomson's definition of an absolute temperature following from the idea of the Carnot cycle (as adapted for the dynamical theory of heat) that there was a completely rigorous basis for working with gases. (See the end of Section 2 of the companion paper to this one [43].)

It is abundantly clear to us today, though, that Carnot had grappled with and considerably illuminated issues of the deepest scientific and technical significance. Sadly it was far from obvious to the engineers and scientists of the today. His book received one excellent review in the *Revue Encylopédique* but was otherwise ignored; probably, it was too abstract and conceptual for the practising engineer, but its subject was too applied for most scientists and mathematicians.

For the remainder of his life, Carnot continued to ponder on these ideas, and it is clear that his belief in the caloric theory faded considerably. Mendoza [16, p. 46] suggests that this process had actually begun before the actual publication of the *Reflections*. At one point, while in the manuscript he wrote: 'The fundamental law that we proposed to confirm seems to us to have been placed beyond doubt, both by the reasoning which served to establish it, and by the calculations which have just been made.' In the published version, however, these words are replaced by: 'The fundamental law that we proposed to confirm seems to us to require, however, in order to be placed beyond doubt, new verifications. It is based upon the theory of heat as it is understood today, and it should be said that this foundation does not appear to be of unquestionable solidity. New experiments alone can decide the question.'

In his later manuscripts which were published posthumously, he went much further. He wrote [16, p. 67]: 'Heat is simply motive power which has changed its form. It is a movement among the particles of bodies. Wherever there is destruction of motive power, there is at the same time production of heat in quantity exactly proportional to the quantity of motive power destroyed,' and he suggested many experiments to check these ideas, including repeating Rumford's experiment, but measuring the motive power consumed as well as the heat produced. However,

Carnot's sad and untimely death prevented any possible substantive work along these lines being performed.

The content of Carnot's *Reflections* was saved from total obscurity only through the labours of Émile Clapeyron [16], another graduate of the École Polytechnique, who took it upon himself to present Carnot's results in a way more accessible to academic physicists and mathematicians. Carnot's work had been non-mathematical, but that of Clapeyron was mathematical throughout. Also Carnot's work had not used any diagrams, but Clapeyron presented the Carnot cycle on the so-called *indicator diagram*, or graph of pressure against volume as we have shown in Fig. 15, which gave a very direct impression of what was involved. Even so, Clapeyron's work, which was published in 1834, received very little attention until it was translated into German in 1843 [16, p. xv]. (It had already been translated into English in 1837 [6, p. 289].)

It appears to have been shortly after the publication of the German translation that news of the ideas of Carnot and Clapeyron reached the Thomson brothers, though it may well have been the English translation that they studied. In a letter to William written in August 1844, James, at that time working at the Fairbairn shipyard, inquired who it was that had proved that a definite quantity of mechanical effort was given out in the passage of heat from one body to another [6, p. 289–290]. It was clearly the work of Carnot and Clapeyron that he was referring to, and James showed that he already had a very good grasp of the issues.

In this letter, James showed himself already highly familiar with the analogy between the water-wheel and the heat engine, which had been heavily used by Carnot and Clapeyron – indeed Smith and Wise [6, p. 841] call it 'the water-wheel analogy of Sadi Carnot'. James, in fact, introduced a new practical point of analogy. In principle, the water should fall from as high a temperature as possible, almost, as James says, at the source of the river; similarly the heat should 'fall' from as high a temperature as possible, and from that point of view it was a pity that the boiler in a steam engine was at a much lower temperature than the actual burning fuel.

In each case, there was a drawback to this ideal maximal situation; for the water case, tributary streams providing water at a lower level would be missed, while, for the steam engine, most of the heat would pass up the chimney and likewise be lost. As we shall see, James would continue to study the Carnot–Clapeyron argument for several years, analysing both its fundamental concepts, and its practical possibilities and difficulties. It will be remembered that it was in this period that, as a semi-invalid in Glasgow, James invented his vortex turbine, the purpose of which was explicitly to maximise the power produced by the device.

Over the next few years, the brothers became steadily more interested and, indeed captivated, by Carnot's argument. While he was in Paris in 1845, William searched in vain for a copy of Carnot's original treatise. For the Paris book-sellers, 'Carnot' meant either Lazare, Sadi's father, or Hippolyte, his brother, author of a book on social studies. However, the brothers were able to locate and read Clapeyron's paper in the original French, and became convinced further of its very great merit. James commented that it was 'a very beautiful piece of reasoning, and of course perfectly satisfactory', while William said that 'nothing in the whole range of Natural Philosophy is more remarkable than the establishment of general laws by such a process of reasoning'.

William was not to obtain a copy of Carnot's original work until late 1848 when he received a copy from Gordon. It was then that he prepared his first important paper on thermodynamics in which he gave a detailed account of Carnot's theory [44]. Rather amusingly, when Clausius presented his own exceptionally significant paper [16] in 1850, he had to admit that he had not yet seen Carnot's paper and was basing his comments on that paper on the descriptions of the paper provided in the works of Clapeyron and Thomson. This explains, at least in part, Cardwell's suggestion that Clausius could be regarded as 'almost a disciple, at one remove, of Kelvin'; much of Clausius' highly admirable work was based on the profound analysis of William Thomson.

James became so enthusiastic about Carnot's work that even earlier in 1848 he had commenced the production of an astounding piece of work which we shall call the Belfast notebooks. These contained a considerable amount of information about his own air-engine, which was very much a result of Carnot's ideas, study of the Stirling engine, the first calculations of the lowering of the freezing point under pressure [1], and, perhaps most important of all from our point of view, important analysis of what would soon be called thermodynamics. This seminal work is discussed in Ref. [43].

Here we shall study the ideas of James Joule and the response to them of the Thomson brothers. As has been said, from the time of their discovery of the work of Carnot and Clapeyron, the brothers were convinced that it should be a crucial component of the sought-after understanding of the relations between heat and work or, as we may say, the future thermodynamics. Yet, they also became steadily, if rather unwillingly, convinced that Joule's work must also provide part of the final understanding, despite the fact that, as they saw it, it was inherently contradictory to that of Carnot.

7 The work of Joule, and the response of James and William Thomson

We saw earlier to what extent the caloric theory had made itself invulnerable to criticism. One important aspect of the theory was the conservation of heat or caloric, but the willingness to include the concept of 'latent caloric' [38] meant that it must be immensely difficult to disprove the theory. One might inquire whether it was *falsifiable* in a Popperian sense [45]. The most natural answer may be that the work of Joule, the Thomsons and Clausius did not actually falsify the caloric theory, but did show that the dynamical theory of heat and the idea of conservation of energy provided a more attractive explanation for the wide range of phenomena studied.

Before we describe the work of Joule himself, we should refer to Julius Mayer, who, as a result of his writings in the 1840s, is sometimes credited with priority over Joule in the idea of the mechanical equivalent of heat. His claim was particularly asserted in the early 1860s by John Tyndall, always an opponent of what Crosbie Smith [9, Chapter 9] calls the North British grouping centred around William Thomson, and including his brother, Joule, Rankine, Tait and the rather younger Maxwell. Smith regards this grouping as keen to assert their own priority in the establishment of what he calls *The Science of Energy* [9], which may be taken to include not only thermodynamics but also, for example, electromagnetism. We do not need to judge the significance of Mayer here, since we are studying specifically the development of the ideas of the Thomsons, and it is certain that their inspiration was Joule and not Mayer.

In view of the difficulty mentioned above of understanding the nature of heat by study of its use in engines of different types in producing work, it is interesting that Joule first approached the question from other angles [9, Chapter 4]. His first significant result, published in 1840 and still important today, was that the heat produced by an electric current was proportional to i^2 and to R, where i and R are the current and resistance, respectively, of the circuit. Today, this heating effect is still known as 'Joule heat'.

Joule also studied the amount of work that might be produced by an electromagnetic engine, and moved on, by 1843, to study of electrochemistry. He was able to show that, in the electrolysis of water, all the 'caloric' produced could be traced back precisely to chemical changes. The use of the word 'caloric' suggested that he was not at the point willing to oppose the caloric theory, but clearly he was at the very least extremely close to a claim of 'conservation of energy'. He noted, for example, that both the mechanical and the heating effects of a current were proportional to the electromotive force, suggesting a relationship between these types of effect. He also remarked that if an electrical generator is inserted into an electrical circuit, there would be a decrease in the heat evolved by the circuit, precisely equal to the mechanical effort produced by the generator.

Soon afterwards he was able to show that an electrical generator was indeed a generator of heat as well as electricity, and suggested that heat should be viewed, not as a 'substance' but as a 'state of vibration', claiming, in this belief, that he was a follower of Rumford. Very importantly, he introduced a 'mechanical value of heat', making it clear his belief that heat was transformed to work, not just related in some way to it, and, on the basis of his experiments, he was able to come up with a range of results for this value. Later experiments included measuring the heat produced by compressing a gas, the heat lost when it expands, and the heat produced by paddle-wheels rotating in water. In all cases, there was a broad though not particularly convincing agreement in the inferred value for the mechanical value of heat. It will be appreciated incidentally that, from the point of view of this paper, the actual physical nature of heat is much less important than the statement that it is not conserved.

Another interesting experiment was to allow a gas to flow into a vacuum, when Joule believed that there should be no increase in temperature because no work was done. This is today known as the *Joule effect*. Joule's expectation of no change in temperature is not quite borne out because the gases are not exactly ideal. In the next few years, Joule and William were to carry out a long series of experiments investigating a related effect, in which a gas moves from one pressure to another through a porous plug. Today, this is called the *Joule–Kelvin* or *Joule–Thom*son effect.

It may be noted that, even at this date, Joule was aware of the Clapeyron paper, and, recognising that Carnot's ideas conflicted substantially with his own, he took the opportunity to criticise them severely. Carnot, he said, stated that the fall of the heat from the temperature of the boiler to that of the condenser was responsible for the production of work. But in that case, the substantial fall of the heat from the temperature of the hot fuel to that of the boiler presented severe difficulties; the work or *vis viva* as Joule calls it, that could have been produced by this fall, appeared to have been lost. 'Believing that the power to destroy belongs to the Creator alone', Joule felt that Carnot's theory must be erroneous [9, p. 69]. He wrote that: 'I conceive that this theory, however ingenious, is opposed to the recognized principles of philosophy, because it leads to the conclusion that *vis* viva may be destroyed by an improper disposition of the apparatus [6, p. 306].' It will be noted that this objection was identical to the worries that the Thomson brothers had shared at Rugby some years before.

For several years, Joule's claims attracted very little attention. From the point of view of the scientific community, he was something of an outsider, coming from a brewing background. Many of the temperature increases he measured were so small as to be unconvincing, and his measured values for the mechanical value of heat covered quite a wide range. More important, perhaps, was the fact that most scientists, very much including the Thomson brothers, found it difficult to move away from the caloric theory with its conservation of heat and its water–wheel analogy. So several of Joule's most important papers were rejected by the *Physical Transactions of the Royal Society*, being published instead in the *Philosophical Magazine*. Even in the June 1947 meeting of the British Association or the Advancement of Science (BAAS) in Oxford, he was only allowed to give an abbreviated presentation of his paper.

However, it was at this meeting that his work attracted the attention of William Thomson, whose support, limited and almost grudging as it was at first, was crucial in, as Joule later put it, rescuing him from obscurity [6, p. 304], and assuring the eventual general acceptance of his views, and his high place in scientific history. Also important was the interest of James Thomson, who was immediately informed of the content of Joule's paper by William, and also of George Stokes, William's close friend and correspondent.

The (extremely) positive component of William's response to Joule concerned the transformation of work to heat, in particular via fluid friction; William and Stokes were greatly concerned with the theory of hydrodynamics, and Joule's ideas and his experiments appeared to have the capacity to make an important contribution to that subject. He later said that he became convinced that 'Joule had certainly a great truth and a great discovery, and a most important measurement [6, p. 302]'.

However, there were also very negative components of William's response, mostly, at least in retrospect, explained by William's adherence, instinctive as much as reasoned, to the idea of caloric, but there was also the point that Joule's ideas, certainly correct in themselves, missed one aspect of Carnot's work – directionality. As William was later to express it, Joule gave science the first law of thermodynamics, but the second was produced by Carnot as elaborated by Clausius.

During Joule's 1847 BAAS talk, William's main disagreement with the speaker came with Joule's statement that the work required to warm water must be proportional to the quantity of heat involved in the process, an obvious result of the assumption of a mechanical value of heat. William was sure, on the basis of Carnot's work, that it must rather be proportional to the square of the quantity of heat.

Smith and Wise [6, pp. 302–304] suggest that William's confidence stemmed from the belief that, when applying heat to a body to increase its temperature, one is both increasing the amount of heat 'in the body' and raising the temperature of the heat – very much a notion inspired by the caloric theory. The situation is at least broadly analogous to the study of the work done against gravity in building a wall steadily higher and higher, the work in this case is definitely proportional to the *square* of the height of the wall. Similarly, the work done in heating the body, on the Carnot–Thomson argument, must be proportional to the square of the temperature difference, or the square of the heat produced. A beautiful picture of this idea was given by John Dalton,

Figure 16: Cylinders of caloric as shown by John Dalton in his book *A New System of Chemical Philosophy* (Manchester, 1808) Vol. 1, p. 217.

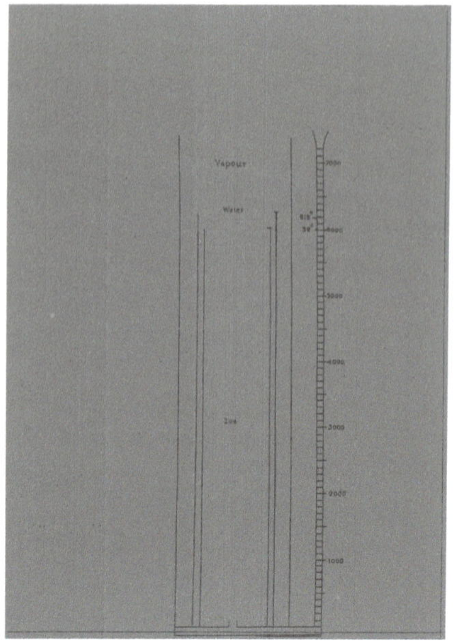

> Figure taken from Ref. [38], Plate 1.
>
> The figure demonstrates the way in which, on a caloric theory, heat or caloric may be said to fill up a hypothetical cylinder in the same way as a liquid, the height of the heat indicating the temperature of the object. The work done in raising the heat is thus proportional to the square of the amount of heat, in close analogy to the work done in building a wall of uniform thickness being proportional to the square of the amount of material in the wall.

a great believer in the caloric theory, in a book published in 1808 and included here as Fig. 16. (Incidentally and ironically, Joule was a disciple of Dalton, both being from Manchester, but Joule's attack on caloric commenced in the period in which Dalton died.)

After Joule's talk, William had fruitful discussions with him and was given copies of several of his published papers. After the meeting, he wrote to his father that he felt Joule was 'wrong in many of his ideas, but he seems to have discovered some facts of extreme importance, as for in stance that heat is developed by the friction of fluids in motion [6, p. 302]'. He promised to write to James about Joule's ideas and to send him a set of Joule's papers. This is yet again evidence that James was at the very least an equal partner in the exchange of ideas between the brothers.

In his letter to James, William [2, p. xxviii] said that

> I enclose Joule's papers which will astonish you. I have only had time to glance through them as yet. I think at present that some great flaws must be found. Look especially to the rarefaction and condensation of air, where something is decidedly neglected, in estimating the total change effected, in some of the cases. Keep all the papers carefully together and give them to me when I return.

In reply, James did indeed raise some issues over Joule's analysis of his experiment on the free expansion of gases. The difference of opinion seems to have been a result of implicit differences in assumptions. In the experiment, the temperature of a vessel at first rose but, by the end of the experiment, had returned to its original value. For Joule, since there was no overall increase in temperature, there could have been no expenditure of work; for James, though, the raising of the temperature of the caloric of itself required the expenditure of work.

However, James was extremely interested in Joule's general position:

> Some of his views have a slight tendency to unsettle one's mind as to the accuracy of Clapeyron's principles. If some of the heat can absolutely be turned into mechanical effect, Clapeyron may be wrong. I think, however, that before coming to a conclusion, we would need to define more accurately what we mean by *a certain quantity of heat* as applied to two bodies at different temperatures. Perhaps Joule would say that if a hot pound of water lose a degree of heat to a cold one, the cold one may receive a greater absolute amount of heat than that lost by the hot one; the increase being due to the mechanical effect which might have been produced during the fall of heat from the high temperature to the low one.

James' suggestion made at the end of the letter has been fleshed out by Smith and Wise [6, pp. 307–309] and by Smith [9, pp. 79–80]. It retains Carnot's conservation of caloric in so far as, for the Carnot cycle case, the heat, Q, deposited in the cold reservoir is equal to that gained from the hot reservoir; work W is also done. However, for the case of conduction of heat directly from hot to cold reservoir, the mechanical work that could have been obtained in the Carnot cycle case is not lost but is effectively transformed to heat, and the heat deposited in the cold reservoir is equal to $Q + W$. (In adding an amount of heat to an amount of work, we are, of course, taking advantage of Joule's mechanical value of heat.)

From the comfort of today's hindsight, James's suggestion may seem a clumsy attempt to splice the ideas of Carnot and Joule. W comes from a fall in caloric, rather than being a transformation of caloric, but if it does not arrive at all, conservation appears to dictate that fresh caloric appears instead. However, the real significance of the remarks is to show how extremely keen both brothers were to bring together the seemingly contrasting ideas of Carnot and Joule in a single theory.

Over the next year or so, this struggle continued. On the one hand, William carried out the kind of experiments that Joule had pioneered – using a paddle-wheel to heat water; indeed, in what might be described as typical Kelvin fashion, he was determined to go much further than Joule and actually boil the water [6, p. 308]! However he and James still had difficulties with both theories. Carnot could not handle the conduction case in anything like a satisfactory way. In one sense, Joule could; heat that was not transformed into work could just remain as heat. But for Joule, conservation of energy implied that there was no reason why that heat could not be transformed to work; for William, heat was an end-product or waste-product of any process, and he could not believe useful work could be obtained from it.

Within a few years, he would realise that low temperature heat is an end-product of a reversible cycle such as the Carnot cycle; since the cycle *is* reversible, it must be possible to transform *this* heat back to work in a straightforward way by running the engine backwards. However, where there is dissipation, the corresponding process must be irreversible and extra heat is produced. Waste, or we may say increase in entropy, is necessarily involved in this process. But for the moment, William could not see his way to solving the problem.

In 1848, when he finally received his copy of Carnot's original treatise, William wrote an account of Carnot's theory for the Royal Society of Edinburgh [44]; he had long promised this to James Forbes, Professor of Natural Philosophy at the University of Edinburgh. In this paper he gave an account of Carnot's theory, providing in addition experimental support from the important experiments of Victor Regnault on the behaviour of steam. He referred to 'Carnot's fundamental principle' whereby 'the quantity of heat ... discharged during a complete revolution (or double stroke) of the engine must be precisely equal to that which enters the water of the boiler', commenting that: 'So generally is Carnot's principle tacitly admitted as an axiom, that its application in this case has never, as far as I am aware, been questioned by practical engineers [44, p. 117].'

In a footnote in this paper, William praised the contributions of Joule and mentions that Joule himself had 'strongly urged' that one should abandon Carnot's fundamental axiom. However, William responded that:

> If we do so, however, we meet with innumerable other difficulties – insuperable without farther experimental investigation, and an entire reconstruction of the theory of heat from its foundation. It is in reality to experiment that we must look – either for a verification of Carnot's axiom, and an explanation of the difficulty [as described above here] we have been considering; or for an entirely new basis of the Theory of Heat.

William and James had made great progress together, but still faced very real difficulties in reaching a theory that would satisfy all their requirements.

Acknowledgements

I thank Deirdre Wildy, Senior Subject Librarian at Queen's University Belfast, for permission to use and quote from the material in the Special Collections in the Library, and Diarmuid Kennedy, Special Collections Librarian and his staff for their courtesy and help in obtaining material from the James Thomson archive. I also thank Joan Whitaker, Physics Subject Librarian, and the other staff of the Science Library at Queen's University Belfast, for considerable support in obtaining books and papers.

I also acknowledge Mark McCartney for many interesting and useful conversations on the topics of James and William Thomson and other aspects of nineteenth century physics.

Finally, I am grateful to Michael Collins for suggesting that I should write this paper.

References

[1] Thomson, J., Theoretical considerations of the effect of pressure in lowering the freezing point of water. *Transactions of the Royal Society of Edinburgh*, **16**, pp. 575–580, 1849; reprinted with some slight but important alterations in Thomson, J., *Cambridge and Dublin Mathematical Journal*, **5**, pp. 248–253, 1850.

[2] Thomson, J., *Collected Papers in Physics and Engineering*, Cambridge University Press: Cambridge, 1912 (eds. J. Larmor & J. Thomson); the Biographical Sketch is on pp. xiii-xci, and also included on pp. xcii-cii is the Obituary Notice by J.T. Bottomley, originally published in 1892 in the *Proceedings of the Royal Society*, **53**, i-x.

[3] Crossland, B., & Moore, J. S., *The Lives of Great Engineers of Ulster*, Vol. 1, NE Consultancy for Belfast Industrial Heritage Ltd: Belfast, pp. 149–156, 2003.

[4] Bowler, P., James Thomson and the culture of a Victorian engineer. *Kelvin: Life, Labours and Legacy*, eds. R. Flood, M. McCartney & A. Whitaker, Oxford University Press: Oxford, pp. 56–63, 316–317, 2008; also in *Queen's Thinkers: Essays on the Intellectual Heritage of a University*, eds. A. Jackson & D.N. Livingstone, Blackstaff: Belfast, pp. 57–68, 2008.

[5] Thompson, S.P., *The Life of William Thomson: Baron Kelvin of Largs*, 2 Vols, Macmillan: London, 1910.

[6] Smith, C. & Wise, M.N., *Energy and Empire: A Biographical Study of Lord Kelvin*, Cambridge University Press: Cambridge, 1989.

[7] Flood, R., McCartney, M. & Whitaker, A., (eds), *Kelvin: Life, Labours and Legacy*, Oxford University Press: Oxford, 2008.

[8] Smith, C.W., William Thomson and the creation of thermodynamics: 1840–1855. *Archives for History of Exact Sciences*, **16**, pp. 231–288, 1977.

[9] Smith, C., *The Science of Energy: The Construction of Energy Physics in the Nineteenth Century*, Athlone: London, 1991.

[10] Crossland, B. & Perry, J., *The Lives of Great Engineers of Ulster*, Vol. 3, eds B. Crossland & J.S. Moore, NE Consultancy and The Bookshop at Queen's: Belfast, pp. 65–74, 2008.

[11] Eastwood, J. & Robert, T., *Australian Dictionary of Biography*, Vol. 6, Melbourne University Press: Victoria, pp. 269–270, 1976, online at http://adbonline.anu.edu.au/biogs/A060290b.htm [Australian National University, 2006].

[12] Thomson, J., Notebook A 14 (B), Thomson Papers, Queen's University Belfast, 1848.

[13] McNeil, I., *An Encyclopaedia of the History of Technology*, Routledge: London, pp. 242–243, 703, 2002.

[14] Moody, T.W. & Beckett, J.C., *Queen's Belfast 1845–1949: The History of a University*, Vol. 1, Faber and Faber: London, 1959.

[15] Thomson, J., Nationalization of public works. *Northern Whig* (21 January, 1869); also in Ref. 2, pp. 472–477.

[16] Carnot, S., *Reflections on the Motive Power of Fire, and on Machines Fitted to Develop That Power* (translated and edited by R.H. Thurston, 1890), Dover: New York, 1960. Carnot's original work was *Réflexions sur la Puissance Motrice du Feu et sur les Machines Propres à Développer cette Puisssance* (Bachelier: Paris) and was published in 1824. The Dover volume is edited by E. Mendoza, and contains an Introduction by Mendoza (pp. ix–xxii); the Carnot paper (pp. 1–59); Selections from the posthumous manuscripts of Carnot (translated by R.H. Thurston and E. Mendoza) (pp. 60–70); Memoir on the motive power of heat, by É. Clapeyron (originally published in French in 1834; translated and edited by E. Mendoza) (pp. 71–106); and On the motive power of heat, and on the laws which can be derived from it for the theory of heat, by R. Clausius (originally published in German in 1850; translated by W.F. Magie, 1899) (pp. 107–152).

[17] Thomson, J., On the vortex water-wheel. *Report of the British Association, Belfast*, 1852; also in Ref. 2, pp. 2–16.

[18] Thomson, J., On a centrifugal pump with exterior whirlpool, constructed for draining land. *Report of the British Association, Glasgow*, 1858; also in Ref. 2, pp. 16–24.

[19] Thomson, J, On the jet pump and its application to the drainage of swampy lands and shallow lakes. Ref. 2, pp. 30–34.

[20] Thomson, J., On experiments on the measurement of water by triangular notches in weir-boards. *Report of the British Association, Leeds*, 1858; also in Ref. 2, pp. 36–42.

[21] Thomson, J., Improved investigations on the flow of water through orifices with objections to the modes of treatment commonly adopted. *Report of the British Association, Glasgow*, 1876; also in Ref. 2, pp. 56–87.

[22] Thomson, J., On the origin of windings of rivers in alluvial plains, with remarks on the flow of water round bends in pipes. *Proceedings of the Royal* Society, **25**, pp. 3–8, 1876; also in Ref. 2, pp. 96–99.
[23] Thomson, J., Bakerian Lecture – On the grand currents of atmospheric circulation. *Philosophical Transactions of the Royal Society*, **183**, pp. 653–684, 1892; also in Ref. 2, pp. 153–195.
[24] Thomson, J., On the plasticity of ice, as manifested in glaciers. *Proceedings of the Royal Society*, **8**, pp. 455–458, 1856–1857; also in Ref. 2, pp. 208–211.
[25] Cunningham, F.F., *James David Forbes: Pioneer Scottish Glaciologist*, Scottish Academic Press: Edinburgh, 1990.
[26] Thomson, J., On ground or anchor ice and its effects in the St Lawrence. Ref. 2, pp. 258–267.
[27] Thomson, J., On the parallel roads of Lochaber. Ref. 2, pp. 407–420.
[28] Rowlinson, J.S., The work of Thomas Andrews and James Thomson on the liquefaction of gases. *Notes and Records of the Royal Society of London*, **57**, pp. 143–159, 2003.
[29] Andrews, T., Bakerian Lecture – on the continuity of the gaseous and liquid states of matter. *Philosophical Transactions of the Royal Society*, **16**, pp. 575–590, 1859.
[30] Thomson, J., Speculation of the continuity of the fluid state of matter, and of relations between the gaseous, the liquid, and the solid states. *Report of the British Association, Edinburgh*, 1871; also in Ref. 2, pp. 286–291.
[31] Thomson, J., On recent theories and experiments regarding ice at or near its melting point. *Proceedings of the Royal Society*, **10**, pp. 152–160, 1859; also in Ref. 2, pp. 222–229.
[32] Thomson, J., On conditions affording freedom for solidification to liquids which tend to solidification, but experience a difficulty of making a beginning of their change of state. Ref. 2, pp. 268–269.
[33] Thomson, J., On the jointed prismatic structure in basaltic rocks, Ref. 2, pp. 422–440.
[34] Thomson, J., An integrating machine, being a new kinematic principle. *Proceedings of the Royal Society*, **24**, pp. 262–265, 1876; also in Ref. 2, pp. 452–458.
[35] Westfall, R.S., *Never at Rest: A Biography of Isaac Newton*, Cambridge University Press: Cambridge, 1993.
[36] Gillispie, C.C., *Pierre-Simon Laplace. 1749–1827: A Life in Exact Science*, Princeton University Press: Princeton, 1997.
[37] Whewell, W., *The Mechanics of Engineering, Intended for Use in Universities and in Colleges of Engineers*, Cambridge University Press: Cambridge, 1841.
[38] Fox, R., *The Caloric Theory of Gases: From Lavoisier to Regnault*, Clarendon: Oxford, 1971.
[39] Tait, P.G., *Sketch of Thermodynamics*, Douglas: Edinburgh, 1877.
[40] Lakatos, I., Falsification and the methodology of scientific research programmes. *Criticism and the Growth of Knowledge*, eds. I. Lakatos & A. Musgrave, Cambridge University Press: Cambridge, 1970.
[41] Cardwell, D.S.L., *From Watt to Clausius: The Rise of Thermodynamics in the Early Industrial Age*, Heinemann: London, 1971.
[42] Truesdell, C., *The Tragicomical History of Thermodynamics 1822–1854*, Springer-Verlag: New York, 1980.
[43] Whitaker, A., *James and William Thomson: The Creation of Thermodynamics*. Following paper in this volume.

[44] Thomson, W., An account of Carnot's theory of the motive power of heat, with numerical results deduced from Regnault's experiments on steam. *Transactions of the Royal Society of Edinburgh*, **16**, pp. 541–574, 1849; also in Thomson, W., *Mathematical and Physical Papers* (MPP), Cambridge University Press: Cambridge, pp. 113–155, 1882.
[45] Popper, K., *The Logic of Scientific Discovery*, Basic Books: New York, 1959.

Chapter 5

Kelvin, Maxwell, Clausius and Tait: the correspondence of James Clerk Maxwell

R.C. Dougal
James Clerk Maxwell Foundation, United Kingdom.

Abstract

The correspondence between Maxwell and Kelvin on many topics of mutual interest during the second half of the nineteenth century was terminated by Maxwell's untimely death in 1879. Much of what little there was on thermodynamics is no longer extant. Peter Guthrie Tait, Professor of Natural Philosophy at Edinburgh University from 1860 to 1901, provided the important link with Kelvin, Professor of Natural Philosophy at Glasgow University, not least because both held office in the Royal Society of Edinburgh. Tait was well aware of Kelvin's crucial contributions to the elucidation of the laws of thermodynamics in the early 1850s, and the need to harmonise them with the insightful contributions by Clausius during that decade; in particular, Clausius's introduction of the thermodynamic property entropy. Processes permitted by the Second Law of Thermodynamics could be charted by changes in the value of entropy. The following decade saw Maxwell introduce his kinetic theory of gases. Thereby Maxwell's interest, especially in the Second Law, was heightened because he felt that he could, overall, enhance Kelvin's dynamics of gases by incorporating explicitly statistical properties of distributions of molecular velocities in assemblies of molecules constituting the gases. Violations of the Second Law by subsets of molecules selected by an artifice soon to become known as Maxwell's Demon would be completely hidden because observations could be made only on a complete assembly. Curiously, Maxwell's first early encounters with entropy were seriously flawed, possibly because he was inadvertently misled by Tait whilst he was writing his own textbook on heat. The strong link between statistics and entropy became apparent later in the century, by which time Maxwell's distributions of molecular velocities had been extended by Boltzmann to distributions of their energies.

1 Introduction

1.1 The correspondence between Maxwell and Kelvin

The correspondence on thermodynamics between William Thomson (1824–1907) – Lord Kelvin from 1892 – and James Clerk Maxwell (1831–1879) was sporadic and not extensive. Losses have left it fragmented as well. Fortunately, some gaps are made good by correspondence between Maxwell and his lifelong friend Peter Guthrie Tait (1831–1901). All that is now available has been collated and presented by Professor Peter Harman in his comprehensive (three volume) *Scientific Letters and Papers of James Clerk Maxwell* [1]. In the introductions and copious footnotes Harman provides continuity and links to the wider literature. These provide a guide to the interactions – direct and indirect – between Maxwell and the main players in the development of classical and, later, statistical thermodynamics in the second half of the nineteenth century. Within it there is to be found much of a nearly self-contained story of Maxwell's particular involvement in the development of the Second Law of Thermodynamics.

In the 1870s, some of the correspondence amongst Thomson, Maxwell and Tait was by postcard. Figure 1 is an example of one of the few which have survived. To save space, Thomson was T, Tait was T′ and Maxwell was dp/dt. The last originated from the equation $dp/dt = JCM$, based on one in Tait's book *Sketch of Thermodynamics* [2] with the factors on the right labelled and arranged to give Maxwell's initials. Thus, Thomson to Maxwell was written 'T 2 dp/dt'.

1.2 Timelines for Kelvin and Maxwell

In the summer of 1871, Maxwell replied to a prompt from Thomson [4] for help with the preparation of his (Thomson's) Presidential Address to the British Association for the Advancement of Science to be held in Edinburgh in the autumn. In June 1871, Thomson had written: '... Where is the Address you undertook to write? It should be in the printers hands soon, so please don't delay longer. Send it to the Athenaeum Club London posting not later than next Wednesday. It must be *complete* (for press) on Dynamical Theory of gases, Diffusion & all who worked meritoriously on it, whether experimentally or theoretically and *their merits carefully weighed*. This last is de rigeur ...'. Maxwell's detailed reply to Thomson was understandably thorough. He had continued to work on fundamentals and applications of the kinetic theory of gases over the preceding 10 years. In addition, he was completing the writing of his *Theory of Heat* for publication in 1871 [5]. The preceding 20 years or so had witnessed the steady development of an understanding of classical thermodynamics arguably begun by Thomson and Rudolf J. E. Clausius (1822–1888), with Maxwell's contributions beginning about half way through.

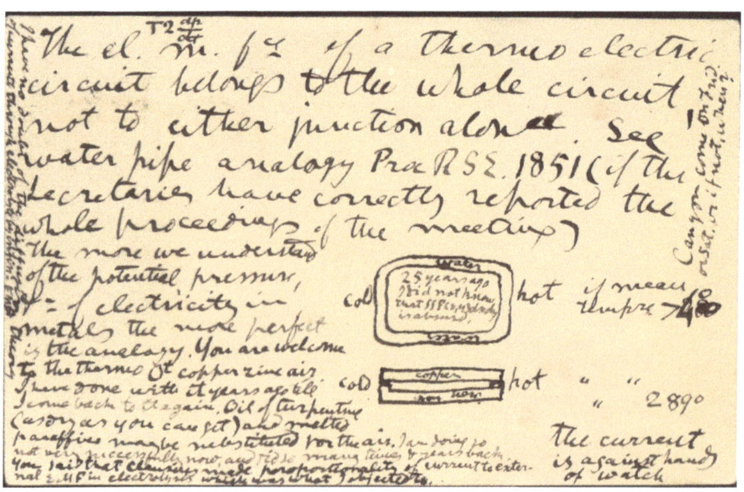

Figure 1: Thomson's reply, dated 4 January 1872, to a query from Maxwell. Reproduced by kind permission of the Syndics of Cambridge University Library, classmark Add. MSS 7655,I,a/24. The transcript in the Appendix is based on the version in Harman [3].

William Thomson was a senior to James Clerk Maxwell by 7 years. From 1846, Thomson was well-established academically and scientifically following his appointment as Professor of Natural Philosophy at Glasgow University. Maxwell began 3 years of study at Edinburgh University in 1847 and moved to Cambridge University in 1850, graduating in 1854. During his undergraduate studies there, Maxwell had kept himself apprised of Thomson's overlapping researches on heat and electricity. These provided an entry to Maxwell's own research on electricity and magnetism. Shortly after graduating from Cambridge University, Maxwell wrote to Thomson on 20 February 1854, on behalf of himself and others, informing him of a 'strong tendency to return to Physical Subjects and several of us here wish to attack Electricity' [6]. He continued 'Suppose a man to have a popular knowledge of electrical show experiments ... how ought he to proceed in reading & working so as to get a little insight into the subject wh[ich] may be of use in further reading?' Thomson's letter of reply was accompanied by a reading list, and encouragement. In his editorial footnote to Maxwell's letter of thanks on 15 May 1855 [7], Harman commented, 'This [Thomson's] letter, like others from Thomson, is not extant.' Maxwell continued that he hoped to make use of Thomson's analogy between steady heat flow and electrostatic force. He added 'By the way do you profess to account for what becomes of the vis viva of heat when it passes thro' a conductor from hot to cold? You must either modify Fourier's laws [8] or give up your theory, at least so it seems to me.' (The concept of vis viva was introduced by Gottfried W Leibniz (1646–1716) [9] to describe a property of certain mechanical systems consisting of masses m each having its own speed v. The sum of terms mv^2 seemed to be conserved. By the middle of the nineteenth century, the name was usually given to $\frac{1}{2}mv^2$, an important quantity in mechanics, referred to as kinetic energy. Harman commented, 'Maxwell probes Thomson's concept of the dissipation of energy.' Maxwell was well aware of contributions by J.P. Joseph Fourier (1768–1830) to the theory of heat. While at Edinburgh University he had borrowed from the library Fourier's *Théorie Analytique de la Chaleur*, and his most recent biographer, Basil Mahon, noted that 'he was so smitten by Fourier's book that he bought his own copy for 25 shillings, a considerable sum in 1849' [10].

The issue had been raised by Thomson himself 6 years earlier in a footnote in his 1849 paper on Carnot's theory of the motive power of heat [11]. 'When 'thermal agency' is thus spent in conducting heat through a solid, what becomes of the mechanical effect which it might produce? Nothing can be lost in the operations of nature – no energy can be destroyed. What effect then is produced in place of the mechanical effect which is lost?' He had gone some way to answering his query in his 1851 paper on the dynamical theory of heat [12]. By that time studies on heat transfer had been extended to include, at least generally, properties of the material supporting the heat flow, intentionally excluded by Fourier in his theory. Energy considerations would lead Thomson a year later [13] to conclude that the conduction of heat through a body be accompanied by such transformations as to leave some of it dissipated and thereby irrecoverable but not lost. If, on this issue, Thomson replied to Maxwell in writing, there is no trace of that letter either.

Thomson had continued his research into the fundamentals and applications of thermodynamics beyond 1851, and his publications would have been available to Maxwell. However, by October 1856, Maxwell had moved to Aberdeen as Professor of Natural Philosophy at Marischal College. His father had died suddenly in the spring of that year, and Maxwell inherited the family estate, Glenlair, in Galloway in southwest Scotland. In 1858, he married Katherine Mary Dewar, daughter of the Principal of Marischal College. Whilst at Marischal College, from 1856 to 1860, the emphases in most of his letters to Thomson as have survived were on electricity, magnetism and optics.

2 Saturn's rings and the Adams Prize: Maxwell and Kelvin

Maxwell's pathway to Thomson's thermodynamics was mainly through the properties of gases in their own right rather than as dynamical systems such as integral parts of heat engines. With the benefit of hindsight, we can surmise that already, at Aberdeen, Maxwell had started an approach to an aspect of the Second Law of Thermodynamics, which can be traced back to his work there on the theory of the structure of the rings of planet Saturn.

The Adams Prize was instituted in 1848 to commemorate the prediction by John Couch Adams (1819–1892) in 1845 of the position of an as yet undiscovered planet. About a year later, in France, calculations by Urbain J.J. Leverrier (1811–1877) had led to essentially the same prediction. Unlike Adams's prediction, Leverrier's was quickly followed up, at the Berlin Observatory. From there, the new planet, to be called Neptune, was discovered late in 1846. In Britain, the Astronomer Royal, Sir George Airy, suggested that Adams be given priority for the prediction, and a squabble followed. The Adams Prize – to acknowledge brilliance in applications of mathematics – can be seen as part of Adams's compensation.

The topic for the fourth Adams Prize, announced in 1855, was 'The Motions of Saturn's Rings and, in particular, the question of their stability'. Maxwell worked intensively on the topic in 1855 and 1856 during his tenure of the Chair of Natural Philosophy at Marischal College. His, the only, submission was outstanding and he was awarded the prize. It was written up and published as a monograph in 1859 titled *On the Stability of the Motion of Saturn's Rings* [14]. This was described by his biographer Ivan Tolstoy [15] as 'ninety quarto pages of mathematics and carefully polished prose ... a bravura performance which, perhaps more than anything he had yet published, made him a star in the eyes of his contemporaries.' While preparing the monograph, Maxwell kept in touch with Thomson (who had been one of the examiners for the prize). In an update on 14 November 1857, Maxwell commented on progress in the consideration of various configurations of matter constituting the rings [16]. He remarked, 'The general case of a fortuitous concourse of atoms each having its own orbit & excentricity is a subject above my powers at present, but if you can give me any hint as to the point of attack I will go at it.' In the monograph, the remark is extended 'When we come to deal with collisions among bodies of unknown number, size, and shape, we can no longer trace the mathematical laws of their motion with any distinctness.' Maxwell returned to this viewpoint when within a few years he turned his attention to the kinetic theory of gases.

3 Molecular model for a gas: Maxwell and Clausius

3.1 Description of model

In the meantime, in 1858 Clausius had published a paper on, amongst other things, the average distance travelled by gas molecules between consecutive collisions, their mean free path [17]. Maxwell noticed the translation in the Philosophical Magazine in 1859. His reworked analysis differs significantly from Clausius's in that he allowed for a distribution of molecular velocities rather than work with just an average value. In Maxwell's 1860 paper [18], he presented what became known as the Maxwell Distribution of molecular velocities.

With a display of audacity rather than bravura, Maxwell kept the derivation brief: four equations derived in 2 pages in Part I of the complete 33-page paper. In the introductory preamble, he remarked: 'In order to lay the foundation of such investigations [of properties of matter, especially when in the gaseous form] on strict mechanical principles, I shall demonstrate the laws of motion of an indefinite number of small, hard, and perfectly elastic spheres acting on one another only during impact.' Extending this basic assumption, he continued: 'Instead of saying that the particles are hard, spherical, and elastic, we may if we please say that the particles are centres of force, of which the action is insensible except at a certain small distance, when it suddenly appears as a repulsive force of very great intensity. It is evident that either assumption will lead to the same results.'

After setting out the key parameters of such collisions – velocities and angles of rebound in particular – Maxwell continued: 'If a great many equal spherical particles were in motion in a perfectly elastic vessel, collisions would take place among the particles, and their velocities would be altered at every collision; so that after a certain time the *vis viva* will be divided among the particles according to some regular law, the average number of particles whose velocity lies between certain limits being ascertainable, though the velocity of each particle changes at every collision.' He had moved to a starting point which was thus a description of an isolated sample of gas that had settled into an equilibrium state.

The paper continued with 'PROP. IV. To find the average number of particles whose velocities lie between given limits, after a great number of collisions among a great number of equal particles'. For a sample of N particles, whose velocity v has components x, y and z in each of three mutually perpendicular directions (Maxwell's notation) with all three components governed by the same function f, the analysis led to, in sequence, four equations. Here, $Nf(x)\mathrm{d}x$ is the number of particles with x in the narrow range x to $x+dx$, and correspondingly for velocity components y and z. The number with actual velocity in the range v to $v+\mathrm{d}v$ is $NF(v)\mathrm{d}v$.

$$f(x) = (1/a\sqrt{\pi})\exp(-(x^2/a^2)) \qquad F(v) = (4/a^3\sqrt{\pi})\, v^2 \exp(-(v^2/a^2))$$

$$\text{mean value of } v \text{ is } 2a/\sqrt{\pi} \qquad \text{mean value of } v^2 \text{ is } (3/2)a^2$$

The factor $-1/a^2$ replaced one of the arbitrary constants in the introductory analysis. Thus, Maxwell excluded the unrealistic steady increase with velocity of the number of particles and made the later analysis immediately amenable to well-known mathematical results. The value of a^2 increases with the temperature and decreases with the molecular mass of the gas. The paper continues: 'It appears from this proposition that the velocities are distributed among the particles according to the same law as the errors are distributed among the observations in the theory of the 'method of least squares'. The velocities range from 0 to ∞, but the number of those having great velocities is comparatively small. In addition to these velocities, which are in all directions equal, there may be a general motion of translation of the entire system of particles which must be compounded with the motion of the particles relatively to one another. We may call one as the motion of translation and the other the motion of agitation'.

3.2 Summary of selected predictions from model

Maxwell summarised the illustrations in the last few paragraphs of the final PART III of the paper, beginning with those in PART I: 'We have now followed the mathematical theory of the collisions of hard elastic particles through various cases, in which there seems to be an analogy with the phenomena of gases.' These began with the relations amongst the thermodynamic coordinates (pressure, temperature and density) and continued with the phenomena of internal friction (viscosity) with the mean free path (the average distance between consecutive collisions) having particular importance. It contained the counter-intuitive prediction that the coefficient of viscosity be independent of density. PART II illustrations were on diffusion for which he had modelled a bulk gas as a large number of parallel strata and treated the net movement of molecules between strata. He added a temperature gradient in the final illustration, the conduction of heat through a gas. Clausius was soon to point out Maxwell's unsatisfactory specifications of the strata [19], which was acknowledged and acted on in Maxwell's preparation [20] of his later paper, in 1867, 'On the Dynamical Theory of Gases [21]. In particular, for thermal conductivity Maxwell had overlooked the variation of kinetic energy with the layer-to-layer variation in temperature. The distributions of velocities could not, therefore, be of the isotropic motion of agitation as Maxwell had assumed. Conduction of heat through a gas was the least satisfactory because it also highlighted a conflict between theory and experiment in thermodynamics, pointing to an inadequacy in modelling of molecular *structure.* Maxwell had replaced molecular mass M in the treatment of diffusion by $\frac{1}{2}Mv^2\beta$, where, following Clausius, β is the unknown factor by which, because of rotational motion, the energy is greater than just the kinetic energy $\frac{1}{2}Mv^2$. However, Maxwell edged the topic one stage further, introducing the final illustration, thus: 'The energy of a single particle is composed of two parts – the *vis viva* of the centre of gravity, and the *vis viva* of the various motions of rotation round that centre, or, if the particle be capable of internal motions, the *vis viva* of these'.

In the concluding PART III of the paper immediately preceding the summary, Maxwell addressed the allocation of energy to the rotational motion of 'perfectly elastic bodies of any form'. He argued that his statistical treatment was valid also for the rotational motion, and so he chose $\beta = 2$ and looked to experiment for judgement. This allocation was part of what became known as equipartition of energy. The crucial test was made using the link between β and the ratio γ of the specific heat of a gas at constant pressure to its specific heat at constant volume: $\beta = 2/(3(\gamma - 1))$ For air $\gamma = 1.408$, hence an experimental value for β of 1.634, in conflict with Maxwell's choice of 2. (Alternatively, $\beta = 2$ leads a predicted value for γ of 1.333 in disagreement with experiment.) On this issue, Maxwell concluded his paper: 'Finally, by establishing a necessary relation between the motions of translations and rotation of all particles not spherical, we proved that a system of such particles could not possibly satisfy the known relation between the two specific heats of all gases.'

3.3 Experimental check and some further developments

It was after he moved to the Chair of Natural Philosophy at King's College in London, in 1860, that he confirmed experimentally the counter-intuitive prediction that the viscosity of a gas would be independent of the pressure [22]. The measurements – a joint effort by Maxwell and his wife Katherine – were made between 1863 and 1865 with apparatus, a viscometer, assembled in the attic of Maxwell's London house in Palace Gardens Terrace, in Kensington. In his 1867 paper, Maxwell addressed the issue of the law of force between colliding molecules. His conclusion is best described from his own abstract [23] of the paper: 'the action between the molecules is supposed to be that of bodies repelling each other at a distance, rather than of hard elastic bodies acting by impact; and the law of force is deduced from [his own] experiments on the viscosity of gases to be that of the inverse fifth power of the distance, any other law of force being at variance with the observed fact that the viscosity is proportional to the absolute temperature [22]. In the mathematical application of the theory, it appears that the assumption of this law of force leads to a great simplification of the results, so that the whole subject can be treated in a more general way than has hitherto been done.'

Also in the paper, he re-cast the final formula for thermal conductivity, rearranging the factors so as to eliminate β in favour of γ, and including the coefficient of viscosity explicitly – thereby winning the security of factors whose values were well known from experiment. He also referred, presciently, to the internal motions as 'oscillation, or vibration'. However, he was never able to fully address their significance. The issue was resolved by the early years of the twentieth century. The nuclear atom provided the lead-in to more realistic atomic and molecular physics, spectroscopic data could be related to frequencies – not directly to but not too distantly from the notion of frequencies of vibrations of Maxwell's 'internal motions' – and rapidly developing quantum theory provided overarching support. However, 'despite the problems with the kinetic theory, Maxwell never seriously doubted its validity or that of the velocity distribution he had derived. Amongst the most compelling of his reasons were the results of his studies of the viscosity of gases.' [24].

In his biographical study of Maxwell, C.W.F. Everitt [25] identifies the derivation of the velocity distribution in the 1860 paper as marking 'the beginning of a new epoch in physics.' In justification, he continues 'Statistical methods had long been used for analyzing observations, both in physics and in the social sciences, but Maxwell's idea of describing actual physical processes by a statistical function was an extraordinary novelty'. The distribution of molecular velocities is still known as the Maxwell Distribution. It quickly became key to further work on mainland Europe, by Clausius and Ludwig E. Boltzmann (1844–1906). There soon emerged the Boltzmann Distribution of molecular energies, crucial to the development of statistical thermodynamics.

4 Two laws of thermodynamics: Maxwell, Clausius, Kelvin and Tait

4.1 The First Law

P.G. Tait had been appointed Professor of Natural Philosophy at Edinburgh University – in preference to Maxwell – in 1860. There he followed assiduously developments in thermodynamics to keep his lectures on the subject up to date. In his book on thermodynamics, *Sketch of Thermodynamics*, whose first edition appeared in 1868 [2], Tait set out the developments historically. He challenged the attribution made in 1862 by John Tyndall (1820–1893) [26] to J. Robert Mayer (1814–1878) of the introduction in 1842 of the idea of conservation of energy, pointing to a mechanical equivalent of heat [27]. Mayer's intuition was strong but his interpretation of available experimental data was suspect and he made little of any link between heat and motion [28]. On the grounds that Mayer had failed to support his hypothesis satisfactorily, Tait insisted that credit should be given to James P. Joule (1818–1889) who had begun reporting on his measurements in 1843 [29]. Joule improved his experimental techniques systematically to measure the rise in temperature of water stirred by rotating paddles. This provided the direct link between mechanical work producing the motion of the paddles and the heat produced, permitting a value for the mechanical equivalent of heat to be found ever more accurately, as reported to the Royal Society in 1850 [30]. The debate became quite bitter [31].

Thomson's substantial and timely paper published in 1851 [12] contains many references to the work of both Joule and Mayer, and had led him to move on from a mechanical equivalent of heat to identify a thermodynamic property of substances, initially called mechanical energy, e, later changed to intrinsic energy and finally to internal energy, denoted by U. A sample heated externally and, whilst expanding, doing mechanical work on its surroundings would experience a change in its internal energy equal to the difference of the energies calculated from the heat and the work, although neither heat nor work on their own could serve as a thermodynamic property.

The year before, Clausius [32] had arrived at essentially the same thermodynamic property, which he represented by (thus introducing the symbol) U, in a treatment which avoided the use of the concept of energy. Instead, simplifying the more detailed discussion by Cropper [33], he argued that the effect of adding heat to a gas is the performance of external work done, if any, and an increase in the value of the property U which could be attributed to an increase in 'sensible heat' together with work done within the gas itself. Represent the net heat input *to* and net external work done *by* a system by ΔQ and ΔW, respectively. The accompanying change ΔU in U is found from $\Delta Q = \Delta U + \Delta W$. By 1865, the status of the concept of energy, and its conservation, having been established, Clausius accepted U as an energy and made his succinct statement of the First Law of Thermodynamics: 'The energy of the universe is constant' [34].

4.2 The Second Law

4.2.1 Background to Kelvin's approach

Tait was well aware of Thomson's abilities. They had collaborated on the writing of the famous 'Thomson and Tait', their *Treatise on Natural Philosophy* [35]. They corresponded and met frequently, not least in connection with the affairs of the Royal Society of Edinburgh. Tait was the General Secretary from 1879 to 1901, and Thomson served as President several times. Tait had followed Thomson's researches on the laws of thermodynamics, and they both acknowledged the clarifications from the contribution [36] in 1850 of W.J. Macquorn Rankine (1820–1872), Professor of Civil Engineering at Glasgow University, though Thomson did not show much enthusiasm [37] for Rankine's speculative modelling of molecular properties in terms of vortices.

Already in 1850 [32], Clausius had sought, independently, to extend conclusions being drawn from the description by N.L. Sadi Carnot (1796–1832) in 1824 [38], rescued by B.P. Emile Clapeyron (1799–1864) 10 years later [39], of an idealised version of a processes common to many heat engines which became known as the Carnot Cycle. This was before the acceptance of the First Law and so indestructible heat, as described in the caloric theory, played a dominant guiding role. The cycle, fully discussed in other chapters, is a closed cycle of four reversible stages, alternately isothermal and adiabatic. A working substance gas absorbed heat Q isothermally at the higher temperature t_1 of a constant temperature heat source, expanding to do work externally. The expansion continued with the gas thermally insulated, doing work adiabatically. Two compression stages followed, with work now being done *on* the gas. The first was isothermal at the lower temperature t_2 of a constant temperature heat sink, to which *all* the heat Q was released. The final compression, with the gas again thermally insulated, returned the gas to its initial state. The net amount of work delivered was associated with the complete throughput of the heat Q, to comply with the indestructibility of caloric.

Further restricted to an infinitesimal temperature difference between source and sink, this reversible cycle would provide the maximum possible work output for a given heat input at the higher temperature. This upper limit on the efficiency of heat engines pointed to a more general law. Thomson [11] had referred to Carnot's admission that the foundations of his theory 'would require a most attentive examination'. Clausius [32] pointed out that accepting that work output was necessarily accompanied by heat flow was compatible with some of the heat input being transformed into the output work without the need to move to a new basic theory of heat. Thomson collated the various such contributions in his 1851 paper [12], in which he introduced the thermodynamic function mechanical energy, e. In the following year, he distinguished reversible from irreversible processes: in the latter, there can be no complete restoration of mechanical energy – part of it having been dissipated. This applied in particular to the conduction of heat in which a transformation of energy leads to part of it becoming irrecoverable. He therefore proposed, in 1852, an overarching law that 'there is at present in the material world a universal tendency to the dissipation of the mechanical energy' [13].

4.2.2 *Alternative approach of Clausius*

By 1854, Clausius – also seeking a general principle – had analysed an enhanced Carnot cycle [40]. The return adiabatic did not restore the working substance to its initial state. Instead it was followed by a reversible isothermal expansion having heat input q at the constant temperature t of a third heat source, with $t > t_1$, and the cycle was closed by a final reversible adiabatic expansion with the temperature decreasing to t_1. The cycle thus comprised two identifiable transformations. One was of heat Q at t_1 to the same amount but now at t_2. This was the necessary accompaniment to the other transformation of the net heat input, just q, to the external work w delivered. Clausius's analysis was based on an 'equivalence value', in terms of heat flow and temperature, for each of these two mutually compatible transformations. Detailed analysis led to $Q(1/T_2 - 1/T_1) - q/T = 0$, where the three temperatures are now represented by their absolute temperature scale values. This he generalised firstly to any number (including infinitesimal) of reversible heat flows dQ at temperatures T: $\int dQ/T = 0$. In a later paper [34], Clausius continued the interpretation introducing a thermodynamic property entropy, denoted S, with $dQ/T = dS$. Allowing for irreversible processes as well, he extended the generalisation to $\int dS \geq 0$, with the equality valid for reversible processes only. Matching the brevity of his statement there of the First Law, he pronounced the Second Law of Thermodynamics in the form 'the entropy of the universe tends to a maximum'. Traced back to its origins – allowed processes identified by changes (if any) in thermodynamic properties – it correlates with Thomson's more easily visualised conclusion of 1852 in terms of dissipation of energy.

4.2.3 *A merging of interpretations*

Maxwell's letter to Tait on 1 December 1873 showed that there was still then some confusion about how entropy was related to available and unavailable energy [41]. Initially, Tait inter preted entropy as available energy, possibly confused by the over-use of the word transformation. Much earlier, in a letter dated 6 December 1867, Tait had sought Maxwell's opinion on his own book, the *Sketch of Thermodynamics* to which Maxwell replied on 11 December 1867 [42]. Later, Maxwell erred by following Tait's lead in the treatment of entropy in the first edition, in 1871, of his own book [5]. It was quickly corrected in later editions.

In particular, Tait had noticed that expressions structurally similar to Clausius's were prominent in Thomson's researches. He insisted on giving Thomson priority over Clausius for the introduction of a thermodynamic function – Clausius's entropy – to which Clausius objected, resulting in a second controversy [43, 44]. When he had had time to consider the matter, Maxwell was more understanding of Clausius's attitude, pointing out in a letter to Tait on 13 October 1876 [45] that in 1854 Rankine [46] had arrived at an expression which he called the thermodynamic function, which was indistinguishable from Clausius's $\int dQ/T$. In that letter Maxwell wrote: 'There are many things in T[homson] which are equivalent to this, because T has worked at the subject and worked correctly, and all mathematical truth is one, but you cannot expect C[lausius] to see this unless it stands very plain in print. In short Rankines statements are *identical* with those of C but T's are only *equivalent*.'

Conventional statements of the Second Law in introductory thermodynamics rule out processes which Thomson and Clausius realised did not happen in isolation. The modern wording of two [47], which can be shown to be equivalent, are the 'Clausius' version 'It is impossible to construct a device that, operating in a cycle, produces no effect other than the transfer of heat from a colder to a hotter body' and the 'Kelvin-Planck' version 'It is impossible to construct a device that, operating in a cycle, will produce no effect other than the extraction of heat from a single body at a uniform temperature and the performance of an equivalent amount of work.'

4.3 Impact of kinetic theory and Maxwell's Demon

4.3.1 The statistical approach of Maxwell

Maxwell was led to suggest that his statistical kinetic theory would fit harmoniously with the Second Law of Thermodynamics. He had to show that the assumption of identifiable velocities of all molecules at all instants could lead to a contradiction of the Second Law. Inverting, the validity of the Second Law would rule out the option of such detailed knowledge particle by particle, but not the option of a statistical specification. In that sense, the Second Law of Thermodynamics had to be regarded as statistical.

In treatments of the Second Law, one thermodynamic system has a heat source and a heat sink defining the ends of a volume filled by a sample gas. They are key to heat in and heat out, compatible with a steady temperature gradient between source and sink. Many observations all indicated that there would be a transfer of heat down the temperature gradient from the source to the sink. In his letter to Tait on 11 December 1867 [42], Maxwell simplified the system to two samples, A and B, containing the same number of molecules of the same gas at different temperatures, with A hotter than B. The samples were in contiguous vessels of the same size, completely closed with a diaphragm forming the side in common. A sliding door covered a hole in the diaphragm. So his system had at the outset a temperature difference, rather than a temperature gradient. He then proceeded 'To pick a hole – say in the 2nd law of thermodynamics, that if two things are in contact the hotter cannot take heat from the colder without external agency.'

He illustrated his reservations to Tait (Fig. 2) by introducing an agency to control energy changes accompanying the movement of molecules – through the hole in the diaphragm – from each vessel to the other. 'Now conceive a finite being who knows the paths and velocities of all the molecules by simple inspection but who can do no work, except to open and close a hole in the diaphragm, by means of a slide without mass. Let him first observe the molecules in A and when he sees one coming the square of whose velocity is less than the mean sq. vel. of the molecules in B let him open the hole & let it go into B. Next let him watch for a molecule in B the square of whose velocity is greater than the mean sq. vel. in A and when it comes to the hole let him draw the slide & let it go into A, keeping the slide shut for all other molecules. Then the number of molecules in A & B are the same as at first but the energy in A is increased and that in B diminished that is the hot system has got hotter and the cold colder yet no work has been done, only the intelligence of a very observant and neat fingered being has been employed.'

Maxwell then summarised his argument, shifting the scenario slightly: 'Or in short if heat is the motion of finite portions of matter and if we can apply tools to such portions of matter so as to deal with them separately then we can take advantage of the different motion of different portions to restore a uniformly hot system to unequal temperatures or to motions of large masses. Only we can't, not being clever enough.'

Tait added a pencilled postscript when he forwarded the letter to Thomson with the request 'to say what occurs to you about the enclosed *which return* speedily.' [48]. Thomson's annotation to the letter was 'Very good. Another way is to reverse the motion of every particle in the universe and preside over the unstable motion thus produced'. He was pointing to the feature of classical mechanics that the solutions to the equations determining the motion of a particle are equally valid for time t decreasing as well as, normally, increasing. He would return to this, and his alert on unstable motion, in a paper to the Royal Society of Edinburgh in 1874.

Figure 2: Peter Guthrie Tait, age 32 (left), courtesy of Scottish National Portrait Gallery, and James Clerk Maxwell, age 30 (right), courtesy of Master and Fellows of Trinity College Cambridge.

Almost immediately after introducing this 'being' to Tait in 1867, Maxwell referred to its role, in severe summary, at the end of a letter on 16 January 1868 to Thomson [49]. Already in 1852 Thomson had pointed out that in natural processes there was a dissipation of energy, which was not recoverable in forwards moving time. Maxwell set out the scenario. 'I believe that if any body consists of finite molecules either solid or fluid knocking about, then this body when of uniform temperature consists of molecules going at different velocities. See my Dynamical Theory of Gases & the Laws of Collision.' There was no mention of diaphragm, hole or sliding door, but the starting point was a system in thermodynamic equilibrium. He continued: 'Now let an intelligent but finite and energy conserving being be appointed pointsman to the flying molecules and when he sees a swift one let him turn it to the place of the swift and when he sees a slow one let him turn it to go among the slow. Then without expenditure of work he will have made one part of the body hotter & the other colder. Hence energy need not always be dizzy pated as in the present wasteful world.' Paraphrasing, the recovery of a prior thermodynamic state in forwards moving time could be done by a sufficiently well apprised being ready to act on molecules individually.

4.3.2 A merging of views and a glance forward

In 1874, Thomson [50] returned to a discussion of dissipation of energy supported by the kinetic theory of gases. His starting point was the evidence that a thermally insulated bar of metal, hot at one end and cold at the other, would eventually assume a uniform temperature by conduction of heat. A spontaneous recovery to the original temperature distribution, with the potential of availability for useful work, is never observed. To illustrate this, he chose the simpler system of a gas consisting of a large number of identical molecules in a leak-proof, thermally insulated vessel. Thomson described a thought experiment in which a large number of Maxwell Demons could control the difference in temperature between two halves of the vessel.

He put them in charge of contiguous areas whose total flat area divided the volume into two equal parts. He expanded on Maxwell's specification of 'turning' molecules to the places of the swift or the slow by allowing the comparatively massive demons to 'bat' the moving molecules accordingly. For the special case of initially uniform temperature throughout the sample, the demons, whilst remaining in their positions, could make their combined area a boundary between hot and cold halves of the volume. They could do this by controlling, in concert, the allowed passages across the boundary of molecules selected according to their energy. If the demons were then made simultaneously inactive, subsequent intermolecular collisions throughout the whole volume of the gas would produce there an equilibrium temperature in a finite time.

He then considered what would happen if at that point, with the demons still inactive, the *velocities* of all molecules were suddenly reversed. In the same time as before, but now negative going, laws of physics would produce, by 'abstract dynamics', a complete re-tracing in reverse of its path by every molecule and the recovery the temperature difference engineered by the demons – which subsequent intermolecular collisions would eventually remove.

Thomson's preference was for 'physical dynamics', in which the motion of molecules would be conditioned by their interactions with the bulk of the sample whether on their 'outward' or 'return' paths. Recovery of the target temperature difference in the time available could be ruled out. A statistical thread in the argument became necessary, providing a bridge to a proposed feature of equilibrium distributions which he now considered in forwards going time. Short-lived non-equilibrium states could happen and persist for a while, more easily the smaller the number of molecules in the sample, before a return to the equilibrium state would happen. Extending the line of reasoning, the larger the fluctuation, the less likely it is to occur, leading naturally to the need for application of the rules of probability theory. Supported by a few numerical estimates from it, Thomson felt justified in aligning his conclusions with his theory of the dissipation of energy. A reinstatement of the original temperature difference across an isolated metal bar, once it had changed irreversibly to uniform temperature throughout, was not possible unless the isolation was relaxed, permitting input from an 'agency of external sources of heat or cold'.

In his own Theory of Heat [5], Maxwell chose the scenario in his summary to Tait, leading to yet another paraphrase of his standpoint, claiming the second law of thermodynamics to be 'undoubtedly true as long as we can deal with bodies only in mass, and have no power of perceiving or handling the separate molecules of which they are made up.' Around this time, in correspondence with others, Maxwell would point to the constraint on the perfect reversibility of classical dynamics posed by the irreversibility of natural processes [51]. This was one facet of a larger issue which in time also had to be faced when the bridge between classical and statistical mechanics was being built. There the probabilistic arguments [52] can be traced back to those presented by Maxwell in his kinetic theory of gases. The imagery associated with the concept of Maxwell's Demon has been used in many areas of science, extending from biology and psychology to computing and information theory, and to the history of science [53].

5 Conclusion

Maxwell's note to Tait 'Concerning Demons', in 1875 [54] was in the form of a catechism, beginning 'Who gave them this name? Thomson.' In his Lecture in London at the Royal Institution in February 1879 [55], Thomson saw fit to juxtapose again energy dissipation and energy distribution changes that could be wrought whenever required by Maxwell's Demon. However, holding to the theme of the Lecture, he changed the order of the key events from that chosen in his 1874 paper – this time dissipation preceded intervention by the demon.

Quoting from the report of the Lecture: 'Clerk Maxwell's 'demon' is a creature of imagination having certain perfectly well-defined powers of action, purely mechanical in their character, invented to help us to understand the 'Dissipation of Energy' in nature. ... By operating selectively on individual atoms he can reverse the natural dissipation of energy ... [a dissipation which] follows in nature from the fortuitous concourse of atoms.' The Lecture was illustrated by a number of experiments, allowing Thomson to remind the audience that 'The conception of the 'sorting demon' is purely mechanical, and is of great value in purely physical science.'

Combined with his 1874 paper Thomson thus effectively completed his elaboration on the annotation he had made on Maxwell's letter to Tait 12 years before [42]. In doing so, he had rounded off, perhaps inadvertently, an important stage in the development of thermodynamics. Maxwell's contribution to it might have at first appeared maverick-like but it now sat more easily – providing clarification – within classical whilst also indicating the way to statistical thermodynamics.

In November of the same year, James Clerk Maxwell died from stomach cancer.

Appendix. Transcript [3] of words on postcard of Fig. 1

$T\,2\,\mathrm{d}p/\mathrm{d}t$

The el. m. f.^{ce} of a thermoelectric circuit belongs to the whole circuit not to either junction alone. See water pipe analogy ProcRSE 1851 (if the Secretaries have correctly reported the whole proceedings of the meeting).

The more we understand of the potential, pressure, &^{c} of electricity in metals the more perfect is the analogy. You are welcome to the thermo O^{t} copper zinc air I have done with it years ago till I come back to it again.

{On the card Thomson sketched a water pipe as a closed rectangular loop with corners rounded and with 'water' inside. The side on left was labelled 'cold' and the other on the right 'hot'. Inside the loop was written '25 years ago I did not know that $\iint \rho(x,y)\,\mathrm{d}x\mathrm{d}y$ is absurd'. Alongside, Thomson added 'if mean temp^{re} > 4°'. Below it he sketched a narrower loop representing two strips of metal, joined at the ends; the upper one 'copper' and the lower one 'iron.' The junctions were labelled 'cold' on left and 'hot' on right. Limit on mean temperature indicated by ' " " 289°'} Below was added 'The current is against hands of watch'.

Oil of turpentine (as dry as you can get) and melted paraffin may be substituted for the air. I am doing so not very successfully now, and did so many times and years back.

You said that Clausius made proportionality of current to external EMF in electrolysis which was what I objected to.

{In what little space was left in the left hand margin Thomson wrote (at right angles to the main text): 'I have no doubt of the diffusion theory of currents through electrolytes in infin. EMF'. Likewise in right hand margin: 'Can you come on Frid. or Sat. or if not, when?'}

Bibliography

Everitt, C.W.F., *James Clerk Maxwell: Physicist and Natural Philosopher*, Charles Scribner's Sons: New York, 1974.

Harman, P.M., *Energy, Force, and Matter: The Conceptual Development of Nineteenth-Century Physics*, Cambridge University Press: Cambridge, 1982.

Knott, K.G., *The Life and Scientific Work of Peter Guthrie Tait*, Cambridge University Press: Cambridge, 1911.

Mahon, B., *The Man Who Changed Everything, the Life of James Clerk Maxwell*, John Wiley & Sons: Chichester, 2004.

Tolstoy, I., *James Clerk Maxwell*, Canongate Publishing Ltd: Edinburgh, 1981.

References

[1] Harman, P.M. (ed), *The Scientific Letters and Papers of James Clerk Maxwell, Vols I (1846–1862), II (1862–1873) and III (1874–1879)*, Cambridge University Press: Cambridge, 1990 (Vol. I), 1995 (Vol. II), 2002 (Vol. III).

[2] Tait P.G., *Sketch of Thermodynamics*, Edmonston and Douglas: Edinburgh, p. 91, 1868.

[3] Harman, P.M. (ed), *The Scientific Letters and Papers of James Clerk Maxwell, Vol. II*, No. 393, Postcard to Peter Guthrie Tait, 7 December 1871, pp. 692–693, footnote (7), Cambridge University Press: Cambridge, 1995.

[4] Harman, P.M. (ed), *The Scientific Letters and Papers of James Clerk Maxwell, Vol. II*, No. 377, Manuscript on the history of the kinetic theory of gases: notes for William Thomson, *circa* summer 1871, pp. 654–660, Cambridge University Press: Cambridge, 1995.

[5] Maxwell, J.C., *Theory of Heat*, Longmans, Green, and Co.: London, 1871.

[6] Harman, P.M. (ed), *The Scientific Letters and Papers of James Clerk Maxwell, Vol. I*, No. 45, Letter to William Thomson, 20 February 1854, pp. 237–238, Cambridge University Press: Cambridge, 1990.

[7] Harman, P.M. (ed), *The Scientific Letters and Papers of James Clerk Maxwell, Vol. I*, No. 66, Letter to William Thomson, 15 May 1855, pp. 305–313, Cambridge University Press: Cambridge, 1990.

[8] Fourier, J.P.J., *Théorie Analytique de la Chaleur*, F. Didot: Paris, 1822.

[9] Article on Vis viva in Wikipedia, https://en.wikipedia.org/wiki/Vis_viva.

[10] Mahon, B., *The Man Who Changed Everything, the Life of James Clerk Maxwell*, John Wiley & Sons: Chichester, p. 29, 2004.

[11] Thomson, W., An account of Carnot's theory of the motive power of heat; with numerical results deduced from Regnault's experiments on steam, *Trans. Roy. Soc. Edinb.*, **16**, pp. 541–575 (1849); reprinted in *Math. Phys. Papers* Vol. I, No. XLI, pp. 113–155, Cambridge University Press: Cambridge, 1882.

[12] Thomson, W., On the dynamical theory of heat, with numerical results deduced from Mr Joule's equivalent of a thermal unit, and M. Regnault's observations on steam, *Trans. Roy. Soc. Edinb.*, **20**, pp. 261–288, 289–298, 475–482 (1851); reprinted in *Math. Phys. Papers*, Vol. I, No. XLVIII, pp. 174–232, Cambridge University Press: Cambridge, 1882.

[13] Thomson, W., On a universal tendency in nature to the dissipation of mechanical energy, *Proc. Roy. Soc. Edinb.*, **3**, pp. 139–142 (1852); reprinted in *Math. Phys. Papers*, Vol. I, No. LIX, pp. 511–514, Cambridge University Press: Cambridge, 1882.

[14] Maxwell, J.C., *On the Stability of the Motion of Saturn's Rings*, Macmillan and Co.: Cambridge, 1859.

[15] Tolstoy, I., *James Clerk Maxwell*, Canongate Publishing Ltd: Edinburgh, pp. 83–84, 1981.

[16] Harman, P.M. (ed), *The Scientific Letters and Papers of James Clerk Maxwell*, Vol. I, No. 134, Letter to William Thomson, 14 November 1857, esp. footnote (17) p. 555, pp. 553–556, Cambridge University Press: Cambridge, 1990.

[17] Clausius, R.J.E., Ueber die mittlere Länge der Wege, welche bei der Molecularbewegung gasförmiger Körper von den einzelnen Molecülen zurückgelegt werden; nebst einigen anderen Bemerkungen über die mechanische Wärmetheorie. *Ann. Phys.*, **105**, pp. 239–258,

1858. Translated as On the mean length of the paths described by the separate molecules of gaseous bodies on the occurrence of molecular motion: together with some other remarks upon the mechanical theory of heat. *Phil. Mag.*, ser. 4, **17**, pp. 81–91, 1859.

[18] Maxwell, J.C., Illustrations of the dynamical theory of gases. *Phil. Mag., Ser. 4*, **19**, pp. 19–32, 1860 (part I) and **20**, pp. 21–37, 1860 (parts II and III).

[19] Clausius, R.J.E., Ueber die Wärmeleitung gasförmiger Körper. *Ann. Phys.*, **115**, pp. 1–56, 1862. Translated as On the conduction of heat by gases, *Phil. Mag., Ser. 4*, **23**, pp. 417–435, 512–534, 1862.

[20] Harman, P.M. (ed), *The Scientific Letters and Papers of James Clerk Maxwell, Vol. II*, No. 207, Draft paper on the conduction of heat in gases, *circa* spring 1863, pp. 72–85, Cambridge University Press: Cambridge, 1995.

[21] Maxwell, J.C., On the dynamical theory of gases. *Phil. Trans. Roy. Soc.*, **157**, pp. 49–88, 1867.

[22] Maxwell, J.C., On the viscosity or internal friction of air and other gases. *Phil. Trans. Roy. Soc.*, **156**, pp. 249–268, 1866.

[23] Maxwell, J.C., On the dynamical theory of gases. *Proc. Roy. Soc.*, **15**, pp. 167–171, 1866 (abstract of [21]). Also in Harman, P.M. (ed), *The Scientific Letters and Papers of James Clerk Maxwell, Vol. II*, No. 263, Abstract of paper on the dynamical theory of gases, 16 May 1866, pp. 279–284, Cambridge University Press: Cambridge, 1995.

[24] Longair, M.S., *Theoretical Concepts in Physics: An Alternative View of Theoretical Reasoning in Physics*, 2nd edn, Cambridge University Press: Cambridge, p. 263, 2003.

[25] Everitt, C.W.F., *James Clerk Maxwell: Physicist and Natural Philosopher*, Charles Scribner's Sons: New York, p. 135, 1974.

[26] Tyndall, J., On force, *Proc. Roy. Institution*, **III**, pp. 527–536, 1862.

[27] Mayer, J.R., Bemerkungen über die Kräfte der unbelbten Natur, *Annalen der Chemie und Pharmacie*, **43**, 233–239, 1842. Translated as Remarks on the forces of inorganic nature in *Phil. Mag., Ser.4*, **24**, pp. 371–377, 1862.

[28] Harman, P.M., *Energy, Force, and Matter: the Conceptual Development of Nineteenth-Century Physics*, Cambridge University Press: Cambridge, pp. 61–64, 1982.

[29] Joule, J.P., On the calorific effects of magneto-electricity, and on the mechanical value of heat, *Phil. Mag., Ser. 3*, **23**, pp. 263–276, 347–355, 435–443, 1843.

[30] Joule, J.P., On the mechanical equivalent of heat. *Phil. Trans. Roy. Soc.*, **140**, pp. 61–82, 1850.

[31] Knott, C.G., *The Life and Scientific Work of Peter Guthrie Tait*, Cambridge University Press: Cambridge, pp. 209–212, 1911.

[32] Clausius, R.J.E., Ueber die bewegende Kraft der Wärme und die Gesetze, welche sich daraus für die Wärmelehre selbst ableiten lassen. *Ann. Phys.*, **79**, pp. 368–397, 500–524, 1850. Translated as on the moving force of heat and the laws of heat which may be deduced therefrom. *Phil. Mag., Ser. 4*, **2**, pp. 1–21, 102–119, 1851.

[33] Cropper, W.H., *Great Physicists: The Life and Times of Leading Physicists from Galileo to Hawking*, Oxford University Press: Oxford, pp. 95–97, 2001.

[34] Clausius, R.J.E., Ueber verschiedene für die Anwendung bequeme Formen der Hauptgleichungen der mechanischen Wärmetheorie. *Ann. Phys.*, **125**, pp. 353–400, 1865. Translated as on several convenient forms of the fundamental equations of the mechanical theory of heat in Clausius, R.J.E., *The Mechanical Theory of Heat with Its Applications to the Steam Engine and to the Physical Properties of Bodies*, Hirst, T.A. (ed), John van Voorst: London, pp. 327–364 (Ninth Memoir), 1867.

[35] Thomson, W. &Tait, P.G., *Treatise on Natural Philosophy*, Clarendon Press: Oxford, 1867.

[36] Rankine, W.J.M., On the mechanical action of heat, especially in gases and vapours, *Trans. Roy. Soc. Edinb.*, **20**, pp. 147–190, 1852. Abstract in *Proc. Roy. Soc. Edinb.*, **2**, pp. 275–288,

1850.
[37] Smith, C. & Wise, M.N., *Energy and Empire: A Biographical Study of Lord Kelvin*, Cambridge University Press: Cambridge, pp. 319–323, 1989.
[38] Carnot, N.L.S., *Réflexions sur la puissance motrice du feu, et sur les machines propres à développer cette puissance*, Bachelier: Paris, 1824.
[39] Clapeyron, B.P.E., Mémoire sur la puissance motrice de la chaleur. *Journal de l'École Polytechnique, cahier 23*, **14**, pp. 153–190, 1834.
[40] Clausius, R.J.E., Ueber eine veränderte Form des zweiten Hauptsatzes der mechanischen Wärmetheorie. *Ann. Phys.*, **93**, 481–506, 1854. Translated as on a modified form of the second fundamental theorem in the mechanical theory of heat, *Phil. Mag. Ser. 4*, **12,** pp. 81–98 1856.
[41] Harman, P.M. (ed), *The Scientific Letters and Papers of James Clerk Maxwell, Vol. II*, No. 483, Letter to Peter Guthrie Tait, 1 December 1873, pp. 944–948, Cambridge University Press: Cambridge, 1995.
[42] Harman, P.M. (ed), *The Scientific Letters and Papers of James Clerk Maxwell, Vol. II*, No. 277, Letter to Peter Guthrie Tait, 11 December 1867, pp. 328–334, Cambridge University Press: Cambridge, 1995.
[43] Knott, C.G., *The Life and Scientific Work of Peter Guthrie Tait*, Cambridge University Press: Cambridge, pp. 218–226, 1911.
[44] Smith, C., *The Science of Energy: A Cultural History of Energy Physics in Victorian Britain*, The Athlone Press: London, pp. 256–258, 1998.
[45] Harman, P.M. (ed), *The Scientific Letters and Papers of James Clerk Maxwell*, Vol. III, No. 623, Letter to Peter Guthrie Tait, 13 October 1876, pp. 397–401, Cambridge University Press: Cambridge, 2002.
[46] Rankine, W.J.M., On the geometrical representation of the expansive action of heat, and the theory of thermo-dynamic engines. *Phil. Trans. Roy. Soc.*, **144**, pp. 115–175, 1854.
[47] Finn, C.B.P., *Thermal Physics*, 2nd edn, Nelson Thornes: Cheltenham, pp. 54–59, 2001.
[48] Smith, C., *The Science of Energy: A Cultural History of Energy Physics in Victorian Britain*, Athlone Press: London, p. 251, 1998.
[49] Harman, P.M. (ed), *The Scientific Letters and Papers of James Clerk Maxwell, Vol. III*, No. 279A, Letter to William Thomson, 16 January 1868, pp. 877–882, Cambridge University Press: Cambridge, 2002.
[50] Thomson, W., The kinetic theory of the dissipation of energy. *Nature*, **9**, pp. 441–444, 1874, also *Proc. Roy. Soc. Edinb.*, **8**, pp. 325–334, 1875 (read 16 February 1874); reprinted in *Math. Phys. Papers*, Vol. V, No. 86, pp. 11–20, Cambridge University Press: Cambridge, 1911.
[51] Harman, P.M. (ed), *The Scientific Letters and Papers of James Clerk Maxwell, Vol. II*, 'The paradoxes of dynamics and thermodynamics', pp. 16–20, Cambridge University Press: Cambridge, 1995.
[52] Ben-Naim, A., *Entropy Demystified: The Second Law Reduced to Plain Common Sense*, World Scientific Publishing Co: Singapore, 2007.
[53] Leff, H.S. & Rex, A.F. (eds), *Maxwell's Demon 2: Entropy, Classical and Quantum Information, Computing*, Institute of Physics Publishing: Bristol, 2003.
[54] Harman, P.M. (ed), *The Scientific Letters and Papers of James Clerk Maxwell, Vol. III*, No. 547, Note to Tait 'Concerning Demons', *circa* 1875, pp. 185–187, Cambridge University Press: Cambridge, 2002.
[55] Thomson, W., The Sorting Demon of Maxwell, *Proc. Roy. Institution*, **IX**, pp. 113–114, 1879; reprinted in *Math. Phys. Papers*, **V(87)**, pp. 21–23, Cambridge University Press: Cambridge, 1911.

Chapter 6

Sir William Thomson, Baron Kelvin of Largs and the Institution of Engineering and Technology

S. Hale
Institution of Engineering and Technology, United Kingdom.

Abstract

Sir William Thomson, Baron Kelvin of Largs, was a member of the Society of Telegraph Engineers and the Institution of Electrical Engineers (the precursors of the current Institution of Engineering and Technology) for 36 years. Elected as the President three times, Lord Kelvin's involvement with the institution has had a lasting effect on the Institution of Engineering and Technology (IET). His bust (Fig. 1) is displayed at the IET's London headquarters at Savoy Place and his portrait hangs in the lecture theatre alongside those of other engineering luminaries such as Alessandro Volta and Michael Faraday. Every year, the IET also hosts a lecture in memory of Kelvin. This chapter will explore the background to the IET, the history of Kelvin's relationship with the Institution and the impact that he had upon the Institution as a whole.

1 Introduction

The history of the IET dates back to the late nineteenth century. It was founded on 17 May 1871 as the Society of Telegraph Engineers (STE), in response to the rapid development of a new communications technology, electrical telegraphy.

1.1 Early days of telegraphy

Before the advent of electrical telegraphy, transmitting communications over long distances without the use of written correspondence relied on visual aids such as fire beacons, smoke signals and flags, and later, mechanised systems such as the semaphore system created by Claude Chappé in 1793 [1]. The idea of using electricity to send messages was first proposed in the 1750s [2] but it was not until the beginning of the nineteenth century that experiments began in earnest. Early electrical telegraph systems made use of the recent invention of the Voltaic pile, such as the electrochemical device invented by the German physician Samuel Thomas Sömmerring in 1809 [3], while Sir Francis Ronalds, a British inventor, used static electricity to create an electrical version of Chappé's semaphore system in his garden in 1816 [4].

In 1820, following Hans Christian Oersted's discovery that a magnetic needle will be deflected by an electric current, André-Marie Ampère proposed an electromagnetic telegraph system.

One of the first people to successfully put this theory into practice was the Russian Baron Pawel Lwowitsch Schilling, who had previously assisted Sömmerring with his electrochemical telegraph in 1809 [5]. Schilling's operational electromagnetic telegraph was exhibited in Europe and Asia in 1832 and attracted much positive attention [6].

1.2 The growth of telegraphy

Following Schilling's achievement, the concept of electrical telegraphy as a reliable communications tool rapidly gathered pace. In 1837, the development of a commercially viable electric telegraph took place on both sides of the Atlantic almost simultaneously, with Samuel Morse in the United States and Charles Wheatstone and William Fothergill Cooke in the United Kingdom creating their own versions [7]. Various institutions in Britain picked up on the utility of the telegraph, especially the railway companies, and the industry subsequently developed rapidly. The Electric Telegraph Company, created by Cooke, was incorporated by the Act of Parliament as the first large undertaking for the electrical telegraph industry in 1846 and by 1868, there were over 4000 telegraph offices and over 22,000 miles of line in the UK [8]. By 1871, therefore, when the Society of Telegraph Engineers was founded, electrical telegraphy was a well-established method of communication.

Figure 1: Bust of Lord Kelvin in the reception of Savoy Place, London (Copyright of the Institution of Engineering and Technology).

1.3 The foundation of the Society of Telegraph Engineers

The society was founded at a meeting held at two Westminster Chambers, Victoria Street, London, and was established for 'the general advancement of Electrical and Telegraphic Science, and more particularly for facilitating the exchange of information and ideas among its Members' [9].

Anecdotal evidence has it that the seeds of the Society were first sown during a conversation between Edward Graves, a civil engineer, and Charles Webber, a captain involved in telegraph matters, beneath a tree on the Uxbridge-Oxford Road [10], but general consensus agrees that the main person behind the practical formation of the Society was Major Frank Bolton. Bolton had been involved in developing a form of visual telegraphy to be used at night by the Royal Artillery via adapted lights to create 'dots and dashes' [11], but by 1869, after having left the army, he became heavily involved with the preliminary work of the Society. He was one of the eight attendees at the inaugural meeting in 1871 and was appointed Honorary Secretary, a post he held until his death in 1887 [12].

At the end of the first year of the Society's existence, there were over a hundred members [13], many of whom were significant figures in the telegraph industry, including Sir Charles Wheatstone and Charles William Siemens, who was also the Society's first President. The cost of membership was two guineas, and members were elected on the provisos that they had been educated as telegraph engineers, were employed as telegraph engineers, or had significant interest in telegraph engineering [14].

Though its name implied that it was primarily concerned with telegraphy, the Society's underlying interest in electricity in general was acknowledged in its earliest days. Cromwell Fleetwood Varley, one of the founding members, stated at the first Ordinary General Meeting held in February 1872 that 'This Society ... will gradually, by natural selection, develop more into an electrical society than into a society of telegraphy proper' [15]. Accordingly, as the practical applications of electricity expanded beyond that of telegraphy, the Society widened its scope, and so in 1880 became the Society of Telegraph Engineers and Electricians and in 1888 the Institution of Electrical Engineers.

2 Lord Kelvin

One of the persons to witness the advances made within the Society was Lord Kelvin. Christened as William Thomson, he was knighted in 1866 and became known as Lord Kelvin in 1892 when he was given the title of Baron Kelvin of Largs in the County of Ayr, Scotland. As this article features his career up to his death, he will be referred to as Lord Kelvin throughout.

When Lord Kelvin joined the Society of Telegraph Engineers in 1871 he was known as Sir William Thomson. He had become a member of the Society within the first 6 months of its being founded; his original membership form (Fig. 2), which the IET is fortunate enough to have in their Archives, is dated 16 November 1871 (on the reverse) [16]. At this time, he was 47 years old and a well-established member of the scientific and engineering community. His addition to the membership of the Society of Telegraph Engineers undoubtedly strengthened the Society's credentials as a professional organisation.

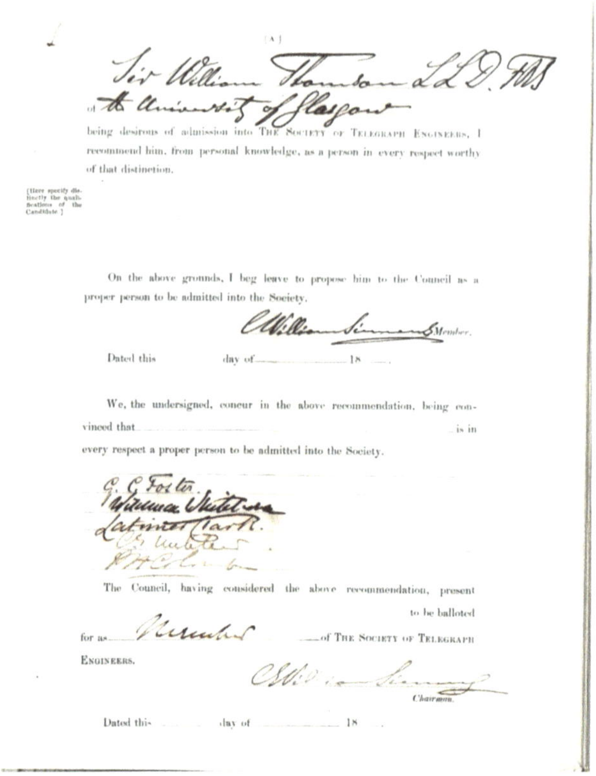

[A]

Sir William Thomson LLD. FRS

of the University of Glasgow

being desirous of admission into THE SOCIETY OF TELEGRAPH ENGINEERS, I recommend him, from personal knowledge, as a person in every respect worthy of that distinction.

[Here specify distinctly the qualifications of the Candidate.]

On the above grounds, I beg leave to propose him to the Council as a proper person to be admitted into the Society.

C William Siemens Member.

Dated this day of 18 .

We, the undersigned, concur in the above recommendation, being convinced that is in every respect a proper person to be admitted into the Society.

G. C. Foster.

Latimer Clark.

The Council, having considered the above recommendation, present to be balloted for as Member of THE SOCIETY OF TELEGRAPH ENGINEERS.

Chairman.

Dated this day of 18 .

Figure 2: Lord Kelvin's application to become a Member of the Society of Telegraph Engineers, 16 November 1871 (Copyright of the Institution of Engineering and Technology).

2.1 Lord Kelvin's early life and career

Lord Kelvin was born in Belfast on 26 June 1824, but from the age of 8 was brought up in Glasgow where his father, Dr James Thomson, was a professor of mathematics at Glasgow University. Kelvin studied at the University of Cambridge, where he graduated as Second Wrangler in the Mathematics Tripos in 1845 [17]. A year later, he returned to Glasgow, where he became the chair of natural philosophy, at the prodigious age of 22, following the death of its previous incumbent, William Meikleham [18].

During the early years of his professorship, Kelvin engaged much of his attention to heat flow and thermodynamics. On realising that the theoretical frameworks of heat and energy could also be applied to electricity and magnetism, Kelvin became interested in the study and practical uses of electricity and in 1853 he formulated the equations for electrical oscillators [19].

Kelvin's burgeoning interest in electricity naturally led him to pursue telegraphy, and particularly, submarine telegraphy. In 1855, he became involved in the laying of telegraph cables across the Atlantic, where he used the inverse-square law to formulate a theory of cable transmission. On the basis of calculations made from this theory, he believed that laying a trans-Atlantic cable might be possible, but only if a heavy conductor and thick insulation was used [20]. Kelvin was invited to accompany a cable-laying expedition by the Atlantic Telegraph Company in 1857.

The cable being used did not conform to Kelvin's specifications, and duly failed, but Kelvin's observations made on this trip allowed him to develop further his ideas about cable laying and resulted in him patenting the instruments he developed to assist in this process, including the marine galvanometer and siphon recorder. Relying on Kelvin's work and inventions, the attempts made to lay the cable across the Atlantic in 1866 were ultimately successful and as a result earned him his knighthood [21].

Submarine telegraphy was not Kelvin's only venture. As a professor of natural philosophy at Glasgow, his interests encompassed a wide range of scientific topics, and it would be impossible to do justice to Kelvin's numerous achievements in this short chapter, but similarly it would be negligent not to mention the areas in which he was most prominent. Developing the first and second laws of thermodynamics was obviously one of his earliest achievements, but Kelvin was also by all accounts an esteemed authority on the science of energy, metrical conventions and electrical standards, calculating the age of the earth, deep-sea sounding and tidal prediction. Alongside all this work, however, Kelvin still found time to devote to electrical engineering, and indeed, even act as the President of the Society of Telegraph Engineers.

2.2 Lord Kelvin and the Society of Telegraph Engineers

Lord Kelvin was first elected as the President of the Society in 1874, just 3 years after becoming a member. He was to be elected twice more during his lifetime: in 1889, the year in which the Society became known as the Institution of Electrical Engineers, and in 1907, the year in which the Institution began taking steps towards making Savoy Place in London its permanent residence.

Lord Kelvin did not limit his association with the Institution to acting as President. His first tenure as President entitled him to be a life-long member of the Council, and he is listed as such in the Institution's journal. In the years between his presidencies, Kelvin was a fully active and committed member of the Institution. Though his home and work life in Glasgow made it difficult for him to attend meetings of the Council and the Ordinary General Meetings of the Institution, which were always held in London, he invariably made the effort to be present. Indeed, while he was President in 1889, some members deemed his ability to seemingly be in two places at once positively miraculous. William Edward Ayrton, during a meeting held 9 January 1890, remarked that 'When one felt certainly that Sir William must be in Glasgow, on entering the Council room at 7 O'clock, one saw him take his seat exactly as if he lived somewhere near Westminster or St. James's Park' [22].

When circumstances permitted him to attend meetings, he often took the opportunity to present his ideas on the papers that had been given. In 1884, William Henry Preece talked to the attendees of the 136th Ordinary General Meeting of the Society about the Electrical Congresses that had been held in Paris since 1881 and the importance of a system of units of measurement for the electrical engineering industry. Kelvin accordingly contributed to the discussion his views on this subject, on which he was a great authority. His presence in turn was regarded as an immense honour to those at the meeting, and President William Grylls Adams expressed his 'great pleasure at having Sir William Thomson with us this evening, and to thank him for his valuable contribution to the discussion' [23]. On the occasions when he was not able to attend, Kelvin often forwarded a note or letter with his comments on a particular issue to the Secretary, to be read out at the meeting in his absence. Thus, in 1903, he communicated a comment on

Alexander Siemens' paper 'Notes on the Metrical System of Weight and Measures', commenting that 'the universal adoption of the French metrical system by electrical engineers and engineers of all classes ... will be a great blessing to every individual person concerned' [24].

The institution frequently followed Kelvin's lead. In the early days, meetings often featured papers by members concerned with practical applications of Kelvin's ideas and inventions regarding submarine telegraphy. Thus, the 1876 volume of the *Journal of the Society of Telegraph Engineers* contains several papers devoted to Kelvin's work, including 'Thomson's Siphon Recorder', by J. A. Ewing, and 'The use of soft iron core in Sir William Thomson's Mirror Galvanometer' by Walter Judd. In later years, as the Institution of Electrical Engineers, the Society became involved in the International Electrical Exhibition at Paris during 1881, in which the latest advances in electrical technology from around the world were presented, as a direct result of Kelvin's association with the International Congress of Electricians [25]. As William Preece put it, in this respect Kelvin 'took us and led us like a gentle shepherd to Paris, and there made us shine before the whole world' [26].

His opinion and approval were also often sought by the institution. In 1896, the newly elected President John Hopkinson implored his fellow electrical engineers to contribute to military service (Britain was involved in the Boer War in the Crimea at the time) and gained Kelvin's support of this cause, who stated 'I cordially approve of your movement to promote the utilisation of the patriotism and abilities of electrical engineers for national defence' [27]. Kelvin's approval clearly held weight, for a year later the Electrical Engineers, Royal Engineers, Volunteers (E.E.R.E.V) was founded.

2.3 Lord Kelvin's later years

In 1899, Lord Kelvin retired from his position as a professor of natural history at the University of Glasgow, having worked there for 53 years. He diligently continued his other work and interests, however, including his involvement with the institution. He continued to attend meetings whenever he could and was delighted to be elected as the President for the third time in November 1907 at the age of 83. Unfortunately, he was unable to give his inaugural address due to his wife being seriously ill, but he sent the institution a letter intimating his gratitude: 'Give the members of the institution my best wishes for all their work, both in the institution and in carrying out its practical objects, and tell them that I hope to be with them before the end of the present Session, and that in the meantime I thank them warmly for their great kindness in allowing me to be their President' [28].

Sadly, this would be the last time Kelvin addressed the Institution. The stress of his wife's illness took a great toll on Kelvin's health, and his own condition rapidly declined: 'Lord Kelvin's misery at her helpless condition was intense ... He wandered distractedly about the corridors of his house unable at last to concentrate his mind on work in hand. A chill seized him and after about a fortnight of prostration he sank slowly and quietly away' [29]. Kelvin died on 14 December 1907, at his house Netherhall in Largs, Ayrshire, in the presence of his wife (who survived her illness), physician and several close friends and relatives [30]. He was buried in the nave of Westminster Abbey on 23 December, next to the tomb of Sir Isaac Newton.

3 The institution after Lord Kelvin

3.1 Public reaction

The public reaction to his death was one of intense mourning, and the national press was flooded with poignant and fond memories of a gentleman who had contributed so much to the development of science and engineering.

Some of these reminiscences were collected in the form of press cuttings by Helen G. Thompson, daughter of Silvanus Phillip Thompson (himself a past president of the IEE and the future biographer of Kelvin) and were pasted into a large scrapbook. This book was later donated to the IET Archives, and it remains an insightful record of how Kelvin was viewed by the public. There are many accounts of his life works, and images abound of a staid old gentleman with an impressive beard, but one cutting, apparently written by an old student of Kelvin's classes in Glasgow, paints a refreshingly endearing portrait of the eminent Victorian scientist. In it, the anonymous student describes how Kelvin used to entertain his students with his ability to contrive a melody using only a piece of wire and a cello bow. The student continues: 'But his greatest feat was the playing of a simple tune on a huge trombone! Conceive, if you can, a learned professor of silvered locks and beard, dressed in the black robes of his office and surrounded by scientific instruments, slowly squeezing from a giant trombone some familiar air, and you will understand why I have said that Lord Kelvin often made too strong appeal to our sense of the humorous' [31].

3.2 The institution's reaction

The institution's reaction to Kelvin's death was no less ardent than that of the general public. The Ordinary General Meeting of the Council, scheduled for 19 December, just 2 days after Kelvin died, was suspended and instead devoted entirely to discussing the ramifications of his death for the institution. He had been a much-loved figure within the institution, and members lamented his death keenly, with Sir William Preece undoubtedly encompassing the feelings of many others when he declared 'I will simply say that, as a man, while he lived I loved him, and now that he is dead I revere him. During the short remainder of my very long experience as an electrical engineer I will take every chance and opportunity of glorifying [him]' [32]. Accordingly, the rest of the Council were adamant that his legacy should be appropriately remembered, and so a Kelvin Memorial Committee was set up to investigate the ways to honour his memory. On 20 February 1908, they submitted a report recommending that the Council 'establish a Kelvin Lectureship, and that a Kelvin Lecture be delivered annually' [33].

3.3 The Kelvin Lecture

Thus, the first Kelvin Lecture was delivered by the past President Silvanus P. Thompson on 30 April 1908, on 'The Life and Work of Lord Kelvin'. In this, he gave an eloquent discourse on Kelvin's long and productive career, as well as some interesting anecdotes that revealed the personal side of his life. This obituary would later become the basis of Thompson's two volume biography of Kelvin. The Kelvin lectures that followed this focused almost exclusively on aspects of Kelvin's work, though from the 1920s onwards the subjects of the lectures became more wide ranging, reflecting the broad scope of Kelvin's own interests.

3.4 Further developments

With the death of Lord Kelvin in 1907, the institution had lost one of their most revered members; but rather than see this as a reason to falter, the Institution entered a new phase in its development. Thus, just a year after Kelvin's death, the IEE purchased the lease on a building in central London that is now the central residence of the institution in the UK, Savoy Place. Originally built in 1889 as a joint Examination Hall for the Royal College of Physicians and the Royal College of Surgeons, and on the site of the ancient manor of the Savoy, Savoy Place became the London home of the institution.

The Institution spent £20,000 to adapt and extend its new residences [34]. The original work can still be seen today, although many rooms have since been updated for modern use. With this prestigious new base to work from, the institution set about establishing themselves as one of the Britain's foremost engineering organisations. This was fulfilled when the institution was granted a Royal Charter of Incorporation in August 1921 [35].

The institution's development continued throughout the twentieth century, achieving charitable status in 1963, and going on to merge and join with a number of similar institutions, such as the Institution of Electronics and Radio Engineers, and the Institution of Manufacturing Engineers. The most recent joining was in 2006 with the Institution of Incorporated Engineers, which led to the creation of a new body, the current Institution of Engineering and Technology.

Conclusion

Despite the numerous institutional changes that have taken place over the years, the IET remains proud of its history and connections to Lord Kelvin. It is perhaps fitting that an organisation which has grown from being a small specialist society to one of the largest engineering institutions in the world, should have such strong links with a man whose interests in engineering ranged far beyond that of electrical telegraphy. The Kelvin lecture is still held annually and is part of the IET's Prestige Lecture series, which includes talks and events named in memory of such esteemed scientists as Sir Edward Appleton and Alan Turing. The legacy of Kelvin's association with the institution, however, can perhaps best be summed up by quoting Silvanus P. Thompson, who concluded the first Kelvin lecture thus: 'We of this Institution may well be proud of him – proud that he was one of our first members, that he was thrice our President, and that as our President he died. We shall not look upon his like again' [36].

References

[1] Beauchamp, K., *History of Telegraphy*, The Institution of Electrical Engineers: London, p. 6, 1999.
[2] Beauchamp, p. 20.
[3] Beauchamp, p. 22–23.
[4] Beauchamp, p. 21.

[5] Beauchamp, p. 25.
[6] Black, R.M., *The History of Electric Wires and Cables*, Peter Peregrinus Ltd: London, pp. 2–3, 1983.
[7] MacLaren, M., *The Rise of the Electrical Industry During the Nineteenth Century*, Princeton University Press: Princeton, pp. 40–44, 1943.
[8] Reader, W.J., *A History of the Institution of Electrical Engineers 1871–1971*, Peter Peregrinus Ltd: London, p. 6, 1987.
[9] Minutes of the first meeting of the Council of the Society of Telegraph Engineers, 17 May 1871, *Signed Minutes of Council Meetings Society of Telegraph Engineers*, 1871–1877, IET Archives reference: IET/ORG/2/1/1.
[10] Reader, p. 18.
[11] Reader, p. 19.
[12] Reader, p. 26.
[13] *The Institution of Electrical Engineers: Founded 1871*, Unwin Brothers: London, p. 15.
[14] Reader, pp. 21–22.
[15] Varley, C.F., Presidential inaugural address. *Journal of the Society of Telegraph Engineers*, **1**, p. 34, 1872.
[16] *Original Membership Form of Sir William Thomson*, IET Archives reference: IET/DEP/9/1/1/105.
[17] Thompson, S.P., *The Life of William Thomson, Baron Kelvin of Largs* (1), Macmillan and Co. Ltd: London, p. 97, 1910.
[18] Wilson, J.P., The life and work of Lord Kelvin, *24th IEE Weekend Meeting on History of Electrical Engineering*, University of York, 5–7 July 1996, IEE: London, p. 2, 1996.
[19] Wilson, pp. 3–4.
[20] Wilson, p. 4.
[21] Wilson, p. 4–5.
[22] Ayrton, W.E., Vote of thanks to retiring President. *Journal of the Institution of Electrical Engineers*, **29**, p. 9, 1890.
[23] Grylls Adams, W., Discussion. *Journal of the Society of Telegraph Engineers*, **13**, p. 388, 1884.
[24] Kelvin, L., Discussion. *Journal of the Institution of Electrical Engineers*, **32**, p. 307, 1903.
[25] Bowers, B., *A History of Electric Light and Power*, Peter Peregrinus Ltd: London, p. 203, 1982.
[26] Preece, W., Vote of thanks to the retiring President. *Journal of the Institution of Electrical Engineers*, **19**, p. 7, 1890.
[27] Hopkinson, J., President's remarks. *Journal of the Institution of Electrical Engineers*, **25**, p. 5, 1896.
[28] Kelvin, Lord, President's Remarks. *Journal of the Institution of Electrical Engineers*, **40**, p. 3, 1907–1908.
[29] Thompson, S.P., The Kelvin lecture: the life of Lord Kelvin. *Journal of the Institution of Electrical Engineers*, **41**, p. 422, 1908.
[30] Thompson, S.P., *The Life of William Thomson, Baron Kelvin of Largs* (2), Macmillan and Co. Ltd.: London, p. 1209, 1910.
[31] *Lord Kelvin Scrapbook*, IET Archives reference: SC MSS 4/4.
[32] Preece, W., The Kelvin lecture: the life of Lord Kelvin. *Journal of the Institution of Electrical Engineers*, **41**, p. 423, 1908.
[33] Minutes of a Meeting of the Council, Thursday 20 February 1908, *Institution of Electrical Engineers Minutes of Council Meetings* 1904–1908, IET Archives reference: IET/ORG/2/1/7.

[34] Appleyard, R., *The History of the Institution of Electrical Engineers (1871–1931)*, The Institution of Electrical Engineers: London, p. 191, 1939.
[35] Appleyard, p. 245.
[36] Thompson, The Kelvin Lecture: The Life of Lord Kelvin, p. 423.

The Laws of Thermodynamics

The University of Glasgow lecture theatre. (Marianne Greated 2014, pencil on paper)

Chapter 7

Engineering Thermodynamics and the Carnot Cycle

M.W. Collins[1], J.A. Stasiek[2] & J. Mikielewicz[3]
[1]*College of Engineering, Design & Physical Sciences, Brunel University London, United Kingdom.*
[2]*Faculty of Mechanical Engineering, Technical University of Gdansk, Gdansk, Poland.*
[3]*Institute of Fluid Flow Machinery of the PASci, Poland.*

Abstract

The Carnot cycle is central to engineering thermodynamics and its teaching. Although on the one hand it is an unattainable ideal, on the other it constitutes a set of concepts to which real heat engine cycles and processes should aspire to and approximate. The Two Property Rule means that cycles may be represented graphically and the principal pairs of properties are ***p, v*** (pressure and specific volume) and ***T,s*** (temperature and specific entropy).

The ***p, v*** diagram is truly practical in the sense that it mirrors the indicator diagram, known to generations of past engineering students from engine laboratory experiments. The indicator and its diagram have a venerable history, dating back to 1796 and the very dawn of thermodynamics. The Scotsman James Watt used them to improve engine design to considerable commercial advantage. In 1824, Carnot was unaware of the diagram, most probably because of its extreme commercial sensitivity for Watt. Not until 1834 did Clapeyron make the first use of the ***p, v*** diagram to describe the Carnot cycle.

The ***T,s*** diagram was proposed by the American J (Josiah) Willard Gibbs in 1873 as part of a more general study of how graphical methods could be used in thermodynamics. Gibbs was a mathematical physicist and his thermodynamic contribution was almost a complete inverse to that of Watt. Despite it dating from well *after* the understanding of the Laws, the ***T,s*** diagram and the Carnot cycle are virtually indistinguishable in present-day understanding. So in representing Sadi Carnot's cycle of 1824 with both p, v and ***T,s*** diagrams (which he was unable to do) almost a century of associated thermodynamics history is involved. In this chapter, we study the historical context of these diagrams and their authors. In particular, James Watt was a massive contributor to the industrial revolution and in certain respects a forerunner to Carnot. We use these diagrams to show how the Carnot concept of a perfect heat engine applies across the thermodynamic board, irrespective of working fluid (gas or vapour), of cycle processes (non-flow or steady flow) or of engine concept (work-producer or refrigerator). In addition, the Carnot cycle was the driving force behind the formulation of the Second Law of Thermodynamics and this aspect is reviewed to highlight the contributions of Kelvin himself. We will find that Kelvin has an enduring high reputation here, based on his absolute scale of temperature and his second law statement.

Finally, in company with Carnot, the greatly contrasting Watt and Gibbs were both committed to the idea of a *perfect* engine. Not only does this demonstrate the underlying commonwealth of thermodynamics understanding, but raises a philosophical question: is thermodynamics Platonic in character?

1 Introduction: the Carnot Cycle

1.1 Engineering thermodynamics

... the methods of converting any form of energy into power are the domain of thermodynamics.

Desmond Winterbone [1, p. ix]

As an undergraduate in the late 1950s, and coming from the non-industrial English county of Dorset, I (MWC) found the subject of 'Engineering Thermodynamics' immediately fascinating. In fact, I still have my First Edition copy of 'Rogers and Mayhew' [2], obviously in its original British Thermal Units (BTU) form. (Anecdotally, it was bought second hand on traditional credit, with the account being rendered by post to my home address and opened in the vacation in the presence of a somewhat disbelieving mother. Those were the University days!) Concepts of work, heat, energy and power seemed so *real*, so *useful*. Even the First Law subtleties of definitions of these gave a warm sense of being a thermodynamics insider. Turbines, compressors, condensers, boilers – all were grist to the Steady Flow Energy Equation's mill. The Second Law showed that power station efficiencies of the low 30% were not a mark of poor engineering ability, but largely a consequence of essential and substantial cyclic heat rejection. The Carnot Cycle, described by Rogers and Mayhew in rather low-key terms as 'one ideal cycle' [3, p. 209], immediately led on, because of its 'low work ratio' and 'practical difficulties associated with compression' [3, p. 214] to the Rankine cycle and thence to the Rankine cycle with superheat, the reheat cycle and regenerative cycle. It all *made sense*.

Inevitably, there were certain intellectual overheads. Firstly, there was the question of working with two sets of steam tables and steam charts in both BTUs and CHUs (Centigrade Heat Units). This was accepted as yet another manifestation of the plethora of systems of units (including MKS, the nascent SI) with which we would-be engineers had to cope. That overhead has evaporated with time. But then there was entropy. Whereas the enthalpy h was introduced by 'the quantity $(u + pv)$ occurs so frequently in thermodynamics...' [3, p. 30] the entropy came in as a bit of a stranger, rather prosaically being introduced as a 'suitable,' a 'useful' property [3, p. 28]. Nonetheless, it was required student good practice to sketch any non-flow cycle as p–v (pressure–specific volume) and T–s (absolute temperature-specific entropy) diagrams and to perform appropriate calculations on them. Flow cycle variants involved the enthalpy and steam charts. Thinking in these terms became part of our thermodynamics DNA.

A further and somewhat disconcerting fact was that with little provocation thermodynamics mutated into the mathematics of the Maxwell relations [3, Chapter 7]. At the back of my 'customer's mind' (to anticipate present-day jargon) was the suspicion that thermodynamics could speak in quite different languages.

Written in retrospect these reactions seem disjointed, but they did represent a sort of personal driving force which led to teaching the subject.

1.2 The Carnot Cycle and the Second Law of Thermodynamics

...Carnot's greatness...

Donald Cardwell [4, p. 201]

...the famous Watt...

Sadi Carnot [5, p. 5]

Sadi Carnot's contribution to the second law of thermodynamics

Don Lemons and Margaret Penner [6, Title]

...since become famous as Kelvin's statement of the Second Law of Thermodynamics...

Crosbie Smith and M. Norton Wise [7, p. 329]

We now know that the above introductory paragraphs form a kind of intellectual closed loop. The very thermodynamic subtleties mentioned above which a twentieth century student took for granted were, historically speaking, very hardly won. In fact, close to two centuries after Carnot's original 1824 publication [8], our presentational view of his ideal heat engine differs from his in two significant ways. Firstly, *caloric* has been replaced by *entropy* as the key thermodynamic property. Secondly, Carnot did not use either of the above property diagrams (p–v and T–s) to explain his cycle. Taking things further, our perception of Carnot's brilliance needs to take into account the heat engine experience existing at the time, especially that due to Watt. Finally, there is the question of the extent of Carnot's contribution to the Second Law of Thermodynamics. This last leads to the similar question, of great interest in the present volume, relating to Kelvin. These issues are now introduced in more detail.

1.3 The caloric controversy

Firstly, according to Wikipedia (please note Acknowledgement with referencing to *Encyclopaedia Britannica*), *caloric* as a thermodynamic theory has disappeared ('an obsolete scientific theory that heat consists of a fluid called caloric that flows from hotter to cooler bodies' [9, p. 1]) in the same way as the earlier *phlogiston* (another 'obsolete scientific theory' [10, p. However, the whole truth is more than this bare statement. Cardwell, in his superb exposition *From Watt to Clausius* [4], points out that the caloric/dynamical debate permitted 'a veritable spectrum of beliefs'[4, p. 142] and in fact there remain ambiguities to this day. Mendoza, in his *Introduction* to [5], convincingly gives the probable reason for this, namely that the term *caloric* was used by Carnot 'interchangeably' for both (what we now know as) 'change of entropy' and 'quantity of heat' [5, p. xv]. On the one hand, we now have evidence that Carnot himself was subsequently convinced that both 'the caloric theory must be false' and 'the whole axiom of conservation of heat must go' ([6, p. 209], referring to a document unpublished during Carnot's lifetime [11]). In fact, Mendoza, commenting on [5, p. 60; 11] refers to Carnot's 'growing doubts about the whole basis of his demonstrations.' This conventional view is confirmed more recently by, for example, Erlichson. In the brief abstract of his paper, he writes of 'the incorrect caloric theory of heat' [12, p. 183] and that 'There can be no doubt that Carnot, after the writing of the Reflexions, abandoned the caloric theory of heat in favour of the mechanical theory of heat' [12, p. 191]. On the other hand, in consistency with Mendoza's abovementioned introductory remarks, caloric has also been interpreted as the simple forerunner of entropy – very recently explained by Newburgh [13] and implied by Thoma and Mocellin in their chapter in this volume. Leaving aside his argument, Newburgh is explicit enough: 'We have then seen that Clausius was able to identify caloric with entropy' [13, p. 726]. Against this, Ehrlichson, in a footnote on p. 190 [12], disagrees with Mendoza's interpretation of Carnot's use. Possibly the ambiguity will never be resolved .

1.4 Cycles and property diagrams

...Clapeyron...made their derivations clear through the use of indicator diagrams

E. Mendoza [5, pp. xii–xiii] – referring to 1834

...Gibbs introduced other diagrams such as temperature-entropy...

Crosbie Smith [14, p. 261] – referring to 1873.

The second question is that of the absence of Carnot's use of property diagrams. He did not use the indicator (p–v) diagram, concludes Cardwell, because 'In all probability he had never heard of it' [4, footnote p. 195]. As we will see in this chapter, Watt kept knowledge of 'his' diagram as covert as possible for commercial reasons. Further, whereas the absence of the p–v diagram causes us to look back from Carnot to Watt, the T–s diagram was absent because it simply did not yet exist. It had to wait for Kelvin's definition of the Absolute Temperature Scale T, for Clausius's formulation of entropy and for Gibbs's formalising of the T–s diagram itself. This leads to our *core thermodynamic sequence in this chapter for assessing the Carnot cycle as 'Watt—to—Gibbs.'* The thermodynamic history involved occupied the best part of a century, from Watt's indicator diagram in 1796 to Gibbs's T–s proposal in 1873. Our approach reflects Crosbie Smith's potted history: 'Just as pressure–volume diagrams (Watt's 'indicator diagram' and Rankine's 'diagram of energy') had long formed part of the science of thermodynamics especially in its relation to practical engineering, so Gibbs introduced other diagrams such as temperature-entropy...' [14].

1.5 The historical context of Carnot's study

The next issue is that of the heat engine and thermodynamic context existing at the time of Carnot's study.

Again, this explicitly focuses on Watt – and Cardwell gives two reasons. 'We have emphasised Carnot's indebtedness to Watt for the invention of and the insight represented by expansive operation; he was also heavily indebted to Watt for the latter's invention of the separate condenser' [4, p. 199]. Just as in order to give a present-day understanding of the Carnot cycle it is necessary to invoke the Watt-to-Gibbs sequence, so in order to appreciate Watt properly we need to understand his thermodynamic 'family tree.' This comprised two lines – that one based on *British steam engine development* (originating with Thomas Savery) and a second based on the *University of Glasgow* (starting with William Cullen).

1.6 The universal Carnot cycle

In this chapter, Carnot is viewed through the eyes of others. Newburgh [13] provides a concise and fitting entrée. Besides describing how Carnot's caloric was replaced by Clausius's entropy, as mentioned above, Newburgh tells us that Carnot studied 'an idealization of an operating steam engine' [13, p. 718]. However, this ideal cycle involved only reversible processes and therefore had a universal maximum efficiency. So Carnot was able 'to show that there is no machine with a greater efficiency, *regardless of the operating substance or design*' [13, p. 722] (our italics). Carnot's own words were 'the maximum of motive power resulting from the employment of steam is also the maximum of motive power realizable by any means whatsoever' [5, p. 12]. This conclusion is of huge significance, and we will show how Carnot cycles are apparent for both vapours and gases, for both flow and non-flow processes and for direct and reversed heat engines. Unlike Carnot himself we will be able to use both p–v and T–s diagrams to describe the variant cycles. In fact, Carnot's focus changed from steam to air ('we have chosen atmospheric air as the instrument,' [5, p. 19] and thence to gases in general [5, pp. 24, 25]. This was taken up by Clapeyron ('let us take any gas whatever,' [5, p. 75] who was the first to use the p, v diagram.

Besides giving a concise exposition of the 'universal Carnot cycle' we attempt to tell the underlying thermodynamic story, and to that end the chapter is arranged approximately historically in two parts. Cardwell is in no doubt of the significance. 'Everything that has happened since 1824 has confirmed Carnot's judgement, and we must therefore divide the history of the heat-engine into two very distinct periods: *before and after Carnot*' [4, p. 208, Cardwell's italics]. Firstly, then, the history of steam power is told up to and somewhat beyond Watt. Then the history of the Carnot cycle itself is told.

1.7 Carnot, Kelvin and Clausius

The climax of this chapter is defined by a key question. Especially from the much later viewpoint of the twenty-first century, what are the nature and extent of Carnot's contribution to the Second Law of Thermodynamics? In seeking to answer this question, we should be able to *address the same question for Kelvin himself*. Of course, we must include Clausius in our consideration. In the chapter itself is explained the participation of Clapeyron and of Kelvin's British contemporaries.

1.8 A philosophy of thermodynamics?

In introducing Carnot's engine Cardwell makes the almost throwaway comment that an essential contributory concept of Carnot 'removes the perfect heat-engine from the world of practical engineering to some Platonic heaven.' He goes on to examine 'how the perfect but wholly impracticable Carnot engine works.' Yet it had 'fundamental and general implications' (all quotes from [4, pp. 193, 194, 200]) for real heat-engine design and operation. So a prima facie case exists for a 'philosophy of thermodynamics' based on Platonism. In fact, Newton himself had an allegiance to Platonism, as do certain present-day mathematicians. Discussion of this forms a rather novel closing section.

2 Timelines of characters

Not one of the engineering and scientific personalities involved in this subject area worked in isolation. They were affected by others in two ways: by inspiration provided by previous workers no longer alive and by interactions with their contemporaries. Since the overall timescale from Savery's patent ('the actual 'breakthrough' came with the...machine devised by...Savery in 1698' [4, p. 13]) to Gibbs's publication is of the order of two centuries, the distinction between them can become blurred. As a consequence a feature of this study is to provide timelines. The personalities involved in the overall story are summarised as a series of Lifetime Sequences:

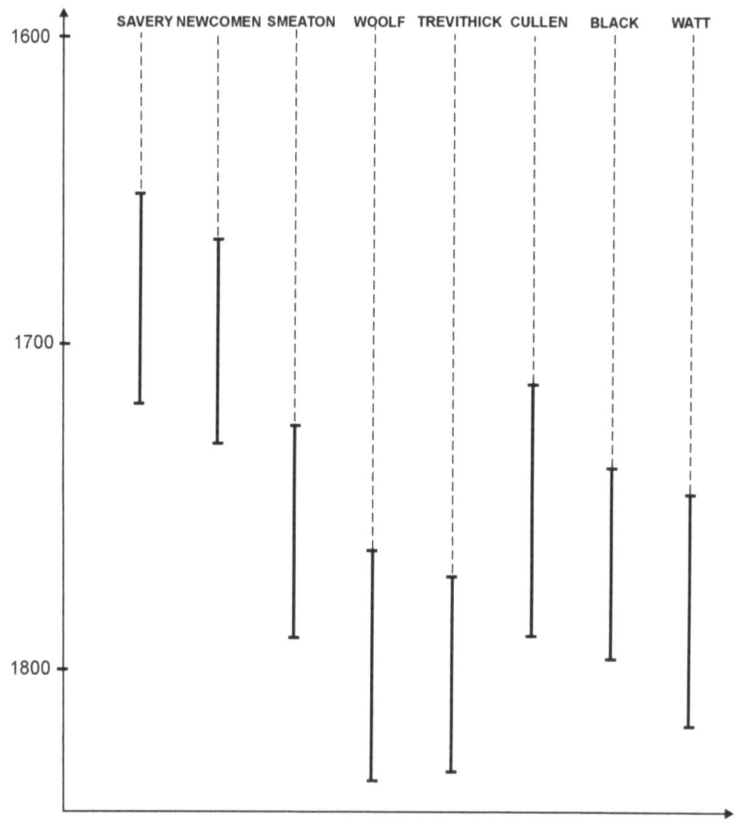

Figure 1: Watt's thermodynamic family tree: left-hand side ~ British steam engine development and right-hand side: Glasgow University.

Thomas Savery	c. 1650–1715	William Cullen	1710–1790
Thomas Newcomen	1664–1729	Joseph Black	1728–1799
John Smeaton	1724–1792		
Jonathan Hornblower	1753–1815		
Arthur Woolf	1766–1837		
Richard Trevithick	1771–1833	James Watt	1736–1819

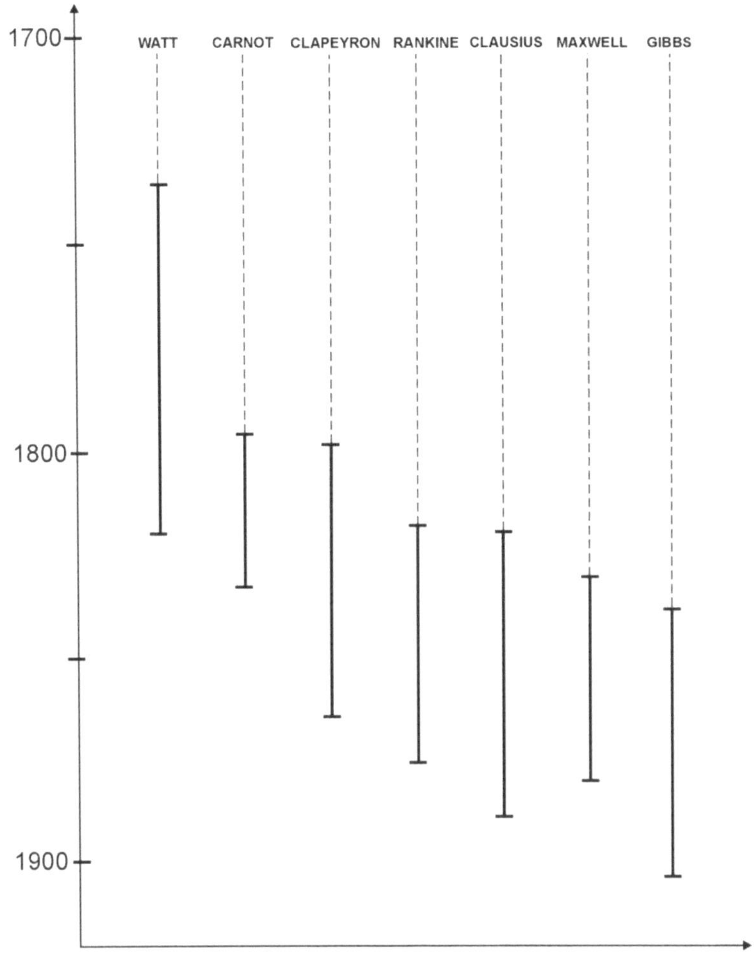

Figure 2: The history of the p–v and T–s diagrams: Watt-to-Gibbs.

James Watt	1710–1790
Nicolas Leonard Sadi Carnot	1796–1832
Benoit Paul Emile Clapeyron	1799–1864
William Macquorn Rankine	1820–1872
Rudolph Julius Emanuel Clausius	1822–1888
James Clerk Maxwell	1831–1879
Josiah Willard Gibbs	1839–1903

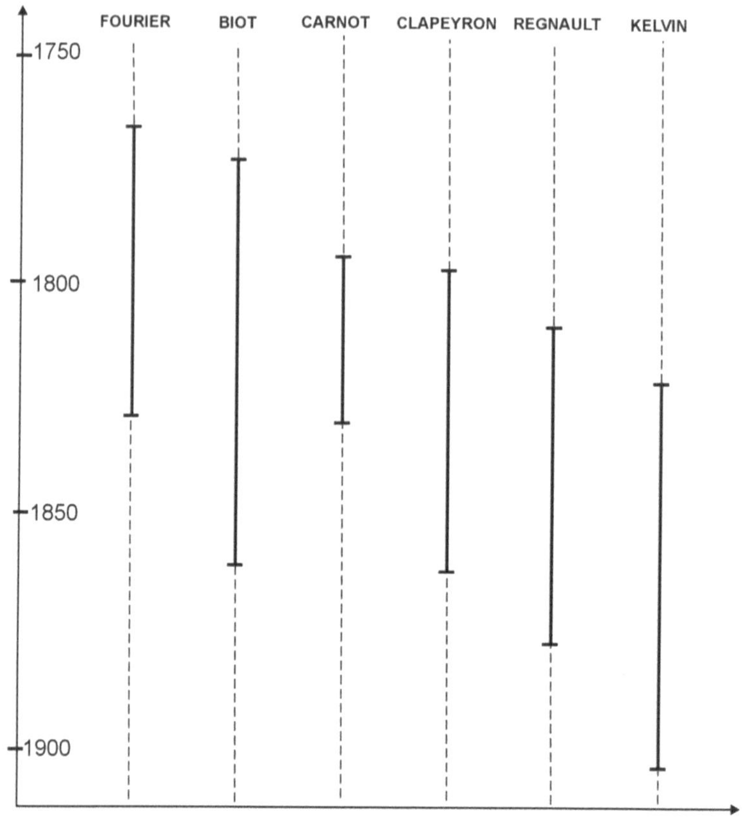

Figure 3: Kelvin's French connections.

Jean Baptiste Joseph Fourier	1768–1830 (Book Read 1840; Use of Law, e.g. 1861 – [7, pp. 167, 566])
Jean-Baptiste Biot	1774–1862 (Met in Paris 1845 – [7, p. 107])
Nicolas Leonard Sadi Carnot	1796–1832 (paper obtained by 1849 – [4, p. 240])
Benoit Paul Emile Clapeyron	1799–1864 (French paper read in Paris 1845 – [7, pp. 293–294])
Henri Victor Regnault	1810–1878 (Worked in Laboratory in Paris 1845 – [7, pp. 107, 243–245])

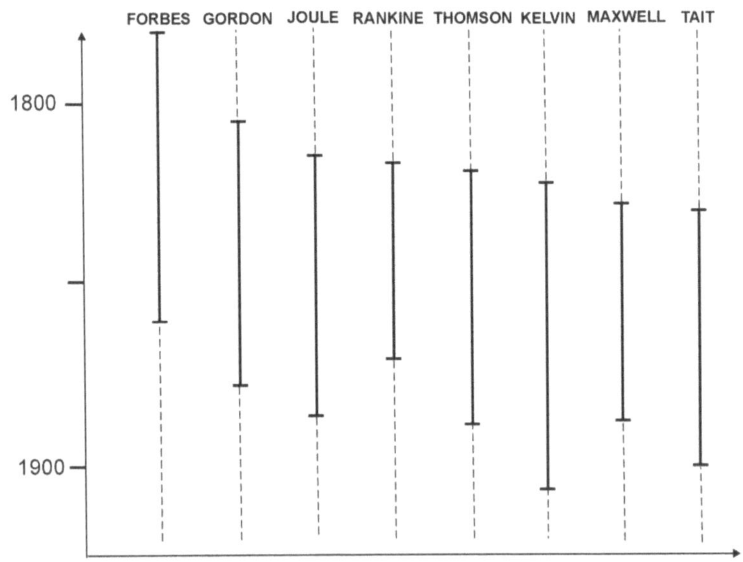

Figure 4: Kelvin and the Second Law: Kelvin and his British contemporaries.

James David Forbes	1785–1868 (Prof. of Natural Philosophy, Edinburgh, Principal, St. Andrews)
Lewis Dunbar Brodie Gordon	1815–1876 (Regius Prof. of Civil Engineering & Mechanics Glasgow, *First UK Professor of Engineering*)
James Prescott Joule	1818–1889 (Brewery Manager & personal researcher – Presentations at British Association of Advancement of Science)
William Macquorn Rankine	1820–1872 (Second Regius Prof. of Civil Engineering, Glasgow)
James Thomson	1822–1892 (Prof. of Civil Engineering, Queen's Belfast, Third Regius Prof. of Civil Engineering, Glasgow)
William Thomson, Lord Kelvin	1824–1907 (Prof. of Natural Philosophy, Glasgow)
James Clerk Maxwell	1831–1879 (Prof. of Natural Philosophy, Aberdeen, Prof. of Natural Philosophy, King's College London, First Cavendish Prof. of Physics, Cambridge)
Peter Guthrie Tait	1831–1901 (Prof of Mathematics, Queen's Belfast, successor to Forbes as Prof. of Natural Philosophy, Edinburgh)

The focuses of Figs 1–4, respectively, are, then

- James Watt: addressed in Section 3,
- Watt to Gibbs: addressed in Sections 4–7, then 9
- Kelvin: addressed in Sections 8 and 12. Joule, James Thomson (Kelvin's brother) and Maxwell feature elsewhere in this volume.

In addition to Figs 1–4, other Timeline sequences are given where appropriate.

3 Watt's thermodynamic family tree and place in history

3.1 Steam engine development: Savery-to-Watt-to-Trevithick

The purpose of this section is to give the reader an overview of the historical context of James Watt's steam engine activities. Two main sources are used: the encyclopaedic biography of Watt by Richard Hills [15–17] and, for ease of access to readers, Wikipedia items for the principal characters: Thomas Savery, Thomas Newcomen, Newcomen steam engine, Newcomen Engine Black Country Living Museum, John Smeaton, Johnathan Hornblower, Arthur Woolf, Compound engine and Richard Trevithick ([18–26], respectively). Additional comments are included from Cardwell's *From Watt to Clausius* [4]. The principal character list, given in Fig. 1(a), summarises the seventeenth and eighteenth centuries' history of steam engine development. Three aspects are noteworthy: it was consistently British in character and preponderantly from South West England; it was achieved through empirical engineering inventiveness and it was applied to the problem of extracting water from metal and coal mines. The first aspect was reflected in the export of Newcomen's engines to France, Belgium, Spain, Hungary and Sweden [20, p. 3] and of Woolf-designed engines 'in many parts of the world' [24, p. 1]. The second aspect expresses the domination of patents on the one hand, and the absence of the Laws of Thermodynamics on the other. The last feature was epitomised by Trevithick who through the use of high pressure steam achieved the best pumping engine – 'the 'Cornish engine' … the most efficient in the world at that time' [26, p. 8]. However, Trevithick went wonderfully further: he 'built the first full-scale working railway steam locomotive' leading to 'the world's first locomotive-hauled railway journey on 21 February 1804' [26, p. 1].

3.2 The dominance of patents

The question of patents is now addressed in more detail, by giving the timeline of Fig. 5. While the question is not of paramount importance to the present study, it is of considerable interest in the overall record of the industrial revolution. It is plain it dominated the whole history. It should be noted that very recently, Boldrin and Levine in Chapter 1 of their *Against Intellectual Monopoly* [27], addressed this by explicitly studying James Watt as a Monopolist. Their specific study can be found on Google [28] and they conclude that 'Quite plainly, the evidence suggests that Boulton and Watt's patent retarded the high-pressure steam engine, and hence economic development, for about 16 years' [28, p. 4]. It is really unfortunate that their sourcing *excludes* Hills's exhaustive study, especially because he discusses this very point in some detail and in a measured manner in [17, pp. 217–220]. He admits that Hornblower might have effected improved designs, but that the then increased pressure was not high enough to have warranted compounding. Otherwise, Hills says, 'the 1775 Act 'does not seem to have stifled inventions and improvements' [17, p. 220].

Year	Patent	Comments
1698	*Savery's* 'early steam engine'[1] ([18], p 1)	14 years granted
1699	Extension by Act of Parliament for England([18], p2)	21 more years granted
1701	Extension by Act of Parliament for Scotland([18], p2) {Several machines built ([18], p3)}	21 more years granted
1712	*Newcomen*'s 'more advanced design'[2] ([18], p 2) {~600 built by 1775 ([19], p 2) often retro-fitted with condensers ([19], p 3)}	Marketed under Savery's patent
1769	*Watt's* 'principle-based' separate condenser [3]([15], pp 389-91)	Covered later developments (p 391)
1775	Extension by Act of Parliament & including Scotland {~450 engines built to 1800([19], p 3)}	New 25 year term ([16], p 46), joint with *Boulton*
1780, 1782 (Feb), 1782 (July), 1784, 1785	Watt's additional patents ([17], p13)	
------	*Smeaton* (lack of patents) Incremental improvements to Newcomen engines ([22], p2) suitable for collieries [4] ([16], p 117)	Always on good terms with Watt -- [15], p 324, [16], p 57
(1781	*Hornblower*, compound engine developed)	
1795-1799	Watt's patent trial against Hornblower: Watt wins	
1802	*Trevithick's* high pressure steam engine ([26], p 4) [5]	Avoided separate condenser
1803	*Woolf's* high-pressure steam boiler ([24], p 2)	
1805	*Woolf's* compound engine ([24], p2)[6]	

Figure 5: Patent History in steam engine development.

Notes

[1] No pistons [18, p. 2] or other moving parts [20, p. 2], a 'thermic syphon' [19, p. 1]. 'Whether any of them worked satisfactorily is uncertain' [4, p. 13]

[2] Used ideas of Savery and Frenchman Denis Papin [18, p. 2; 19, p. 1] to invent and build the first practical steam engine' [19, p. 1]. 'The Newcomen engine was extremely important. It was perhaps the first major invention to be made by an Englishman, and it ushered in a period of British technological supremacy that was to last about 150 years' [4, p. 15].

[3] See [15, pp. 367–370; 16, p. 30ff]. '…first … The engine he envisaged was, in a sense, an ideal engine: one that was to develop the maximum power with the maximum efficiency … Second

... Watt's engine was unambiguously a *heat*-engine ... the presence of the condenser indicates that *heat* is the basic principle ... Watt's intuitive insight into thermodynamics...' [4, p. 54]

4 Also see [16, pp. 168–169]. 'In Britain John Smeaton was undoubtedly the leading engineer, and ... was able to approximately double the efficiency of a group of Newcomen engines on the north-east coast of England'[4, p. 33].

5 'Trevithick, an intuitive genius if ever there was one, ...intuited very early on that high-pressure engines might be inherently more economical than low-pressure ones ... he was one of the main sources of British wealth and power during the nineteenth century' [4, p. 83, 154/5]

6 Steam expanded in two (or more) stages in high- and low- pressure cylinders [25, p. 1ff]. 'The first Woolf compound expansive engine ... in 1814 ... something like a 100% improvement on the best performance of the Watt low-pressure engine ... the Woolf compound engine found its greatest market in France' [4, pp. 154–157]'

3.3 Glasgow University: Cullen-to-Watt

This sequence is simpler than in Section 3.1 and involves only three characters. The same sourcing method is used: for William Cullen, Joseph Black and James Watt, respectively, the Wikipedia references [29–31]. Similarly, in addition there is the comprehensive material of Hills [15–17] and Cardwell [4]. Figure 6 gives the Glasgow timelines with positions/activities, hence the interactions between them. This helps in understanding how Watt entered into Glasgow University's nascent tradition of chemical and heat studies.

3.4 Watt's place in history

3.4.1 *A contributor to world history*

...Watt belonged to one of the most active scientific groups in the world...

D.S.L. Cardwell [4, p. 54]

Boulton has perhaps been most widely known for his steam-engine partnership with James Watt

R. McLean [32, p. 1]

We now consider a number of aspects of Watt's enduring reputation, but with two introductory provisos.

Firstly in this section, we have been unable to give a true overview of Watt's activities, especially that relating to Matthew Boulton. Our understatement 'joint with *Boulton*' in Fig. 4 is the main focus of Hills's biography, from Volume 1, Chapter 6 ([15, p. 414, '*Enter Matthew Boulton*) onwards. More significant is the place in history of Boulton himself. In 2009 Birmingham (UK) celebrated Boulton's Bicentenary with an Exhibition organised by the Birmingham Museum & Art Gallery. This was accompanied by the elegant and authoritative volume *Matthew Boulton. Selling what all the world desires* [32]. Mike Whitby, the Leader of Birmingham City Council describes

	Year	Position/Activities
Cullen[1]	1744	Informal lecturing in physiology, botany, medical materials, chemistry ([29], p 3)
	1747	Lectureship in Chemistry—Britain's first ([29], p 3)
	1751	Professor of Medicine ([29], p 3)
	1757	Leaves for Edinburgh in Jan ([15], p 104) to become Professor of Chemistry & Medicine there ([29], p 3)
Black[2]	1744	Student in Arts Course-becomes laboratory assistant to Cullen ([15], p 76/77)
	1752	Medical student at Edinburgh
	1753	MD degree at Edinburgh
	1756	Leaves Edinburgh for London
	1757	Regius Professor of Medicine and lecturer in Chemistry ([30], p 1)
	1762	Latent heat of steam
	1766	Leaves for Edinburgh
Watt[3]	1753[4]	Arrives in Glasgow ([15], p 46) to be his father's agent (p 48)
	1755-6	In London: trained as mathematical instrument maker ([15], p 55 ff)
	1756-7	In Greenock: paid work for University ([15], p 64)
	1757-9	Mathematical instrument maker at University, Making/repairing lecture apparatus for Black ([15], p 78)
	1761	Start of steam power experiments at University ([30], p 3)
	1763	Repair of model Newcomen engine ([15], p 329)
	~1764-	Measurements of boiling point of water, kettle experiment & latent heat ([15], pp 329-340),
	1765	Working model of condenser ([31], p 4)
	1766-	Construction of full-scale engine—help from Black and Roebuck ([31], p 4) &
	-1769	Continuing experiments at Glasgow ([15], p 397)
	1769	First patent ([15], p 389)

Figure 6: Glasgow University timelines for Cullen, Black and Watt.

[1] Cullen was a significant figure in the Scottish Enlightenment, Scottish medicine and in the formation of the Royal Society of Edinburgh [29, p. 1,2]. '…he actually published very little. His single paper on heat ¼ influenced the course of James Watt's researches' [4, p. 34].

[2] Black was an eminent Scottish physician and chemist, and, like Cullen, had Scottish Enlightenment associations [30, p. 1, 2]. He 'was one of the most influential scientific thinkers of the eighteenth century' [4, p. 34]. Watt closely collaborated with him. Indeed, they were friends, and Black provided substantial financial support for Watt [30, p. 2].

[3] The three volumes of Hills's biography, especially the first, show how multi-faceted were Watt's abilities. He had substantial civil engineering accomplishments [15, *Chapter 4. Civil Engineer*, pp. 180–293] and his links with Charles Smeaton largely stemmed from this. Such engineering adaptability was reciprocated. Smeaton himself, although being 'the first self-proclaimed civil engineer' [22, p. 1], also contributed not insubstantially to steam engine development, as already noted in Fig. 4.

[4] Watt's life resume is given in full in [15, pp. 12–17], and in [16, pp. 6–8], reflecting note [3].

Boulton's reputation in these words: 'Matthew Boulton…helped to establish Birmingham as one of England's greatest manufacturing centres, and England as the industrial leader of the known world' … 'the father of Birmingham' [32, p. vii]. In commemorating 'one of the great manufacturers and entrepreneurs of his age,' though, 'Boulton has perhaps been most widely known for his steam-engine partnership with James Watt' [32, p. 1]. In other words, the key to Boulton's enduring reputation itself was 'the Boulton & Watt steam engines exported all over the world' [32, again: dustjacket blurb]. So even this proviso gives ample evidence (especially in Jim Andrew's'*The Soho Steam Engine Business*' Chapter 8 of [32]) to show that Watt was a key contributor to steam engine development and hence to the significance of the industrial revolution in world history – '…whose improvements … were fundamental to the changes brought by the Industrial Revolution in both his native Great Britain and the rest of the world' [31, p. 1].

Secondly, we have not yet considered the place of Watt's indicator diagram within engineering thermodynamic history. This will be addressed in the following section.

Bearing in mind the provisos, we will consider four broad aspects of his reputation:

- with Carnot himself,
- at the start of Kelvin's career,
- at the present day in thermodynamic thinking,
- and at the present day in public perception

3.4.2 With Carnot

…he was heavily indebted to Watt…

D.S.L. Cardwell [4, p. 199]

Watt, to whom we owe almost all the great achievements in steam engines….

Sadi Carnot [5, p. 50(footnote)]

In an essential way Watt anticipated Carnot, in that he promulgated the idea of the perfect engine. He was highly complimented by Carnot throughout the Reflections [5]. In fact, Carnot, who mentioned virtually all the personages of Fig. 4, referred to him alone as *famous* [5, p. 5]. Further, Carnot, while being fully aware of the work of Woolf and Trevithick, said regarding Watt's engines: 'Wheal Abraham … somewhat exceptional, for it was only temporary … is generally regarded as an excellent result for steam-engines. It is sometimes attained by engines of the Watt type, but very rarely surpassed [5, p. 58 (footnote)].

In fact the generally fulsome way Carnot refers to Watt, does two things. It justifies Watt's enduring reputation on the one hand, and on the other throws into light Carnot's own superlative genius. Regarding Carnot, Cardwell quotes Sir Joseph Larmor: '…perhaps the most original in physical science' [4, p. 191]. We will return to a comparison of Watt with Carnot later.

3.4.3 Glasgow University and the Thomson brothers

...gave pride of place to the steam-engine model, which had inspired James Watt's discoveries...

Crosbie Smith and M. Norton Wise [7, p. 53]

In 1763 Glasgow University commissioned Watt to make its model Newcomen engine work properly [15, p. 312]. This he did [15, pp. 312–318] and it was the defects that he encountered that led him to the idea of a 'perfect engine' (p 318). (A rather amusing digression – of some comfort to present-day researchers--is to do with how good Watt's memory was. This story features on Richard Hills himself. He had staff member experience of the 1/3rd scale model of the Dudley Castle Newcomen engine [15, p. 316], of which the current full-scale replica is described in [21]. The model was built by the Mechanical Engineering Department at UMIST, now part of the integrated University of Manchester. Because of *his* knowledge of making the 1/3rd scale model work, Hills came to the conclusion that in later life Watt must have forgotten what he exactly did to make the Glasgow engine perform properly!). For the intervening centuries since Watt Glasgow University have been justly proud of their model. Hills mentions that it is in the Hunterian Museum at Glasgow and he devotes a number of Plate photographs to it [15, Plates 11–14).

Returning to history, in 1840 Glasgow hosted the 10th Annual Meeting of BAAS (the British Association for the Advancement of Science). The meeting was originally proposed by the Glasgow Philosophical Society of which James and William Thomson (then 18 and 16 years old) were senior and junior members, respectively. James Thomson became secretary of the Model and Manufactures Committee and organised an exhibition for this international Meeting, at which the Newcomen/Watt model engine, together with the engine for Bell's *Comet* steamboat, took pride of place. The future Lord Kelvin gave secretarial assistance.

The whole story is told by Crosbie Smith and Norton Wise in their (like Hills's) encyclopaedic biography of Kelvin, *Energy and Empire* [7, pp. 51–55]: they point out that even when so young '...the Thomson brothers were already moving to the centre stage of British science.'

3.4.5 Watt and Black in current thermodynamic thinking

...having thus resolved the dilemma in a moment of insight, Watt was immediately able...

D.S.L. Cardwell [4, p. 48]

James Watt, the most famous of Joseph Black's pupils...

this invention brought unimagined power and speed to everything...

Dilip Kondepudi and Ilya Prigogine [34, p. 67,68]

Here there is some controversy. Cardwell does not so much raise an important question, as lucidly expose a problem. Certain aspects of Black's record he criticises, notably his failure to publish [4, p. 40]. While Black is 'able' he 'did not play the game according to the rules: his friends played it for him.' So 'The exact relationship between the reticent, able but academically indolent professor and the inventor and applied scientist of genius has been a subject of discussion ever since … The legend is that Black told Watt about his discovery of latent heat, whereupon Watt perceived the importance of economising the consumption of steam in the Newcomen engine, and to this end invented the separate condenser' (all quotes from [4, pp. 40,41]. Hailing like Hills from Manchester, Cardwell refers to him as 'my friend Richard Hills' [4, p. xv]. Despite this friendship their views do not quite converge. For the record, Cardwell's [4, pp. 40–42] and onwards should be compared with Hills's [15, pp. 303–310]. Hills, taking into account his usual measured style, tends to give rather more credit to Black.

What of the authors of thermodynamics texts? We take two examples. Myron Tribus, in the Introduction to his *Thermostatics and Thermodynamics* (a text which uses information-theory-oriented entropy as the entrée into the Laws of Thermodynamics),gives Black credit for the initial distinction between 'temperature' and 'heat,' his definition of 'heat capacity' and for his 'method of mixtures' ('first clearly understood by Black'). For Watt, Tribus says: 'One of Black's pupils was James Watt who applied Black's ideas to the improvement of the steam engine' (all quotes from [33, p. xiii]). This is just the 'legend' Cardwell describes! The second example is the text by Dilip Kondepudi which is co-authored by the Nobel Laureate Ilya Prigogine – '*Modern Thermodynamics. From Heat Engines to Dissipative Structures*' [34]. Again Black gets full credit for his ideas on temperature and heat, together with his discovery of specific heat [34, pp. 8, 9] and that credit is accompanied by a portrait reproduction on p 8. There Watt is mentioned as giving 'enthusiastic help.' Significantly unlike Tribus, however, in introducing the Second Law, Kondepudi and Prigogine elevate Watt (together with portrait) to the initial historical spot followed by Carnot [34, pp. 67, 68].

3.4.6 *Watt and the Age of Enlightenment*

… 'when Watt had so far perfected the Steam Engine…

he invented the units…on a safe commercial principle'

Crosbie Smith and M. Norton Wise [7, p. 53] (quoting Kelvin)

Adam Smith knew and helped Watt

Richard Hills [15, p. 44]

Boulton was…Britain's leading Enlightenment entrepreneur

Matthew Boulton. *Selling what all the world desires* [32] (Dustjacket blurb)

Further to the discussion immediately above, Kondepudi and Prigogine (K & P) start their very first chapter with a telling comparison of Adam Smith with Watt [33, p. 3]. Adam Smith was 'a pioneer of political economy and one of the key figures of the Scottish enlightenment' [35, p. 1] and being at Glasgow was well known to Watt. Hills points out that 'Adam Smith…knew and helped Watt and could have influenced the way Watt organised his workshop' [15, p. 44]. Despite the involvement that Adam Smith had in Watt's activities, it was clear that Smith failed to recognise the potential of steam power. K & P quote Smith in his iconic *Wealth of Nations* (published in 1776, seven years after Watt's patent [35, p. 1]) as saying that coal was only for 'providing heat for the workers.' So in an application which could barely be more important industrially K & P have noted an instance of myopia: 'Adam Smith did not see in coal this hidden wealth of nations (quotes from [33, p. 3]). We should take this matter further, since this was a key issue for Kelvin himself. As we might expect, in contrast to Adam Smith, Kelvin fully appreciated the significance of coal-fuelled steam power, and that moreover in relation to Watt. In *Energy and Empire* the story is told of the projected *Principia Mathematica* of Tait and Kelvin [7, Chapter 11, T & T' or Treatise on Natural Philosophy, starting around 1861, p. 348 ff]. Reducing the quoted material to the minimum: 'wherever the steam-engine entered Thomson's (Kelvin's) thinking, there also the concept of work as labour value entered ... when Watt had so far perfected the Steam Engine ... he invented the units of horsepower as a comparative measure of its effectiveness 'on a safe commercial principle.' Thomson's national and economic concerns with the measure of work thus went hand in hand as concerns with the wealth of the nation. They were in fact concerns with political economy' [7, p. 358]. In other words, without mentioning Adam Smith, Kelvin in 1862 was essentially making the same point as K & P some 140 years later.

In fact, Kelvin was not being partisan on behalf of Watt. While Adam Smith does not figure largely in *Energy and Empire*, Smith and Wise point out that probably both Thomson brothers were influenced by the 1841 Cambridge publication of William Whewell [7, pp. 285–287] *The mechanics of engineering*. In it Whewell used the expression 'Labouring Force is the labour that we pay for,' a whole section being quoted in [32, p. 286]. Following this Smith and Wise make the comment that 'These remarks could largely have been quoted from Adam Smith or virtually any other political economist. But Whewell went further...' [7, p. 287]. This is in the context of a more diffuse discussion on what was the best terminology to use for engine output – 'labouring force,' 'work' (Whewell introduced the term (p 286)) or 'mechanical effect' as preferred by the Thomsons (p 290). This latter harked back to Watt's use of the term 'effect' (pp 290,291). The above words 'went further' showed that the nascent University discipline of engineering sought to contribute to the understanding of political economy. The probable implication is that Kelvin and his brother devoted attention to this issue from about 1842 (p 286) at least to the above 1862.

Smith and Wise tell the story of the formation in 1811 in Cambridge of the Analytical Society leading to the 'revitalization of British mathematics' and forming part of 'the much larger movement to reform British science.' Dr. Thomson (Kelvin's father) 'led his own component of this reform movement.' Now 'The call of the Analytical Society was the call of ... the Enlightenment' (all quotes from [32, pp. 151/152]). This hugely significant movement was in Britain as a whole, and indeed 'across Europe and the New World.' A fine potted history entitled *The Eighteenth-Century Intellectual Climate* is given by Chris Rice/Shena Mason/Laura Cox in the context of Boulton's activities. Their first descriptor of the 'Age of Enlightenment' is a 'surge of interest in science and empirical reason' (all subsequent quotes from [32, p. 128]). Adam Smith, James Watt and Matthew Boulton all qualify, therefore, with Kelvin and James Thomson not far behind.

3.4.7 Watt and the current UK scene

... ' the invaluable contribution that Boulton and Watt made to the advancement of engineering'...

Sir Mervyn King, Governor of the Bank of England, News Release, 2011 [37]

We will see in this volume's concluding chapter (*Conclusion: honoured by banknotes*) that being featured on paper currency represents one of the most important ways a country expresses honour to its past notable citizens. A fairly full story is told there, but for the sake of the current line of thought some details are given here. For the UK Adam Smith is the *only* person so honoured on both Scottish (Clydesdale Bank £50 notes from 1981 to 2008 [36, pp. 833–836]) and English notes (Bank of England £20 notes from 2006 onwards [36, p. 408]). In other words for several years this was being done *simultaneously*. In contrast to Adam Smith, however, the very recent Bank of England £50 note features Boulton and Watt themselves. Has the record been straightened? Has Watt caught up with Adam Smith in official esteem? Given that the Bank of England series has only four denominations – £5, £10, £20 and £50 notes – it means that half of the current notes feature the very three characters mentioned immediately above: Adam Smith, James Watt and Matthew Boulton. This shows that the issues addressed here resonate with current UK political priorities. With reference to the new £50 note the Bank of England News release of 2 November 2011 [37] quotes the governor Sir Mervyn King, as saying additionally to the heading quote:

'The partnership of an innovator and an entrepreneur created exactly the kind of commercial success that we will need in this country as we rebalance our economy over the years ahead.'

As this chapter is being written, the UK is indeed undergoing a refocus of economic activity towards engineering. James Watt, together with Matthew Boulton, would have been extremely proud to think that their achievements would be so highly regarded a quarter of a millennium later!

An intriguing but key final point is that whereas we have introduced Boulton via Watt, hence giving precedence to Watt, in fact the duo is *always called* Boulton and Watt. The reasons for this are also given in the *Conclusion*.

4 The *p–v* diagram: James Watt and the Indicator Diagram

4.1 Watt: inventor of genius and perfectionist

Watt's activities and achievements wholly predated the formal statements of the Laws of vThermodynamics. However, he anticipated them in two ways. Firstly, he was wedded to the concept of the 'perfect engine.' Hills frequently refers to this, but especially in [15, pp. 318, 319]. As a consequence of his initial series of tests on the Glasgow model Newcomen engine Watt concluded that it was 'far from perfect.' However, whereas (for instance) Smeaton successfully made improvements on an incremental basis (Fig. 4), from the start Watt made the intellectual leap to 'the 'perfect engine' [15, p. 318]. Cardwell elaborates on this. Watt 'the inventor of genius' [4, pp. 40, 41] conceived of an 'ideal' engine and then construed from it the closest approximation to it that he could, that is his 'perfect' 1765 design. Cardwell illustrates this combination of conceptiveness and superior engineering design as Fig. 4 (The physically impossible 'ideal' engine) and 5 (The Watt Engine) on [4, pp. 48, 49].

In this striving after a perfect engine Watt anticipated Carnot by *over 50 years*. In fact it could well be argued that any appreciation of Sadi Carnot's greatness should also include an appreciation of James Watt as an authentic forerunner, a sort of 'brother under the skin.'

Hills seeks to answer the question 'from where did Watt derive this idea of perfectability?' by referring to 'the common concept during the eighteenth century in many different spheres such as some aspects of religion or architecture with Greek temples' [15, p. 318]. We will address the matter of there being a distinctive philosophy of thermodynamics later in this chapter.

The second anticipation was due to the *empirical* nature of the Laws of Thermodynamics. Rogers and Mayhew express it this way: 'The structure of thermodynamics rests upon two important principles called the *First and Second Laws of Thermodynamics*. These cannot be proved and are treated as axioms; their validity rests upon the fact that neither they, nor any deductions made from them, have ever been disproved by experience' [3, pp. 5, 6]. Hence any reliable experimental data will always be consistent with the Laws and *therefore sharp intuition (in this case on the part of Watt) could give insight into them.* The crucial objective for Watt was overall engine performance, and he made measurements which expressed this for standard Newcomen engines and for his own designs [16, pp. 168–170]. Energy input was expressed as 'coal consumed per hour' and work output in terms of 'pounds of water raised over a foot.' This gave output / input figures of 'lbs of water raised over a foot per bushel of coal,' that is in the precise terms which would be of interest to a mine-owner. In these terms a Newcomen engine, a Watt engine without expansion and one with expansion produced respectively 10.5 million, 19.8 million and 26.6 million [16, p. 169]. So Watt engines were at least twice as efficient as the Newcomen ones. Now the expansion engine was one where the steam supply to the cylinder was cut off part way through the stroke and then allowed to expand over the remainder [16, pp. 129,130]. Moreover, if the steam admission could be varied using an additional valve the power output of the engine could be varied, a clear operational advantage recognised by Boulton. Different mines meant that depths of water extraction varied and an engine design which accommodated this factor was distinctly advantageous [16, p. 131]. For the case quoted the expansion engine gave a *further* improvement in performance of some 70% of the original Newcomen value.

Such achievements on the part of Watt stemmed, then, from his intuitive thermodynamic understanding and hence inventiveness-'inventor and applied scientist of genius' [4, pp. 40, 41]. Moreover Watt was a superlative experimentalist and 'a perfectionist' [4, p. 48]. The notebook facsimile given in [16, p. 132] illustrates this. It could be a model for the working notebook of a present-day PhD experimentalist (that is, some 230 years later!). More importantly, it recorded the pressure-volume progression over the stroke and Hills points out that a preceding pressure fall-off diagram 'is the earliest representation to have been found of what later became the *steam engine indicator diagram*' (in 1776 [16, p. 131], our's in italics). The experimental information represented the rate of energy transfer in the engine cylinder, that is intermediate between the gross 'energy input' (coal consumption) and 'power output' (water extracted). To have knowledge of this would give Watt an ability to optimise operating conditions and deduce improvements in design.

The above gives the historical background to the invention of the indicator and the p–v diagram.

4.2 Watt's indicator diagram

Hills describes the development of this in some detail in [17, pp. 83–87]. Watt's abovementioned primitive antecedent of the indicator diagram depended on using a mercury barometer to record the highest and lowest pressures over a stroke [17, p. 83]. 'Soon after 1790' a small steam cylinder was developed with a spring-balanced piston/pointer arrangement (i.e. no mercury) which Watt called an indicator. The fully-fledged device, 'fitted with a pencil' and board –mounted paper which moved in time with the piston, was essentially the same as used in undergraduate engine tests (at City University, London for example) at least until the 1960s. The invention is attributed to John Southern; he was originally employed as a draughtsman, (on the condition he gave up music as a 'source of idleness'!) and became 'a very valuable assistant' and then manager and partner [16, pp. 157,158]. Hills makes the telling comment 'It is not known who recognised that this diagram could be used to calculate the power output of the engine' [17, p. 84]. This would have been a crucial step – as it required measuring the net positive cyclic area of the p–v plot. In the 1960s experiments referred to above this was achieved using a planimeter.

An example of an indicator diagram of that time is given as Fig. 7

4.3 Keeping it quiet

Boulton and Watt traced four other pirate engines by this concern … the numbers of pirates

R.L. Hills [17, p. 175, 218]

Carnot did not make use of the indicator diagram. In all probability he had never heard of it

D.S.L. Cardwell [4, p. 199]

Watt and Boulton had perpetual trouble with 'pirates' stealing their designs. This was because, according to Hills, 'rival types of engines were less successful' [17, p. 218]. However, the invention of the indicator enabled Watt and Boulton to determine 'how much power their engines produced' [17, p. 72]. It was therefore of very considerable commercial advantage and Watt and Boulton successfully kept things secret, at least probably until 1826 [4, p. 220], which

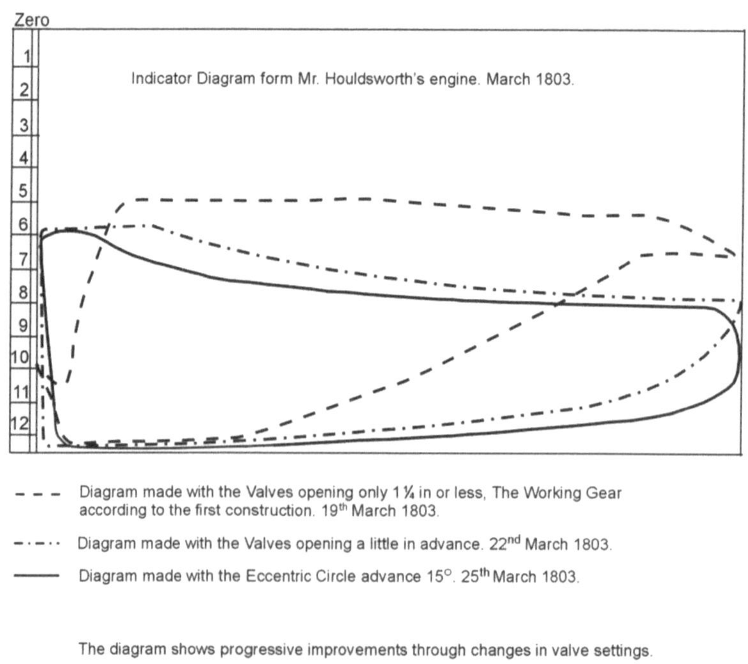

Figure 7: An example of Watt's Indicator Diagram (after Hills [17, p. 86]).

is *after* Carnot's Reflections was published. However, and the exact reason is not known [4, p. 220], Clapeyron used the p, v diagram freely to explain the Carnot cycle in his important 1834 publication. This will be described below.

5 The Carnot cycle: Carnot's great achievement

5.1 Introduction

In Section 12 of this chapter, we will assess in some detail – from our twenty-first century perspective – the relative contributions made by Carnot, Kelvin and Clausius to the Second Law of Thermodynamics. Here we make an assessment of Carnot's greatness in terms of:

- the debt he owed, and how his achievement can be assessed by comparison with, Watt
- how he was ignored by the scientific establishment
- how Clapeyron and Clausius took up his theory and
- how Kelvin and his brother viewed his work

5.2 Comparison with Watt: Watt's ideal and perfect engines

The engine he (Watt) envisaged was, in a sense, an ideal engine

D.S.L. Cardwell [4, p. 54]

Watt claimed that he had invented the perfect engine

R.L. Hills [16, p. 31]

Just as Kelvin and his contemporaries followed Carnot, so others, notably Watt, preceded him. Now to this point we have given an appreciation of Watt in *his* historical context of steam-engine development and the Industrial Revolution. The succeeding sections give credit enough to Carnot for us to appreciate the sheer force of his achievement (but, of course, ignoring the caloric misconception, common at the time). We have made the point that Watt predated Carnot with the *idea* of an 'ideal' (or conceptual) engine, on which was based his 'perfect' (and achieved) engine. Here, by comparing Watt's perfect engine with the Carnot cycle, we can appreciate Carnot's achievement more fully. We use as a source for the former Cardwell's description [4, pp. 47–49], together with diagrams.

Figures 8–10 compare Watt's ideal and perfect engines with the Carnot engine and its cycle, where Figs 8 and 9 address the engine arrangements, and Fig. 10 the processes involved.

It is important to realise that the motives of Watt and Carnot were quite different. In one respect Watt achieved more than Carnot in that he successfully built an engine.

5.3 Comparison with Watt: Watt's conceptual steps

Watt's intuitive insight into thermodynamics

D.S.L. Cardwell [4, p. 54]

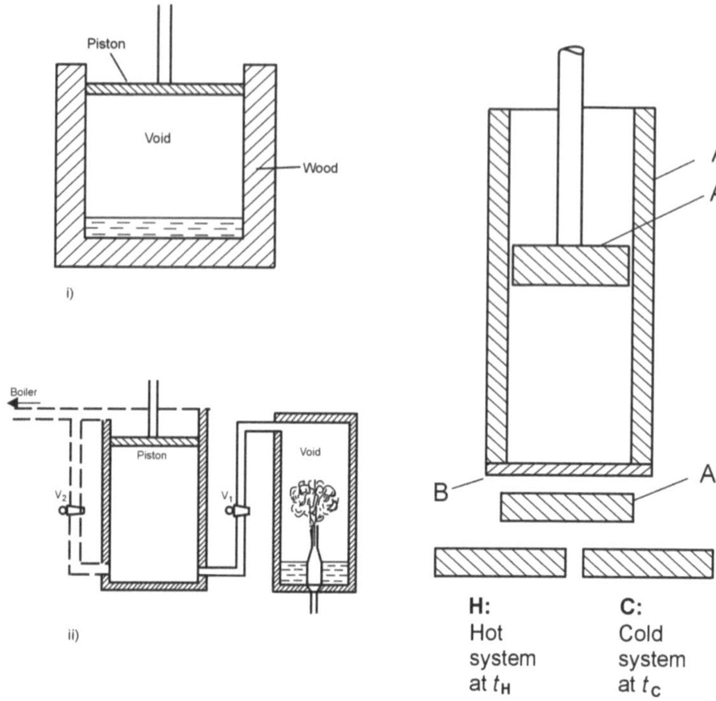

Figure 8: Comparison of Watt's perfect engine with the Carnot Cycle: 1. The engine arrangements: (a) Watt's perfect engine: (i) the impossible ideal engine and (ii) the 1765 Watt engine (after Cardwell [4, pp. 48/49] and (b) Carnot cycle (adapted from Spalding & Cole [38, p. 231]).

	Watt's impossible **ideal engine**	Carnot's engine
Practicality:	'Impossible' ([4], p 48)	'Wholly impracticable' ([4], p 194)
Structure:	Single piston-cylinder	Single piston-cylinder
Material:	Wood (like Newcomen)	Perfect thermal insulator
Fluid:	Steam/water Watt's 1765 engine	Generic: example air
Thermal separation of source and sink temperatures:	By using a separate condenser (that is: practicable)	By having Hot & Cold perfectly conducting caps (that is: impracticable)

Figure 9: Comparison of Watt's perfect engine with the Carnot cycle. (1) The engine arrangements (details).

Watt carried out two conceptual steps. *Combined, they form an example of superb engineering achievement and justify his enduring reputation.*

1. Watt's ideal engine. He extrapolated from the Newcomen engine a physically impossible 'ideal' engine given as Fig. 8(a, i) (Cardwell [4, p. 48]. The notional operation of this was based on the premise 'that for maximum economy the cylinder must be kept hot all the time, but for maximum power it must be cooled down once every cycle. These desiderata are quite incompatible, for a cylinder cannot be both hot and cold at one and the same time' [4, pp. 46/48]. This was Watt's first conceptual step: he identified the two key design requirements, incompatible though they were.
2. Watt's perfect engine. He then approximated to this as best as he possibly could, resulting in his 'perfect engine' design, Fig. 8(a, ii), termed 'The Watt Engine' of Cardwell [4, p. 49]. The essence of this second step, was his inspiration of having have *two* cylinders, which facilitated and drove the design process.

5.4 Comparison with Watt: Carnot makes heat-engine history

...sixty years were to elapse before ... Sadi Carnot ... one great synthesis

Everything that has happened since 1824 has confirmed Carnot's judgment

D.S.L. Cardwell [4, pp. 55/208]

Carnot's concept, on the other hand, was truly *ex novo*; he did not extrapolate from practice, nor had he an immediate intention to construct an 'approximating Carnot engine.' It was true that Carnot 'abstracted from the highly complicated mechanical contrivance that was the steam engine (even as early as 1824)' [4, p. 201] but that abstraction was of 'the essentials, and the essentials alone, of his argument. Nothing unnecessary is included and nothing essential is missed out.' Carnot's hugely important creative engineering leap can to some extent be appreciated from Fig. 9 which extends the comparison of Fig. 8. Various aspects of Watt's two engines, Fig. 8(a, i and ii), are detailed and compared with the same aspects of Carnot's 'abstraction' Fig. 8(b).

It was not just the engines themselves, but the processes which they could carry out. In Fig. 10, we compare these. Firstly, Fig. 10(a) gives the twentieth century conceptual indicator diagrams for the Carnot cycle for (a) gas and (b) steam, and then in Fig. 10(b) details of Watt's achieved processes are compared with these Carnot processes

The two noted aspects (* and +) of Fig. 10 reflect Carnot's greatness. Further to *, 'It was Carnot and Carnot alone, who saw that only if the working substance is brought back at the end of each cycle, *exactly* to its original condition, can we be sure that all the heat ... has been used to perform work' [4, p. 200], or as put by Mendoza 'Above all, he invented the closed cycle of operations' [5, p. x]. Then for + 'Again, it was only Carnot who saw the importance of reversible changes' [4, p. 200], '...he saw ... the need to avoid irreversible changes' [5, p. x].

Cardwell's crowning compliment to Carnot is: 'Finally, Carnot was unique in that he perceived that a cold body-or at any rate a temperature difference was as necessary for the performance of work by heat as was a hot body, or a source of heat' ... the essence of the second law of thermodynamics' [4, p. 201]. Now Fig. 9 shows that Watt essentially achieved this in 1765 with his separate condenser, and *to some extent* (our italics) in his striving after a perfect engine Watt

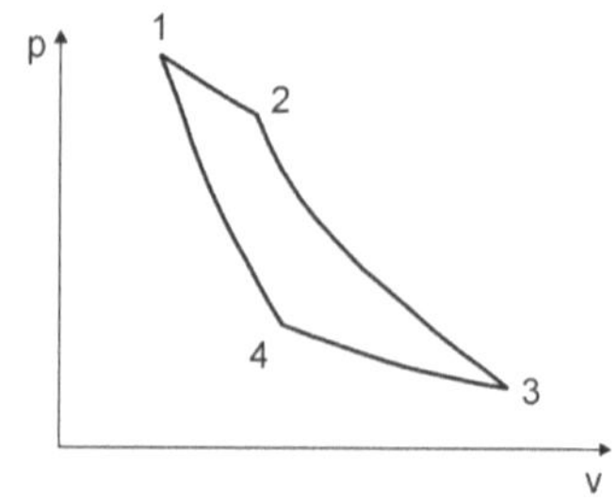

(a) Carnot cycle for gas (adapted from Cardwell, [4], Figure 16, p 194)

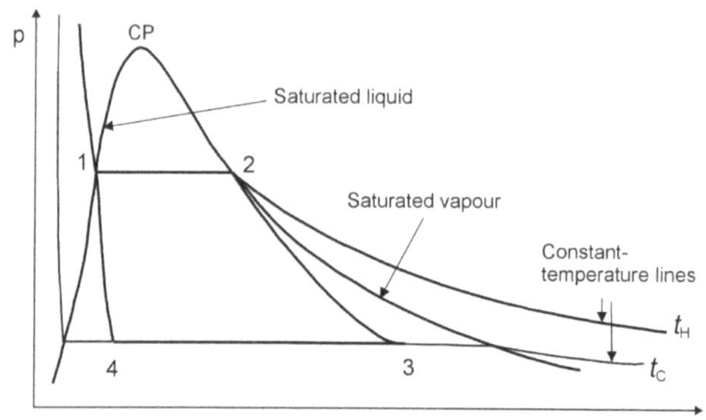

(b) Carnot cycle for steam (after Spalding and Cole [38], Figure 11.9, p 231)

	Watt's 1765 engine	**Carnot's engine**
Working fluid:	Steam/water: necessity for condensing/boiling is focus for processes	Generic/gas: focus is on ensuring fluid is in same thermodynamic state at end as at beginning, a *cycle**
Control of processes:	via automatic valves	via three ideal caps
Movement of piston:	external atmospheric steam/internal vacuum	driven by internal pressure
Individual processes:	includes necessity for inter-cylinder movement	defined by necessity for avoiding heat losses +

Figure 10: (a) The Carnot cycle for (a) gas and (b) steam: twentieth century conceptual indicator diagrams. (b) Comparison of Watt's perfect engine with Carnot's cycle engine. (2) Processes.

* Cardwell notes the 'implicit recognition' of this [4, p. 198] by steam engine designers, including Trevithick.

+ Without using the term 'reversibility' [3, p. 9] at that point, Cardwell discusses the requirement very nicely: 'there must be no contact between bodies at different temperatures during the operation of a heat engine' [4, p. 193].

anticipated Carnot by over 50 years. Could it be that in complimenting Watt so much Carnot was recognising him as a sort of authentic forerunner, a 'brother under the skin'?

5.5 Ignored by the scientific establishment

For would-be scientific discoverers the following fact can be taken either as a steely encouragement or as a sheer discouragement: Carnot could barely have had a worse scientific response. It may have achieved 'one long and excellent review' [5, p. xi], 'a long and favourable review' in the *Revue Encyclopaedique* and a mention in *Bulletin des Sciences Technologiques* [4, p. 211], but for the rest – unalloyed bleakness. Mendoza's description has a dreadful beauty: 'Carnot's *Reflexions* had been received by the official world of science in utter silence. No contemporary reference to his work can be traced in any published paper; his personal friends seem to have been the only people who had read his book' [5, p. xii]. Cardwell records Kelvin's efforts as scientific archaeology! 'When, many years later, the young and enthusiastic student William Thomson (Kelvin) tried to buy a copy of the book he found that the Paris booksellers had never heard of it, or its author' [4, pp. 211/212].

It was Clapeyron who rescued Carnot, and that posthumously. 'The only person who kept the memory of Carnot's results alive was Emile Clapeyron' [5, p. xii]: 'he rescued Carnot's ideas from the oblivion into which they had fallen and restated them with such conviction and clarity that they aroused…interest…' and gave him in turn a 'posthumous reputation' [4, p 220].

5.6 Clapeyron and Clausius on Carnot

Elegantly combined by Mendoza [5], with the translation of the *Reflexions* [8], are translations of the key associated papers by Clapeyron in 1834 [5, p. 71; 40]) and Clausius in 1850 [5, p. 107; 41]. In Clapeyron's words 'The idea…seems to me to be both fertile and beyond question…it seems to me to be free of all objections…S. Carnot, avoiding the use of mathematical analysis, arrives by a chain of difficult and elusive arguments at results which can be deduced easily from a more general law which I shall attempt to prove' [5, p. 74]. In relation to Clapeyron's stated attempt, Mendoza makes the comment that 'Practically all his equations are correct, whereas many of Carnot's are incorrect, however they are construed' [5, p. xiii].

Turning from Clapeyron to Clausius: 'The most important of the researches here referred to was that of S. Carnot, and the ideas of this author were afterwards given analytical form in a very skilful way by Clapeyron' [5, p. 107]. Also, towards the end of the paper, Clausius demonstrates that the 'work equivalent of the unit of heat' (i.e. the 'replacement' of caloric initiated by Joule) was indeed constant.

Quantitatively the expenditure of a unit of heat (that quantity of heat which will raise the temperature of 1 kg of water from 0° to 1°) will lift 'something over 400 kilograms to the height of $1^{m'}$ [5, p. 150,151]. Clausius compares his theoretical value of 421 with no less than *three* experimental values obtained by Joule, of 460 (magneto-electricity), 438 (expansion of atmospheric air) and 425 (a large number of friction experiments') [5, p. 151]. The telling point is Clausius's conclusion, which we quote in full: 'The agreement of these three numbers, in spite of the difficulty of the experiments, leaves really no further doubt of the correctness of the fundamental principle of the equivalence of heat and work, and their agreement with the number 421 confirms in a similar way the correctness of Carnot's principle, in the form which it takes when combined with the first principle.' Leaving aside the reflection this gives of the genius of *Clausius*, this superb example of concentrated thermodynamic logic both buries the caloric issue for Carnot, and via Clapeyron's mathematics, justifies Carnot's essential thesis.

5.7 James and William Thomson on Carnot

Mendoza makes the point that, 'After 1850, a number of papers were published which rewrote Carnot's results in the new terminology.' Kelvin is included in this, but Mendoza selects Clausius's paper over that of Kelvin because 'its terseness is more attractive than Kelvin's rather pompous style' (quotes from [5, p. xix]. What of Kelvin then? We take up the story a decade after Clapeyron's Carnot-related paper. 'The discussions in James's letter of 1844 are the earliest record of implicit references to Carnot and Clapeyron by the Thomsons. They show a considerable understanding of the basic principles involved in Carnot's theory' [7, p. 289]. This dialogue continued and, James in 1846, like Clausius in 1850, expressed admiration for Clapeyron's analysis : 'a very beautiful piece of reasoning, and of course perfectly satisfactory.' This was followed by Kelvin's key compliment to Carnot in 1849: 'nothing in the whole range of Natural Philosophy is more remarkable than the establishment of general laws by such a process of reasoning' (all from [7, p. 294]. *In other words, Kelvin's praise for Carnot was unqualified.*

6 The Carnot cycle: diagrams of Clapeyron and Clausius

6.1 Clapeyron and Clausius

Clapeyron deduced new results and made their derivation clear through the use of the indicator diagram

E. Mendoza [5, p. xii]

He represented the Carnot cycle in terms of a Watt indicator diagram

D.S.L. Cardwell [4, p. 220]

...given analytical form in a very skilful way by Clapeyron...

R. Clausius [41, p. 109]

After 1850, a number of papers were published which rewrote Carnot's results in the new terminology. Of these, Clausius' has been selected for this volume.

E. Mendoza [5, p. xix]

Leaving aside the details of Clapeyron's analysis, for the very first time in 1834 the Carnot cycle appears as a two-dimensional graphical plot: the p–v indicator diagram (in consistency with the Two Property Rule ([38, pp. 118–119] – two properties normally sufficient to define the thermodynamic state of a system). Now Clapeyron and Clausius were thorough: *both of them addressed both classes of working fluid* – permanent gases and steam – and gave the diagrams corresponding to them.

6.2 p–v diagrams for gases

Figure 11 gives the complete sequence for Clapeyron's and Clausius' p–v diagrams. Essentially, Clausius' versions are repeats.

Process in (b), (c)	Cap on end of Cylinder in Figure 8 (b)	
CE	H	HEAT INPUT: Isothermal expansion (Boyle's Law)
EF	A1	Adiabatic expansion (unknown Law)
FK	C	HEAT REJECTION: Isothermal compression (Boyle's Law)
KC	A1	Adiabatic expansion (unknown Law)

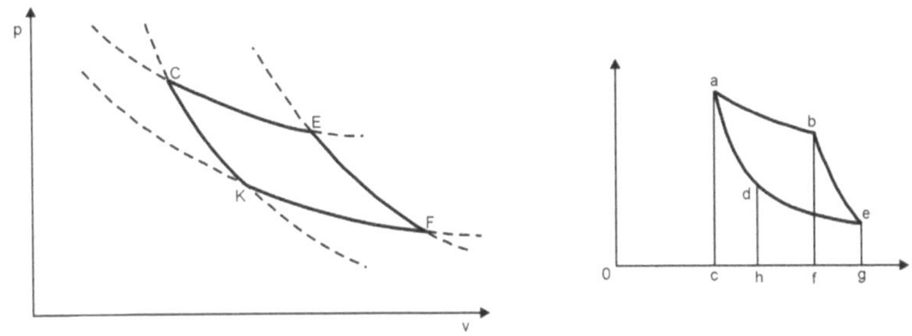

(b) Basic diagram (Clapeyron)

(d) Basic diagram (Clausius)

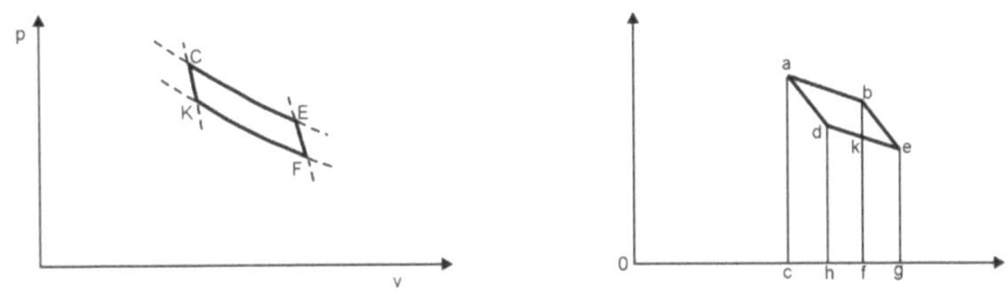

(c) Parallelogram (Clapeyron)

(e) Parallelogram (Clausius)

Figure 11: p–v diagrams of Clapeyron and Clausius of the Carnot cycle for a gas: (a) description of processes: after Cardwell [4, p. 220] and with reference to Fig. 8; (b) basic diagram for a gas: after Cardwell [4, Fig. 19, p. 220]; (c) 'small parallelogram' model: after Cardwell [4, Fig. 20, p. 222]; (d) Clausius' version of (b) –abcd replaces CEFK– after Clausius' [41, p. 118]; and (e) Clausius' version of (c) – abcd replaces CEFK – after Clausius' [41, p. 119].

Notes: Cardwell's versions of (a), (b) and (c) are more reader-friendly, partly because of Clapeyron's use of the term 'caloric.' Clapeyron's original versions may be found in [40] as (a) pp. 75–77, (b) Fig. 1, p. 75 and (c) Fig. 3, p. 83.

In interpreting Fig. 11, the idea behind (c) of the 'infinitesimal quadrilateral' [40, p. 83] becoming a 'small parallelogram' [4, p. 222] is that it enabled Clapeyron to compute the integrated p–v diagram, that is the cyclic work done. Clausius, starting from Clapeyron's analysis [41, p. 109], using data from Kelvin (p 137) and Joule (p 151), ditching 'caloric' and systematising the mathematics, gave essentially the same p–v diagrams.

6.3 p–v diagrams for steam

Figure 12 gives an overview, similar to Fig. 11, of how Clapeyron and Clausius interpreted the Carnot cycle for the fluid used overwhelmingly in the then current heat engines. What is missing is the vapour-liquid phase boundary.

Process in (b)	Cap on end of Cylinder in Figure 8 (b)	Description
CE	H	HEAT INPUT: Partial boiling takes place
EG	A1	Adiabatic expansion (unknown Law)
GK	C	HEAT REJECTION: Partial condensation takes place
KC	A1	Adiabatic compression (unknown Law)

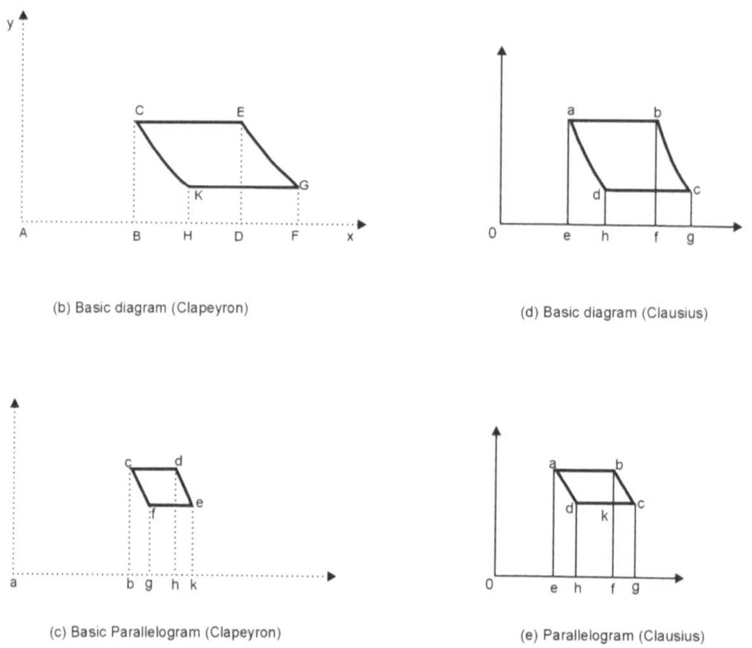

Figure 12: p–v diagrams of Clapeyron and Clausius of the Carnot cycle for steam. (a) description of processes with reference to Fig. 8; (b) basic diagram for steam: after Clapeyron [40, Fig. 2, p. 79]; (c) ' small parallelogram' model – cdef replaces CEGK – after Clapeyron [40, Fig. 4, p 89]; (d) Clausius' version of (b) – abcd replaces CEGK – after Clausius' [41, Fig. 3, p. 123]; and (e) Clausius' version of (c) – abcd replaces cdef – after Clausius' [41, Fig. 4, p. 124].

6.4 Reversed cycle heat engines

It was part of Carnot's conceptual framework that the whole heat-engine cycle could be exactly reversed. By putting work *into* such a device it would then be possible to transfer heat from a lower to a higher temperature. The practical significance is very substantial – not only could heat-engine engineers deliver to humankind *useful work* from the chemical energy of a fuel, they could also deliver *refrigeration* and *pumping of heat* (useful heat from a low temperature source). Clapeyron and Clausius also incorporated this concept, as shown by their quotes given *as* Fig. 13. The mathematically simple logic belies the engineering importance.

<u>*Gas*</u>

The reverse operation is equally possible............

Clapeyron [40], p 78

If the process above described is carried out in the reverse order

Clausius [41], p 118

<u>*Steam*</u>

In the same way as for gases, it may be proved that by repeating the same operation in the reverse order....

Clapeyron [40] p 80

Figure 13: Reversed Carnot cycles, after Clapeyron and Clausius: diagrams not given, only above quotes are used.

6.5 Realistic simulated versions of the *p–v* diagrams

Clapeyron and Clausius anticipated the practice of generations of authors and students with respect to their *p–v* diagrams. Figures 11 and 12 are constructed in illustrative, rather than true scale mode, as are almost all of the property diagrams in this chapter. However, examples of more true-to-scale versions of the Carnot gas and steam cycles are given in Fig. 14.

6.6 Carnot, Clapeyron and Clausius

Mendoza's three papers fit well together and effectively summarise one of the main messages of this chapter. Clapeyron and Clausius confidently promulgate, in company with Carnot, the universality of his cycle: *it applies equally to all fluids and to direct and reversed heat engines*. Cardwell makes several striking assessments of Carnot: the approach of the present chapter, in identifying the particular legacy of Watt, reflects the following quote.

The image of Carnot as a wholly isolated figure has been dissolved. Nevertheless in doing this we have not explained the thought processes whereby he came to make his extraordinary synthesis. Indeed we cannot do so, for he was one of the most truly original thinkers in the whole history of science [4, p. 211].

7 The Carnot cycle for a gas

7.1 Introduction

There have been made, it is said, recently in England, successful attempts to develop motive power through the action of heat on atmospheric air. We are entirely ignorant in what these attempts have consisted.

Sadi Carnot [5, footnote p. 56]

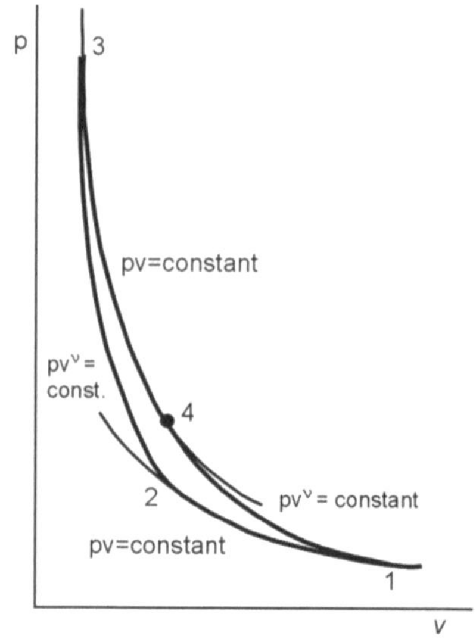

(a) For gas

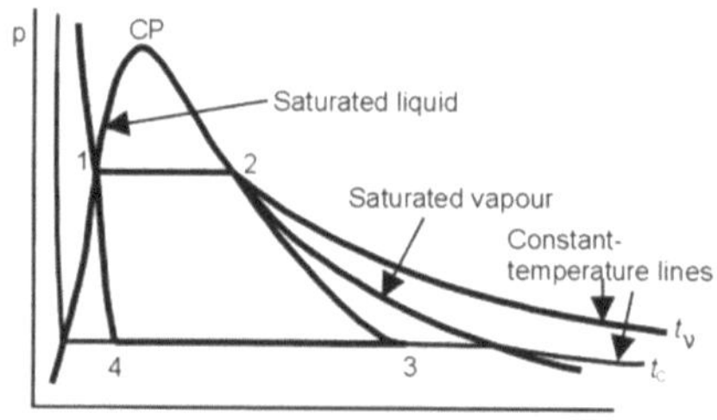

(b) For steam

Figure 14: Simulated (more true-to-scale) p–v diagrams of the Carnot cycles for gas and steam: (a) Carnot cycle for a gas after Rogers and Mayhew [3, Fig. 12.19, p. 270] and (b) Carnot cycle for steam: after Spalding and Cole [38, Fig. 11.9, p. 231].

Notes: in (a) the unknown relationship for adiabatic processes has been replaced and in (b) CP indicates critical point.

The doyen of Scottish air-engine inventors, Robert Stirling … John Ericsson's ambitious project

Crosbie Smith [15, p. 157/8]

As noted immediately above, part of the genius of Carnot was his insight that his ideal cycle was independent of working fluid. In fact, he comments that if certain 'difficulties' can be overcome the use of air 'would doubtless offer a notable advantage over vapor of water' [5, p. 5]. In Mendoza's gallery of eight engines 'discussed by Carnot' [5, Plate 1 wording], Robert Stirling's (1790–1878) air-engine is the sixth. But in fact, Carnot had no direct knowledge of this ('we are entirely ignorant' [5, footnote, p. 5]. Now historically, as Figs 1 and 5 show, heat-engine development was originally based exclusively on steam, with the final principal strategic invention, that of Trevithick's first locomotive, dating at 1804 [26, p. 1]. This gives the historical context for the invention of Stirling's air-engine. In fact it is a tribute to the fertility of human inventiveness, and that of Robert Stirling in particular, that an alternative working engine using air should be produced, being patented in 1816 [42].

The Stirling engine invention and patent was based on and specified a heat exchanger of considerable conceptual originality. Stirling himself termed it a *Heat Economiser*, and Rankine a *regenerator* [45, p. 140]. Rather later, the Swedish-American John Ericsson (1803–1889) [43] independently invented a somewhat similar air engine,-'the caloric or hot air engine.' Given the prevailing confusion about caloric it was not altogether surprising that Kelvin's brother James concluded that *even by 1848* Stirling 'does not understand his own engine'[44]. John Ericsson had the additional difficulty that his engines were unsuccessful. His *Novelty* locomotive, while fastest in the 1829 Rainhill Trials, had boiler problems, and the Trial was won by the Stephenson brothers [43, p. 2]. His 2000 ton *Caloric Ship Ericsson* [44, Fig. 5 and accompanying text] was sunk by a torpedo, raised and had its engine replaced with one using steam.

And yet, despite all these shortcomings, Stirling and Ericsson invented better than they knew. This section will show their theoretical cycles had Carnot efficiencies. This was achieved by incorporating a perfect heat exchanger between the pair of non-isothermal processes for either cycle. Stirling invented and patented his *Heat Economiser* in 1816, on which his engine was based, but today's practice follows Rankine in terming the device a regenerator [3, p. 271].

Two other air engine concepts should also be noted. James Thomson – and it is part of his contribution to the development of thermodynamics – had already conceptualised an air engine of his own by the time he met Stirling in 1848 [44]. He was convinced that his engine cycle was superior in certain respects to those of Stirling and Ericsson and in fact differed from the Carnot cycle. Despite his drawing up an actual design the 'Thomson air-engine' was never built. Another air-engine design – illustrated in Fig. 8.2, p 161 of [15] … was patented by Rankine and Napier, but again construction was not achieved. This was due to the negative effects of the above-mentioned Ericsson disaster.

7.2 Enter Rankine

7.2.1 Rankine and the p–v diagram

….the diagram described by Watt's Indicator is the type of all diagrams representing the expansive action of heat…as the principles of expansive action of heat are capable of being presented to the mind more clearly by the aid of diagrams of energy than by means of words and algebraical symbols alone, I purpose in the present paper to apply those diagrams…chiefly to the solution of new questions

W.J. Macquorn Rankine [45, pp. 115/6]

7.2.2 Rankine: the thermodynamic basis

William Macquorn Rankine (1820–1872) was a very significant figure in early thermodynamics. At Glasgow he was the successor to Lewis Gordon, the very first UK Professor of Engineering (Fig. 4). Here we focus on his integrated analysis of the Carnot and Stirling/Ericsson cycles., given in his iconic Royal Society of London paper of 1854 [45]. He had the advantage of decades of appreciation of Carnot's and Clapeyron's works of, respectively, 1824 and 1834 and of the recent publications of Kelvin and Clausius, all of which he acknowledged. Above all, however, is the foundation provided by Watt's 'well-known Steam-engine Indicator' as explained in the quote above.

We now summarise his logic. From a generalized p–v diagram as in Fig. 15(a) the integrated area of the curve gives what Rankine termed the 'motive power or 'Potential Energy' (in modern terms 'work output') during a complete stroke' [45, p. 115]. Although not making the point himself, *he has started to identify the clarity of thermodynamic understanding given by the p–v diagram. It is a measure of the essential performance of a working engine.* As already noted, Watt's secrecy about the diagram was because of its perceived vital commercial value.

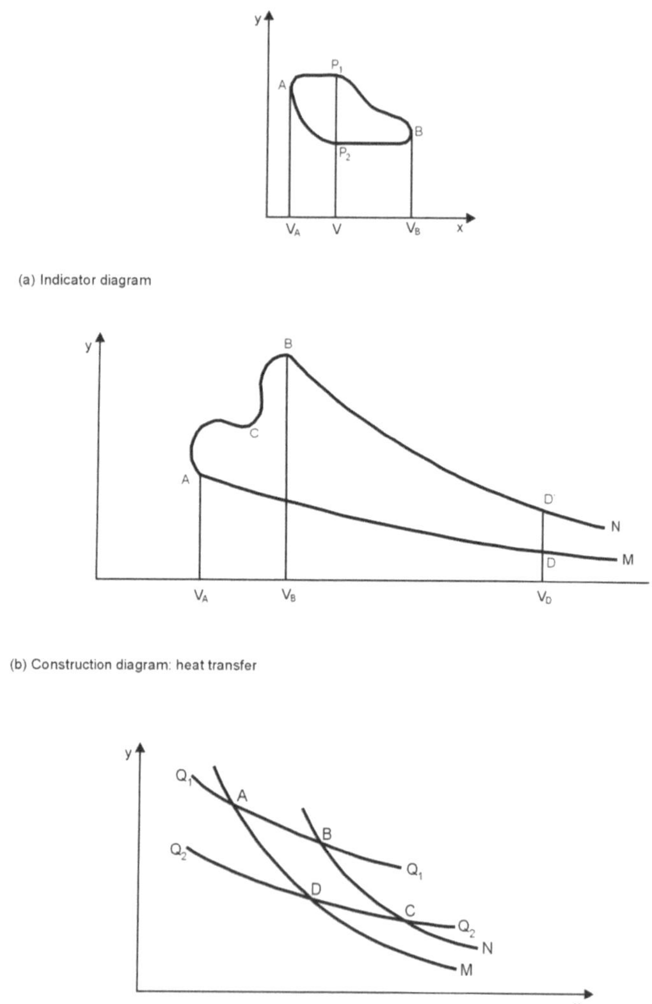

Figure 15: Rankine and the p–v diagram. I: Thermodynamic understanding: (a) generic indicator diagram: after [45, Fig. 1, p. 115]; (b) construction diagram to obtain heat transfer: after [45, Fig. 3, p. 119]; and (c) cyclic construction diagram giving thermal efficiency for a perfect engine: after [45, Fig. 6, p. 125];

Rankine then defines two kinds of curves on the *p–v* diagram as 'isothermal' and 'No transmission of heat' (our term adiabatic) and makes the penetrating deduction that at absolute cold (our absolute zero, c.f. p 149) the two sets coincide.

With reference to Fig. 15(b), he then shows how the *heat transfer* may be found for any non-adiabatic (hence including isothermal) process ACB by way of a Theorem. If from the end states A and B, adiabatic lines are drawn indefinitely in the direction of increasing volume to M and N, respectively, then the resultant area so generated represents the heat transferred. Not only so, but if the arbitrary line ACB lies partly above and partly below AM, the respective sub-areas represent heat absorbed and given out. Significantly, this enables the *p–v* diagram also to be used to determine the *thermal efficiency* for a cycle. Figure 15(c) shows a cycle for a 'perfect engine' consisting of two isothermal processes AB and DC and two adiabatic processes BC and DA (that is a Carnot cycle but not named as such). The 'two quantities of actual heat' Q_1 and Q_2 (our internal energy) stay the same for the isothermal processes AB and DC. The adiabatic curves BC and DA (maintaining direction) are the first part of the construction lines for the heat transfer deduction Theorem above. This means that the two areas MABN and MDCN represent the heat transfers at the upper (heat in) and lower (heat out) temperatures of the cycle, with ABCD the *net heat absorbed.*

Hence, the thermal efficiency $(Q_1 - Q_2)/Q_1$ = ABCD/MABN (his equation (10), p 125)

This led him on to the concept of *entropy*, but instead of accepting (or even acknowledging) Clausius's prior proposal of the expression, Rankine called it 'Thermodynamic function' (p 126).

7.2 Rankine: the Carnot cycle

Rankine then repeats the cycle of two isothermals + two adiabatics as in Fig. 16(a), defining it as a volume-limited cycle operating between v at A and C. He describes it as having the 'greatest efficiency of a Thermo-dynamic engine without a Regenerator' (p 140). Joule is credited for a (pressure-limited) air cycle where p at A and C define the cycle, but as with entropy and Clausius, so Carnot is not given credit here.

7.3 Rankine: the Stirling/Ericsson cycle

Rankine then makes the incisive assertion that, for the given isothermal lines Q_1 and Q_2 of our Fig. 16(a), a range of cycles can be constructed, where the adiabatic curves AD, BC of Fig. 16(a) are replaced with 'a pair of curves of equal transmission of any form,' where c and d represent the Carnot points C and D respectively. Therefore, Fig. 16(b) is the ideal indicator diagram for both Stirling and Ericsson cycles, where the AD and BC curves are for equal volume and equal pressure, respectively.

Rankine's adaptation of the Carnot statement for a non-Carnot cycle reads: 'The greatest efficiency of a thermo-dynamic engine, working between given limits of actual heat, with a perfect regenerator, is equal to the greatest efficiency of a thermo-dynamic engine, working between the same limits of actual heat, without a regenerator' [45, p. 142].

Despite avoiding crediting Carnot, this is a superb thermodynamic statement, its significance being identified almost 20 years later by Gibbs.

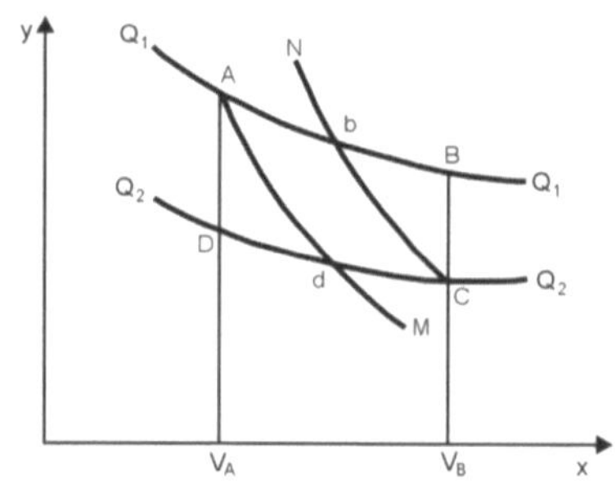

a) Carnot cycle

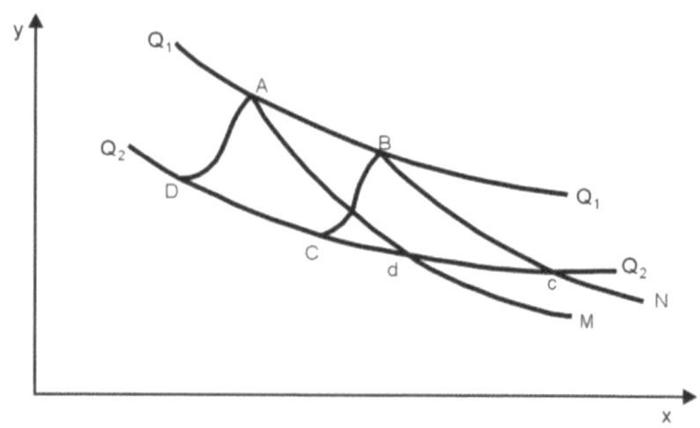

b) Stirling/Ericsson cycle

Figure 16: Rankine and the *p–v* diagram II: Cycles for Carnot and other perfect engines: (a) Carnot cycle: after [45, Fig. 16, p. 140] and (b) Stirling Ericsson cycle after [45, Fig. 18, p. 143].

7.4 *p–v* diagrams for the Stirling and Ericsson cycles

In present day terms, the *p–v* diagrams for the two cycles are shown in Fig. 17. This includes an illustration of the perfect regenerator matrix, necessary for achievement of the Carnot efficiency.

7.5 Engine history: the Stirling cycle

Stirling not only had an engine pumping water from a quarry by 1818 but his brother James built a large engine in the 1840s which drove his Dundee Foundry machinery [42, p. 2]. Also a working model was discovered by Kelvin at Glasgow University in 1847 and used in his lectures in 1850 [7, pp. 294–297].

Today, Stirling engines are commercially available, from model size to 500 W output [46]. Also the reversed Stirling cycle is 'widely used' in air liquefaction [3, p. 271].

7.6 Engine history: the Ericsson cycle

The initial lack of success has already been noted, but subsequently Ericsson's 'caloric engine' successfully competed with steam engines for lower power demands and made Ericsson 'comfortably wealthy.' However it is viewed as so similar to the Stirling engine,

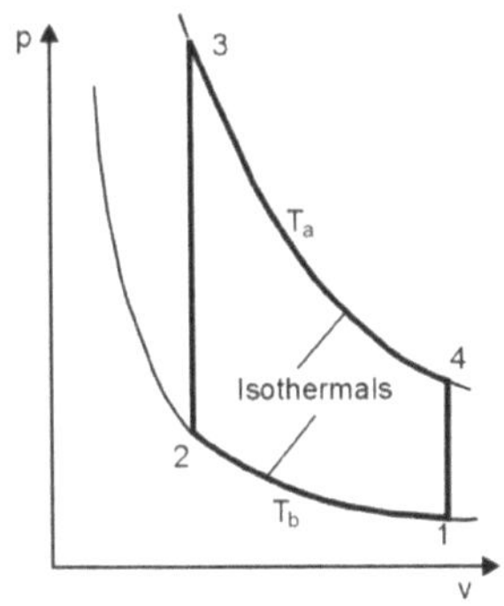

a) Stirling cycle

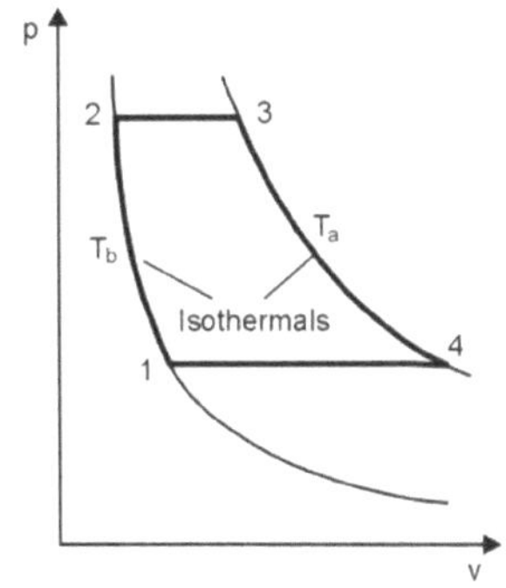

b) Ericsson cycle

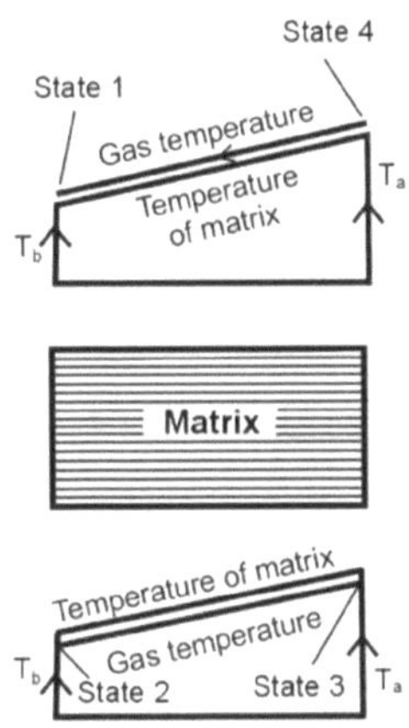

c) Regenerator structure

Figure 17: *p–v* diagrams for gas cycles having the Carnot efficiency: (a) Stirling cycle (heat input at T_a and heat rejection at T_b); (b) Ericsson cycle (heat input at T_a and heat rejection at T_b) and (c) Regenerator design for both cycles (all after Rogers and Mayhew [3, Figs 12.21 and 12.22, pp. 271/272].

Note: in the regenerator, as the hot air cools from T_a to T_b, it heats the cold air from T_b to T_a. Since the closed cycle mass flow rate is the same, an infinitesimal temperature difference may be postulated across the entire temperature range, assuming a perfect regenerator matrix.

that the fact of prior invention gives the 'Stirling Engine' title to the common concept (all quotes [43, p. 4]).

7.7 The *p–v* diagram: from Watt to Rankine

This concludes the fascinating story of the *p–v* diagram, spanning over 50 years. Originally kept as secret as possible by Watt for commercial advantage, it was eventually publicized by Rankine as a supreme thermodynamic descriptor. We now turn to its brother – the *T–s* diagram.

8 From p–v to T–s: Kelvin's absolute scale of temperature

8.1 A historical progression: the sequence in quotes

Watt called it an 'Indicator' because it showed the maximum and minimum pressures … it is not known who recognized that this diagram could be used to calculate the power output of the engine.

Richard Hills on Watt, Southern and the p–v diagram [17, pp. 83/84, date: 1796]

Thus he saw that in an engine – any engine – an amount of caloric fell from a high to a low temperature.

E. Mendoza on Carnot [5, p. x, date: 1824]

The principal result of Carnot's theory … had already become the basis of his famous absolute scale of temperature

Crosbie Smith and M.Norton Wise on Kelvin [7, p. 299, date: 1847]

Lord Kelvin introduced the absolute scale of temperature… which is independent of any material properties

Dilip Kondepudi and Ilya Prigogine on Kelvin [34, p. 76, date: 1847]

Clausius introduced this new quantity S in 1865 saying, 'I propose to call the magnitude S the entropy of the body…'

Dilip Kondepudi and Ilya Prigogine on Clausius [34, p. 80, date: 1865]

Just as pressure–volume diagrams … had long formed part of the science of thermodynamics …so Gibbs introduced other diagrams such as temperature-entropy

Crosbie Smith on Gibbs [14, p. 261, date: 1873]

We have to consider the following quantities … the (absolute) temperature … the entropy

Gibbs's use of *absolute* temperature and entropy [47, p. 1, date: 1873]

The unit of thermodynamic temperature is the fraction 1/273.16 of the thermodynamic temperature of the triple point of water. This unit is called the *kelvin,* denoted by the unit symbol K

Gordon Rogers & Yon Mayhew on the internationally adopted thermodynamic scale [3, p. 133, date: 1954]

8.2 A historical progression: Watt's experiments 1758–1765

This short section is a historical summary of the entire chapter, in that it traces how the heat engine focus progresses from the original p–v diagram to the T–s diagram. The above quotes explicitly demonstrate this progress, which dates from Watt's original series of experiments.

In 1758 Watt 'had no idea of the true principles of the steam engine' [15, p. 318] but his experiments from 1763 onwards resulted in a 'new engine' constructed in 1765 [15, p. 340]. The story of these iconic experiments on the model Newcomen engine is fully told by Hills ([15, pp. 329–340, more historical]) and by Cardwell ([4, pp. 40–52, general but more interpretive, as given in Fig. 8 (a)]. We are interested here in the parts pressure and temperature played. Over this period Watt perfected his method of measurements (the 'balance for weights, the barometer for atmospheric and steam pressures and the thermometer' p 330) and identified the crucial importance of latent heat (with Black's input p 334). The 'uncommon genius' (p 334) of Watt resulted in him being able to quantify the latent heat of steam and the heat capacity of the cylinder, and thereby *accurately* calculate the steam consumption (all quotes in [15, p. 339]). Temperature measurements were crucial in that research, in the process of which he measured the boiling temperature of water at different pressures, as given in Fig. 18. We have already mentioned his notebook facsimile [16, p. 132]; a facsimile of Watt's original graph is given in [15, p. 336].

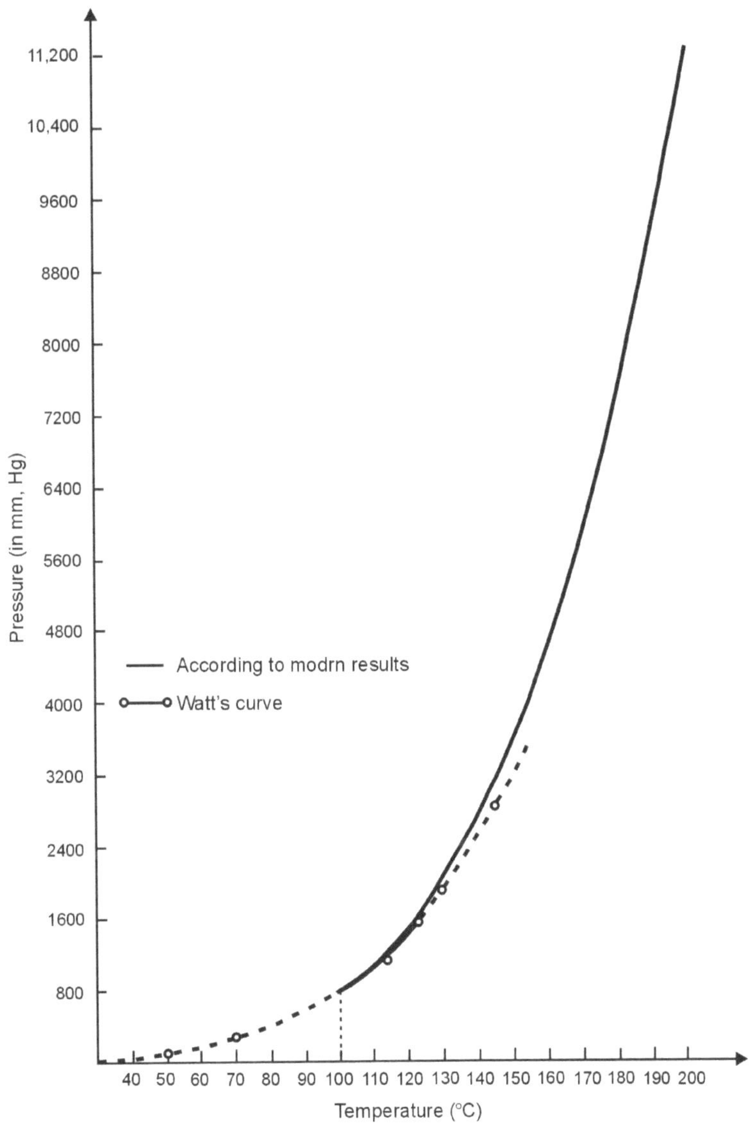

Figure 18: Watt's pressure-temperature measurements (after Cardwell [4, p. 47]).

Watt's engine followed the 'atmospheric' design, although instead of air at atmospheric pressure driving the piston against the vacuum in the cylinder, steam was used (Fig. 8(a, ii)). He could not construct a high-pressure boiler, despite its much greater output potential [4, p. 42]. Subsequently his successful patenting managed to fight off the development of high pressure steam engines, as discussed in Section 3.2 of this chapter.

This brings us to temperature, because hand in hand with a higher pressure would have been a higher temperature, as shown in Fig. 18. So Watt, by retaining the atmospheric design, limited the fundamental thermal efficiency and had to concentrate on improving engine output as represented by the *p–v* diagram.

Hence pressure, rather than temperature, became the key property. It took the whole sequence of Carnot's theorem highlighting temperature in 1824, of Kelvin's definition of the absolute temperature scale from 1847, of Clausius defining entropy in 1865 and then Gibbs's introduction of the *T–s* diagram in 1873 to put forward temperature as a dominant property. In consistency with this rather slow progress, up to this stage Fig. 8 (a, ii) of the chapter the *p–v* diagram has been used exclusively, reflecting the fact that *despite Carnot* Clapeyron, Clausius and Rankine all used the *p–v* diagram. From now on the *T–s* diagram will predominate.

Two major conclusions can be drawn from this sequence. The lack of understanding of the laws of thermodynamics and of latent heat, only serves to emphasise more strongly Watt's intuition and experimental ability. It was these which facilitated his ultimate engineering goal of manufacturing and improving working heat engines. Not only so but his *p–v* diagram formed a permanent contribution to the understanding of thermodynamics, being used to this day. *As previously noted Watt's was a truly brilliant achievement.*

Secondly, there is Kelvin's concept of the absolute temperature scale. The story is told, for example, in [7, pp. 249/250 with 299/301 and 324/325 and will be returned to in Section 12]. Whereas the extent of Kelvin's contribution to the Second Law is a matter of some debate, no such argument is needed to recognise his *superlative thermodynamic insight* in proposing the absolute thermodynamic scale. The internationally agreed use in 1954 of the degree Kelvin is no more than he deserved.

9 The 'representational' approach of J Willard Gibbs and the *T–s* diagram

> *'...it is the general if not the universal practice to use diagrams in which the rectilinear co-ordinates represent volume and pressure. The object of this article is to call attention to certain diagrams of different construction ... and preferable to it in many cases...'*
>
> **J. Willard Gibbs [47, p. 1]**

9.1 Introduction: J. Willard Gibbs

J(osiah) Willard Gibbs (1839–1903) was from a New World family with a history dating back to 1658, when his ancestor Robert Gibbs emigrated from Warwickshire, England [48, p. xi]. J Willard Gibbs epitomised the way in which a tradition of academic excellence can be systematically achieved both for an individual and for a University. His own CV was enviable: *Yale*: undergraduate (1854–1858),

New Jersey: postgraduate and tutor (1858–1866), one year each at Paris, Berlin and Heidelberg (1867–1879), *New Jersey* (1869–1871), returning to *Yale* as Professor of Mathematical Physics (1871 onwards). Before him, his father was a Professor (of Sacred Literature) at Yale for almost 40 years and 6 generations of the Gibbs family graduated either from Harvard or Yale. Unlike Kelvin, he was not a teenage prodigy, being 34 by the time he displayed in 1873 'evidence of his extraordinary powers as an investigator in mathematical physics' [48, p. 12]. At the time he wrote, almost all of Kelvin's British contemporaries were still alive (Fig. 4) and in fact Kelvin outlived Gibbs.

9.2 The *T–s* diagram

The main focus of our interest is in Gibbs's introduction of the Temperature-Entropy diagram within a general analysis of 'graphical methods in thermodynamics' [47, Title, p. 1]. His basis was the five properties of Volume (V), pressure (p), absolute temperature (T), internal energy (U) and entropy (S) – using our notation which is consistent with Crosbie Smith's [14, p. 260]. He obtained (his equation (4) [47, p. 1, 2] in differentials the interproperty relationship for the combined first and second laws as:

$$dU = TdS - pdV \qquad [14, p. 261]$$

Hence a state is 'capable of two independent variations' [47, p. 2] and may be represented as a point on a (two-dimensional) graph. This is otherwise termed the Two Property Rule, mentioned in the Abstract of this chapter (Spalding and Cole, [38, pp. 118–119]). Further, a cycle may be represented as a closed loop [48, p..35],

$$\text{with the cyclic work } W = \int_1^2 p\, dV \qquad (5)\ [47, p. 3]$$

$$\text{and cyclic heat } Q = \int_1^2 T\, dS \qquad (6)\ [47, p. 3]$$

$$\text{where } W = Q \qquad [47, p. 3]$$

Based on the above, Gibbs promulgates the *T–s* diagram as being able to define the two perfect thermodynamic engines of Carnot and Rankine, the latter being the 'perfect regenerator' adaptation. These diagrams are as shown in Fig. 19.

9.3 Understanding the thermodynamics

At this stage we should not miss the thermodynamic wood for the trees. Despite the mathematical nature of Gibbs's approach in [46], and that of Rankine before him [45], both thermodynamicists adopted the heat engine rationale of *expressing their thermodynamics in terms of diagrams.* The following extended quote from Gibbs has an unexpected ending: 'The method in which the co-ordinates represent volume and pressure has a certain advantage … and its analogy with Watt's indicator diagram has doubtless contributed to render it popular. On the other hand a method involving *entropy* … will doubtless seem to many far-fetched and may repel beginners…this inconvenience is more than counterbalanced by the advantages of a method which makes the second law of thermodynamics so prominent … the ordinates shall represent the temperature … the use of the entropy-temperature diagram may serve a useful purpose in the

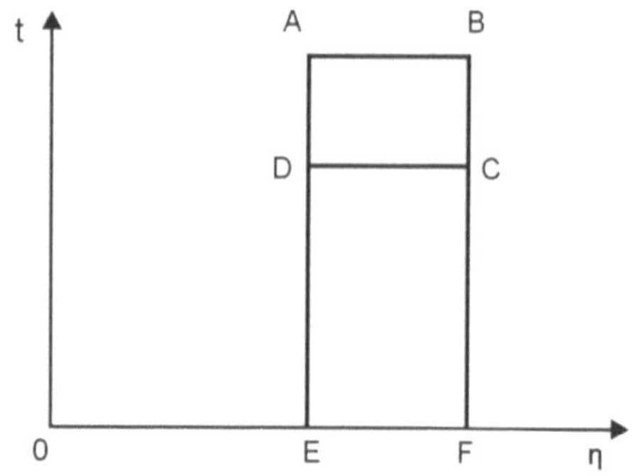

(a) Carnot cycle

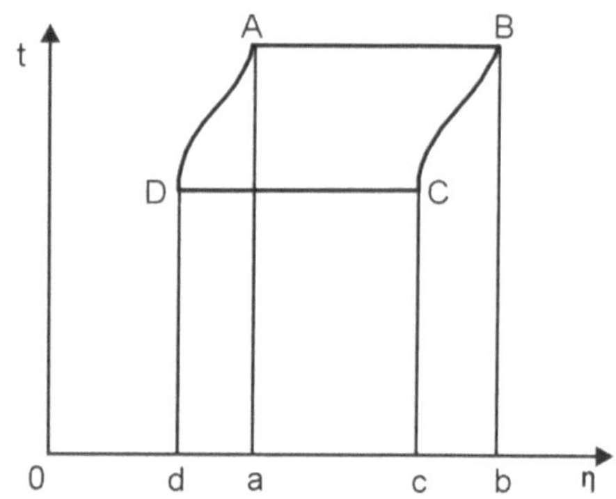

(b) Stirling/Ericson cycle

Figure 19: T–s diagrams of Gibbs for perfect thermodynamic engines: (a) Carnot cycle [47, Fig. 3, p. 10] and (b) Stirling/Ericsson cycle [47, Fig. 3, p. 10)].

popularizing of this science' [47, p. 11]. So Gibbs has the motive (unusual for a mathematician?) of *marketing* thermodynamics.

This lateralising approach by a pioneer mathematical thermodynamicist reflected to some extent Kelvin's masterly breadth of vision of himself:, the epitome of the mathematician turned experimentalist.

9.4 Three-dimensional diagrams: James Thomson and Maxwell

In his representational studies, Gibbs also considered three-dimensional surfaces [47]. Gibbs strongly preferred the volume/energy/entropy diagram , where simple differentiation yields the p–v -T relations [48, p. xiv]. His proposal 'of exceptional importance and beauty' inspired Maxwell to spend a 'considerable time, in the last years of his life in carefully constructing…a model of this surface (with) a cast' being sent to Gibbs [48, p. xiv]. Here we see Maxwell's alter ego of practicality, here catalysed by Gibbs's desire to make thermodynamics understandable.

Bumstead's introduction also mentions James Thomson's three-dimensionalising of the 2-D volume-pressure diagram by including temperature as a third co-ordinate [48, p. xiii]. In dismissing this surface Gibbs's thermodynamics sophistication overcame his marketing drive.

James Thomson's p–v -T surface has stood the test of time, especially since it involves the key complexity of expanding/contracting on freezing. The undergraduate text of Rogers and Mayhew, for example, gives the p–v -T surface for water [3, p. 141] and the statistical thermodynamics oriented text of Tribus reproduces Zemansky's p–v -T surfaces for the two expanding/contracting freezing alternatives [33, p. IX-23].

Could it possibly be that the different languages of thermodynamics which came across to a twentieth-century undergraduate (Section 1.1 of this chapter) are in fact an essential feature of the subject – namely a combination of high mathematics and engineering-oriented practicality?

10 The Carnot cycle for steam

It is necessary to establish principles applicable … to all imaginable heat- engines, whatever the working substance and whatever the method by which it is operated.

[5, p. 6]

10.1 Non-flow and steady-flow Carnot cycles

Carnot's immensely inspired objective, when fully quoted, extrapolates *from* steam engines, whereas in this chapter we have addressed gas first and steam second. None the less, here we show how the Carnot cycle applies to steam (hence fulfilling Carnot's 'whatever the working substance') and both to non-flow and steady-flow processes (additionally fulfilling his 'whatever the method by which it is operated'). Figure 20 gives the overall information. 20(a) for non-flow, refers back to the engine arrangement of Fig. 8 (b) and the p–v or 'indicator diagram' of Fig. 10(a). Figure 20(b) gives the Gibbs-style T–s diagram, which is the equivalent of Fig. 19(a) but with the dry-steam phase boundary added. In fact 20 (b) holds for *both* non-flow and steady flow, where the steady-flow cycle arrangement is given in 20 (c). The steady-flow steam version of the Carnot cycle, then, comprises the notional set of boiler + turbine + condenser + compressor.

We have already shown in Fig. 9 that the thermodynamic advantage of the Carnot proposal was not in its practicality – it was no more practicable than 'Watt's impossible ideal enginee.' To use that term, though, its thermodynamic advantage was in its transcendent 'degree of idealness.' For future exploitation with steam, the engineering challenge would be to devise a practical version of the steady-flow 20 (c). Here Rankine reappears.

Refer back

to Figure 8 (b)
engine arrangement

& Figure 10A (b)
p-v diagram

(a)

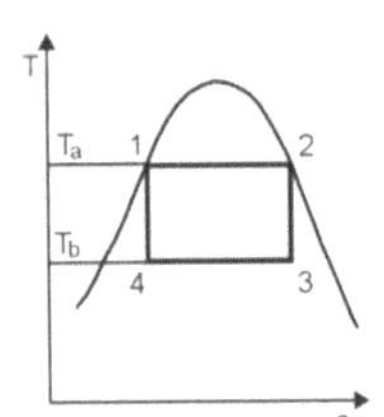

(b)

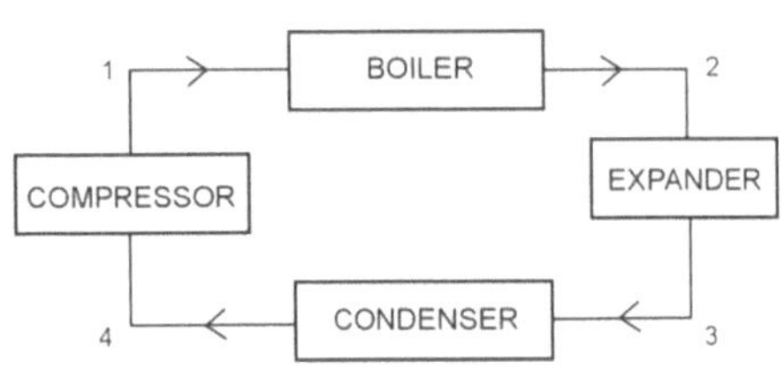

(c)

Figure 20: The Carnot cycle for steam: (a) Non-flow arrangement – refer back to Figs 8 (b) & 10A (b), (after Spalding and Cole [38, p. 231]); (b) common T–s sketch diagram (after Rogers and Mayhew [3, p. 210]); (c) steady-flow circuit diagram (adapted from Rogers and Mayhew [3, p. 210]).

10.2 The Rankine cycle

The more practical 'adapted Carnot cycle,' attributed to Rankine, is shown in Fig. 21. It (i) extends the impractical state point 4 of Fig. 20(b) to the phase boundary point 4 of Fig. 21(b), so giving true liquid water of zero dryness and (ii) carries out compression of the water by the feed pump of Fig. 21(a) rather than by the notional compressor of Fig. 20(c). The pump merely compresses the liquid to the (upper cycle) boiler pressure.

It should, however, be appreciated that this is still a notional cycle, in that the processes 2–3 and 4–5 are assumed to be isentropic or loss-free.

However, the Rankine cycle of Fig. 21 can be thermodynamically conceptually improved by *superheating*. This increases the thermodynamic efficiency of the cycle by adding heat at higher temperatures, as shown in Fig. 22. The superheating takes place away from the presence of liquid in the boiler but at the same pressure. This reflects the core Carnot condition that the efficiency of a perfect cycle is defined by the absolute temperature of the heat addition.

The logic can be taken a stage further by reheating, where the expansion is in two stages of high and low pressure, with intervening reheating at the intermediate pressure. This cycle is shown in Fig. 23.

The sequence of ideal cycles in Figs 21–23 reflects Carno *T*-style pointers to improvements. Whether or not they are worth carrying out is purely a question of cost: does the improvement in efficiency (so reducing the cost of generation) outweigh the additional capital cost of the extra complexity?

10.3 Carnot and Rankine cycles compared

By comparing Figs 20(b) with 21(b), 22(b) and 23(b) it is possible to appreciate the design process needed to conceptualise practical engineering plant rather than the ideal steady-flow Carnot circuit of Fig. 20(c). Such a comparison is made directly in Fig. 24, where the Rankine cycle with superheat is used as a basis. In the ideal Carnot thought-world the equivalent (but wholly impractical) Carnot cycle has all the heat added at the extreme superheat temperature. It has a much larger cyclic T–s area than the Rankine cycle. As we have seen in the discussion on Gibbs, this represents the net cyclic heat transfer, hence the net cyclic work transfer. So the Carnot cycle has a conceptual but wholly unattainable much higher work output. Also, the comparison

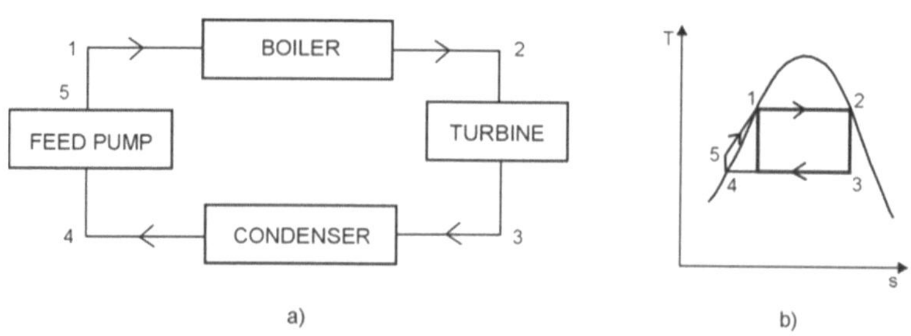

Figure 21: The Rankine cycle (adapted from Rogers and Mayhew [3, p. 214]): (a) circuit diagram and (b) *T*–*s* diagram.

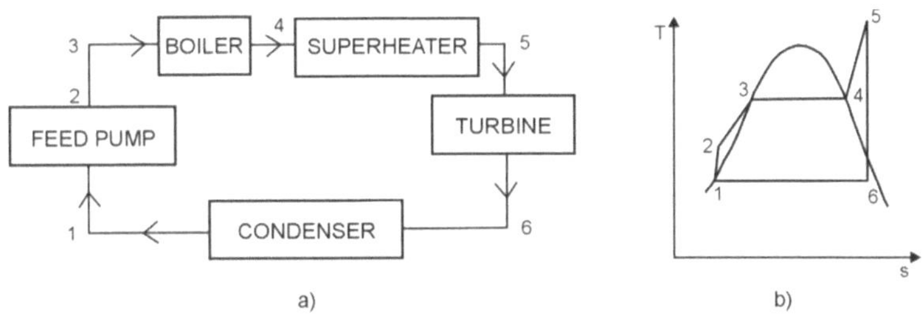

Figure 22: The Rankine cycle with superheat (adapted from Rogers and Mayhew [3, p. 217]): (a) circuit diagram and (b) *T–s* diagram.

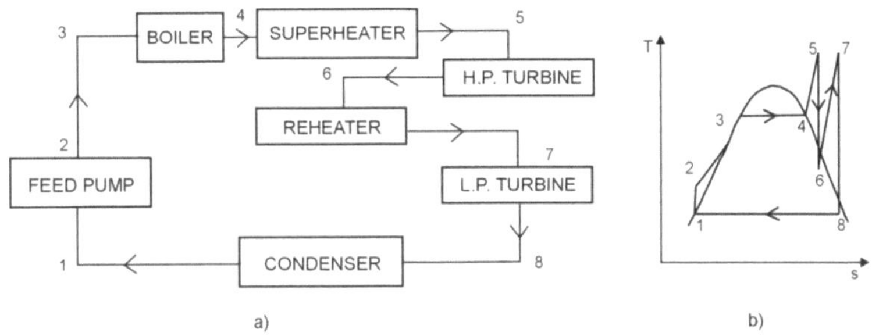

Figure 23: The reheat cycle (adapted from Rogers and Mayhew [3, p. 219]): (a) circuit diagram and (b) *T–s* diagram.

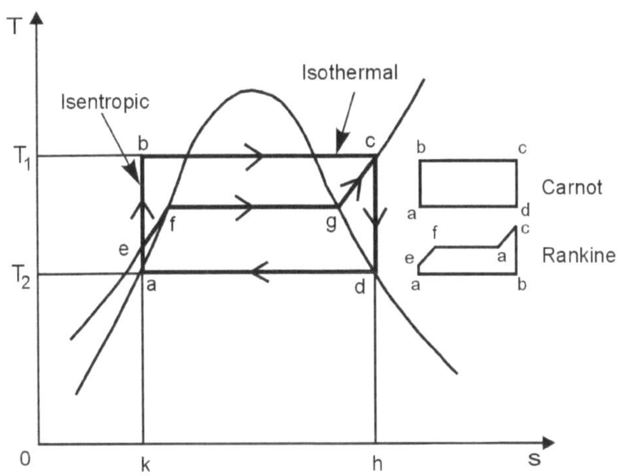

Figure 24: Comparison of cycles: Carnot with Rankine + superheat (after Spalding and Cole [38, p. 270]).

can be interpreted in reverse. The Rankine cycle with superheat is the best possible practical approximation to the Carnot cycle for steam between the temperatures in question. In other words, the process of going from the Carnot to the Rankine cycle is a similar engineering conceptual step that Watt achieved in going from his 'impossible ideal engine' to his '1765 engine' (Figs 8 and 9).

In general, given that historically so much mains-generated electricity has been via steam plant, the key factor is the generation cost of electricity over the lifetime of the power station. Generally speaking, for large output plant, the increase in efficiency resulting from an identifiable thermodynamic improvement outweighs the additional cost of the equipment. This may be seen, for example in the comparative calculations for Rankine cycle without and with superheating in [1, pp. 69–76] and in the sequence Rankine cycle, Reheat cycle, Regenerative cycle in [3, pp. 214–229].

11 The reversed Carnot cycle: refrigerators and heat pumps

11.1 The pure Carnot concept

In Section 6.3, we have noted the recognition by both Clapeyron and Clausius that the Carnot cycle may be exactly reversed. In acting in this way a reversed heat engine can become either: a *refrigerator*, where the reversed heat input is the cooling effect desired, or a *heat pump*, where the reversed heat rejection is the heating effect desired. Figure 25 shows the fundamental thermodynamic concepts, where 25 (a) is the reversed T–s diagram, and 25 (b) & 25 (c) the equivalent refrigeration and heat pump circuit diagrams respectively.

11.2 Practical cycles and circuits

We have shown how the impracticable direct Carnot cycle was replaced with the progressively more efficient steam cycle sequence of Figs 21–23. In a similar manner a basic refrigerator circuit and cycle are given as Fig. 26(a) and (b), respectively.

In a rather similar manner to the Carnot compressor (Fig. 20, 4–1) being replaced with the feed pump of Fig. 21 for the direct cycle, so the notional expander from 3 to 4 of Fig. 25(a) is replaced

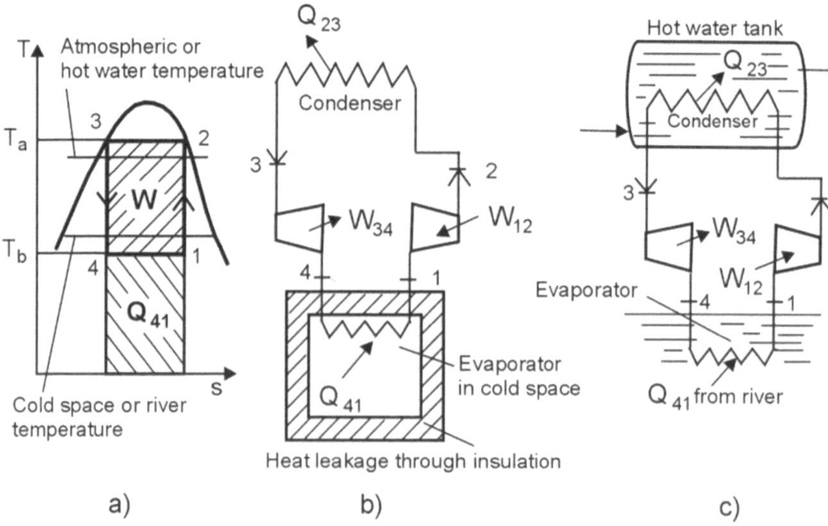

Figure 25: The reversed Carnot cycle (after Rogers and Mayhew [3, p. 274]): (a) *T–s* diagram and (b) conceptual Carnot refrigeration cycle, where the desired energy effect is Q_{41}; and (c) conceptual Carnot heat pump cycle, where the desired energy effect is Q_{23}.

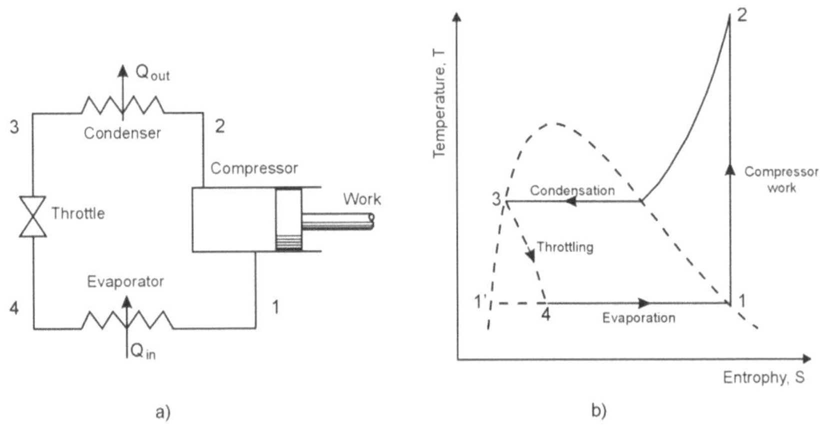

Figure 26: A simple refrigeration cycle (after Winterbone [1, p. 136]): (a) circuit diagram, where the desired energy effect is Q_{in} at the evaporator and (b) *T*–*s* diagram.

with the (non-isentropic) throttle valve of Fig. 26. Again comparing direct and reversed cycles, the reverse of the turbine work of Fig. 22 (b) is the compressor work of Fig. 26 (b): both lie completely, or almost so, in the superheat region.

12 Contributions to the Second Law of Thermodynamics: Carnot, Kelvin and Clausius

12.1 Introduction

The focus of this chapter has been on Carnot and the unique scientific character of his *Reflexions* [8]. His thesis led to the formal statements of the second law of thermodynamics some quarter of a century afterwards. *So what was the extent of Carnot's contribution to the second law*? And in addressing that question can it help us to understand better *Kelvin's contribution* to the Second Law? In so doing, we must, of course, also include Clausius.

12.2 Carnot: marks of a genius

How did Carnot come upon such beautiful, general and powerful arguments?

J.S. Dugdale [51, p. 34]

Carnot's contribution has been cited as one of the greatest intellectual achievements of the human mind

Myron Tribus [33, p. VI-20]

Until recently, engineers had a mixed record of paying strong compliments to Carnot. We take some representative examples. For Rogers and Mayhew: 'the original conception of a reversible cycle was due to Sadi Carnot,' and 'One ideal cycle … is that suggested by Sadi Carnot' ([3], originally 1957, pp. 60, 209). Similarly, Tribus pays compliment in his Introduction as 'the demonstration of the Second Law as given by Carnot and Clausius has been unchanged , in essence, in over one hundred years' [33, 1962, p. xv]. Then later there is a full presentation of Carnot's analysis including a proposed demonstration experiment (pp VI-17-VI-21) culminating in the unalloyed praise of the heading quote. Also we have Hatsopoulos and Keenan's 'Carnot introduced several new concepts. Carnot's principle that the reversible cyclic engine is the most efficient engine possible was in reality the second law of thermodynamics' [39, 1965, p. xviii]. Finally, Kondepudi and Prigogine are unstinting: '... Carnot ... a brilliant French military engineer ... Carnot's great insight ... Clapeyron ... realized its fundamental importance' [34, 1999, pp. 68, 69].

We turn to two specialised books on entropy. In *Entropy* by Fast [50] there is a full description of the Carnot cycle in pages 27-31, with one or two other uses of his name. Dugdale, in his more recent *Entropy and its Physical Meaning* by [51], however, accompanies a full explanation of the cycle (pp 27-35) with an expression of praise for Carnot himself: 'Carnot's analogy ... provided an astonishingly clear insight into the working of heat engines and from that to a profound understanding of thermodynamics' (p 34). This is high praise indeed, reflecting previous quotes in this chapter.

In addition a number of Journal articles address Carnot explicitly [6, 12,13], previously referenced in the context of caloric's meaning. The title of Ehrlichson's 1999 publication is quite explicit: *Sadi Carnot, 'Founder of the Second Law of Thermodynamics* [12]. Apart from the title, which is repeated in the paper, Ehrlichson uses such expressions as 'brilliantly evaded' (p 188) 'more amazing still' (p 189) and 'prophetically foreseen' (p 190). The fuller 2008 *Sadi Carnot's contribution to the second law of thermodynamics* by Lemons and Penner [6] has the same message. Their conclusion reads '¼what we call Carnot's statement of the second law is.... logically equivalent to Clausius's version of the second law (p 24). Finally, in 2009 we have Newburgh's *Carnot to Clausius: caloric to entropy* [13]. This covers the same ground, especially focusing on the meaning of caloric. It is wider in scope (Clapeyron's p–v diagram is used as Fig. 1) and describes the bound and free caloric of the cycle (Fig. 2). His explicit compliment is that 'Carnot conceived of an ingenious device' [13, p. 718]. In summary, then, Carnot's reputation is extremely high. Also, it has to be borne in mind that personal assessments in scientific publications are not common. The mere use of such expressions as Carnot cycle and degree Kelvin take for granted their contributions to the subject. Having said that, however, with the passage of time, a more open appreciation is becoming apparent. So it is the 1998 text Modern Thermodynamics [34], co-authored by the Nobel prize-winner Prigogine, which makes a virtue of including descriptions and portraits/photographs of the pioneers.

12.3 Kelvin, Clausius and the second law

...since become famous as Kelvin's statement of the Second Law of Thermodynamics...

Crosbie Smith and M.Norton Wise [7, p. 329]

...Clausius in 1850 and Thomson (Lord Kelvin) in 1851 proposed distinct versions of the second law of thermodynamics

D.S. Lemons and M.K. Penner [6, p. 23]

Detailed analysis of second law statements are in the relevant companion chapter in this volume: here we focus on Kelvin's (and necessarily Clausius's) enduring reputation, using the same approach as for Carnot.

To introduce this, we note that if compliments to Carnot fell short until recently, much the more so held for Kelvin. So in Rogers and Mayhew [3, p. 51] we read that their form of the second law 'is essentially the statement given by Planck.' Also, the Rogers and Mayhew First Corollary of the Second Law [3, p. 53] is presented as the Clausius statement. From this information the reader would not consider Kelvin as having contributed. In fact, their Reference 7 (p 692) is the same as Reference 21 of Lemons and Penner [6]. The latter refer to the Planck statement as only (p 23) that 'Planck corrected Thomson's original 1851 statement of the second law by insisting that the word 'only' be included,' thus giving primary credit to Kelvin. In fact, if we insert [6] into [3] then the preferred second law statement of Rogers and Mayhew is *sourced from Kelvin.*

The text of Hatsopoulos and Keenan [39] gives weak praise to Kelvin. Having read that in 1851 Kelvin 'credited Carnot and Clausius with the second law' (p xix), we find that Planck's 1897 'statement seems to have been one of several statements made by Kelvin' (p xx). Their second law focus, like Rogers and Mayhew, is on Planck's statement (p xxiv/xxv), but for H & K his (and Kelvin's in the background) 'proves to be a tautology' (p xxv). This is because the essential thesis of H & K is that the laws of thermodynamics may all be proven from three fundamental principles. As shown in their logic diagram (p 368) the First and Second Laws can be deduced from their Law of stable equilibrium and on those grounds the significance of the customary statements is discounted. Winterbone, in his 1997 *Advanced Thermodynamics for Engineers* [1] inadvertently gives the same impression. Despite being 'a strong believer in the ...necessity of introducing thermodynamics through the traditional route of the Zeroth, First, Second and Third Laws, rather than the Single-Axiom Theorem of Hatsopoulos and Keenan or The Law of Stable Equilibrium of Haywood'[1, p. x], it is Keenan's general criterion of equilibrium that has immediate focus as a boxed statement , on p 4. This compares with inconspicuous statements of the Laws themselves ('...the First Law of Thermodynamics, which is a law of energy conservation...the Second Law stated that is impossible to construct a heat engine that does not reject energy to its surroundings,' given in the simple text of introductory material for Chapter 2, p 13. However, ignoring this lack of emphasis, the *actual statement Winterbone uses* stems from Kelvin.

For Tribus [33] Clausius predominates: 'Carnot's contribution was expanded upon by Kelvin and later by Clausius, who in 1850 made it one of the two principles...the Second Law as given by Carnot and Clausius...' (p xv).

Kondepudi and Prigogine [34], make the same presentation as Rogers and Mayhew. Their first statement [34, p. 83] coincides with [3, p. 51] but neither Planck nor Kelvin are mentioned. Again, like R & M they give Clausius's statement as an alternative [34, p. 84; cf. 3, p. 53].

In their specialized books on entropy, both Fast and Dugdale give equal weight to the Clausius and Kelvin statements and [50, p. 25; 51, p. 33]. This is repeated in the Journal publications of Lemons & Penner and Newburgh [6, p. 23; 13, p. 724].

Cardwell's comprehensive discussion of the issue gives a rather cooler view of Kelvin on this point. 'Kelvin was fair in ascribing the credit for the two main laws of thermodynamics to Joule and to Carnot and Clausius. But he does not accept Clausius' axiom on which the second law rests...it is surely more likely that Kelvin wanted to express his own ideas in his own way. Hence he put forward the second law of thermodynamics in the form...' [4, p. 254/255]. Whatever the truth about Kelvin's motives, we can conclude from the above that the independence and broad value of his second law statement has general acceptance today. However, commonly this acceptance is not reflected in explicit scientific credit.

12.4 Kelvin's absolute temperature scale

Kelvin...was the inventor of the absolute Kelvin temperature scale. In this paper we review his achievement.

Herman Ehrlichson [52, p. 325]

Thomson (Kelvin) recognized even before the general acceptance of the present theory of heat, that the fact expressed in Carnot's theorem could provide the basis for an absolute scale of temperature. He first proposed such a scale in 1848...In 1854 Thomson defined a second absolute scale, chosen this time to agree with ... effectively the gas scale of temperature

J.S. Dugdale [51, p. 35]

In contrast to the discussion involving Kelvin and the second law, there is no ambiguity about Kelvin's inspired proposal for an absolute scale of temperature independent of thermometric substance. Putting to one side the notable input of Joule and James Thomson, Kelvin, and Kelvin alone, was responsible.

The enduring reputation of Kelvin in his definition of the absolute scale is shown in Fig. 27. The dates quoted are for original publication years, which give the historical sequence of his reputation.

The above representational sample covers approximately the last 50 years. The varied publications give clear and undiluted credit to Kelvin, the material consistently re-appearing. Even his 1848 scale still merits attention.

Of course, as with so many of the topics addressed in this chapter, there is a caveat, though quite mild. We have already mentioned Winterbone's 'downplay' of formal statements of the first and second laws [1]. Curiously, the absolute temperature scale is not treated in [1] either. However, the caveat doesn't dilute the explicit appreciation of the other publications.

12.5 Carnot, Clausius and Kelvin: worthy pioneers

In this section we have assessed the reputations of Carnot and Kelvin, together with Clausius. With regards to the second law, there is strong support for Carnot's 1824 thesis to be interpreted as a proto-statement. Clausius then takes historical precedence with his formal 1850 statement combined with exclusive credit for the proposal of entropy in 1865. Kelvin's 1851 statement, is regarded as independent and of broad equivalent value, with a subsequent refinement by Planck, so making a triumvirate. As far as the absolute scale of temperature goes, it is Kelvin alone of the trio who takes enduring credit for its introduction in the two forms of 1848 and 1854.

Reference (Type, Year)	**1848 Exponential Scale** Mention/ Explanation?	**1854 Gas Thermometer Scale** Mention/ Explanation?
Tribus[33] (undergraduate)(1961)	No	Yes p V-9/No
Fast [50] (entropy) (1962)	No	No (absolute scale, pp 31-34)
Hatsopoulos & Keenan [39] (undergraduate)(1965)	Yes (p 152)	Yes/Yes (thermodynamic scales, pp 145-152)
Rogers & Mayhew [3] (undergraduate) (1967)	Yes p 66/No	Yes/Yes pp 130-133 (+ Int'l Temperature Scale)
Dugdale [51] (entropy) (1996)	Yes/Yes, p 35	Yes/Yes pp 35/36
Kondepudi & Prigogine [34] (introductory) (1998)	No	Yes/Yes pp 76/77
Ehrlichson [52] (journal article) (2001)	Yes/Yes pp 326/327	Yes/Yes p 328

Figure 27: Kelvin's absolute temperature scales: his enduring reputation.

13 The perfect heat engine: is thermodynamics Platonic in character?

13.1 A summary in quotes

On Watt's engine

The engine he envisaged was, in a sense, an ideal engine … to develop the maximum power with the maximum efficiency

D.S.L. Cardwell [4, p. 54]

Watt had at last discovered the basis for his perfect engine

Richard L. Hills [15, p. 340]

On Rankine's engine

Rankine's 'diagram of energy' for the 'perfect' air engine

Crosbie Smith [14, p. 157]

A perfect regenerator

W.J. MacQuorn Rankine [45, p. 141]

There is another form of the perfect thermo-dynamic engine, viz: one with a perfect regenerator as defined by Rankine

J. Willard Gibbs [47, p. 10]

On philosophical interpretations

...William Thomson's concept of a 'perfect thermodynamic engine'... I suggest had cosmological and spiritual ...foundations

Crosbie Smith [14, p. 93]

Watt's Concept of a Perfect Engine ... from where did Watt derive this idea of perfectability? It seems to have been a common concept during the eighteenth century in many different spheres such as some aspects of religion or architecture with Greek temples.

Richard L. Hills [15, p. 318]

Of course this removes the perfect heat-engine from the world of practical engineering to some Platonic heaven...the perfect but wholly impracticable Carnot engine

D.S.L. Cardwell [4, pp. 193/194]

On current views

This concept of an ideal, perfect form behind the messy particulars of reality is ... generally attributed to Plato

Philip Ball [53, p. 10]

...sympathies lie strongly with Platonistic views that mathematics truth is absolute, external and eternal...

Sir Roger Penrose [54, pp. 146–151]

Many physicists share this Platonic vision of mathematics

Paul Davies [55, p. 145]

13.2 Platonism: the essential argument

The question of whether there is an underlying philosophical rationale behind design and manufacture is not a current issue among engineers. However, the striking antithesis between Carnot's concept of 'perfection' (that is extreme desirability) which is at the same time totally 'impracticable' does raise this very point; has Cardwell's 'Platonic heaven' any meaning, or is it just an intriguing aside? The subject is addressed more fully in the *Editorial Introduction* to this volume, but here we have given a concise entrée by means of a series of quotes.

Firstly, we have focused on Watt's engine and Rankine's air cycle. In fact, as Fig. 8 shows, Watt's achievement had two stages: an 'ideal' (or purely conceptual) engine and the closest practical approximation to it, which he none the less termed his perfect engine. Rankine , on the other hand, was unable to build his patented design of engine based on a perfect regenerator

[14, pp. 158–164]. Despite that, Rankine's underlying theory was unerringly correct, as Gibbs recognized some 20 years later. The first series of quotes contain a further significance. Watt and Gibbs contrasted in almost every sense of the word. Both were extremely able thermodynamicists, but their abilities could barely be more different. Despite such a contrast both readily thought in terms of perfection. This leads on to the second series of quotes. There was indeed a sense of searching for perfection during the times of our study, and this coincided with the explicit beliefs of many of the British characters. In fact, again as described in the introduction, there is a continuous timeline for a Platonic-related conviction from the Renaissance onwards.

Finally, while this question may not be a subject of interest for today's *engineers*, it is in science generally. The third set of quotes are given by the internationally known Philip Ball (chemistry and physics), Sir Roger Penrose (mathematics) and Paul Davies (physics).

14 Conclusion

14.1 The wider picture: public perception

An essential theme of this volume is how the public and scientific worlds *perceive* Kelvin and the thermodynamic pioneers today. For this we use Reference [56]. This is a popular publication, giving concise information in as far as possible a non-technical manner: 'Compiled by a panel of expert contributors, **The Complete Family Encyclopedia** is the essential companion for the home, the school or the office' [56]. We are interested, not in the contents of the items per se (generally reliable), but in the pre-assessed importance for the readership. *The number of lines allocated to the main characters of this chapter* can be regarded as a rule of thumb for this marketing-based assessment. (By so doing we admit we are ignoring other effects on length, such as complexity of material). This is given in Fig. 28.

Broadly speaking, Fig. 28 reflects both the chapter in general, and section 12 in particular. Perhaps Carnot's uniqueness is underplayed, although he is referred to as having 'founded thermodynamics' [56, p. 154]. The omissions of James Thomson and Rankine should be noted..

Featured in chapter		*Omitted*	
Newcomen	5 lines	Trevithick	
Black	10 lines	Clapeyron	
Boulton	7 lines	Rankine	
Carnot + Carnot Cycle	15 lines	James Thomson	
Clausius + entropy	16 lines		
Ericsson	6 lines		
Gibbs + Gibbs function	14 lines	*Comparison with Biology*	
Joule + Joule SI unit	16 lines	thermodynamics	16 lines
Kelvin + Kelvin Scale	18 lines	evolution +	
+ absolute zero	11 lines	natural selection	48 lines
+ Joule-Thomson effect	11 lines	Darwin	14 lines
Mayer	4 lines	T H Huxley	15 lines
Adam Smith	13 lines		
Stirling engine	3 lines		
Watt + Watt SI uni	22 lines		

Figure 28: Guide to perceived significance for general population.

Also, in comparison with Ericsson, Stirling merits more than his 'scraping in' via his engine! Finally, Kelvin is portrayed as a dominating thermodynamic figure, with the Kelvin scale and the Joule–Thomson effect.

Finally, this book relates to Kelvin and the natural world, so the comparisons of Fig. 28 include Darwin, Huxley and evolution/natural selection. Again, broadly speaking the character lengths are comparable, as are the overall subject lengths. However, the long section on evolution outbids thermodynamics and this fairly certainly reflects *the relative impact of thermodynamics and biology on the public consciousness.*

14.2 The wider picture: scientific perception

In the world of scientific and engineering research, an essential feature is the bread-and-butter activity of reporting the work via peer-reviewed publications. The ability to transmit original information accurately, economically and in an interesting manner is a key attribute of the researcher. Perception and impact are due to (i) the text itself, (ii) photographs and diagrams of equipment and (iii) graphical results. So there are both *direct visual and intellectual* aspects. In this section we show that current perception of the individual thermodynamic pioneers has both aspects.

14.2.1 In portrait: Kondepudi and Prigogine's 'Hall of Fame' [34]

One aspect of perception of individuals is the visual compliment by way of portrait reproduction. This has already been noted for Kelvin, Boulton & Watt, and Adam Smith, in the context of paper currency. This is also beginning to happen in thermodynamics texts, and that by Kondepudi and Prigogene [34] includes portrait reproductions of a range of pioneers of thermodynamics. Given Prigogine's Nobel Laureate status, their selection has considerable authority. Figure 29 references the sources in K & P's 'Hall of Fame' for the characters of our chapter. Interestingly, James Thomson, Clapeyron and Rankine, the three omitted in [56], similarly do not rank for a portrait in [34].

14.2.2 In writing: individual reputations

We turn to the more familiar world of scientific perception by way of writing. This part of the study focuses on individual reputations, mainly for the sequence of Fig. 2. In view of their absences in [56] and [34] we regard James Thomson, Clapeyron and Rankine as special cases to be considered separately later.

James Watt: already extensively addressed in this chapter.
Scientific perception: of very high rank.

Sadi Carnot: similarly addressed in this chapter.
Scientific perception: of the highest rank

Joseph Black p 8,	James Watt p 67,	James Prescott Joule p 37,
Sadi Carnot p 68,	Lord Kelvin p 76,	Rudolph Clausius p 79,
	J Willard Gibbs p 104	

Figure 29: Kondepudi and Prigogine's Hall of Fame (after [34]).

Rudolph Clausius: also addressed in this chapter. His statements relating to the universe have achieved almost cult status:

1. the energy of the universe is constant
2. the entropy of the universe tends toward a maximum

They were quoted early on by Gibbs (in German, Bumstead [48, p. xiv]),and appear in the historical texts by Bruce Hunt [57, p. 41] and Cardwell ([4, p. 273], in the thermodynamics texts of Kondepudi and Prigogine [34, p. 84]) and Tribus [33, p. X-1], in German) and in other contexts—for example the chaos and complexity study by Cembel [58, p. 134].

Scientific perception: of the highest rank

James Clerk Maxwell: Maxwell's achievements, mentioned in this chapter, are fully addressed in the chapters on his correspondence with Kelvin and on the Second Law. *Of all the scientists considered here, Maxwell has the highest present-day reputation.* Two examples suffice. Stuart Kauffman, of the Santa Fe Institute for complexity studies puts it thus: 'Arguably James Clerk Maxwell was the greatest scientist of the nineteenth century, notwithstanding giants such as Carnot, Boltzmann and Darwin' [59, p. 85]. Bruce Hunt, in his very recent historical study, echoes this: '...now widely regarded as the greatest physicist of the nineteenth century' [57, p. 53]. This is high praise indeed!

Scientific perception: of *the* highest rank

J Willard Gibbs: Gibbs's Free Energy Function. 'The great American scientist' [58, p. 140] Gibbs made a wide contribution to thermodynamics [57, p. 66] of which the *T–s* diagram was only one. His quantification of the *free energy* concept, that is 'the maximum useful work that can be obtained from a system' [1, p. 6] is of little less significance than entropy itself. Unfortunately, *for long-term perception of Gibbs's genius,* and unlike Clausius with entropy, Gibbs did not see any need of proposing a distinctive name. In thermodynamic circles this has acted to his advantage, and it is variously termed Gibbs energy, Gibbs function or Gibbs potential (Winterbone [1, p. 6], Rogers and Mayhew [2,3, p. 87]) or more commonly *Gibbs free energy* [39, p. 270; 33, p. XIII-14; ; 34, p. 127; 58, p. 140; 60], in an important recent paper on waste management). This latter work [60] involved operating experience with an engineering demonstration plant. Of the above expressions *Gibbs free energy,* unfortunately, contracts quite readily to *free energy* itself. This has occurred not only in engineering thermodynamics [3, p. 87] but also in thermodynamics applied to biology, where the meaning is identical [61, p. 36]. The loss of name is especially evident in the superbly produced American biological text *Life.* Whereas in the 1995 4th Edition [62, p. 117], the expression *Gibbs function* is used, by the 2001 6th Edition [63] it has become *free energy.* It is no surprise that the popular encyclopaedia item for *Gibbs function* [56] gives *free energy* as the same alternative.

Apart from the loss of Gibbs's name there is an important wider issue. As discussed in the *Editorial Introduction,* entropy and the Second Law of Thermodynamics have generated a quite wide cultural significance. By comparison the wider public is unlikely to appreciate the significance of *free energy* particularly as it is a 'two-word' expression involving the commonplace adjective *free.* Entropy, by contrast, has impact.

Scientific perception: slightly below Clausius?

14.2.2 *James Thomson, Clapeyron and Rankine*

We now address (1) James Thomson, (2) Clapeyron and (3) Rankine.

Portraits.

Firstly they *do* have portraits in the historical texts, as given in figure :

1. James Thomson—*Energy and Empire* [7], portraits as a teenager (p 52) and as professor at Glasgow University (p. 284).
2. Emile Clapeyron—via Mendoza [5, p. 72]
3. William MacQuorn Rankine--*Energy and Empire* [7], portrait as professor at Glasgow University (p. 319).

Scientific achievements and reputations

1. James Thomson. There is little doubt that Kelvin's elder brother James should be regarded as a leading pioneer. In the footing notes to the portrait on p 284 in [7] we read 'James played a central role in the creation of his brother's thermodynamics.' Current perception has tended to subsume James's contributions in those of his brother. Quite apart from the extensive coverage James receives in [7] two chapters in this volume seek to set the record straight.
   Scientific perception: insufficiently high
2. Emile Clapeyron. Of the pioneers, Clapeyron seems to have come off badly. He was the first person to publish the use of the p–v diagram and 'Practically all his equations are correct whereas many of Carnot's are incorrect however they are construed' [5, p. xiii]. However, his name *is* commemorated in another context: (1) the *Clapeyron equation,* relating phase change heat capacities and latent heat, and (2) the consequent *Clausius-Clapeyron Equation* involving the entropy. The references are given in order of originally published dates as:

   Tribus [33, pp. X-20 to X-22] (for (1) and enthalpy/volume form of (2))

   Rogers and Mayhew [3, p. 113] (enthalpy /volume form of (2))

   Dugdale [50, pp. 48 (for (1)) and 51 (entropy/volume form of (2)])

   Winterbone [1, p. 117 (for (1) and enthalpy/volume form of (2))]

   Kondepudi and Prigogine [34, pp. 179/180 (as for Winterbone)].

   Unlike James Thomson, then, Clapeyron has two equations credited to him.
   Scientific perception: secure but not in this chapter's context
3. William MacQuorn Rankine. Like Clapeyron, Rankine has suffered badly.

His superb thermodynamic grip and analysis of the perfect cycle [45] was recognized by Gibbs but by no-one else. Moreover, the basic Rankine cycle of Fig. 21 was like doing 'half a James Watt' on the steady flow version of the Carnot cycle. Of course, unlike Watt there was no consequent Rankine engine.

However, Rankine didn't help himself. He was significantly involved in Kelvin's progression of thought towards the latter's formulation of the Second Law [7, pp. 318–327]:'...his interaction with Rankine during 1850 was crucial' [7, p. 327]. However, 'while Thomson owed much to Rankine...for molecular vortices' [7, p. 469 of the actual paper on the subject], he said' The title is of more importance than anything else in the work' [7, p. 470]. So Rankine failed to secure any stake in the debate by way of publications. Maxwell criticized his numerous 'assumptions' and wrote 'Rankine's Second Law of Thermodynamics we find...inscrutable' [14, p. 165] and P.G. Tait was dismissive, too [14, p. 166]. A second area where Rankine's name has now been lost was in the 1954 internationally agreed set of four Thermodynamic Temperature Scales, where the Rankine scale gave absolute Fahrenheit temperatures, paralleling the Kelvin scale with Celsius temperatures [39, pp. 147–148]. The Fahrenheit/Rankine scales are living on borrowed time: Rogers and Mayhew, for instance, give their personal death warrant as 'In this book ,Fahrenheit and Rankine scales will never be used' [3, p. 134]. *However, we are now living in a newer thermodynamic world.* The combined drives towards environmentally-friendly and lower-cost energy solutions are resulting in highly imaginative thermodynamic cycles. On an even larger scale (300kW output from turbine) than the demonstration plant of [60] is the novel utility-exploitable plant of [64, 65], based on cryogenic energy storage. The significance here is that the plant thermodynamics is portrayed quantitatively on *T–s* diagrams and the performance compared with that of both Carnot and Rankine cycles [65].This constitutes a return of thermodynamic thinking to the heat engine behaviour of this chapter and Rankine's work in particular.

Scientific perception: Rankine's contributions at the critical stage of formulation of the Laws is in danger of being forgotten. Very recent developments in mains electricity production may help.

14.3 The wider picture: Kelvin

14.3.1 Kelvin and Joule

James Prescott Joule has barely featured in this chapter, although he does appear in Fig. 4. His story is yet another told elsewhere in this volume, but here we give a potted appreciation. He is a very important originator of thermodynamics, featuring in Figs 28 and 29:'his series of experiments established the equivalence between mechanical energy and heat' (Kondepudi and Prigogine [34, p. 13)]. Over the whole period from their first meeting at Oxford in June 1847 [7, p. 294] to February 1851 [7, p. 327], Kelvin opposed Joule's weddedness to that equivalence. The latter date identifies Kelvin's burying of the caloric theory and it is quite clear from [41] that Clausius preceded him in doing so.

A further aspect was the Joule-Thomson effect, explained in [56] as 'the fall in temperature of a gas as it expands adiabatically through a narrow jet...it is the basic principle of most refrigerators.' Rather like Maxwell's equations (which is not an item in [56]) it forms a standard part of the thermodynamics corpus: [3, p. 98]; 1, pp. 141–145; 39, p. 114; 33, p. XII-27].

Scientific perception of Joule, leaving aside the issue with Mayer: of consistently high reputation, from Kelvin and Clausius onwards.

14.3 Kelvin

Appropriately arrowed by the Joule-Thomson effect, we note in conclusion the reputation of Kelvin himself. His overall reputation has been addressed in the *Editorial Introduction*, with echoes of Leonardo da Vinci [66]. In the context of engineering thermodynamics it has been fully addressed in Section 12.

14.4 In summary

To understand more fully the achievements of Kelvin, this chapter has focused on his most brilliant contemporary, Sadi Carnot, and the latter's conceptual cycle and engine. The universality of Carnot's thesis has been demonstrated in its application to any fluid (steam or air), any process (nonflow or steady flow) and in either direction (direct or reversed). His misconceptions regarding caloric did not prevent either his peers or those who followed recognizing his genius at the highest level.

Throughout, though, we have used the p–v and then T–s diagrams of Watt and Gibbs, neither of which were available to Carnot. So both these pioneers are also studied, especially the former. Watt, like Carnot, was a man of genius and enduring reputation, and it was only Watt that Carnot described as 'famous.' To understand Watt properly needs an appreciation of steam engine development stretching back to the origins of the industrial revolution. A feature of this chapter is the provision of a comprehensive set of timelines for all the principle characters.

In terms of pre-Carnot history the achievements of Watt and Carnot are compared, and then the way is studied in which the second law of thermodynamics followed Carnot. This enabled us to make assessments of the contributions of Carnot, Clausius and Kelvin himself. Kelvin not only is a co-discoverer of the second law, but was also solely responsible for the absolute scale of temperature. So Kelvin too is shown to have demonstrated his genius during the critical quarter-century after Carnot.

Acknowledgements

The background material for this chapter covers persons and events of wide significance in energy-related world history. The sheer amount of information involved is awesome. The books dedicated to Watt, [15–17], and Kelvin [7] alone are 1023 and (xxvi + 866) pages long, respectively. It is hoped that readers will wish to follow up some of the themes discussed, and to facilitate access to the very extensive material, Wikipedia sources have been frequently referenced in this chapter. For formal referencing the matching *Encyclopaedia Britannica* sources are also given.

Finally, we are indebted to Teresa Figurska-Stempa, who was responsible for producing the diagrams.

References

[1] Winterbone, D.E., *Advanced Thermodynamics for Engineers*, Arnold: London, 1997.
[2] Rogers, G.F.C. & Mayhew, Y.R., *Engineering Thermodynamics. Work and Heat Transfer*, 1st edn, Longmans: London, 1957.

[3] Rogers, G.F.C. & Mayhew, Y.R., *Engineering Thermodynamics. Work and Heat Transfer*, 4th Edition, Prentice Hall: Harlow, UK, 1992.
[4] Cardwell, D.S.L., *From Watt to Clausius*, Heinemann: London, 1971.
[5] Carnot, S., *Reflections on the Motive Power of Fire, and Other Papers on the Second Law of Thermodynamics by* E. Clapeyron and R. Clausius, with Introduction by E. Mendoza (ed.), New Edition Dover: New York, 1988.
[6] Lemons, D.S. & Penner, M., Sadi Carnot's contribution to the second law of thermodynamics, *Am. J. Phys.*, **76(1)**, pp. 21–25, January, 2008.
[7] Smith C., & Wise, M.N., *Energy & Empire*, Cambridge University Press: Cambridge, UK, 1989.
[8] Carnot, S., *Reflexions sur la puissance motrice du feu et sur les machines propres a developper cette puissance*, Bibliotheque du Conservatoire National des Arts et Metiers: Paris, 1824.
[9] Wikipedia item for *Caloric Theory*, 2011//Article for *Caloric Theory*, Encyclopaedia Britannica Online, Web, 20 August 2012.
[10] Wikipedia item for *Phlogiston Theory*, 2011// Article for *Phlogiston*, Encyclopaedia Britannica Online, Web, 20 August 2012.
[11] Carnot, S., (in English, labelled as [5, p. 60]) *Notes on mathematics, physics and other subjects*, and translated as *Appendix: Selections from the Posthumous Manuscripts of* Carnot [5, p. 60], 1988.
[12] Ehrlichson, H., Sadi Carnot, 'Founder of the Second Law of Thermodynamics.' *Eur. J. Phys.*, **20**, pp. 183–192, 1999.
[13] Newburgh, R., Carnot to Clausius: caloric to entropy, *Eur. J. Phys.*, **30**, pp. 713–728, 2009.
[14] Smith, C., The Science of Energy, The Athlone Press: London, UK, 1998.
[15] Hills, R.L., *James Watt. Volume 1: His Time in Scotland, 1736–1774*, Landmark Publishing: Ashbourne, Derbyshire, UK, 2002.
[16] Hills, R.L., *James Watt. Volume 2: The Years of Toil, 1775–1785*, Landmark Publishing: Ashbourne, Derbyshire, UK, 2005.
[17] Hills, R.L., *James Watt. Volume 3: Triumph through adversity, 1785–1819*, Landmark Publishing: Ashbourne, Derbyshire, UK, 2006.
[18] Wikipedia item for *Thomas Savery*, 2012//Article for *Thomas Savery*, Encyclopaedia Britannica Online, Web, 20 August 2012.
[19] Wikipedia item for *Thomas Newcomen*, 2012//Article for *Thomas Newcomen*, Encyclopaedia Britannica Online, Web, 20 August 2012.
[20] Wikipedia item for *Newcomen steam engine*, 2012//Article for *Thomas Newcomen*, Encyclopaedia Britannica Online, Web, 20 August 2012.
[21] Wikipedia item for *Newcomen Engine Black Country Living Museum*, 2012.
[22] Wikipedia item for *John Smeaton*, 2012//Article for *John Smeaton*, Encyclopaedia Britannica Online, Web, 20 August 2012.
[23] Wikipedia item for *Jonathan Hornblower*, 2012//Article for *Jonathan Hornblower*, Encyclopaedia Britannica Online, Web, 20 August 2012.
[24] Wikipedia item for *Arthur Woolf*, 2012//Article for *Arthur Woolf*, Encyclopaedia Britannica Online, Web, 20 August 2012.
[25] Wikipedia item for *Compound engine*, 2012//Articles for *Jonathan Hornblower and Arthur Woolf* , Encyclopaedia Britannica Online, Web, 20 August 2012.
[26] Wikipedia item for *Richard Trevithick*, 2012//Article for *Richard Trevithick*, Encyclopaedia Britannica Online, Web, 20 August 2012.
[27] Boldrin, M. & Levine, D.K., *Against Intellectual Monopoly*, Cambridge University Press: Cambridge, UK, 2008.

[28] Boldrin, M. & Levine, D.K, James Watt: Monopolist: (Google on *James Watt Monopolist*, 2012).
[29] Wikipedia item for *William Cullen*, 2012//Article for *William Cullen*, Encyclopaedia Britannica Online, Web, 20 August 2012.
[30] Wikipedia item for *Joseph Black*, 2012//Article for *Joseph Black*, Encyclopaedia Britannica Online, Web, 20 August 2012.
[31] Wikipedia item for *James Watt*, 2012//Article for *James Watt*, Encyclopaedia Britannica Online, Web, 20 August 2012.
[32] Mason, S., (ed.), *Matthew Boulton*, Birmingham City Council: UK/Yale University Press: New Haven, CT, USA, 2009.
[33] Tribus, M., *Thermostatics and Thermodynamics*, Van Nostrand: Princeton, NY, 1961.
[34] Kondepudi, D. & Prigogine, I., *Modern Thermodynamics*, John Wiley: Chichester, UK, 1999.
[35] Wikipedia item for *Adam Smith*, 2012//Article for *Adam Smith*, Encyclopaedia Britannica Online, Web, 20 August 2012.
[36] Cuhaj, J.S. (ed.), *Standard Catalog of World Paper Money Modern Issues 1961–Present, 18th Edition*, Krause Publications: Iola, WI, USA, 2012.
[37] Bank of England News Release. New-style £50 banknote issued today, 2 November 2011.
[38] Spalding, D.B. & Cole, E.H., *Engineering Thermodynamics* (3rd edn), SI Units, Arnold: London, UK, 1973.
[39] Hatsopoulos, G.N. and Keenan, J.H, *Principles of General Thermodynamics*, John Wiley: New York, USA, 1965.
[40] Clapeyron, E., *Memoir on the Motive Power of Heat*, Translated and Edited by E. Mendoza from *Journal de l'Ecole Polytechnique*, **XIV**, 153, 1834; Poggendorff's Annalen der Physik, **LIX**, 446, 566, 1843. in Carnot, S., *Reflections on the Motive Power of Fire, and other papers on the Second Law of Thermodynamics by* E. Clapeyron and R. Clausius, with Introduction by E. Mendoza, (ed.), New Edition Dover: New York, pp. 71–105, 1988,
[41] Clausius, R., On the motive power of heat, and on the laws which can be deduced from it for the theory of heat, translated by W.F. Magie, (*Poggendorff's Annalen der Physik*, **LXXIX**, p. 368, 500, 1850}. In Carnot, S., *Reflections on the Motive Power of Fire, and Other Papers on the Second Law of Thermodynamics by* E. Clapeyron and R. Clausius, with Introduction by E. Mendoza, (Ed.), New Edition Dover: New York, pp. 107–152, 1988.
[42] Wikipedia item for *Robert Stirling*, 2012//Article for *Robert Stirling*,Encyclopaedia Britannica Online, Web, 20 August 2012.
[43] Wikipedia item for *John Ericsson*, 2012//Article for *John Ericsson*,Encyclopaedia Britannica Online, Web, 20 August 2012.
[44] Whitaker A., *James and William Thomson: The Creation of Thermodynamics*, Chapter 3, 2015.
[45] Rankine, W. J.Mac Q., On the geometrical representation of the expansive action of heat, and the theory of thermodynamic engines. *Phil. Trans. R. Soc. Lond.*, 144, pp. 115–175, 1854.
[46] http://greenpowerscience.com/SHOPSTIRLINGENGINES.html, 23 January 2012.
[47] Gibbs, J.W., Graphical methods in the thermodynamics of fluids, Paper I, & A Method of Geometrical Representation of the Thermodynamic Properties of Substances by means of Surfaces, Paper II, in *The Scientific Papers of J Willard Gibbs, Volume II, Thermodynamics*, Dover Publications: New York, 1961.
[48] Bumstead, H.A. & Gibbs, J.W., reprinted from *American Journal of Science*, as pp. xi–xxvi, in *The Scientific Papers of J. Willard Gibbs, Volume II, Thermodynamics*, Dover Publications: New York, 1961.

[49] Mikielewicz, J., Stasiek, J.A. & Collins, M.W., The laws of thermodynamics: cell energy transfer, Chapter 3 in *Nature and Design*, eds. M.W. Collins, M.A. Atherton & J.A. Bryant, Vol. 1, International Series on Design and Nature, pp. 29–62, WIT Press: Southampton, 2005.

[50] Fast,J.D., *Entropy*, 2nd edn (Revised and Enlarged), MacMillan and Co: London, UK, 1968.

[51] Dugdale, J.S., *Entropy and Its Physical Meaning*, Taylor & Francis, London: UK, 1996.

[52] Ehrlichson, H., Kelvin and the absolute temperature scale, *Eur. J. Phys.*, **22**, pp. 325–328, 2001.

[53] Ball, P., *The Self-Made Tapestry*, Oxford University Press: UK, 1999.

[54] Penrose, R., *The Emperor's New Mind*, Oxford University Press: UK, 1989/1999.

[55] Davies, P., *The Mind of God*, Penguin: London, UK, 1993.

[56] The Complete Family Encyclopedia, Fraser Stewart Book Wholesale Ltd/Helicon.

[57] Hunt, B.J., *Pursuing Power and Light*, Johns Hopkins University Press: USA, 2010.

[58] Cambel, A.B., *Applied Chaos Theory*, Academic Press: San Diego, CA, 1993.

[59] Kauffman, S., *Investigations*, Oxford University Press: UK, 2000.

[60] Morrin, S., Lettieri, P., Chapman, C. & Mazzei, L., Two stage fluid bed-plasma gasification process for solid waste valorization: technical review and preliminary thermodynamic modeling of suphur emissions. *Waste Management*, **32(4)**, pp. 676–684, April 2011.

[61] Wicken, J.S., *Evolution, Thermodynamics and Information*, Oxford University Press: UK, 1987.

[62] Purves, W.K., Orians, G.H. & Heller, H.C., *Life*, 4th edn., Sinnauer Associates/W.H. Freeman and Co., Ltd: New York, NY, USA, 1995.

[63] Purves, W.K., Sadava, D., Orians, G.H. & Heller, H.C., *Life*, 6th edn., Sinnauer Associates/ W.H.Freeman and Co., Ltd: New York, NY, USA, 2001.

[64] http://www. highview-power.com.

[65] Morgan, R., *Cryogenic Energy Storage. A Cost Effective Utility Scale Power Storage Solution.* Presentation at the Institution of Mechanical Engineers, 2 February 2012.

[66] Grassi,W. & Collins, M.W., Leonardo da Vinci, Chapter 12 in *Nature and Design*, eds. M.W. Collins, M.A. Atherton & J.A. Bryant, Vol. 1, International Series on Design and Nature, pp. 29–62, WIT Press: Southampton, 2005.

Chapter 8a

The first law of thermodynamics:

1. Kelvin's relationship with Joule

A.K. Wróblewski
Physics Department, University of Warsaw, Poland.

Abstract

A concise account is given for the gradual transition of William Thomson, Lord Kelvin, from acceptance of Carnot's theory of indestructible caloric to the dynamical theory of heat.

1 Introduction

One may say that William Thomson (Lord Kelvin) was destined to study the phenomena of heat. Being a child prodigy, in 1834, at the age of 10, he matriculated at the University of Glasgow where his father James was a professor of mathematics. In 1840, the teenage William read the famous book *Analytical Theory of Heat* by Jean Baptiste Joseph Fourier. This event to a large extent formed his career as a mathematical physicist.

Fourier showed that one could develop the theory of heat without committing oneself on the issue of the nature of heat. He laid out his program in the first two sentences of the first chapter of his book.

> "The effects of heat are subject to constant laws which cannot be discovered without the aid of mathematical analysis. The object of the theory which we are about to explain is to demonstrate these laws; it reduces all physical researches on the propagation of heat, to problems of the integral calculus whose elements are given by experiment' [1].

Thomson was impressed by the fact that Fourier's theory was not based on hypothetical entities such as caloric which at that time was accepted by practically every physicist. Fourier's work served as a model for Thomson in his further studies.

2 William Thomson's fascination with Carnot's theory of heat

The evidence is that the first acquaintance William Thomson had with the radical theory of Sadi Carnot (1796–1832) was third hand. The overall historical sequence appears to be as follows. In 1824 Carnot, still only a young engineer, wrote the original booklet [2], entitled *Réflexions sur la puissance motrice du feu et sur les machines propres a développer cette puissance* (*Reflexions on the motive power of heat and on machines capable of developing that power*). As explained in this volume's chapter *Engineering Thermodynamics and the Carnot Cycle*, Carnot's original publication had a cold reception in Paris. It was really revived only by Emile Clapeyron's 1834 paper entitled 'Memoire sur la puissance motrice du feu'[3], in which the author gave a quantitative analysis of the work produced by a steam engine based on Carnot's analysis. In expounding the famous doctrine familiar to every engineer under the name of Carnot's Cycle, Clapeyron was the first, and possibly the only one, of Carnot's compatriots to recognise its significance. So far, then, the story was French and in the French language but from then on it would become English. Three years later, Richard Taylor (who was to join with William Francis in 1852 to form the publishing company Taylor and Francis) republished Clapeyron's paper in English [4] in the first volume of his *Scientific Memoirs Selected from the Transactions of Foreign Academies of Science*.

It was this *third* publication that apparently attracted William Thomson's attention. Smith and Wise, in their exhaustive biography of Kelvin, *Energy & Empire*, explained in detail [5, p. 289] how in their correspondence the brothers Thomson started to discuss the matter. In a letter to William in 1844, James referred to 'the paper you mentioned to me'. Smith and Wise comment that 'This paper was almost certainly the 1837 translation for Taylor's *Scientific Memoirs* of Clapeyron's 'Memoir on the motive power of heat'. They conclude that the 1844 letter of James's is '... the earliest record of implicit references to Clapeyron and Carnot by the Thomsons' since it showed 'a considerable understanding of the basic principles involved in Carnot's theory' (all quotes from [5, p. 289]).

It is on the grounds of this 'considerable understanding', then, that Smith and Wise use the expression 'almost certainly' for their interpretation of the Thomson brothers' link with Carnot's original publication. Two aspects should be noted. Most importantly, James used the expression 'whatever gaseous substances are acted on'. The independence of the Carnot cycle on working substance was a key concept and barely something James could have referred to without the knowledge of Carnot's work. Also 'James exhibited his familiarity with the waterfall analogy employed by Carnot'. In all, James Thomson had a sure grasp of 'the fundamental principles of Carnot and Clapeyron'. By including Clapeyron, Smith and Wise are, of course. identifying Clapeyron's paper as the source. (All quotes are from [5, p. 289/290].)

James's letter was sent to William during his 4 years, 1841–1845, in Cambridge. In the latter part of 1845, William Thomson went to Paris (Smith and Wise [5, p. 300/301]) and in the laboratory of Victor Regnault he learned about heat as a source of motive power. Thomson, while being evidently fascinated by Clapeyron's account had no direct knowledge of Carnot's work. Being in Paris gave him the opportunity to look for an 'original Carnot'.

However, Carnot's booklet was exceedingly rare, very few copies having been printed. Thomson inquired for it in the Library of the College de France. No one could tell him even where a copy might be seen. Years afterwards, he narrated how in 1845, when in Paris, he sought in vain for this work.

'I went to every book-shop I could think of, asking for the *Puissance motrice du feu*, by Carnot. 'Caino ? Je ne connais pas cet auteur'. With much difficulty I managed to explain that it was 'r' not 'i' I meant. 'Ah ! Ca-rrr-not ! Oui, voici son ouvrage,' producing a volume on some social question by Hippolyte Carnot; but the *Puissance motrice du feu* was quite unknown' [6].

Thomson was gripped by the French scientist's argumentation. In his analysis of the motive power of heat Carnot believed, as was commonly assumed at that time, that heat is a substance, a subtle fluid named caloric. Then, he also employed the analogy between the fall of water from a higher to a lower level and the 'fall' of caloric from higher to lower temperature:

'According to established principles at the present time, we can compare with sufficient accuracy the motive power of heat to that of a waterfall. Each has a maximum that we cannot exceed, whatever may be, on the one hand, the machine which is acted upon by the water, and whatever, on the other hand, the substance acted upon by the heat. The motive power of a waterfall depends on its height and on the quantity of the liquid; the motive power of heat depends also on the quantity of caloric used, and on what may be termed, on what in fact we will call, the height of its fall, that is to say, the difference of temperature of the bodies between which the exchange of caloric is made' [7]. It is worth pointing out here that water power was a key contributor to the industrial, especially the textile, scene. Cardwell describes the water-steam rivalry in Chapter 3 of Ref. [8]. It was logical, therefore, for Carnot to use such an analogy. James Thomson did exactly the same – [5, p. 289/290]).

After a detailed analysis of what we now call Carnot's cycle, Sadi Carnot concluded in his work that 'The production of motive power is then due in steam-engines not to an actual consumption of caloric, but to its transportation from a warm body to a cold body, that is, to its re-establishment of equilibrium – an equilibrium considered as destroyed by any cause whatever, by chemical action such as combustion, or by any other. We shall see shortly that this principle is applicable to any machine set in motion by heat' [9].

3 Enter James Prescott Joule

William Thomson was unaware that in the 1840s there were other people studying the problem of heat. James Prescott Joule (1818–1889) from Manchester was educated at home, partly by the famous John Dalton. He was quite well-off because of the prospering family brewery and built himself a small laboratory at home. There he performed various experiments on the production of heat by mechanical actions.

Joule read his first paper on the mechanical equivalent of heat on 21 August 1843, during the annual meeting of the British Association for the Advancement of Science (BAAS) at Cork, in southern Ireland. He presented it to the Chemistry Section because the Physics Section was full. Joule's conclusion was that 'By a dynamometrical apparatus attached to his machine, the author has ascertained that .... a quantity of heat, capable of increasing the temperature of a pound of water by one degree of Fahrenheit's scale is equal to a mechanical force capable of raising a weight of about 838 pounds to the height of one foot' [10].

Joule's paper attracted little attention. He, however, continued experiments. In 1845, at the BAAS meeting in Cambridge, he presented the results of his next experiment in which a pair of falling weights drove, by means of two light cords, a vertical drum mounted on a shaft at the lower end of which was a brass paddle. This rotated in a copper drum full of water. The weights fell through 36 feet at a speed of one foot per second. The mechanical equivalent of heat determined from the rise of the temperature of water and the easily calculable expense of the potential energy of the weights was found to be between 774 and 890 foot×pounds. Joule's presentation again did not attract attention.

William Thomson met James Joule for the first time during the BAAS meeting in Oxford, in 1847. On Thursday 24th June, Joule presented his paper on the new theory of heat. He was told by the chairman to be brief in his presentation because it was late in the day and the subject was not interesting to the audience. Joule reported again on experiments with his paddle-wheel apparatus. He gave 781.5 and 782.1 foot×pounds for the mechanical equivalent of heat measured using water and sperm oil, respectively. Joule's presentation would have passed without notice but a young man at the back of the lecture hall asked some penetrating questions about it. That created an interest in the paper. The young man was William Thomson, then 22 years old, who later presented another version of the incident, saying that he remained seated and put his questions to Joule after the meeting. Whichever version was true, this encounter marked the beginning of a life-long friendship between the two great scientists.

After the meeting, Thomson wrote to his father from Cambridge on 1 July 1847. He described the meeting as 'delightful' and wrote: 'I am going to write to James [his brother] about [Joule's paper on the dynamical theory of heat] and enclose him a set of papers I have received from Joule, whose acquaintance I made.... Joule is, I am sure, wrong in many of his ideas, but he seems to have discovered some facts of extreme importance, as for instance, that heat is developed by the friction of fluids...' [11].

The meeting with Joule marked the beginning of Thomson's very slow conversion to the dynamical theory of heat. Still, in 1848 he published a paper [12] explaining how Carnot's theory [9] may be used to define an absolute temperature scale.

At the end of 1848 Thomson obtained a copy of Carnot's booklet from Professor Lewis Gordon. He realised that Carnot's wordy discussion can be expressed better with the use of mathematics. Thus, he wrote a paper [13] in which he presented an improvement of Carnot's calculations by using more precise experimental data obtained by Regnault. Joule was disappointed and surprised that Thomson still adhered to Carnot's heat theory.

4 Rudolf Clausius finds the solution

Meanwhile, another important player appeared on the scene. The German physicist Rudolf Clausius (1822–1888) was only 2 years older than William Thomson. He received his doctorate at the University in Halle in 1847 and was teaching at the Royal School of Artillery and Engineering in Berlin. He did not know Carnot's booklet but learned about his ideas from Clapeyron [2] and Thomson [13].

In an important paper 'On the moving force of heat' [14] published in 1850 Clausius realised that, contrary to Thomson's belief, there was no need to make a choice between Carnot's statement that a loss of heat never occurs in the production of heat by heat engines, and Joule's experimental finding of the equivalence between heat and mechanical work.

Clausius argued that Carnot's and Joule's ideas could be reconciled if it was assumed that in the production of work a certain portion of heat was consumed and a further portion transmitted from a warm body to a cold one. He further proposed that both portions stand in a certain definite relation to the quantity of work produced. Thus, Clausius stated two important principles, the equivalence of heat and work, and the principle that heat always shows a tendency to equalise temperature differences and therefore pass from hotter to colder bodies. He obtained an explicit form of the first law of thermodynamics and also formulated the second law.

5 William Thomson's answer

Thomson read Clausius' paper and regretted that he did not see this point earlier. Then he wrote his own version of the solution of the motive power of heat.

'The whole theory of the motive power of heat is founded on the two following propositions, due respectively to Joule, and to Carnot and Clausius.

Prop. I (Joule) – When equal quantities of mechanical effect are produced by any means whatever from purely thermal sources, or lost in purely thermal effects, equal quantities of heat are put out of existence or are generated.

Prop. II (Carnot and Clausius) – If an engine be such that, when it is worked backwards, the physical and mechanical agencies in every part of its motions are all reversed, it produces as much mechanical effect as can be produced by any thermo-dynamic engine, with the same temperatures of source and refrigerator, from a given quantity of heat' [15].

In the same paper Thomson also proposed his own version of the second law of thermodynamics. This paper marked the final conversion of William Thomson to the dynamical theory of heat.

6 Concluding remarks

It is worth remembering that in the middle of the nineteenth century force had not been clearly distinguished from what is now called energy. Thus, for example, Hermann Helmholtz's important treatise on the conservation of energy was entitled 'Über die Erhaltung der Kraft' (On the conservation of force). Helmholtz was speaking about living force (*Lebendige Kraft*) and tension force (*Spannkraft*).

Thomson first used the terms 'statistical energy' and 'dynamical energy'. The terms 'potential' (hidden) energy and 'actual' (visible) energy were introduced by W.J. Macquorn Rankine. Finally, in 1867 William Thomson and Peter Guthrie Tait introduced 'potential' energy and 'kinetic' energy in their famous textbook '*Treatise on Natural Philosophy*' [16].

References

[1] Fourier, J.B.J., *Analytical Theory of Heat*, translated by Alexander Freeman, Cambridge University Press: Cambridge, p. 14, 1878.

[2] Carnot, S., *Réflexions sur la puissance motrice du feu et sur les machines propres a développer cette puissance*, Bibliotheque du Conservatoire National des Arts et Metiers: Paris, 1824.

[3] Clapeyron, E., Memoire sur la puissance motrice de la chaleur. *Journal de l'Ėcole Polytechnique*, **14**, pp. 153–190, 1834.

[4] Clapeyron, E., Memoir on the motive power of heat, Richard Taylor (ed.). *Scientific Memoirs*, **1**, pp. 347–376, 1837.

[5] Smith, C., & Wise, M.N., *Energy & Empire*, Cambridge University Press: Cambridge, UK, 1989.

[6] Thompson, S.P., *The Life of William Thomson, Baron Kelvin of Largs*, vol. 1, Cambridge University Press: Cambridge, p. 133, 1910.

[7] Carnot, S., *Reflections on the Motive Power of Heat*, ed. R.H. Thurston, John Wiley & Sons: New York, Chapman & Hall, London, p. 61, 1897.

[8] Cardwell, D.S.L., *From Watt to Clausius*, Heinemann: London, 1971.

[9] Carnot, S., *Reflections on the Motive Power of Heat*, ed. R.H. Thurston, John Wiley & Sons: New York, Chapman & Hall, London, p. 46, 1897

[10] Joule, J.P., Report of the BAAS meeting, Cork, 1843, Transactions of the Sessions, p. 33. His full report was published as On the Calorific Effects of Magneto-Electricity, and on the Mechanical Value of Heat, *Phil. Mag. Series*, **3(23)**, pp. 263–276, 347–355, 435–443, 1843.

[11] Cardwell, D.S.L., *James Joule. A Biography*, Manchester University Press: Manchester and New York, p. 85, 1989.

[12] Thomson, W., On an absolute thermometric scale, founded on Carnot's theory of the motive power of heat, and calculated from the results of Regnault's experiments on the pressure and latent heat of steam. *Phil. Mag.*, Series 3, **33**, pp. 313–317, 1848; reprinted in Mathematical and Physical Papers by Sir William Thomson, Vol. 1, p. 100, Cambridge, 1882.

[13] Thomson, W., An account of Carnot's theory of the motive power of heat with numerical results deduced from Regnault's experiments on steam. *Transactions of the Royal Society of Edinburgh*, **16**, pp. 541–574, 1849, and *Annales de Chimie*, **25**, pp. 248–255, 1852; reprinted in *Mathematical and Physical Papers* by Sir William Thomson, vol. **1**, p. 113, Cambridge, 1882.

[14] Clausius, R., Über die bewegende Kraft der Wärme, und die Gesetze, welche sich daraus für die Wärmelehre selbst ableiten lassen. *Annalen der Physik*, **79**, pp. 368–397, 500–524, 1850; English translation published in *Phil. Mag., Series 4*, **2**, pp. 1–21, 102–119, 1851.

[15] Thomson, W., On the dynamical theory of heat, with numerical results deduced from the Joule's equivalent of a thermal unit, and M. Regnault's observations on Steam [1851], *Transactions of the Royal Society of Edinburgh*, **20**, pp. 261–288, 1853; reprinted in *Mathematical and Physical Papers* by Sir William Thomson, Vol. **1**, p. 183, Cambridge 1882.

[16] Thomson, W. & Tait, P.G., *Treatise on Natural Philosophy* (Oxford 1867); new edn. (2 vols), Cambridge, pp. 1879–1883.

Chapter 8b

The first law of thermodynamics:

2. The Joule–Mayer Controversy

M.W. Collins
College of Engineering, Design & Physical Sciences, Brunel University London, United Kingdom.

Abstract

Two pioneering scientists are associated with the formulation of the first law of thermodynamics, James Prescott Joule and Julius Robert Mayer. There are two distinct aspects to their contributions: the actual statement of the equivalence of heat and work, and the experimental justification for that statement. In terms of the former, Mayer and Joule independently made that statement, Mayer a year or so earlier than Joule. In terms of the latter, Joule carried out a scientifically superb range of experiments allowing him to quantify accurately the mechanical equivalent of heat.

To begin with, both had very great problems in gaining acceptance from their peers. Joule was ultimately successful, in large measure due to the strong support of Kelvin. Mayer, on the other hand, had the bitter experience of never gaining that acceptance in his lifetime. The situation was exacerbated by the heated controversy initiated by John Tyndall and the biologically oriented X-Club. In strong opposition to the thermodynamic community, they not only maintained the historical precedence of Mayer over Joule but were derogatory to Joule's contributions.

In this chapter, it will be shown that at the time the controversy was unnecessary.

Moreover, scientific history, while fully crediting Joule, has been kind to Mayer, in the present-day use of the expression 'the Mayer–Joule Principle'.

1 Introduction

The scientific community's ethos includes correctly recognising the originality of researchers' work. The time-honoured way of doing this is by peer review of papers submitted to scientific journals. This achieves further protection for the researcher since the act of publishing is a defence against plagiarism. In the case of the First Law of Thermodynamics, there was the question, even at the time, of Julius Mayer's contribution to the debate, and as a consequence and more importantly whether he should be given precedence over Joule in the development of the First Law. This issue arose at the time and it had two quite separate aspects: peer assessment by scientific contemporaries and the way in which the matter was inflamed by John Tyndall of the X-Club (the group founded in 1864 by T.H. Huxley and eight associates) [1, p. 171]. These are addressed in this chapter.

2 Joule, Mayer and the First Law: their peer assessment problems

With reference to the first aspect, there was a nascent peer assessment in place in those times, from which both Joule and Mayer suffered. '...both ... Mayer ... and ... Joule had, in different ways, been largely ignored and even scorned by their respective German and British scientific peers' [1, p. 9].

Joule had a great struggle to achieve scientific credibility, epitomised by his striving after publication in the Royal Society's *Proceedings*. In 1844, 'Once again, however, the Royal Society's *Phil. Trans* rejected the full paper ...' [1, p. 68], which was published instead in the *Philosophical Magazine*. Cardwell gives Joule's paper the highest praise: 'This splendid paper was rejected by the Royal Society' [2, p. 234], but via *Phil Mag* 'the scientific world learned not only of Joule's fundamental insights ... but also of his penetrating criticisms of Clapeyron and Carnot'. It was a scientific 'first' – 'for the first time the fundamental ideas of thermodynamics were subject to the (constructive) criticism of a first-class scientific intellect'. Recounting the story as told by Cardwell, Joule achieved another 'first' in 1847. In 1845, as fully described in Section 3 above, he presented at the Cambridge meeting of the BAAS his 'new method of determining the mechanical equivalent of heat' using a paddle-wheel. Again, with the exception of Helmholtz, in 'his important paper *On the conservation of energy* in 1847 'no one seemed very interested'. It was in that year, at the Oxford meeting of the BAAS that he presented 'an improved paddle-wheel apparatus' which aroused 'the keen appreciation' of Kelvin. Again, more details of this meeting have been given in Section 3. Cardwell described Joule's second 'first' as 'an historic stage in the history of thermodynamics: the real beginning of the acceptance of the first law' (all quotes from [2, p. 235]. Returning to the question of Joule's publication endeavours, he was ultimately successful, with his definitive Royal Society paper of 1850 'On the mechanical equivalent of heat' [3, p. 312].

What of Mayer? While both Joule and Mayer struggled, Joule's success contrasted strongly with Mayer's unfortunate experience. Mayer, a German physician, was initially unsuited to scientific publication and Crosbie Smith tells how in a number of regards his submission to the 'prestigious' Poggendorff's Journal was unacceptable. His paper was not just rejected – it was not even returned to him. However, like Joule, Mayer did not give up and (like many successful researchers today?) 'undertook a thorough revision of strategy'. In 1842, he achieved his sole success – 'his only paper submitted directly to, and accepted for publication by, a national and international journal'. It was 'On the Forces of Inorganic Nature' in Liebig's *Annalen der Chemie und Pharmazie*' (all quotes from [1, p. 75]). Cardwell paints a vivid picture of Mayer's despair. Despite his sole success 'he was almost entirely ignored' especially by his fellow-countrymen Helmholtz and Clausius who instead referenced Joule's work. The 'tragic irony' of this epitomising 'the evident scorn of his peers led to mental collapse' (quotes from [2, p. 231]).

In the late 1840s, Mayer became aware of Joule's papers and in 1848 'wrote to the French Academy of Sciences pointing out his claims to priority' [1, p. 73]. *This was an extremely significant matter – in effect Mayer was claiming precedence over Joule in being credited with the concept of the mechanical equivalent of heat*. Joule's approach, agreed with Kelvin, was to acknowledge the fact of Mayer's historical priority, but to claim that the scientific substance was his (Joule's). What Joule wrote is worth quoting in full 'I shall be quite content to leave M. Meyer in the enjoyment of the credit of having predicted the law of the equivalency. But it would certainly be absurd to say that he has established it' [1, pp. 73, 74 from Smith's Reference 70, [1, p. 325]. *The chronological order (the years are given) has considerable significance,* because it was in 1850 that Clausius published the key paper (Section 4) re-expressing Carnot's theory in non-caloric terms [4, 5]. Sections 4 and 5 explain the significance of Clausius's paper for Kelvin's progress towards the 'dynamical theory of heat'. Here, the focus is on the relevance to Joule and Mayer. Mayer's publication was indeed referenced by Clausius [5, p. 111] in a well-focused introductory discussion which needs some explanation.

Clausius identified the problem clearly: Did the 'quantity of heat remain unchanged' (as promulgated by Carnot and Clapeyron) or could there be 'an increase in the quantity of heat' as 'the careful investigations of Joule ... have almost certainly proved'? He then cites Holtzmann who, using Mayer's method both agrees with Joule's equivalent but then in developing the equations 'proceeds as Clapeyron' and 'tacitly assumes that the quantity of heat is constant' (all quotes from [5, pp. 110, 111]). In contrast to identifying the ambiguity of Holtzmann (and, therefore, Mayer), Clausius, in the climactic conclusion of his paper, hugely complimented Joule. He quoted no less than three of Joule's experimental methods 'The agreement of these three numbers, in spite of the difficulty of the experiments, leaves really no further doubt of the correctness of the fundamental principle of the equivalence of heat and work, and their agreement with the number 421 confirms in a similar way the correctness of Carnot's principle' [5, p. 152]. Such praise, by a peer of both Mayer and Joule, must be regarded as justifying Joule's words of the following year. There are two further points, though. First, Ref. [5] is part of the Dover publication edited by E. Mendoza, who also gives a thorough introduction in [6]. Referring to Mayer's paper, Mendoza comments 'the famous paper by Mayer ... is almost unreadable today, the physical ideas being obscured by layers of philosophical verbiage' [6, p. xix]. Secondly, the story does not end there. In 1854, Rankine confirmed a prediction of Mayer in his (Rankine's) very masterly Royal Society publication [7]. It related to the possible deviation of high-density saturated steam from the perfect gas law. Rankine pointed out that Clausius's calculations showed either that 'the supposition of Mayer is erroneous, or steam deviates very much .....' [7, p. 158]. Rankine then calculated a table of steam volumes concluding that it added to the evidence for such deviation. [7, pp. 159, 160].

It would have been great comfort to Mayer who had been able to foresee that history would treat him kindly, in fact very kindly, reflecting the 'famous' compliment of Mendoza rather than his 'unreadability'. Today, this kindness goes hand-in-hand with due regard to the genius of Joule. Two examples are given. In the text, *Modern Thermodynamics* of 1999, co-authored by the thermodynamics Nobel Laureate Ilya Prigogine [8], we first read of Joule's 'series of experiments that established the equivalence between mechanical energy and heat' [8, p. 13], and then 'the most important contributions to the idea of conservation of energy as a universal law of Nature came from Julius Robert von Mayer ... James Prescott Joule and Hermann von Helmhotz. Two landmarks ... are a paper by Robert Mayer ... published in 1842 ...' [8, p. 36]. Overall, the relative credits to Joule and Mayer could barely be clearer, and virtually coincident with the assessment of Joule himself – Kondepudi and Prigogine using Joule's own word –'established'.

An even more recent publication is that by Newburgh and Leff [9] – *The Mayer–Joule Principle: The Foundation of the First Law of Thermodynamics*. In the summary we read: 'Independently, Julius Robert Mayer and James Prescott Joule found the connection between heat and work, the Mayer–Joule principle'.

This unexpectedly long treatment represents an attempt to assess objectively the contribution of Mayer. We will now examine how, one decade after the original historical events, the well-connected researcher John Tyndall 'unleashed a massive extension of hostilities to North Britain … .ostensibly over the historical claims of Joule and Mayer' [1, p. 171].

3 Joule, Mayer and the First Law: the X-club controversy

The whole story is told by Crosbie Smith in Chapter 9 of [1] from the point of view of the 'North British' thermodynamicists. The controversy involved John Tyndall on the one side and P.G. Tait on the other. The latter was yet another ex-Cambridge Wrangler thermodynamicist to add to Kelvin and Maxwell. (To digress, the mathematical abilities of these three were bewildering: Kelvin in 1845 (senior wrangler and first Smith's prizeman [1, p. 80]), Tait in 1852 (senior wrangler and first Smith's prizeman [3, p. 174]) and Maxwell in 1854 (second wrangler and first-equal Smith's prizeman [1, p. 214]). To cap it all, Tait was credited with the remark 'I could coach a coal scuttle to be Senior Wrangler' [1, p. 174]. The complexities take Crosbie Smith over 20 pages (170–191) to describe. The X-club was the group founded in 1864 by T.H. Huxley and eight associates to promote a gospel of 'scientific naturalism'. It used Darwin's evolutionary theory as a principal weapon. Crosbie Smith has two main themes of interpretation – scientific authority morphing into Christian belief. 'I argue that the Tait–Tyndall controversy should be construed as a contest for scientific authority' and 'some of the fundamental differences between scientific naturalism (united by a stand against Christian doctrine) and the science of energy (promoted as a natural philosophy in harmony with, though not subservient to, Christian belief)' [1, pp. 171, 172]. The arguments need not be repeated as the key facts about Joule and Mayer have already been discussed. Both Tyndall and Tait tended to downplay the achievements of Joule and Mayer, respectively: and, so Smith concludes, 'Tait sought to secure not only the reputation of Joule but the …basis of a North British science of energy against … its appropriation by the scientific naturalists' [1, p. 191].

Cardwell [2] addresses this matter, too, in two sections in pages 229–238 and page 282 onwards. In the first section, Tyndall is interpreted as someone whom today we would call a white knight. 'Mayer was eventually rescued from his undeserved oblivion through the good offices and by the generous impulses of John Tyndall [2, p. 231]. It is in the second section that the interchanges are more fully addressed and Cardwell does this *personally*, that is without mention of the X-club. Using the source [1] in addition, the events may be summarised as follows: (i) in 1862, Tyndall gave an address to the Royal Institution in which he suggested 'it had all been the work of a comparatively unknown … doctor' [2, p. 283); (ii) Joule then wrote to the Philosophical Magazine 'to set the history … right' [1, p. 181, 283]); (iii) still in 1862, Tyndall replies that 'to Mr Joule … we are almost wholly indebted for the experimental treatment of this subject' ([1, p. 182, Smith's Reference 42 of Chapter 9, p. 337]). Cardwell describes (ii) and (iii) in terms of quite complimentary to Tyndall: 'Joule … felt his particular claim to have established, independently and scientifically, the doctrine of energy conversion could not be denied'. 'With this Tyndall entirely agreed' (both quotes [2, p 283]).

From the above, Cardwell concludes:'. ... the whole matter could – and certainly should – have ended amicably with Joule's credit unimpaired and justice at last done to the unhappy Mayer'. He lays the blame for 'this unnecessary and harmful squabble' (both quotes from [2, p 283]) largely with P.G. Tait, the professor of natural philosophy at Edinburgh. In the more diffuse discussion which follows Cardwell paints Tait rather unattractively. However, in 1875 Tyndall wrote a rather intriguing letter to Tait, which includes 'I would rather see you and Clausius friends than Zollner and myself. Trust me, C. is through and through an honest high-minded man' [2, p. 288]. Public conflict and personal good relations are not unknown today in British political life!

4 Conclusion: honour satisfied

The realisation that heat and work were equivalent formed a crucial watershed in the history of thermodynamics and led to the first law. Our appreciation of this may have dulled with time and familiarity, but two of the references remind us the strongest terms. Cardwell describes the situation thus: '... after 1850 a developed, established science ... had to be re-established on a new basis: that of the axiom of the conservation of energy'. *This was the first time in history that such a thing had happened* ([2, p. 291], Cardwell's italics). Newburgh and Leff have very recently given the same message: 'The path-breaking work of Mayer and Joule fusing heat and work is as fundamental to physics as the ideas of James Clerk Maxwell and Albert Einstein fusing electricity and magnetism' [9, p. 484].

Identification of the relative contributions of Mayer and Joule is therefore a matter of correspondingly high scientific import. Broadly speaking, there was no disagreement, even at the time from Joule himself, that Mayer had historical priority in *predicting* the law. What Joule claimed was that he had *established* it, and again, at the time this was accepted by his peers and led ultimately to the SI Joule unit of energy. For Mayer, if the expression 'the Mayer–Joule Principle' becomes generally accepted honour will be more than satisfied.

References

[1] Smith, C., *The Science of Energy*, Athlone Press: London, 1998.

[2] Cardwell, D.S.L., *From Watt to Clausius*, Heinemann: London, 1971.

[3] Smith, C. & Wise, M.N., *Energy & Empire*, Cambridge University Press: Cambridge, UK, 1989.

[4] Clausius, R., Über die bewegende Kraft der Wärme, und die Gesetze, welche sich daraus für die Wärmelehre selbst ableiten lassen. *Annalen der Physik*, **79**, pp. 368–397, 500–524, 1850; English translation published in *Phil. Mag. Series* 4, **2**, pp. 1–21, 102–119, 1851.

[5] Clausius, R., On the motive power of heat, and on the laws which can be deduced from it for the theory of heat, translated by W.F. Magie, from *Poggendorff's Annalen der Physik*, **LXXIX**, p. 368, 500, 1850. Carnot, S., *Reflections on the Motive Power of Fire, and Other Papers on the Second Law of Thermodynamics by E. Clapeyron and R. Clausius, with Introduction by E. Mendoza, (ed.)*, New Edition Dover: New York, pp. 107–152, 1988.

[6] Mendoza, E., Introduction, in Carnot, S., *Reflections on the Motive Power of Fire, and Other Papers on the Second Law of Thermodynamics by E. Clapeyron and R. Clausius, with Introduction by E. Mendoza*, (ed.), New Edition Dover: New York, 1988.

[7] Rankine, W.J., Mac, Q., On the geometrical representation of the expansive action of heat, and the theory of thermo-dynamic engines. *Phil. Trans. R. Soc. Lond.*, **144**, pp. 115–175, 1854.
[8] Kondepudi, D. & Prigogine, I., *Modern Thermodynamics*, John Wiley: Chichester, UK, 1999.
[9] Newburgh, R. & Leff, H.S., The Mayer–Joule principle: the foundation of the first law of thermodynamics. *The Physics Teacher*, **49(8)**, p. 484, 2011.

Chapter 9

The emergence and evolution of the Second Law of Thermodynamics

D. Wilkie[1], R.C. Dougal[2], M.W. Collins[3] & A. Whitaker[4]
[1]*Formerly of British Nuclear Fuels, United Kingdom.*
[2]*James Clerk Maxwell Foundation, United Kingdom.*
[3]*College of Engineering, Design and Physical Sciences, Brunel University London, United Kingdom.*
[4]*Department of Physics, Queen's University of Belfast, United Kingdom.*

Abstract

The results of measurements on thermal properties of common substances provided a backdrop to the successful development of heat engines in the eighteenth and nineteenth centuries. Early in the nineteenth century, Carnot's theory of an idealised heat engine provided the stimulus to the ongoing studies, both practical and theoretical, of the properties of materials and of mechanical and thermal processes. By mid-century, the First and Second Laws of Thermodynamics had been established, in various equivalent forms. These clarified the role of energy and its transformations and led to the introduction of a new thermodynamic function, entropy, to complement energy and, from its thermodynamic properties, to clarify the limitations imposed by the Second Law. The understanding of thermal properties of matter from the laws and techniques of classical thermodynamics was enhanced by the introduction of and alignment with statistical thermodynamics. This provided an understanding of properties of assemblies of large numbers of molecules by incorporating the rules of probability theory, leading to a statistical interpretation of entropy. A merger with the well-known laws and processes of chemistry led to substantial advances in chemical thermodynamics, permitting an insightful subsequent reassessment of thermodynamics as a whole.

1 Introduction

Classical thermodynamics, emerging from studies of heat and bulk properties of matter, is underpinned by the First and Second Laws of Thermodynamics. The First Law is a special case of the principle of conservation of energy. The Second Law provides a criterion for the selection from processes permitted by the First Law only of those which occur naturally. These laws were formulated from a wide range of observations and measurements, and have a correspondingly wide range of applicability. In this chapter, we show how tactically the formulation of the Second Law benefitted from the formulation of the First Law, how both demonstrate the interplay of engineering and mathematics and how they led endeavours, which might have seemed peripheral, into the mainstream of science.

2 Heat and temperature: conceptualised and applied

2.1 Black and Watt

The science of energy began with Joseph Black (1728–1799) at Glasgow University in the early 1750s with his definition of the distinction between heat and temperature. Black stated that to assume two bodies at the same temperature would possess the same quantity of heat would be 'taking a very hasty view of the subject. It is confounding the quantity of heat in different bodies with its general strength or intensity, though it is plain that these are two different things and should always be distinguished, when we are thinking of the distribution of heat'. Black showed that the temperatures of different substances increased at different rates when they were heated equally, implying different 'capacities' for absorbing heat. He was able to compile a comparative list of heat capacities per unit mass ('specific heats') for a number of common materials. He went on to introduce the concept of latent heat by heating ice and finding that no change in temperature occurred until all the ice had melted and likewise for water, which remained at boiling point until all the water had been converted to steam. He concluded that the heat supplied in these two events was latent in the melted ice and in the steam and was available for recovery in the reverse processes of freezing and condensation.

James Watt (1736–1819), born in Greenock, was engaged by Black as an instrument mechanic to repair and develop instruments. Watt had spent a year in London working for an instrument maker but his health suffered from overwork. As Watt had not served a full apprenticeship he was barred from obtaining employment in the trade, but the University had no such restrictions.

Watt repaired the department's demonstration model Newcomen engine, which had been returned from London as beyond repair. This was an atmospheric rather than a steam engine, with the piston powered by the air above it, making it move down on a partial vacuum produced by the condensing of steam on the other side. During this work, he became aware of its extremely poor performance. This led him to invent a steam condenser separate from the cylinder thus avoiding the need to cool the piston on each stroke and re-raise steam resulting in a saving of time and fuel. It provided a fivefold improvement. In his early experiments with steam, Watt realised that raising steam required a lot of heat and deduced that the steam contained this heat.

2.2 Watt and Boulton

Black financed Watt's development work initially but the industrialist John Roebuck (1718–1794), founder of the Carron ironworks (near Falkirk in Central Scotland), took over the financing in return for a two-thirds share in the patent rights. Roebuck became bankrupt in 1773 when his own businesses failed and Matthew Boulton (1728–1809), a versatile manufacturer from Birmingham, who had already met Watt and had been trying for some time to engage him, acquired Roebuck's share. Watt went into partnership with Boulton in 1775.

In Watt's engine, steam entered through an aperture in the side of the cylinder, above the piston. Early improvements included letting the steam enter alternately through opposite sides, making it a double-acting engine. Heat losses were reduced by the addition of a steam jacket, and insulation. A pressure indicator was modified by the works manager, John Southern

(c. 1755–c. 1825) to show the variation of pressure with piston displacement. From a plot of pressure versus piston displacement, a closed figure was produced, whose area represented the power developed by the engine. This allowed Watt to assess the effectiveness of developments to improve the power delivery. Among other things, Watt found that stopping steam entry to the cylinder before the end of the stroke improved the efficiency.

Being fully conversant with the operation of his steam engine Watt must have realised the importance of latent heat and therefore of stored energy. He also knew of the importance of attaining a maximum temperature difference, as did his extremely inventive assistant, William Murdock (1754–1839), who was responsible for many of the improvements to the Watt engine. For instance, he suggested using high pressure steam to increase the upper temperature, but this was dismissed by Watt because of the dangers of exploding boilers.

3 Towards the laws of thermodynamics

3.1 Carnot, Clapeyron and the original Carnot cycle

By the turn of the nineteenth century developments in the construction and operation of the steam engine had led to an appreciation of both its potential and its limitations. In 1824, the French engineer N.L. Sadi Carnot (1796–1832) stepped back from matters of detail of steam engines so as to identify their basic features and skilfully assemble them into a coherent whole. He then presented a description of the basic theory of an idealised steam engine, followed by a simpler, similarly idealised, air engine [1]. Carnot's memoir, his only publication, was largely ignored until B.P. Emile Clapeyron (1799–1864) interpreted and re-presented it more analytically in 1834 [2,3] in a version that was the essential input to later researches of others, including William Thomson (1824–1907) [4], appointed as the professor of natural philosophy at Glasgow University in 1846. Clapeyron's treatment led to a specification of 'the Carnot cycle'. (Thomson did eventually, in 1848, get a copy of Carnot's memoir from his friend, Lewis Gordon, Professor of Engineering at Glasgow University [5].

3.1.1 Carnot engines and Carnot's theorem

The 'structure' of the idealised steam engine was chosen so that it could provide external mechanical work broadly in line with that from existing steam engines. This allowed Carnot to best place effort in identifying those features of the ideal engine which, once clarified, would allow an interpretation of relevance to other engines operating in repeated cycles. Optimum performance meant a tight control of heat flow: no waste from heat leaks down temperature gradients which would impair performance. Frictional effects were also ignored. He held to the caloric theory of heat, which obscured the link between heat flow, in and out, and external work done, per cycle, in the operation of the device, but he was still led to propose that – for the idealised engine – the operation of all stages in the cycle could be reversed. The performance of a reversible engine would then be optimum. He argued that if there was an 'other' engine having superior performance for the same heat flow, its output, being more than enough to drive the cycle of the reversible engine backwards, would result in net mechanical power from them connected as a composite system as from a perpetual motion machine, which was inadmissible. Carnot extended the argument to propose (Carnot's theorem) that the output power from a reversible idealised cyclic engine had to be the maximum, or the process was at its optimum efficiency, whatever its working substance.

Carnot's modelling (adapted) and analysis (essentially intact) were then transferred to an air engine, whose simpler working substance made it easier to visualise. In the resulting idealised cycle, heat Q_1 entered the working substance from heat source or reservoir (boiler, for steam) at a constant high temperature t_1 and caused the working substance to expand isothermally at that particular temperature and do mechanical work externally. This was followed by continued expansion with the working substance thermally isolated (later called adiabatic), and with its temperature dropping from t_1 to t_2. Two compression stages, now with work being done *on* the working substance, followed: isothermal at the temperature t_2 of a heat sink or reservoir (condenser) accompanied by a necessary release there of heat, Q_2, then a closing return adiabatic with temperature increasing to the initial t_1. The same *net* amount of mechanical work was delivered *by* the working substance each cycle. If the four stages of the closed loop are plotted clockwise on a p–v diagram, like Watt's indicator diagram, the area of the loop is a measure of the work done by the working substance per cycle. The caloric theory of indestructible heat meant choosing $Q_2 = Q_1$.

For the steam engine, Carnot delayed addressing a potentially problematic return stage to the initial state, as described here, of the steam. Raising steam could have required heat input through finite temperature differences, for which reversibility would not pertain. This led him to consider an infinitesimal cycle for the working substance, working between an infinitesimal temperature difference. The loop area was also infinitesimal, meaning an infinitesimal work output per cycle. Undeterred, Carnot proposed that for an engine operating between a finite temperature difference $t_1 - t_2$, its cycle could be modelled as a composite in the form of a stack of infinitesimal cycles with the heat output from any one serving as the heat input to the next one down in the stack [3].

3.1.2 The Carnot function

All the stages of idealisation led to the independence of the efficiency of the idealised engine on everything except temperature. Engineering practice pointed to an improvement of mechanical output W with increasing temperature *difference* $t_1 - t_2$, moderated by some improvement for lower actual temperatures. Analogy with the mechanical output from a fall of water – proportionality to the height of the fall – suggested proportionality to $t_1 - t_2$ for a heat engine, moderated by a factor whose value increased as temperatures decreased. This is written here as C, called the Carnot function. (Clapeyron used $1/C$ for the moderating factor, and he called this C the Carnot function.) Heat throughput Q could serve as the analogue of mass of water falling. For an infinitesimal cycle, with heat throughput dq and temperatures $t + dt$ and t, the Carnot function is easier to visualise, since some kind of averaging over temperature of it is not required. The work output dw can then be written $dw = dq \times C \times dt$. The infinitesimal cycle and the Carnot function both had key roles in later developments of thermodynamics (Section 5.1), buttressing further Carnot's remarkable insight.

3.1.3 Postscript

Carnot's working notes after 1824 show that he moved from caloric towards accepting a mechanical theory of heat [3]. His premature death in 1832 prevented them from being recognised as a contribution to the formulation of the First Law of Thermodynamics: in 1824, there was no First Law, with little of direct relevance from the theoretical side on its precursor, transformations between heat and work. Carnot had, however, provided a convincing association of heat flow with mechanical work which – however, tenuous – was supported – however, loosely – by a conservation principle: conservation of caloric thereby permitting the working substance to return to its starting state at the end of every cycle. It was Clausius who would provide significant additions to and replacements of – and consequential realignment of – the various pieces (Section 3.4); and eventually follow Thomson in accepting energy as the unifying concept (Section 3.5).

Remarkably, Carnot's memoir is recognised as pointing the way to the more subtle Second Law of Thermodynamics (Section 4.2), with a *universal* Carnot function providing a clue to a fundamental interdependence of heat flow and temperature (Section 4.4) which was crucial to the establishment of that law.

3.2 Davy, Rumford, Mayer and Joule: transformations and conservation

Observations had been reported from just before 1800 onwards of a direct link between mechanics and heat, casting doubt on the caloric theory. Sir Humphrey Davy (1778–1829) produced heat from mechanical energy by melting the surfaces of contact between two blocks of ice by rubbing them together [6]. Count Rumford (1753–1814) observed the large amount of heat generated in the process of boring cannons which led him to deduce a relationship between mechanical work and heat which he published in 1798 [7]. Energy would appear later – in parallel with developments in language – as the enabling concept for the successful merging of heat and mechanics, building on to the notion that heat could be related to some kind of motion. In the 1840s, Robert J. Mayer (1814–1878) in Germany had speculatively broadened his interpretation, in chemical and thermal terms, of certain changes occurring in physiological processes. He suggested, in 1842, that conservation of some kind of 'force' was involved [8, 9], to parallel conservation of mass in chemistry. His treatment there of a relationship between mechanical work and heat prevented him from quantifying it convincingly, but history credits him with having recognised but not having driven forward adequately a principle of conservation of 'energy' (replacing 'force'). In 1845, his more detailed account in a follow-up paper was rejected and he published it privately [10]. By that time measurements in England by James Prescott Joule (1818–1889), and his interpretation of them, had begun to dominate the field. In a sequence of remarkably precise direct measurements on the transformation from work to heat, he established an accurate value for a mechanical equivalent of heat including, early in the sequence, observations which had also shown a similar transformation of electrical energy [11]. The outcomes from these observations became crucial to the understanding of the parts of the Carnot cycle involving transformations between heat and work.

3.3 Thomson's assessment of Carnot's theory

In 1849, Thomson [4] presented an account of Carnot's theory, putting a focus on the return of the working substance to its original state at the end of each cycle with heat gains by and heat losses from the working substance in balance. He noted Carnot's own remarks that 'to deny [it] would be to overturn the whole theory of heat, of which it is the fundamental principle', followed by 'several experimental facts appear nearly inexplicable in the actual state of this [caloric-based] theory'. Thus prepared, Thomson then gave an authoritative and sympathetic interpretation of the basis of Carnot's theory, both steam and air engines, together with data from measurements, mainly at the laboratories of Henri Victor Regnault (1810–1878), on thermometry and associated thermal properties of materials, seeking compatibility. (Following graduation from Cambridge

University in 1845, Thomson had spent several months in Regnault's laboratories in Paris. Results of measurements being made there provided a complementary addition to theoretical work [12] of J.B. Joseph Fourier (1768–1830) on heat transfer, already well known to him.)

In particular, he developed analytical expressions linking M (our W) and H (our Q): M = H∫μdt, with μ denoting 'Carnot's coefficient'. From experimental values of μ at one degree intervals from 1°C to 230°C, he calculated and presented as a look-up table, the value of M (in foot-pounds) per 'thermic unit centigrade' of heat transferring from 'a body at any temperature less than 230°C to a body at 0°C'. These show that Thomson's μ corresponds to $J \times C$, where J, the mechanical equivalent of heat, converts thermal to mechanical values, and C is the Carnot function (Section 3.1.2). He showed that there was good agreement between the few values (of μ) quoted by Clapeyron and his own derived from Regnault's observations. However, he cautioned that '[Carnot's theory of the motive power of heat] and with it every other branch of the theory of heat, may ultimately require to be constructed upon another foundation, when our experimental data are more complete'.

3.4 Clausius's extension of Carnot's theory

In 1850, Rudolf J.E. Clausius (1822–1888) presented an analysis [13] of the assumptions in Carnot's paper. He accepted as a fundamental principle 'that whenever work is produced by heat and a permanent alteration of the body [working substance] in action does not at the same time take place, a certain quantity of heat passes from a warm body to a cold one'. However, he doubted that 'the assertion, that in the production of work loss of heat never occurs, is sufficiently established by experiment'. He continued, 'The difference between the two ways of regarding the subject has been seized with much greater clearness by W. Thomson, who has applied the recent investigations of Regnault, on the pressure and latent heat of steam, to the completing of the memoir of Carnot'.

Clausius sought to remove the conflict. Introducing a small alteration to Carnot's model, he claimed 'it is quite possible that in the production of work … a certain portion of heat may be consumed, and a further portion transmitted from a warm body to a cold one; and both portions may stand in a certain definite relation to the quantity of work produced'. Clausius elaborated on the 'portion of heat consumed', treating it as the combination of heat retained (by a working substance) and work done internally (whilst the substance expands) and work done externally. Shortening Clausius's argument, in a reversible cyclic process the heat retained and the internal work could be recovered when the working substance regains its initial state but not so the external work. The first two together could be regarded as a thermodynamic function of any two appropriate thermodynamic coordinates – say temperature t and volume v – just as, from the 'gas laws', the pressure p of a gas can be related to t and v to a good approximation via the ideal gas equation. He gave it the symbol U, a function which limited whilst enabling the conversion of heat to external work. For small changes in heat Q, U and volume v, he wrote '$dQ = dU + Ap\,dv$ … where A denotes a constant which expresses the equivalent of heat for the unit of work'. He thus improved the tools with which to build a law suggested by the researches of Mayer and Joule taken together.

Throughout, Clausius was guided by 'the general deportment of heat, which everywhere exhibits the tendency to annul differences of temperature, and therefore to pass from a warmer body to a colder one'. He was able to adapt Carnot's reasoning on the composite of two reversible engines, one driving the other in reverse (Section 3.1.1), coupled this time so that the conjectured more efficient was tuned to provide, from a smaller throughput of heat, exactly the work needed to drive the Carnot engine in reverse. The resulting *net* flow of heat from cold to hot, unaccompanied by a net amount of work, was contrary to the general deportment of heat and the conjectured more efficient engine rejected.

3.5 Thomson and Clausius: the First Law of Thermodynamics stated

Within Thomson's dynamical theory of heat, presented in 1851 [14], he associated heat with molecular motion (rather than caloric) and identified a *mechanical energy*, e, of a body as its means of participating in transformations and transmissions of heat. Thus, the increase Δe in mechanical energy of a body resulting from heat input ΔQ and performance by it of external work ΔW could be written $\Delta e = \Delta Q - \Delta W$. Evaluating e from a sufficiently extensive database of values of thermodynamic coordinates and properties was feasible. Later, Thomson referred to e as *intrinsic energy* before it became widely accepted as *internal energy* and given Clausius's symbol U. Much later, in a paper in 1865 on the dynamical theory of gases [15], Clausius, referred back to his introduction of U 15 years earlier, and remarked that 'since then, however, W. Thomson has proposed the term *energy* of the body for this magnitude, which mode of designation I have adopted as one very appropriately chosen'.

With heat Q and external work W both now accepted as manifestations of energy, the tracking of energy changes, without loss, led to a formulation of the First Law of Thermodynamics in the form of a tally. Following Clausius by starting with ΔQ, it is: $\Delta Q = \Delta U + \Delta W$. He also gave there a succinct statement of the First Law: 'the energy of the universe is constant'.

4 The Second Law of Thermodynamics

4.1 Interlude: an absolute temperature scale

Thomson [16] recognised the need for an absolute scale of temperature in the science of energy and conceived one which he published in 1848. He had identified a difficulty arising from the variation with temperature of thermal properties of materials. If such a property were used in the design of a thermometer, its responses to the same heat input at different temperatures would be different temperature rises as indicated by the increase in number of degrees or scale divisions of the chosen scale. The device would be unable to record heat input correctly (i.e. *absolutely*) at all temperatures.

In his search for an absolute thermometric scale, Thomson drew attention to the properties, following Carnot, of an ideal reversible heat engine operating between temperatures t_1 and t_2, (as described above). He wrote, using mechanical work output (the 'mechanical effect') as an intermediary: 'The characteristic property of the scale which I now propose is, that all degrees have the same value; that is, that a unit of heat descending from a body A at the temperature T° of this scale, to a body B at the temperature $(T\text{-}1)^\circ$, would give out the same mechanical effect, whatever be the number T. This may justly be termed an absolute scale, since its characteristic is quite independent of the physical properties of any specific substance'. He explained how the scale could be realised, approximately, based on existing data from Regnault's measurements on thermal properties of gases to access the mechanical effect, via the resulting values of his 'Carnot coefficient' μ (Section 3.3). He anticipated that more extensive data would improve the calculation of absolute temperature by correlation with temperature measured with an air thermometer [17].

Thomson later simplified the specification of absolute temperature in 'Fundamental Principles of General Thermo-dynamics recapitulated', his introduction to a paper in 1854 on thermo-electric currents [18]. This time he summarised the argument: 'the absolute values of two temperatures are to one another in the proportion of heat taken in to the heat rejected in a perfect thermo-dynamic engine working with a source and refrigerator at the higher and lower of the temperatures respectively'. Thus, for a Carnot cycle working between heat reservoirs at absolute temperatures T_1 and $T_2<T_1$, and where the work output per cycle, W, is equal to the difference, $Q_1 - Q_2$, between heat in and heat out, $Q_1/Q_2 = T_1/T_2$. The efficiency for such Carnot engine, $W/Q_1 = (Q_1 - Q_2)/Q_1 = (T_1 - T_2)/T_1$, proportional to t emperature difference and with $1/T_1$ as the factor which moderates the efficiency in line with Carnot's interpretation (Section 3.1.2).

4.2 Axioms and statements

In the course of clarifying Carnot's theory by input from the First Law of Thermodynamics (Section 3.4), Clausius had been led to propose in 1850 an overarching principle: heat flows spontaneously only to reduce a temperature difference, not to increase it. For a Carnot engine working in reverse, heat flow from cold to hot had to be accompanied by work input. Thomson also identified a principle by a path that held closer to temperature-dependent restrictions on the transformation of heat to mechanical work. He presented both, as axioms, in the first section (March 1851) of his long paper [14] on a dynamical theory of heat.

Section 1 included a key paragraph comprising two propositions which together provided the foundations for 'the whole theory of the motive power of heat'. Briefly, 'PROP. I', acknowledging Joule, was that the same amount of heat is used up (or generated) when equal amounts of mechanical effect are produced (or lost) purely thermally. The two axioms were then presented so as to validate the paper's 'PROP. II', attributed to Carnot and Clausius, that a fully reversible engine is the most efficient of all engines extracting heat from the same constant temperature heat source whilst delivering heat to the same (lower) constant temperature heat sink:

Clausius's axiom: *It is impossible for a self-acting machine, unaided by any external agency, to convey heat from one body to another at a higher temperature.*

Thomson's axiom: *It is impossible, by means of inanimate material agency, to derive mechanical effect from any portion of matter by cooling it below the temperature of the coldest of surrounding objects.*

Thomson remarked that 'It is easily shown, that ... either [axiom] is a consequence of the other'. This kind of exercise is still standard in the presentation of the two typical statements, below, of the Second Law in textbooks on thermodynamics [19].

In Parts I and II of Section 1, on the 'motive power of heat', he amended Carnot's treatment replacing the caloric theory with interconversion of heat and mechanical work as demonstrated by Joule. The infinitesimal Carnot cycle retained the moderating factor, with μ now called Carnot's function. Although it had the optimum efficiency, a large proportion of the heat input was 'irrecoverably lost to man, and therefore 'wasted', although not annihilated'. He gave an analytical treatment of the relationship between work output and heat input for heat reservoirs having a finite temperature difference, also modelling the system as a stack of infinitesimal cycles. The work output was less than would be got from Carnot's theory, but tended to it as the temperature difference decreased. He also noted Joule's suggestion that μ might be inversely proportional to absolute temperature. Part III demonstrated the ease with which the dynamical theory relationships between physical *properties* of substances – both those whose experimental values were well known and others – which had a bearing on thermodynamic *processes*. The single Part IV of Section 2 (April 1851) dealt with possible measurements on the relationship between work done and heat released during compression of gases, exploiting Joule's proven instrumentation. Section 3 (December 1851) comprised Part V in which Thomson introduced the concept of the mechanical energy, e, of a fluid (Section 3.5).

The axioms provide the starting points for typical statements of the Second Law of Thermodynamics:

The Clausius statement: *It is impossible to construct a device that, operating in a cycle, will produce no effect other than the transfer of heat from a cooler to a hotter body.*

Thomson's axiom was developed more substantially, following this re-statement of the law by Max Planck (1858–1947): *It is impossible to construct an engine which will work in a complete cycle, and produce no effect except the raising of a weight and the cooling of a heat-reservoir* [20]. It led to the compromise:

The Kelvin–Planck statement: *It is impossible to construct an engine that, operating in a cycle, will produce no effect other than the extraction of heat from a reservoir and the performance of an equivalent amount of work.*

4.3 Thomson's insight

Thomson's short paper [21] in 1852, just a few months after his major dynamical theory, was a response to Carnot's proposition that there is 'absolute waste of mechanical energy' accompanying heat flow from a hot to a cold body by any process incompatible with the constraints satisfied by an ideal engine. As in his dynamical theory, he chose to replace waste by a transformation, rather than an annihilation of energy. That led him to consider the possibility of restoration 'to its primitive condition' of mechanical (in present day terms, potential and kinetic) energy that had, in part, been transformed to heat. He introduced the term dissipation of energy as a means of distinguishing irreversible from reversible processes.

Thomson set out four propositions which were 'necessary consequences' of his own axiom (Section 4.2). The first proposition put a condition on heat produced *reversibly* by transformation of mechanical energy. His choice of illustrating that the transformation could not be accounted for in isolation was: 'When heat is created by a reversible process (so that the mechanical energy thus spent may be *restored* to its primitive condition), there is also a transference from a cold body to a hot body of a quantity of heat bearing to the quantity created a definite proportion depending on the temperatures of the two bodies'. The next three propositions, all involving dissipation of mechanical energy thereby ruling out its full restoration, were much more general, reflecting the abundance of irreversible processes. The examples were: any heat produced *irreversibly* (such as frictionally), heat *diffused* by conduction and the *absorption* of radiant heat or light.

Setting aside his few brief remarks on their relevance in general terms to biology and chemistry, his main conclusion from the propositions was that 'there is at present in the material world a universal tendency to the dissipation of mechanical energy'. This overarching statement has to be acceptable as an informative, even if qualitative, guide to the Second Law of Thermodynamics.

4.4 Clausius's initiative

By 1854, Clausius [22] had moved on from combining the Carnot cycle with his 'first fundamental theorem', essentially the First Law of Thermodynamics, extending his treatment from that set out four years before [13], which he included by way of introduction in a form which he claimed was 'at once more general and more concise'. He claimed that the four-stage Carnot cycle was too restrictive. In particular, a higher temperature body (the boiler) had to provide both the heat to be converted to external work and the heat released to the lower temperature body (the condenser). It was, however, important to retain the link identified by Carnot between the conversion of heat to external work and the accompanying throughput of heat – both per cycle with all stages of the cycle reversible.

Clausius therefore introduced a six-stage reversible cycle. The first three stages are chosen here as in the Carnot cycle described above (Section 3.1.1) starting with an isothermal expansion at the higher temperature. The return adiabatic compression does not close the cycle. Instead it takes the working substance to a temperature of $t > t_1$, followed by an isothermal expansion at t of such duration that the cycle can be closed by an adiabatic expansion, during which the temperature falls from t to t_1. This provided Clausius with a combination of two mutually supportive heat transformations per cycle: heat flow Q in at the temperature t_1 *transformed* to become the same amount Q out at a lower temperature t_2; and heat flow q in at a temperature $t > t_1$ which, being the net heat in, was *transformed completely* into external work w. In completing the cycle, the working substance returned to its initial thermodynamic state.

He argued that these two particular transformations were 'phenomena of the same nature'. Retaining the simplicity (dependence on temperatures only) of the expression for efficiency (Section 3.1.2), he gave them 'equivalence values': $Q \times F(t_1,t_2)$ for the transformation of Q, and $q \times f(t)$ for the 'heat-to-work' transformation of q. He noted that the transformation of Q occurred in the direction of spontaneous heat flow, but that the transformation of q was in the opposite direction to what could be chosen to be the spontaneous direction for transformations between heat and work, namely work to heat. He sought the implications of a balance between the two heat transformations, now defining the cycle by combining their equivalence values algebraically: $Q \times F(t_1,t_2) - q \times f(t) = 0$.

Clausius then showed how to extract a relationship between the two functions of temperature. The strategy was similar to that used by Carnot: combine two Carnot engines, one driving the other in reverse (Section 3.1.1). Clausius's second cycle, performed in reverse, otherwise differed only in having the 'third' reservoir at a temperature $t' > t$, and $q' > q$. For it, $Q \times F(t_2,t_1) + q' \times f(t') = 0$. Because $F(t_2,t_1) = - F(t_1,t_2)$, the two heat flows Q cancelled in the composite system, leaving only q at t and q' at t', and $q' \times f(t') = q \times f(t)$ which became an input to his analysis by the equivalence value method to a now two-reservoir system – essentially a Carnot cycle. The upshot was $F(t_1,t_2) = f(t_2) - f(t_1)$. The equivalence value for heat throughput could thus be replaced by the combination of two equivalence values; one, $Qf(t_1)$, for the release and the other, $Qf(t_2)$, for the absorption of heat, by reservoirs at the denoted temperatures. The algebraic sum of the equivalence values in his reversible six-stage cycle could now be rewritten $Q \times f(t_2) - Q \times f(t_1) - q \times f(t) = 0$ without losing sight of the balance of the two processes involved. This was the basis of several immediately ensuing stages of generalisation and interpretation.

The return to the same thermodynamic state in each cycle suggested that the product $f(t)Q$ could, like U (Section 3.4), form the basis of another thermodynamic function.

Then for *any* multistage reversible cycle with any number of *transformations of heat Q at temperature t*, the summation, now written $\Sigma f(t)Q$, could even be replaced by an integration $\int f(t)\,dQ$ to allow for infinitesimal heat flows, each for its corresponding temperature. Clausius proposed that $\Sigma f(t)Q = 0$ generalised to $\int f(t)\,dQ = 0$ could be 'the analytical expression for all *reversible cyclical processes*, of the second fundamental theorem in the mechanical theory of heat'.

Furthermore, the temperatures could equally well be taken to be those of the working substance during reversible, isothermal (the only) heat flows. This gave the option of a change of focus from heat reservoirs to working substance. Heat leaving a reservoir became heat entering the working substance but $\Sigma f(t)Q = 0$ still pertained.

Clausius then introduced a descriptive term 'compensation' to distinguish reversible from irreversible processes. His selective list of the latter overlapped those in Thomson's introduction of dissipation (Section 4.3). Clausius's description then parallels Thomson's. The sum of all the positive terms in $\Sigma f(t)Q$ would be compensated by the sum of all the negative terms in a *reversible* cycle. In proceeding to address *irreversibility*, he proposed that heat thereby generated could only be included as an *uncompensated transformation*. Thus 'for all cyclical processes, those that are reversible forming the limit: the algebraical sum of all transformations occurring in a cyclical process can only be positive'. In the notation being used here, $\Sigma f(t)Q \geq 0$.

Finally, an appeal to the gas laws provided the path to the simplest explicit function of temperature for $f(t)$, namely $1/T$, with T the absolute temperature. A rare detailed account of Clausius's 1854 paper is given by Cropper [23].

Clausius's subsequent papers trace his systematic development of the issues and concepts he had recognised, removing the restriction to cyclic processes and extending a consideration of thermodynamic properties of substances. The final paper [11] in the sequence included writing $dQ/T = dS$ and choosing S and the name entropy for this now well-established thermodynamic function. The more general version of his second fundamental theorem could be written $\int dS \geq 0$, its brevity hiding its massive content. As he did with his statement of the First Law, he gave his succinct alternative statement of the Second Law of Thermodynamics: 'the entropy of the universe tends to a maximum'.

4.5 Interpretation

One attraction of Clausius's general statement (of the Second Law) is that it implies the importance of *changes*, if any, in the value of a thermodynamic function entropy. In introductory thermodynamics, a range of illustrative examples can be collated and a generalisation made:

entropy *increase* can be associated with a *decrease* of order that can be related to structure in its widest sense.

From the extreme of bulk properties of materials is the historically challenging change from an initial spatially structured (or ordered) temperature gradient across a metal bar to an equilibrium structureless (thereby less ordered) uniform temperature throughout, accompanied by no work output. This example serves to illustrate the increase in entropy in irreversible processes culminating in an equilibrium state. The alternative 'disordered' for 'less ordered' is not very helpful here.

Topics from condensed matter physics appear in the centre ground. Elementary examples are associated with differences in order below and above some transition temperature, itself determined by *energy* considerations. The 50/50 copper/zinc alloy, β-brass, shows large peak in its specific heat capacity around 730 K. Below that temperature, the copper and zinc atoms have an ordered arrangement on a body-centred-cubic, crystalline lattice. The arrangement can be visualised as having a zinc atom at the centre of each cube, and a copper atom at those points where eight contiguous cubes meet, so that each copper atom is shared equally amongst the eight cubes, one-eighth to each. Above the transition temperature, both sets of constituent atoms are randomly distributed, hence disordered, on the same complete set of lattice sites. The transition is accompanied by an increase in *entropy* [24]. For this topic-specific transition, the brief description 'order-disorder' *is* appropriate.

5 Thermodynamic properties and processes

5.1 Overview and Maxwell's relations

In classical thermodynamics, thermodynamic *properties* of gases can be described in terms of those two coordinates, chosen from pressure, volume and temperature, which match them to reversible thermodynamic *processes* whose outcomes they determine. Internal energy U and the more sophisticated entropy S were augmented by other thermodynamic functions of classical thermodynamics such as enthalpy H, free energy E and the Gibbs function G. They are interrelated and their utility is topic-specific. Combined with inputs from engineering, there was a steady extension of the understanding – in breadth and depth – of basic thermodynamic properties. Some were essentially thermal, such as heat capacities for heat transfer and, for phase transitions, the heat storage and release in terms of latent heat. Others were essentially mechanical, building on to an empirical base of coefficients determining properties of solids such as compressibility and ductility. Inevitably, with heat regarded as energy and entropy change frequently acting as a kind of proxy for heat transfer, this theory provided thermal and mechanical interdependence.

Also in this period, descriptions of relationships amongst thermodynamic coordinates moved to the use of the language of differential calculus from that of geometry, the latter as exemplified in their thermodynamics books by Maxwell [25] and Tait [26]. For example, they chose a geometrical scenario of the extreme theoretical Carnot cycle (Section 3.1.2) with an infinitesimal difference in temperature Δt between heat source and heat sink, and the corresponding infinitesimal difference in pressure ΔP between the two isothermal stages. The geometry of this cycle was nearly a rectangle of sides ΔP and, also infinitesimal, ΔV for the isothermal expansion and compression stages. Two expressions for the efficiency were equated. One was largely a matter of definition: $\Delta W/(J\times\Delta Q)$ for work output ΔW, heat input ΔQ at the higher temperature with the factor J, the mechanical equivalent of heat, converting heat input from thermal to mechanical units. The other was Carnot's deduction that the efficiency was a function of temperature, independent of the choice of working substance, and was eventually rewritten $C\times\Delta t$, with C representing the universal Carnot function.

The development, outlined in detail by Klein [27], centred on rearranging the factors. ΔW was replaced by $\Delta P\times\Delta V$, a good approximation to the area of the cycle plotted on a P–V diagram. Thus, $\Delta P/\Delta t = J\times C\times\Delta Q/\Delta V$. Tait chose to put $\Delta Q/\Delta V = M$, with the order of the three factors being Maxwell's initials, JCM. In the three-way correspondence amongst Maxwell, Thomson and Tait, Maxwell became $\mathrm{d}p/\mathrm{d}t$, with Thomson and Tait adopting the simpler T and T', respectively.

Returning to thermodynamics, that equation (together with another three, all similarly derived, concisely, in Maxwell's book) was later re-presented. The absolute temperature T replaced t, for consistency with the replacement $1/T$ for the Carnot function and J was presumed to have been merged with ΔQ. In the transition to differential calculus, the constant volume $\Delta P/\Delta T$ was re-cast as $(\partial P/\partial T)_V$ and the isothermal $(\Delta Q/\Delta V)/T$, firstly written $(\Delta S/\Delta V)$ with the entropy difference ΔS replacing $\Delta Q/T$, was re-cast as $(\partial S/\partial V)_T$ to give the first of the four Maxwell's relations:

$$(\partial P/\partial T)_V = (\partial S/\partial V)_T$$

$$(\partial T/\partial P)_S = (\partial V/\partial S)_P$$

$$(\partial V/\partial T)_P = -(\partial S/\partial P)_T$$

$$(\partial T/\partial V)_S = -(\partial P/\partial S)_V$$

5.2 Illustrative example: the Joule–Kelvin process

This process consists of pushing a column of gas steadily through a large number of tiny holes in a barrier, settling beyond the barrier at a lower pressure. Its historical importance is twofold. It stems from a collaboration of Joule and Thomson on measuring the temperature change of a gas made to flow through a barrier consisting of a thermally insulated plug of either tightly packed cotton or a tightly rolled piece of silk. Its popular name, the 'porous plug' process, was introduced by Joule and Thomson [28]. A typical analysis, *many* years later [29], exemplifies some of the strategy of classical thermodynamics.

It is assumed that the gas can be tracked to the extent that during some arbitrary time interval an amount of mass M in volume V_1 at uniform pressure P_1 and temperature T_1 is pushed by the rest of the column further upstream to emerge to occupy a larger volume V_2 and so have the smaller uniform pressure P_2 and the common temperature T_2 of all the downstream gas. The barrier and the gas column are thermally insulated and so the difference in internal energies of the mass M before and after, $U_2 - U_1$, is equal to the work done on the mass M by the gas column also at pressure P_1 upstream (pressure × volume moved through) minus the corresponding work done by M pushing on the gas ahead of it downstream: $P_1V_1 - P_2V_2$. Hence, $U_2 + P_2V_2 = U_1 + P_1V_1$. The combination $U + PV$ is called the enthalpy H (Section 5.1). Although H cannot be tracked during the irreversible passage of the gas through the holes, the initial and final states of the mass M are modelled as equilibrium states and for them the change in any other thermodynamic function, here the temperature T, can be calculated for a constant enthalpy process between the same end states, this being optimally chosen to ease the analysis. Thus, the temperature T is chosen to be a function of pressure P and enthalpy H. This leaves T varying with P only because H does not change. The particular variation of T with P can therefore be written as $(\partial T/\partial P)_H = \mu_{JK}$, the Joule–Kelvin coefficient. For a change in pressure from P_1 to P_2, the change in temperature $T_2 - T_1 = \int \mu_{JK}\, dP$, integration being between the limits P_1 and P_2.

The now traditional tactics in thermodynamic analyses include replacing H in terms of other thermodynamic functions and parameters. Start with $(\partial T/\partial P)_H = -(\partial T/\partial H)_P(\partial H/\partial P)_T$. $(\partial T/\partial H)_P$ is readily shown to be equal to the reciprocal of the heat capacity, C_P, of the mass M of gas at constant pressure. Entropy enters by incorporating a version of the First Law written $TdS = dU + PdV$, leading to $(\partial H/\partial P)_T = T(\partial S/\partial P)_T + V = -T(\partial V/\partial T)_P + V$, using the Maxwell Relation $-(\partial S/\partial P)_T = (\partial V/\partial T)_P$. The outcome is $\mu_{JK} = \{T(\partial V/\partial T)_P - V\}/C_P$, which can be positive or negative, depending on the values of the thermodynamic coordinates and their mutual variation. From it, the choices making $T_2 - T_1$ negative point to the design details of devices using the effect for gas cooling and liquefaction, with the great advantage that, especially as the final stage of the cooling, it has no moving parts.

Joule and Thomson arrived at the same formula for $T_2 - T_1$ within their extensive papers. These merge a detailed account of Joule's meticulous measurements of temperature with a steady development of immediately applicable theory 'founded on elementary thermodynamic principles' over a 10-year period, together matching the inspired design of the apparatus. The set of named functions such as enthalpy and entropy, and the packaging of equivalences of properties as exemplified by Maxwell's Relations, came later.

6 Statistical thermodynamics

6.1 An early sighting: dissipation, recoverability and Maxwell's Demon

In his 1874 paper on the kinetic theory of dissipation of energy, Thomson [30] returned to the much discussed conduction of heat through a thermally insulated bar of metal initially warmer at one end than the other. The relevant details he extracted from a detailed treatment of a similar – but more easily visualised – process of heat diffusion between two halves, each initially in equilibrium, of a sample of gas, one half warmer than the other, in a closed, insulated vessel. As with the bar, the state of that system would change, tending to equilibrium. He concluded that for both bar and gas, a return to the initial temperature difference could only happen following input 'by the agency of external sources of heat and cold'. This was the basis of his 1851 version of Clausius's axiom clarified a year later by his introduction of dissipation, and subsequent unavailability, of energy (Section 4.3).

In the intervening years, interest had grown in acquiring an enhanced understanding of thermodynamics from the dynamics of gases treated as purely mechanical systems. Almost by way of expressing caution, Thomson began his paper by contrasting 'abstract dynamics' with 'physical dynamics'. In the former, the motion of every part of a system could be reversed exactly – and still satisfy the laws of motion – if the velocity of each part were to be instantaneously reversed: 'in mathematical language, any solution remains a solution when

[time] t is changed into $-t'$. He asserted that such perfect reversibility – by implication helpful in the study of some of the properties of gases – fails in the physical dynamics of responses to a wide variety of forces. These could be attributed to friction, imperfections in properties of materials in bulk – electrical and chemical as well as mechanical – and including absorption of radiation. Thomson addressed the properties of the simpler system, a gas consisting of a large number of identical molecules. He used much of the imagery of abstract dynamics. However, the impact of physical dynamics led him to replace certainties with probabilities.

He embedded in his discussion a thought experiment involving a large number of identical Maxwell Demons. Maxwell had pointed out the properties of such a 'being' in a letter to his lifelong friend Professor P.G. Tait (1831–1901), in Edinburgh [31] in December 1867 followed within a few weeks – in February 1868 – by a much briefer version to Thomson in Glasgow [32]. The context was Maxwell's comments on the draft version of Tait's textbook on thermodynamics. A detailed account is included in Chapter 5, 'Kelvin, Maxwell, Clausius and Tait: the correspondence of James Clerk Maxwell'. Suitably provided with a sorting device and, significantly, with a complete knowledge of positions and velocities of each of a very large number of molecules constituting the sample, a demon (Thomson's name for the being) could select molecules according to their energy and produce a non-uniform distribution of energy, and thus of temperature, within the sample. The letter to Tait had the demon in charge of a lightweight trapdoor in a partition so as not to disturb with the normal energies of the molecules, of which the more energetic could be chosen – by the demon briefly opening the trapdoor – to become more numerous on one side of the barrier and the number of less energetic made more numerous on the other (see also the *Maxwell's Demon* Section in the Appendix). This procedure, continued, would enable a continuous transfer of heat from cold to hot in circumstances which violate the Second Law of Thermodynamics. Maxwell dispensed with such a structure in his letter to Thomson, who chose to extend Maxwell's concept to his own approach to describing changes in, and stability of, energy distributions as might be monitored by distributions of temperature regarded as a bulk property, in Thomson's phrase, 'in the gross'. Outcomes from a three-stage thought experiment, also detailed in Chapter 5, included a much enhanced understanding of the equilibrium state, as well as the intended support for his theory of the dissipation of energy.

An army of Maxwell Demons was put in positions in a vessel containing a sample of gas in equilibrium so as to define the position of an eventual interface between two halves of the vessel. Acting to order, they allowed passage in either direction to enough molecules, identified by energy, to produce a disequalisation (Thomson's word) in energy between the two halves. With the army of demons suddenly stood down, the dynamics of the interactions amongst the moving molecules would lead to an equilibrium state in some finite time interval. Then, with the velocities of all molecules suddenly reversed, the system would seek to recover the demons-produced disequalisation by having every molecule retrace its path. Thomson argued that this might begin but in the same finite time interval would be inhibited from happening. Thereafter, the system, still left to itself, would seek some equilibrium state.

Thomson moved to propose that, in reality, without external influence an equilibrium state would persist (essentially Clausius's axiom), whilst subject continually to fluctuations about properties such as averages of distributions are amenable to description at the molecular level. Thomson used the language of probability theory to discuss these. The number of ways, from the mathematics of permutations, of making any allocation of molecules to locations in the vessel had to be multiplied by the probability of occurrence of that allocation. This is an extension of predicting the *numbers*, H and T, of 'heads' and 'tails', respectively, in a repeated exercise of tossing a number, n, of identical, perfectly symmetrical coins. Different H's and T's are the more likely if n is small. If n is large, many repeats are needed for a significantly different H's and T's to occur. Translating to a very large number of molecules in the vessel, a prohibitively long time is needed for any substantial disequalisation to happen. Thomson confirmed this with calculations of 'the number of chances against one' that a mixture of oxygen and nitrogen molecules would redistribute themselves to different volumes in their containing vessel.

6.2 From Maxwell's kinetic theory to the Maxwell–Boltzmann distribution

Maxwell's kinetic theory [33] and his follow-up dynamical theory of gases (guided by experimental results) [34] established their roles in the description and interpretation of properties of fluids. It was not feasible to track the movements of all molecules in a sample of gas in equilibrium. However, it was possible to predict the average value of velocity v, and of v^2, of the molecules whose collisions with one another were governed by very short range intermolecular forces, but which otherwise moved freely.

For an equilibrium sample of N molecules, the number with velocity in the range v to $v+dv$ is $NF(v)\mathrm{d}v$, where $F(v) = (4/a^3 \sqrt{\pi}))\ v^2 \exp(-(v^2/a^2))$. Here, a is a constant, whose value is determined by the molecular mass and the sample temperature. The analysis led to $(3/2)a^2$ for the average value of v^2. The function $F(v)$ can be transformed to a function $F(w)$ of kinetic energy $w = ½\ mv^2$. Algebraic manipulation leads to $F(w) = (2/\sqrt{\pi})(1/kT)(\sqrt{(w/kT)})\exp(-w/kT)$. Now, the number of molecules with kinetic energy in the range w to $w+dw$ is $NF(w)\ \mathrm{d}w$, where $F(w)$ is usually called a Maxwell–Boltzmann distribution function. The factor kT, a kind of reference energy, replaces $½\ ma^2$ and uses a link from the kinetic theory of gases between average molecular kinetic energy and temperature T: $½\mathrm{m}(3/2)a^2 = 3kT/2$. 'Boltzmann's constant' $k = R/N_0$, where R is the universal gas constant and N_0 is Avogadro's number.

6.3 Statistical mechanics: the Second Law and entropy

Klein [27] has provided an excellent commentary in his summary of early developments in the search for an account of principles of thermodynamics solely in terms of the dynamics of particles. It soon became apparent that statistical methods would have to be included, leading to 'statistical mechanics'. From about 1865 onwards, the path of what amounted to a sequel to Maxwell's kinetic theory was developed by Ludwig Boltzmann (1844–1906), with substantial contributions on matters of interpretation and substance from Clausius and others. The mathematics evolved around an exponential energy factor, structurally a generalisation of the kinetic energy factor in the Maxwell–Boltzmann distribution (Section 6.2). Once a gas has reached an equilibrium state, there could be continual reallocations of energy amongst the molecules whilst always retaining the equilibrium distribution. However, there was the option – of solutions in keeping with laws of motion – of having all molecular movements repeated in reverse. Thomson had discussed a very special case of this (Section 6.1) in the context of an inadequacy of 'abstract mechanics' to provide a complete description of changes in *energy* distributions, and was led to introduce probability arguments in support. Now, the issue centred on changes in *entropy*. The reversal of all molecular motions would move the system steadily away from equilibrium accompanied by an entropy decrease, thereby challenging the Second Law. Boltzmann countered this with an argument amounting to a general acceptance of a probabilistic strand in the Second Law. Simplifying, a system reaching its most probable state – the equilibrium state – would then never stray far from it.

6.4 Statistical entropy and chemical thermodynamics

The parameters specifying the state of a system at molecular level must be compatible with molecular properties, typically position and velocity. Guided by the mathematics of permutations and of probability, as was Thomson (Section 6.1), an analysis leads to that set of parameter values which will keep recurring most often during their continual reallocation amongst the molecules as a result of intermolecular collisions – subject to appropriate conditions. These are typically that the sum of the molecular energies is equal the total energy, and that the number of molecules is constant. That particular set is chosen to define the equilibrium distribution. Others that occur less often will not persist and will revert to the equilibrium one.

A convenient link to entropy is provided by seeking an analogy, between classical and statistical thermodynamics, in describing the approach to equilibrium of a gas sample. The resulting increase in classical entropy to an ultimate final value can be thought of as being matched – in statistical terms – by a path, through intermediate states, settling at the equilibrium state. It is the most probable by having the maximum number of ways (Ω) of finding the molecules with the equilibrium set of parameter values across their ranges. [35]. The analogue of the state's classical entropy is a statistical entropy defined as $k \times \log\Omega$, where k is Boltzmann's constant and log denotes logarithm to the base e. It is a feature of statistical thermodynamics that $\log \Omega$ acts as a valid, and mathematically very convenient, proxy for Ω. Throughout this period, there remained a gap between the developing statistical theory of molecular motions and the laws of thermodynamics. Professor J. Willard Gibbs (1839–1903), of Yale University, pointed the way to narrowing the gap with a paper in 1876 [36] dealing with the equilibrium of multicomponent systems including chemical effects explicitly. This led to the use of changes in entropy and energy in tracking chemical processes. A special thermodynamic function (to be called the Gibbs function G) was soon identified. Its constant value for reversible processes occurring at constant temperature and pressure was particularly useful in describing basic phase transitions. He led the subsequent systematic development of chemical thermodynamics jointly with the establishment of statistical mechanics as a new branch of theoretical physics.

7 Later developments: Keenan

In this section, we take a brief look at the work of Joseph Henry Keenan [37], an engineer who introduced entirely new ideas into the conceptual structure of thermodynamics. In particular, he endeavoured to construct a totally coherent account of the fundamental aspects of the subject, removing all the remaining ambiguities.

Keenan, who was born in 1900, was involved with every aspect of engineering, both in industry and in education. His first position, which he took up in 1922, was as a turbine design engineered for the General Electric Company in Schenectady. This was the start of a lifelong concern for the properties of steam and, of course, for thermodynamics. In 1928, he moved into academia with a position at the Stevens Institute of Technology. Here, he was a delegate to the First International Conference on the Properties of Steam in 1929; he was to maintain attendance at the following conferences on this area of engineering for the rest of his life. He also made a major contribution to the Steam Tables of 1930. In successive editions of 1936, 1939 and 1969, to which he also made major contributions, these tables would be indispensable for generations of students and also for every practitioner in the steam-power industry for decades to come. In 1934, he moved to MIT, where he headed the Department of Mechanical Engineering for a number of years. Here, his work on thermodynamics became increasingly well known and he developed what is often called the Keenan School of thermodynamics or the MIT School of thermodynamics.

At MIT, he followed up his work on the Steam Tables by co-authoring the Air Tables and the Gas Tables, which were compulsory reading for those developing the gas-turbine industry. He also carried out research that was virtually always ground breaking. This included evaluation of the performance of steam-turbine nozzles, studies on the operation of the free-piston compressor in gas turbines, and the development of equipment to process coffee and cocoa, and for dust separation. He also worked on jet and rocket propulsion, carrying out research on the friction coefficients of air at supersonic speeds, and on injectors and on heat transfer at high speeds.

Keenan was equally devoted to teaching, where he was determined to achieve a simple approach to thermodynamics, which was nevertheless totally rigorous, and would eliminate those unnecessary and awkward features and ambiguities in the then current accounts that were a residue of historical struggles and misunderstandings. It was this desire in his teaching that led to his development of entirely new conceptual foundations of thermodynamics. In the 1930s, he adopted many of the views of J. Willard Gibbs, in particular, the use of the quantity of thermodynamic availability, sometimes called exergy. This is the maximum useful work that is possible during a process bringing a system into equilibrium with a heat reservoir. In the case that the reservoir is just the surroundings, the thermodynamic availability is the ability of the system to cause a change as it comes to equilibrium with its environment. It is effectively the energy that is available for use. The concept of thermodynamic availability is of great use in the steady-flow processes of engineering, being crucial in the allocation of fuel costs in processes that have many outputs in chemical and power-plant engineering.

His two books, *Thermodynamics* [38], published in 1941, and *Principles of General Thermodynamics* [39], published with George Hatsopoulos in 1965, have been hugely influential in the study of thermodynamics, and the second in particular has revolutionized our ideas of the subject. It shows that the second law of thermodynamics can be expressed in terms of the existence of stable equilibrium states. In fact, the authors claim that the first and second laws may be combined into a single law, which they call the law of stable equilibrium. Their statement is that: 'A system having specified allowed states and an upper bound in volume can reach from any given state one and only one stable state and leave no effect on its environment'.

The law has been restated [40] as: 'When an isolated system performs a process, after the removal of a series of internal constraints, it will always reach a unique state of equilibrium; this state of equilibrium is independent of the order in which the constraints are removed and is characterized by a maximum value of entropy'. This principle entails the Kelvin–Planck and Clausius statements of the second law but also allows the traditional definition of entropy to be extended to non-equilibrium situations. It describes the behaviour of systems that are some distance from equilibrium, and which, when certain constraints are removed, will move to a local equilibrium. It includes a formal statement of the second law that has all the attributes of the concept of entropy but does not use it in its definition. Thus, the statement implies a direction and an equilibrium end point for all processes [41].

It may also be noted that this restatement of thermodynamics has considerable potential for reaching out from its home of classical physics to the quantum realm, as well as coping with relativistic situations. The authors describe their intentions as follows: 'Thermodynamics has been known as a science which is effective in getting directly to the heart of a problem with a minimum of interpretation. It was considered, however, to be limited in its application to non-relativistic non-fluctuating systems ... The extensive developments in the book show that by relatively minor modifications of definitions and principles and by adoption of the Gibbs concept of an ensemble to describe a system, most of the limitations previously assumed cease to exist. Thermodynamics can now be applied to obtain quite easily familiar and unfamiliar relations from the special theory of relativity. It can also be applied to fluctuating systems to provide a firm new basis for what was previously a basically incompatible mechanical theory of heat. This application will lead quite naturally to other developments of which information theory is one. Thermodynamics appears to be a renascent science on the threshold of a new era'.

8 Myron Tribus's thermodynamics: an alternative understanding

Myron Tribus [42], who was born in 1921, may be regarded as a polymath. In the second worldwar, he worked as a design-development officer for the United States Air Force, and after graduating with a PhD in 1949 he began work as a gas turbine design engineer at General Electric, but was unhappy in industry and moved back to education at the University of California at Los Angeles. In 1960, he moved to Dartmouth College as Dean of Engineering. Here, he developed an updated curriculum based on engineering design and entrepreneurship.

In 1969, he joined the Nixon administration as assistant secretary of Commerce for Science and Technology, but in the following year he became the senior vice principal for Research and Engineering in Xerox. Then, from 1974 to 1986 he was the director of the Centre for Advanced Engineering Study at MIT. He is a co-founder of Exergy, Inc., a company that specializes in designing high-efficiency power production systems. (As explained in the previous section, exergy is another word for thermodynamic availability.)

He has carried out research in a wide range of areas including heat transfer, fluid mechanics, probability theory, statistical inference, sea water demineralisation, aircraft heating and ice prevention on aircraft. He is a great advocate of Bayesian methods of probability and, in connection with his work on the economic considerations for converting sea water into fresh water, coined the term thermoeconomics.

As something of a guru, he has produced his famous Perversity Principle, which says that: 'If you try to improve the performance of a system of people, machines, and procedures by setting numerical goals for the improvement of individual parts of the system, the system will defeat your efforts and you will pay a price where you least expect it'.

Like Keenan, Tribus regarded himself as a follower of J. Willard Gibbs, though in his case, the ideas of Gibbs were filtered through the work of Claude Shannon (1916–2001) and Edwin T. Jaynes (1922–1998). Shannon is regarded as the founder of information theory. His wartime experience in cryptography and communication theory led to his celebrated paper of

1948 [43] which initiated this topic. In this paper, he produced an expression for the amount of information contained in a sequence of numbers, and broadly it could be said that the more unlikely an event is, the higher the related information. Shannon was able to build up a rigorous mathematical account of the relationship between probability and information, perhaps better expressed as the relationship between uncertainty and information. In the course of this he coined the term information entropy, though the connection between this quantity and thermodynamic entropy was not immediately obvious.

The work of Jaynes [44] centred around the relationship between thermodynamics, statistical mechanics, probability and statistical inference, and in a paper of 1957 he introduced the so-called MaxEnt or maximum entropy interpretation of thermodynamics. This principle is based on inference techniques and is based on Shannon information theory, Bayesian probability and the principle of maximum entropy. While Jaynes' work has been criticized by many, Tribus believed that it revolutionized the foundations of thermodynamics. In his famous book *Thermostatics and Thermodynamics* [45], he wrote that the work of Gibbs was certainly correct but mathematically complicated.

In contrast, when he was directed to the work of Jaynes, he realized that it showed how, starting from the work of Shannon 'one could obtain in an elementary way all the results given by Gibbs. It is difficult to overestimate the value to science education of this contribution. At one fell swoop the enormous conceptual difficulty of Gibbs' approach has been swept away. It now becomes possible to begin by considering the properties of the smallest particles and by simple mathematical methods to deduce the properties of large systems. The results obtained by Joule, Mayer, Carnot, Maxwell, Boltzmann and Gibbs are seen in a new light. It is no longer necessary to cover the development of thermodynamics in the same way as the subject developed in history. New generations of students can, in shorter time, reach a state of understanding heretofore impossible for more than a small number of persons'.

9 Conclusions

The development of thermodynamics by the aforementioned formidable scientists was geared to the recognition and refinement of the First and Second Laws of Thermodynamics. These laws emerged from the empirical correlation of relevant observations and measurements via the postulation of equivalents ('mechanical equivalent of heat') to transformations. Once the overarching concept of energy was accepted, the quantitative content of the laws became clear. Tallying served the First Law well enough. However, some sophistication was needed for the Second Law. The interpretation of the energy tally had to be supported by an additional thermodynamic property, entropy, whose value always increased or, in the limit for reversible processes, remained constant. For various specialized processes other properties, such as enthalpy (H) and the Gibbs function (G), were identified. Mathematics provided links amongst the properties and its own increasing sophistication was essential. These supported the subject, theoretical and applied, as it expanded across the sciences, with a statistical component crucial to the successful incorporation of information theory, fortified by developments in computing generally. Thus, the last 50 years or so have completed a platform for current and future developments of which some will almost certainly continue the expansion into the realm of nanoscience.

References

[1] Carnot, N.L.S., *Réflexions sur la puissance motrice du feu, et sur les machines propres à développer cette puissance*, Bachelier: Paris, 1824.

[2] Clapeyron, B.P.E., Mémoire sur la puissance motrice de la chaleur. *Journal de l'École Polytechnique*, cahier 23, **14**, pp. 153–190, 1834.

[3] Erlichson, H., Sadi Carnot, founder of the Second Law of Thermodynamics. *Eur. J. Phys.*, **20**, pp. 183–192, 1999.

[4] Thomson, W., An account of Carnot's theory of the motive power of heat; with numerical results deduced from Regnault's experiments on steam. *Trans. Roy. Soc. Edinb.*, **16**, 541–575, 1849; reprinted in *Math. and Phys. Papers*, **I(XLI)**, pp. 113–155, 1882.

[5] Thompson, S.P., *Life of Lord Kelvin*, vol. I, Macmillan: London, 1910.

[6] Davy, H., Experimental essay on heat, light, and the combinations of light, in *Contributions to Physical and Medical Knowledge principally from the West of England*, collected by Thomas Beddoes, M.D., pp. 5–147, especially 'Experiment II' pp. 16–17, Longman and Rees: London, 1799.

[7] Thompson, B (Count Rumford), An enquiry concerning the source of the heat which is excited by friction. *Phil. Trans. Roy. Soc.*, **88**, pp. 80 and 99, 1798.

[8] Mayer, J.R., Bemerkungen über die Kräfte der unbelbten Natur. *Annalen der Chemie und Pharmacie*, **43**, pp. 233–239, 1842. Translated as Remarks on the forces of inorganic nature. *Phil. Mag., Ser.* 4, **24**, pp. 371–377, 1862.

[9] Newburgh, E., Carnot to Clausius: caloric to entropy. *Eur. J. Phys.*, **30**, pp. 713–728, 2009.

[10] Mayer, J.R., *Die organische Bewegung in ihrem zusammenhange mit dem Stoffwechsel (Organic Movement in Connection with Metabolism)*, C. Drechsler: Heilbronn, 1845.

[11] Joule, J.P., On the mechanical equivalent of heat. *Phil. Trans. Roy. Society*, **140**, pp. 61–82, 1850.

[12] Fourier, J.P.J., *Théorie Analytique de la Chaleur*, F. Didot: Paris, 1822.

[13] Clausius, R.J.E., Ueber die bewegende Kraft der Wärme und die Gesetze, welche sich daraus für die Wärmelehre selbst ableiten lassen. *Ann. Phys.*, **79**, pp. 368–397, 500–524, 1850. Translated as On the moving force of heat and the laws of heat which may be deduced therefrom. *Phil. Mag.*, Ser. 4, **2**, pp. 1–21, 102–119, 1851 and in Clausius, R.J.E., *The Mechanical Theory of Heat With Its Applications to the Steam Engine and to the Physical Properties of Bodies*, Hirst, T.A. (ed.), pp. 327–364 (First Memoir), John van Voorst: London, 1867.

[14] Thomson, W., On the dynamical theory of heat, with numerical results deduced from Mr Joule's equivalent of a thermal unit, and M. Regnault's Observations on steam. *Trans. Roy Soc. Edinb.*, **20**, pp. 261–288, 289–298 and 475–482, 1851; reprinted in *Math. and Phys. Papers*, **I(XLVIII)**, pp. 174–232, 1882.

[15] Clausius, R.J.E., Ueber verschiedene für die Anwendung bequeme Formen der Hauptgleichungen der mechanischen Wärmetheorie. *Ann. Phys.*, **125**, pp. 353–400, 1865. Translated as On several convenient forms of the fundamental equations of the mechanical theory of heat, in: Clausius, R.J.E., *The Mechanical Theory of Heat with Its Applications to the Steam Engine and to the Physical Properties of Bodies*, Hirst, T.A. (ed.), pp. 327–364 (Ninth Memoir), John van Voorst: London, 1867.

[16] Thomson, W., On an absolute thermometric scale founded on Carnot's theory of the motive power of heat, and calculated from Regnault's observation. *Phil. Mag. Series 3*, **33**, pp. 313–317, 1848; reprinted in *Math. and Phys. Papers*, **I(XXXIX)**, pp. 100–106, Cambridge, 1882.

[17] Erlichson, H., Kelvin and the absolute temperature scale. *Eur. J. Phys*, **22**, pp. 325–328, 2001.
[18] Thomson, W., Thermo-electric currents. *Trans. Roy. Soc. Edinb.*, **21**, pp. 123–172, 1854; reprinted in *Math. and Phys. Papers*, **I(XLVIII)**, pp. 232–291, 1882.
[19] Finn, C.B.P., *Thermal Physics* (Chapter 7), 2nd edn, Nelson Thornes: Cheltenham, 2001.
[20] Planck, M., *Vorlesungen über Thermodynamik*, Walter de Gruyter: Berlin, 1897, Translated as *Treatise on Thermodynamics*, 3rd edn, Dover: New York, p. 89, 1945.
[21] Thomson, W., On a universal tendency in nature to the dissipation of mechanical energy. *Proc. Roy. Soc. Edinb.*, **3**, pp. 139–142, 1852; reprinted in *Math. and Phys. Papers*, **I(LIX)**, pp. 511–514, 1882.
[22] Clausius, R.J.E., Ueber eine veränderte Form des zweiten Hauptsatzs der mechanischen Wärmetheorie. *Ann. Phys.*, **93**, pp. 481–506, 1854. Translated as On a modified form of the second fundamental theorem in the mechanical theory of heat. *Phil. Mag. Ser. 4*, **12,** pp. 81–98, 1856; Clausius, R.J.E., *The mechanical theory of heat with its applications to the steam engine and to the physical properties of bodies*, pp. 327–364 (Fourth Memoir), Hirst, T.A. (ed.), John van Voorst: London, 1867.
[23] Cropper, W.L., Rudolf Clausius and the road to entropy. *Am. J. Phys.*, **54**, pp. 1068–1074, 1986.
[24] Adkins, C.J., *Equilibrium Thermodynamics*, McGraw-Hill Book Co.: London, 1968.
[25] Maxwell, J.C., *Theory of Heat*, Longmans, Green, and Co.: London, 1871.
[26] Tait, P.G., *Sketch of Thermodynamics*, Edmonston and Douglas: Edinburgh, 1868.
[27] Klein, M.J., Maxwell, his demon, and the Second Law of Thermodynamics. *American Scientist*, **58**, pp. 84–97, 1970.
[28] Joule, J.P. & Thomson, W., On the thermal effects of fluids in motion. *Phil. Mag.*, **4**, pp. 481–492, 1852 (Preliminary), *Phil. Trans. Roy. Soc.*, **143**, pp. 357–365, 1853 (Part I), **144**, pp. 321–364, 1854 (Part II), **150**, pp. 325–336, 1860 (Part III), **152**, pp. 579–589, 1862 (Part IV). Reprinted in *Math. and Phys. Papers*, **V(XLIX)**, pp. 333–455, 1882.
[29] Finn, C.B.P., *Thermal Physics*, Chapter 7, 2nd edn, Nelson Thornes: Cheltenham, 2001.
[30] Thomson, W., The kinetic theory of the dissipation of energy. *Nature*, **9**, pp. 441–444, 1874, also *Proc. Roy. Soc. Edinb.*, **8**, pp. 325–334, 1875 (read 16 February 1874); reprinted in *Math. and Phys. Papers*, **I(XLIX)**, pp. 333–455, 1882.
[31] Harman, P.M., (ed.). *The Scientific Letters and Papers of James Clerk Maxwell, Vol. II*, No. 277, Letter to Peter Guthrie Tait, 11 December 1867, pp. 328-334, Cambridge University Press: Cambridge, 1995.
[32] Harman, P.M., (ed.). *The Scientific Letters and Papers of James Clerk Maxwell, Vol. III*, No. 279A, Letter to William Thomson, 16 January 1868, pp. 877–882, Cambridge University Press: Cambridge, 2002.
[33] Maxwell, J.C., Illustrations of the dynamical theory of gases. *Phil. Mag. Ser. 4*, **19**, pp. 19–32, 1860 (part I) and **20**, pp. 21–37, 1860 (parts II and III).
[34] Maxwell, J.C., On the dynamical theory of gases. *Phil. Trans. Roy. Soc.*, **157**, pp. 49–88, 1867.
[35] Boltzmann, L., Ueber die Beziehung zwischen dem zweiten Hauptsatze der mechanischen Wärmetheorie and der Wahrscheinlichkeitsrechnung respektive den Sätzen über das Wärmegleichgewicht. *Wiener Berichte*, **76**, pp. 373–435, 1877.
[36] Gibbs, J.W., On the equilibrium of heterogeneous substances. *Trans. Connecticut Acad. of Arts and Sciences*, **3**, pp. 108–248, 1876.
[37] Biography of Joseph Henry Keenan, in: Meeting the Entropy Challenge (MIT Symposium); http://web.mit.edu/keenansymposium/aboutjhkeenan/biography/index.html.
[38] Keenan, J.K., *Thermodynamics*, Wiley: New York, 1941.

[39] George, N., Hatsopoulos, G.N. & Keenan, J.K., *Principles of General Thermodynamics*, Wiley: New York, 1965.

[40] Schneider, E.D. & Sagan, D., *Into the Cool: Energy Flow, Thermodynamics and Life*, University of Chicago Press: Chicago, 2005; article on Joseph Keenan in Hmolpedia, an Encyclopedia of Human Thermodynamics: http://www.eoht.info/page/Joseph+Keenan?t=anon.

[41] Schneider, E.D., Towards a thermodynamics of life, in *Scientists Debate Gaia: The Next Century*, Schneider, S.H., Miller, J.R., Crust, E. & Boston, P.J. (eds). MIT: Boston, pp. 45–56, 2004.

[42] Article on Myron Tribus in Hmolpedia, an Encyclopedia of Human Thermodynamics: http://www.eoht.info/page/Myron+Tribus.

[43] Shannon, C., A mathematical theory of communication, *Bell Systems Tech. J.*, **27**, pp. 379–423, 623–656, 1948.

[44] Jaynes, E.T., Information theory and statistical mechanics. *Phys. Rev.*, **106**, pp. 620–630, and 108, 171–190, 1957.

[45] Tribus, M., *Thermostatics and Thermodynamics: An Introduction to Energy, Information and States of Matter, With Engineering Applications*, Van Nostrand: Princeton, 1961.

Chapter 10

The teaching of thermodynamics today

R. Bartlett & J. Kubie
Edinburgh Napier University, United Kingdom.

Abstract

Although there are several robust theories of 'thermodynamics' operating very successfully over a range of different disciplines, for example, biology, chemistry, geoscience, quantum theory, astrophysics and numerous associated technologies, there are still no unifying foundations for this most important branch of science. Yet, most elementary thermodynamics is taught from an unjustifiable assumption of consistency and certainty. The current inconsistencies that permeate the science community's thinking about 'heat' first appear at the time of Kelvin. This chapter traces some of the early attempts to formulate a theory of 'heat' and indicates a way in which we can improve our approach to the teaching of elementary thermodynamics by attempting to empathise with Kelvin's perspective.

1 Introduction

Heat is a very tricky thing. During the first half of the nineteenth century, despite an industrial revolution powered by coal and the steam engine, the study of heat (thermodynamics) was in very poor shape. Most modern periods of intense technological changes – such as those caused by the introduction of the motor car, the microwave oven, the Internet or the mobile phone – have a science which precedes and drives invention but, peculiarly, this was not true during the development of the steam engine.

Natural Philosophy (Science) was floundering to find a convincing account of the steam engine, and the threads of an understanding only began to come together around 1850. William Thomson brilliantly wove them into the first 'complete theory of the motive power of heat' in his paper of March 1851, 'On the Dynamical Theory of Heat' [1].

2 The technological context of Thomson's 1851 paper

It might be useful to remember that previously, in the sixteenth and seventeenth centuries, the ground water that drained into mining excavations was the major problem facing European industrial development. A typical excavation would have to raise ten times as much water as it did coal or ore. Particularly where coal was easily to hand, the question arose: maybe coal as fuel could be used to raise water?

In this respect, three significant technological developments took place between 1700 and Thomson's 1851 paper: First, in 1702, Thomas Savery [2] introduced his 'Miner's Friend'. Steam was introduced into a large egg-like chamber which, when externally dowsed with cold water, sucked mine water up from the floor of the mine. More steam was then pressured into the chamber, forcing out the mine water up to the surface. Unfortunately, it used a lot of coal and it was prone to explode.

Thomas Newcomen [3], on seeing the shortcomings of the Miner's Friend, introduced an enormously improved engine in 1712. It rocked an 8-m lever attached to a conventional stirrup pump. Steam pushed a piston which moved the lever up. Cold water was then flushed into the steam-filled piston chamber, causing the cylinder to contract and the lever to be drawn down. It consumed only a third as much coal as did the Miner's Friend.

Technical improvements continued but a third (and revolutionary) breakthrough came when James Watt [4] realised that the repeated heating and cooling of a chamber was somehow wasting heat. In 1769, he patented an engine that kept the hot chamber warm by flushing the steam into a second chamber before dowsing it to pull the piston back. The scientific principle underlying this leap forward was not really understood for another 50 years or recognised for a further 30 years but, more importantly at the time, this engine used only half the fuel of a Newcomen engine (one-sixth as much as the Miner's Friend).

3 The 'scientific' context of Thomson's 1851 paper

Throughout the eighteenth century, there was still no science or theory to help move the technology along. In fact, the scientific community was still arguing about the nature of heat. In the seventeenth century, Francis Bacon had described heat as quiddity in motion [5] ('quiddity' being used with its Aristotelian meaning as the 'what-it-is-ness' or essence of matter) and both Galileo and Newton, in the eighteenth century, thought heat to be the frictions and movements of tiny particles of which matter is made. However, many scientists/thinkers of the time believed heat to be an indestructible substance – The Caloric – which flowed like a liquid through all matter and from warmer bodies to cooler ones. Wood and coal were believed to have plenty of Caloric, somehow 'frozen' within them, which was released only when they burned. In 1799 Humphrey Davy, arguing for the Bacon camp, described an experiment where he melted two blocks of ice by rubbing them together [6]. Since ice was supposed to be water with its supply of Caloric largely depleted – what could be melting the ice?

But it was Sadi Carnot, a theorist in the Caloric camp, who presented the first incisive analysis of the workings of the motive power of heat in 1824 [7]. His idea was to analyse the changes which took place in a piston chamber as some combination of two separate processes. The first process was the addition of heat (whilst allowing the expansion to push the piston out enough to keep a constant temperature in the chamber) – and, when harnessing the resulting piston movement – to get some work done. The second process was allowing the chamber to expand and reduce its pressure (but allowing no heat to leak in or out of the chamber) – and again getting some work done.

Both these processes (heat change/pressure constant and pressure change/heat constant) are 'reversible' – an important concept in understanding thermodynamics. The first can be reversed by gently pushing on the piston while allowing heat to drain out of the chamber. This gets everything back to where it started. More importantly, it requires exactly the same amount of 'work' to reverse the process as we could have harnessed from the process in the first place. The second process can be reversed by compressing the chamber and increasing its pressure, but not allowing heat to leak out – again it requires exactly the same work to reverse this process as could be harnessed from it in the first place.

Carnot went on to design an engine 'on paper' that completely separated these processes in time. It cycled continuously through four stages: (i) allowing heat into the piston chamber; (ii) allowing the expansion to continue; (iii) cooling the chamber and (iv) allowing the contraction to continue. His engine worked between two heat 'reservoirs' – the heat source and the cooler environment – and it would produce work by virtue of the transfer of heat from the source to the environment.

He then imagined a second engine of identical manufacture driven at the same speed but driven in reverse – it would 'consume' work to carry heat from the environment and back to the heat source. This was an ingenious thought experiment that had the work output of a heat engine harnessed to and driving a reversed engine; together maintaining some kind of heat equilibrium with nil overall work output.

Now comes a surprising implication: If it were possible to have an engine more efficient than a reversible engine – then only a part of its output would be necessary to power Carnot's reversed engine – thus, the heat lost from the source could be infinitely replenished by the reversed engine while the surplus work from the more efficient engine could be infinitely harnessed. Carnot sensibly decided this to be impossible and concluded that a reversible engine must be the most efficient possible. He also concluded that for an engine to work, it is necessary for heat to go out into the environment (be 'dumped' or wasted). His work explained why James Watt's engine was so successful.

However, Carnot's masterpiece of scientific thought was not immediately recognised as such. The French Revolution had given rise to an elite republican generation of mathematicians and scientists (Ampere, Cauchy, Fourier, Gay-Lusac, Poisson, etc.) through the Ecole Polytechnique [8], but they were seen from the other side of the English Channel as somewhat arrogant and over theoretical – especially in the ultra-practical field of engineering.

4 Enter Thomson

William Thomson, on the other hand, admired much of France's achievements. He started at Glasgow University (at the age of 10) where there was a strong interest in the French approach to mathematics and science. When he entered Cambridge (age 17) in 1841, his first published paper was on 'Fourier's expansions of functions in trigonometrical series' [9]. After he left Cambridge, he studied in Paris where he began to work in the physical laboratory of Henri-Victor Regnault and where he began to publish his mathematical works in the *Liouville Mathematical Journal*. When he was elected as a professor of natural philosophy at Glasgow University in 1846 (age 22), he continued to keep an eye on developments in mainland Europe.

Meanwhile, in Britain, James Joule had demonstrated that work could be converted to heat in a consistent way [10]. Although he had published his early results in 1843, he was finding it difficult to be taken seriously. It would have been especially difficult for Thomson to accept that heat and work were somehow equivalent, for which he would have had to abandon his understanding of Carnot's work based upon the theory of indestructible Caloric.

Then, in 1848, Thomson publishes an astounding paper – 'On an Absolute Thermometric Scale' [11]. In this paper, Thomson explains that the problem of measuring temperatures may have been largely solved to all practical purposes. However, because every material used in thermometry behaves with a different heat uniformity, there is no sound theoretical basis for choosing one thermometer over another. He suggests that instead of measuring material expansion, thermometers should be calibrated by determining the amount of work that an idealised Carnot engine can produce from a one degree temperature difference.

In this paper, Thomson feels the need to educate his erudite audience (the Cambridge Philosophical Society) with a description of a Carnot's work. Therefore, it would seem that Thomson had not only come down on the side of Carnot and the Caloric, but that he was using the theory to shore up structural defects in current understanding. Of particular note is his comment that 'the conservation of heat (or caloric) into mechanical effect is probably impossible'. Much as no actual water is lost when a water–wheel produces energy, it would be Carnot's fundamental assumption that the movement of Caloric from a warm reservoir to a cool one would conserve Caloric, even though 'mechanical effect' could be obtained.

However, in a footnote to this comment Thomson states, 'This opinion seems to be nearly universally held by those who have written on the subject. A contrary opinion, however, has been advocated by Mr Joule of Manchester; some very remarkable discoveries which he has made with reference to the generation of heat by the friction of fluids in motion …'. In Joule's view, a steam engine worked by converting some of the heat into 'mechanical effect'. The two theories the Caloric and Joule's 'dynamical theory of heat' were quite incompatible. But this footnote indicates that Thomson was actively, cognitively challenged by the juxtaposition of two seemingly irreconcilable points of view.

5 Death of the Caloric

In early 1850, Rudolph Clausius published a paper 'On the Moving Force of Heat' [12] in Germany, abandoning the Caloric theory. Clausius accepted Joule's results and rejected 'the conservation of Caloric' as a false assumption. He replaced it with 'the conservation of energy' and proclaimed it as the first law of thermodynamics. And he proclaimed the tendency of heat to flow from hot to cold bodies as the second law. However, and significantly, he shows that Carnot's analysis – that no engine is more efficient than a reversible engine – is completely compatible with both these laws.

It is difficult to know at what point Thomson came to the same conclusion – that it was possible to ditch the Caloric but still to keep Carnot. In terms of 'publication', he was clearly beaten to it by Clausius. In his ground-breaking 'On the Dynamical Theory of Heat' [1] presented to the Royal Society of Edinburgh in March 1851, Thomson says 'I may be allowed to add that I have given the demonstration exactly as it occurred to me before I knew that Clausius had either enunciated or demonstrated the proposition'.

Thomson's 'dynamical theory of heat' is founded upon two propositions. First, he attributes to Joule that when work (or mechanical effect) is lost, an equal quantity of heat must be generated somewhere. If work is gained, an equal quantity of heat must be 'put out of existence'. This is an exact (but still clumsy) expression that 'heat is energy' and that 'energy is indestructible'. Attributing to this first principle, his new theory of heat to Joule was an act of fine diplomacy.

Such diplomacy was not lost on Joule who was at last able to join the top echelon; he and Thomson went on to collaborate fruitfully with one another over the next few years.

The second proposition he attributes to Carnot and Clausius that no engine can be more efficient than a reversible engine. This was Carnot's *deduction* from his analysis of heat reservoirs and reversible processes – he did not consider it to be a fundamental principle. Neither did Clausius, who was able to make the same deduction from his two laws.

Even if this was 'exactly as it occurred' to him, his decision to publish this in a quite way can be seen as an act of finely judged politics. Thomson had nailed his colours to Carnot's mast when he proposed his Absolute Thermometric Scale [11] and declared himself a 'Calorist'. Clausius had rescued Carnot from the wreckage of Caloric theory and Thomson was in danger of appearing to need to cling to the same life-ring. Thomson's paper pre-empted any such problems. What it did was to reconcile the new dynamical theory with the best of the old – to make sense of all previous findings – and to raise Carnot's insight above that of Clausius' and to reprieve himself.

Clausius, however, was convinced that he alone had founded thermodynamics and went on over the next two decades to make further ground-breaking contributions, insights and clarifications to a new and successful theory of heat.

6 The second law of thermodynamics

Both these original formulations – by Clausius and by Thomson – of 'the second law of thermodynamics', apparently quite different and unleashed onto the world within months of one another, have weaknesses or drawbacks (and we look at them more closely below). First, however, it is worth making a few comments – from a modern perspective – about that physical phenomenon we call 'the second law'. A 'law' is really only a belief (strongly supported by observation) which, by virtue of its fundamental nature, is given the working status of 'absolute truth' by the scientific community.

In very loose terms, the second law promotes the belief that although the energy around us can be found in concentrations, over time, any such concentrations will tend to disperse. If we (or natural forces) do succeed in collecting energy together by some kind of process, this must be at the expense of some greater dispersal of energy elsewhere. Energy is never destroyed – it just becomes more and more 'dispersed'. For instance, oil or coal is a concentration of stored energy, but the biological and geological processes involved in its formation – of photosynthesis, of building sugars and proteins, of subsequent rotting and geological compression – must (according to this law) disperse more energy than can be stored.

Scientists use the term 'work' for energy employed to lift a weight or to set something in motion. Work is directed energy and, therefore, considered to be equivalent to concentrated, well-ordered energy. When an 'engine' converts heat energy into (concentrated) work, an even greater dispersal of energy must also be taking place, somewhere in the vicinity. Basically, some quantity of unconverted heat energy must be allowed to 'waste away' into the cooler environment – and no matter how inventive we become, any further attempt to 'harness' that wasted energy will result in some yet greater dispersal. It is natural for engineers to see this 'throwing heat away' as an unfortunate reality but, if it were possible to convert heat into energy without waste, then, as Carnot's own proof demonstrated, it would be possible endlessly to concentrate energy and, indeed, to create energy and matter.

Shortly after the events we have described, Thomson published 'On a Universal Tendency in Nature to the Dissipation of Mechanical Energy' [13] in 1852 (where in the introductory passage, Thomson again invokes Carnot as the primary source of thermodynamic thinking). This paper begins a discussion of startling implication; that the universe is irrevocably heading towards a state of total dispersion, which was to be known by the 1860s as 'the heat death of the universe'.

One way to quantify this 'dispersion of energy' during a transfer of heat is with some quite simple arithmetic – by dividing the transferred heat by any temperature change. This measure of dispersal is called 'entropy' (as named by Clausius in 1865 [14]). Indeed, any physical object that has measurable properties like 'volume', 'mass' and 'temperature' can also have measured 'entropy'. When we first pour cool cream into hot coffee, both the cream and the coffee have a distinct temperature and entropy. Adding the entropies gives the combined pre-mix value – but as the cream diffuses into the coffee, and some heat leaves the coffee to enter the cream, the measurable entropy of the mix increases. The total 'dispersion of energy' in the universe has increased – according to its natural tendency.

7 Flawed lawmaking

Clausius' original formulation of the second law, which heat will never spontaneously move from cool to warm bodies, is superbly clear and very easy to understand – but it hides a circular assumption. The relative warmth of two bodies is measured by their temperatures – and a temperature is a measure of a body's propensity to disperse energy. This version of the law can, therefore, be read as 'the body with the greater propensity to spread energy will spread its energy' – which really says nothing at all. By 1865, Clausius had overcome this difficulty with a new statement of the law; that 'the total entropy of the universe tends towards a maximum' [14]. The heat energy in the mug of coffee, as described above, very quickly reaches its maximum dispersion.

Thomson's original formulation (that no engine can be more efficient than a reversible engine) is not quite as simple as Clausius' original formulation. (For the novice student, it would require them to already know about the workings of a heat engine and to understand the concept of reversible process.)

Thomson's formulation is also blemished by a logical problem. It is based upon an ideal (the reversible process) which, according to the law it attempts to define, cannot exist in the real world. Of course, many physical laws only describe behaviour in an ideal situation (if only we had frictionless surfaces, perfectly elastic bodies or 100% vacuums). However, it is the presence of the second law of thermodynamics, solely, which 'spoils' the perfect operating conditions for all these other laws (for instance, if we strike an elastic body it causes an energy transfer and some energy must be lost to dispersal; if we try to create a vacuum there will be a tendency for energy to disperse itself into the vacuum, riding on material bumped off the container walls). To say, 'this particular law works perfectly – if we disregard the effects of the second law of thermodynamics' is quite reasonable because we specify laws in order to help us analyse and separate physical effects. But, to say 'the second law of thermodynamics works perfectly – if we disregard the second law of thermodynamics' is suspect.

Some workers in the field, notably Max Planck [15] in 1897, have tried to keep something of Thomson's formulation, but with the reliance on 'reversibility' removed, by going back to the 1851 paper 'On the Dynamical Theory of Heat' [1]. In this paper, Thomson presented a common sense notion: that no engine could produce work solely by taking heat from the coolest surrounding objects. He then showed that his (reversibility) formulation is the equivalent of this notion. Planck, in 1897, tweaks Thomson's 'common sense' and provides a 'new' formulation of the second law which he credits to Thomson: it is not possible to take heat from a single reservoir and produce work (without dumping some heat at a lower temperature). This became known as the Kelvin-Planck formulation – despite the fact that Carnot made exactly this deduction in 1824.

Today, we have a veritable plethora of ways to view the second law. In 1877, Ludwig Boltzmann gave us a new and completely different perspective on thermodynamics [16]. By treating the heat energy in an object as the aggregate of the small individual energies belonging to a large collection of components, it is possible to use statistical measures of dispersion to represent energy dispersion. In the same way that if a coin is tossed 100 times, getting around 50 'heads' is far more likely than getting near to 100, the most likely distributions of energy are those where the energy is spread evenly through the object. Boltzmann saw the universe's tendency for 'entropy to increase to a maximum' as a tendency for things to shift towards their most likely statistical outcome. Needless to say, in common with the other formulations discussed, a logically satisfying link between statistical dispersion and heat dispersion has yet to be found.

Strangely, despite the situation today where we have no secure or unifying formulation of the second law, there is complete agreement that, however, we might eventually come to formulate it; the second law is a basic foundation stone upon which we must build our understanding of the physical universe. Therefore, imagine a bottle of milk left to stand. Many physicists will see that, as the milk organises itself into cream and whey (losing entropy in the process by 'un-dispersing'), there must be a greater dispersal of heat energy (which is probably radiated away into the environment). Astrophysicists, on the other hand, will choose Boltzmann's perspective and see the gravitational clustering of particles of cream as a more likely outcome, and see the bottle gaining entropy. Their observations and energy calculations will agree. They will both believe that, according to their own standards, the total entropy in the universe will have increased. But they will not agree about whether the new arrangement within the bottle has greater or less entropy.

8 Returning to Thomson's formulation

Therefore, in this context, let us turn to the question how thermodynamics should be taught today, and to what extent Thomson's formulation is of value. We feel that a modern approach to elementary thermodynamics, which looks to the need to educate (rather than being over-concerned to rationalise its choice of foundations), should be based upon the following rules of thumb (each of which we justify below):

- Keep (or reinstate) the heat engine
- Discard attempts at logical consistency
- Use reversibility only as a model

8.1 Keep the heat engine

As we have indicated above, Thomson must have thought long and hard before selecting and presenting his two 'propositions' for the 1851 paper [1]. He would have been acutely aware that his second proposition (that no engine can be more efficient than Carnot's ideal engine) was not very 'elementary'. Clausius' earlier, parallel proposition (that heat energy naturally moves from hot to cold) does not refer to 'engines' and is clearly a statement about heat, pure and simple. It would seem that Thomson wanted the new science of heat to incorporate some notion of 'transformation' – between heat and work. Indeed, he described his theory as one of 'the motive power of heat'.

Today, at least one school of thought believes that, in an attempt to clarify thinking, students should be told that physical objects have intrinsic energy, one aspect of which is their 'internal energy'. They should not refer to this 'internal energy' as 'heat'; instead, that word should be reserved to measure a 'transfer of internal energy' (when the process of heating/cooling takes place'). An engine may transform 'internal energy' into 'mechanical energy' – but not into 'work'; that word should be reserved to measure a 'transfer of mechanical energy'. All four concepts are measured in 'Joules'.

Everyday language usage frequently militates against applying this refinement with precision. However, whether we use language with precision or not, it is important that we do not lose sight of the magic that happens when heat is turned into work. Many of us see the pistons in an internal combustion engine (and the steam engine before it) as being driven by a series of small explosions – clever, but quite prosaic. If it were as simple as that, these engines would be about as efficient as the Miner's Friend. A piston under pressure moves in only one direction and, therefore, is a kind of filter. It acts to extract a single directional component (work) from all of the randomised (heat) energy on either side the piston. Thomson's principle tells us something about maximising this process.

With today's economic and environmental concerns about energy, it is even more important that our technologists understand its nature. In temperate climes, our individual energy needs tend to be twofold; we need randomised energy (warmth) and organised energy (work). Work is even needed to drive our computers, TVs and music systems, in order to create a voltage. Individually, we tend to spend more money on heat than on work. But when heat is turned into work (usually to generate electrical power) a significant quantity of heat has to be dumped directly out into the environment. Our motorcars dump *all* of their heat.

Some of us employ electricity (which has come from heat, and much heat was dumped in its production) to generate heat in our homes. We could employ a cheap-to-run 'reversed engine' to retrieve energy from the cooler environment and use that to heat our homes – but, in general, we do not. We do not recognise, as Carnot did, that any 'temperature gradient', where we allow heat to spread of its own accord down our corridors or out through our walls and roofs, is a lost opportunity.

In short, the magic of the heat engine should be emphasised, not only to challenge and inspire students, but to throw a stark perspective upon the way in which energy around us is utilised.

Thomson, standing at the leading edge of thermodynamic understanding, was aware of the miraculous nature of the heat engine. He chose to keep it as a central principle for understanding the ways of energy and we, maybe for quite different reasons, require that perspective even more today.

8.2 Discard attempts at logical consistency

It is partly because Thomson introduced his second proposition in the face of Clausius that modern thermodynamics is still pre-occupied by the logical equivalence of alternative propositions. A greater reason for this pre-occupation is that we still do not understand the nature of the universe's predisposition to spread its energy and, in denial of this, we cling to pet rationalisations that appeal to some of us, but rarely have more than conventional acceptance. And we visit this rehearsal of historical argument and preferred treatments on to our students.

We frequently run the risk of losing students by introducing pedantry before any need for language precision is clear and then insisting on the performance of logical exercises to satisfy some kind of external standard, usually before immature young minds can grasp their sophisticated abstraction. Thermodynamics can be a lot easier than we make it, and a lot more fun.

It is interesting that Thomson did not try to present a logically complete structure when he presented his two propositions. Mainland European science is often characterised by British commentators as being over-theoretical and it is true that, in this case, Clausius had attempted to present a grand theory of thermodynamics, looking rather like Euclidean geometry, based upon 'laws' and their logical progeny of theorems and demonstrations.

Thomson, with his admiration of French mathematics and science, was well aware of this tradition. Indeed, Carnot's beautifully created arguments and his theoretical heat engine were, in this sense, typically European. However, Thomson's real genius depended upon a quite different approach. He recognised that real progress could be made by facilitating conversation between theory and practice, and he had a special gift for this. He was principally a theorist, but he did not despise practical experiment – he positively encouraged it. His collaboration with James Joule is one example of his approach but, well before that, he had been instrumental in ensuring that laboratory work was known to all 'natural philosophers' coming through the University of Glasgow.

His 1851 paper [1], could be summarised as: 'here are two propositions, which common sense should persuade us to accept, upon which the whole of our understanding of motive power can be based – and here is the experimental data to support this assertion'. He does not call them 'laws' but he does present Carnot's arguments to reinforce our 'common sense'. He uses the power of argument but does not strive for any kind of axiomatic structure or logical completeness.

The second law of thermodynamics still hides a deep truth about our universe which we have not yet penetrated. Until we do, we should not pretend that we have a complete theory of thermodynamics and should not cover for our own lack of understanding by pretending sympathy for the student's confusion. If we present weak, unsubstantiated arguments, then good students should be confused! Instead, let us find ways to talk about second law phenomena, which can improve a student's understanding of what we do know, and instil a genuine fascination for a scientific mystery which is yet to be solved.

8.3 Use reversibility only as a model

In his 1851 paper, Thomson also uses an alternative form of his second proposition (namely, that it is impossible to obtain work from heat by cooling the coolest reservoir in the vicinity). In later years, as we described above, Max Planck developed this idea to formulate what was to become known as the Kelvin–Planck statement of the second law (that it is impossible to obtain work from heat without the existence of a cooler reservoir – in which to dump some of that heat).

Planck went much further and developed a theory which placed the concept of 'reversibility' at the centre of understanding the nature of thermodynamics. Much of British thermodynamics teaching throughout the twentieth century has had this concept at its heart. It might be possible to characterise the difference between Planck and Kelvin in this respect by saying that Planck envisaged circumstances where reversibility was possible (and, consequently, this is still the position of many working engineers today), whereas Thomson saw Carnot's engine more as an ideal to indicate the limits placed upon heat phenomena.

There is some justification for making this characterisation when we look at Thomson's earlier 1849 paper on thermometry [11]. He makes it very clear from the outset that his 'absolute temperature calibration' (which could be determined by placing Carnot engines between a series of heat reservoirs one degree apart) would not provide a better thermometer than already existed. Thomson saw the Carnot engine (and the reversible process) as a tool for thinking about and understanding heat – not as a realistic proposition. Considerable experimentation with various gas thermometers had already provided us with a high degree of confidence in our measurement of temperature; what Thomson did was to provide a theoretical basis for that confidence. In that paper, Thomson suggests experiments to test his theory, but not to develop a better thermometer.

This is 'science as model-making' at its best. Most of us, when we use the term 'degree Kelvin', or 'absolute temperature', see it as interchangeable with the 'degree Celsius' (or 'centigrade' – as it was known in Thomson's time); it is just a question of where the zero lies. However, it can be argued that, whereas the degree Celsius represents temperature seen in terms related to the mean behaviour of expanding substances, the degree Kelvin is a measure which stands outside physical analogue and one which is based upon a Carnot model of the world.

In our teaching, we should celebrate these kinds of opportunities to uncover the nature of scientific thinking. Instead of presenting the second law as 'given', and routinely expounding on its implications, we should say 'this model represents what we think is going on – it is well tested but, even now, we do not have certain proof'.

References

[1] Thomson, W., On the dynamical theory of heat; with numerical results deduced from Mr. Joule's equivalent of a thermal unit and M. Regnault's observations on steam. *Math. and Phys. Papers*, **1**, pp. 175–183, 1851.

[2] Savery, T., *The Miner's Friend: Or, an Engine to Raise Water by Fire*, S. Crouch: London, 1827.

[3] Rolt, L.T.C. & Allen, J.S., *The Steam Engine of Thomas Newcomen*, Landmark: Ashbourne, 1997.

[4] Dickenson, H.W., *James Watt: Craftsman and Engineer*, Cambridge University Press: Cambridge, 1935.
[5] Broad, C.D., *The Philosophy of Francis Bacon*, Cambridge University Press: Cambridge, 1926.
[6] Davy, H., *Researches, Chemical and Philosophical*, Biggs and Cottle: Bristol, 1800.
[7] Carnot, N.L.S., *Réflexions sur la Puissance Motrice du Feu et sur les Machines Propres a Développer cette Puissance*, Bachelier: Paris, 1824.
[8] Suleiman, E.N., *Elites in French Society*, Princeton University Press: Princeton, 1979.
[9] Thomson. W., On Fourier's expansions of functions in trigonometric series. *Cambridge Mathematical Journal*, **2**, pp. 258–259, 1841.
[10] Joule, J.P., On the caloric effects of magneto-electricity, and on the mechanical value of heat, *Philosophical Magazine*, **23**, pp. 263-276, 1843.
[11] Thomson, W., On an absolute thermometric scale founded on Carnot's theory of the motive power of heat, and calculated from Regnault's observations. *Math. and Phys. Papers*, **1**, pp. 100–106, 1848.
[12] Clausius, R., Über die bewegende Kraft der Wärme. *Annalen der Physik*, **79**, pp. 368–397, 500–524, 1850.
[13] Thomson, W., On a universal tendency in nature to the dissipation of mechanical energy. *Math. and Phys. Papers*, **1**, pp. 511–514, 1852.
[14] Clausius, R., Über die Wärmeleitung gasförmiger Körper. *Annalen der Physik*, **125**, pp. 353–400, 1865.
[15] Planck, M., *Vorlesungen über Thermodynamik*, Leipzig: Von Veit & Comp., 1897.
[16] Boltzmann, L., Über die Beziehung zwischen dem zweiten Hauptsatz der mechanischen Wärmetheorie und der Wahrscheinlichkeitsrechnung respektive den Sätzen über das Warmegleichgewicht, *Wien. Ber.*, **76**, pp. 373–435, 1877.

Chapter 11

Entropy as thermal charge: an application of bond graphs inspired by Carnot and his cycle

J. Thoma[1] & G. Mocellin[2]

[1]*Department of Systems Design, University of Waterloo, Canada*
[2]*Straco Consulting, Switzerland*

1 Sadi Carnot

Sadi Carnot (1796–1832) is recognised as the founder of thermodynamics. He formalised the Carnot cycle which we now call the second law of thermodynamics, the principle of the irreversibility of physical phenomena. This invention was published at his own expense in 1824 in his book *Reflections on the Motive Power of Fire and on Machines Fitted to Develop That Power* [1].

A precursor of Lord Kelvin (1824–1907), he also discovered towards the end of his life the first law of thermodynamics, that of the conservation of energy, as shown in his posthumous articles published in 1878. This principle was rediscovered and stated in 1842 by Mayer (1814–1878) following in the footsteps of Clapeyron who had already published his 'Memoir on the Motive Power of Heat' in 1834, in which he developed the ideas of Carnot who had died 2 years before.

Carnot's small book is at the root of all thermodynamics, the science of energy exchanges, thermal ones in particular, in all mediums whatsoever, particularly steam and gases.

It was published well after:

- the invention of the first cylinder and piston steam engine (1690) by Denis Papin (1647–1712),
- the separation of boiler and piston (1712) by Thomas Newcomen (1664–1729), and
- the introduction of the separate condenser (1763) by James Watt (1736–1819),

at a time, therefore, when steam technology was available to everyone, and after the invention of the first internal combustion engine (1804) by brothers Claude and especially Nicephore Niepce (1765–1833).

Although the main principles of work in a heat engine had been identified and applied, it still remained to produce a theory of thermal machines, which Carnot did. Carnot's book has been reprinted several times both in French, the original, and English. As a reference for this study, we use principally Mendoza's English edition (1960), reprinted in 2005 [2]. All things considered, Carnot's little book contains many new ideas despite some inconsistencies in terminology: Carnot often refers to 'calorique', what we would now call 'entropy', and to 'chaleur' which today we still call 'chaleur' or heat. The book, however, must be considered in the context of its times when all thermal effects were thought of in terms of 'calorique', a substance without mass that enters matter when the latter is heated, increasing its temperature.

Then, the theory of calorique was replaced by the kinetic theory of heat by Clausius (1822–1888) and Joule (1818–1889), a theory in which the notion of entropy was completely absent. It was only introduced afterwards as a factor of integration: a somewhat obscure procedure. Carnot's idea is much better: heat flow is equal to the product of entropy flow times the absolute temperature, just as in electricity the flow of power is the product of the flow of charge times the voltage. In this sense, entropy is a very simple concept, a kind of thermal charge. Moreover, entropy is a simpler concept than electric charge as it does not generate a field as electric charges do. In short, entropy has numerous properties of electric charges but differs from them in other respects. The idea of considering entropy as a thermal charge is by no means a new one. Apart from Carnot himself, authors such as Callendar [3], Thoma [4], Job [5], Falk [6,7] and Fuchs [8] all use the same point of view.

Carnot's book contains other new ideas for the time, such as that of integrating in a repetitive process the various phenomena at work in a heat engine, in particular those occurring in a cylinder closed by a piston (although in reality steam was replaced at the end of each cycle in the steam engines of the day). In addition, Carnot compares the power of steam engines to that of hydraulic machines, in which the pressure of water is the intensive variable and the quantity of water the extensive variable, the product of one times the flow of the other giving the available power. Water is not affected by the process of extracting power, just as entropy is not itself affected in a thermal process. In other words, it is the transport of 'calorique' from the heat source to the cold source that produces motive power, just like the movement of water in a hydroelectric power station. In modern hydraulic terms, pressure is the intensive variable and the flow of water is the extensive flow variable. Transcribed to the field of heat engines, temperature difference is the intensive variable and entropy flow is the extensive flow variable. Carnot explicitly said that a cold source is needed to eliminate 'calorique' once it has performed its task of producing motive power.

On the other hand, he also mentions that zones of different levels of 'calorique', that is, of different temperatures, can balance each other out without producing motive power and that to obtain good heat engines such contacts should be avoided. Moreover, one of Carnot's recommendations is that a change in temperature must only take place by a change in the volume or quantity of gas and not by conduction, which he calls the pointless establishing a balance of 'calorique'. To sum up, entropy must not be exchanged by conduction, which is of no use, but by a change in the volume of gas, which allows an adjustment to high and low temperatures, and vice versa. He further adds that 'calorique' is not consumed, but transported from a warm body to a cold body, which produces motive power. Finally, Carnot mentions an inverse cycle: by providing motive power, we can pump 'calorique' from a cold source to a warm source.

2 Bond graphs

Bond graphs are a tool for studying dynamic systems, thermodynamic systems in particular, developed by Paynter [9] at MIT in the 1960s and still appearing (2009) in the MIT course by Hogan [10]. This tool, which provides an easy method to show Carnot's ideas, is now documented in a large body of literature, developed in particular by Thoma [4,11–13], a former team from MIT [14], and extended into French by Dauphin-Tanguy [15].

One of the main ideas of bond graphs is to show the reality studied as a network of elements in energy interaction, that is to say, to show by arrows the flows of energy (therefore, of power) circulating between the elements making up the dynamic reality that one seeks to understand.

According to the maximum abstraction principle that Lazare Carnot (1753–1823) had instilled in his son Sadi when teaching him differential equations, another fundamental idea of bond graphs is to represent simply by letters the constituent elements of the reality studied that can produce, consume, accumulate or dissipate energy. The consequence of this idea is that the elements of representation are selected on the basis of the constraints limiting the observation and action of the scientist studying his problem. Using this principle, the particle of matter and the environment of Fig. 1 are reduced to simple letters in bond graphs, as shown in Fig. 2.

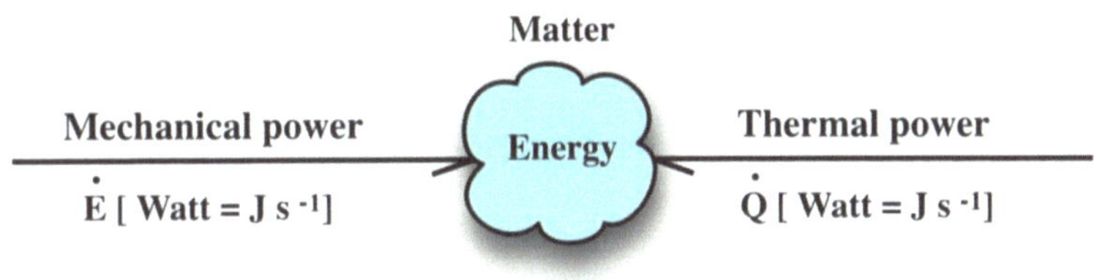

Figure 1: Graphical representation of a domain of study consisting of a particle of matter interacting mechanically and thermally with its environment. (i) The two modes of energy interaction (mechanical and thermal) are represented by thick semi-headed arrows to indicate clearly that these arrows represent an energy flow, i.e. power. (ii) The position of the arrowhead, determined by the researcher, indicates to him the conventional positive sign for energy circulation in the bond considered.

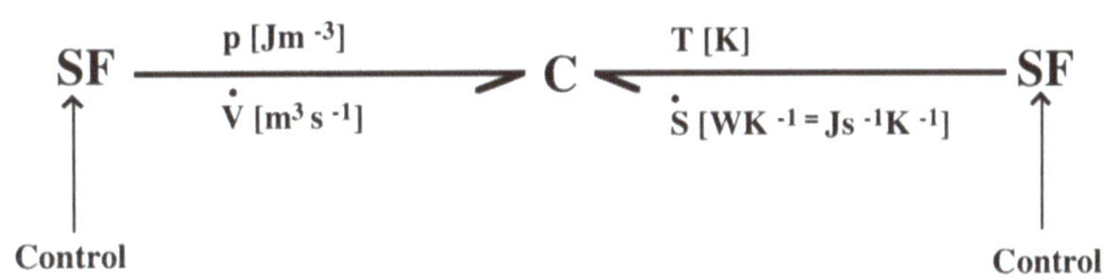

Figure 2: Representation of a particle of matter C interacting hydraulically and thermally with its environment represented by the two sources of flow SF. (i) The letter C represents a particle of matter and, specifically, its capacity to accumulate the extensive magnitudes of the energy domains considered: volume for the hydraulic domain (left) and entropy for the thermal domain (right). (ii) The letters SF (Source of Flow) represent sources (or sinks) of extensive flow variable. Indeed, it represents the conditions imposed at the boundaries of the system considered, the so-called conditions at the limits of partial differential equations. (iii) These conditions at the limits can be modulated, which is indicated by the narrower vertical arrows terminated by a small arrowhead. These arrows, which differ from the thick ones of the horizontal energy bonds, represent signals and thus enable the representation of reality in the form of bond graphs to be used immediately to make automatic adjustments. When the sources are not modulated by a control signal, they are simply considered as constants. (iv) The names of the conjugate variables of the energy domain considered appear on the energy bonds: pressure and volume variation for the hydraulic domain and absolute temperature *T* and entropy flow rate for the thermal domain.

We can show here the power of representation of bond graphs, which can be used to describe, in a highly compact and calculable manner (as they can be directly implemented as computer programmes to carry out simulations) phenomena encompassing a multitude of energy domains. This arises from the fact that each energy bond represents a power considered as the product of an effort (the conventional word in bond graphs for a potential) times a flow:

$$Power = Effort \times Flow$$

in which the efforts (potentials) and flows can belong to different scientific disciplines. In classic thermodynamics, the conservation of the power that represents the variations of energy in a system is expressed by the equation:

The relation between a representation of thermodynamics in the form of equations and one in the form of a bond graph becomes apparent in Fig. 3.

For the purpose of synthesis, we can construct a table of the principal domains of energy interaction:

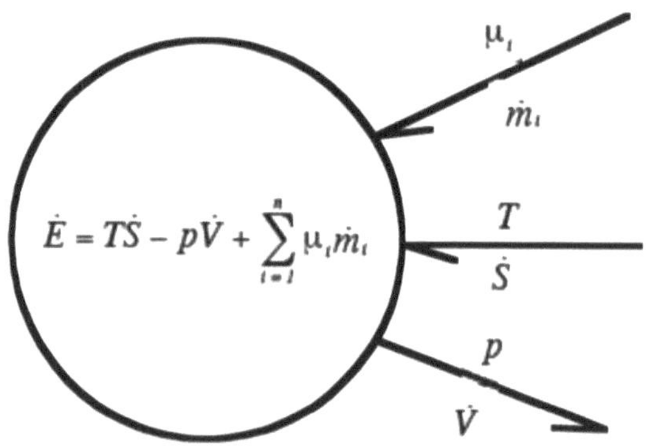

Figure 3: Representation of a thermodynamic reality in the form of a formula and a graph. The sign of the powers in the formula is represented by the direction of the arrows, the heads of which correspond to the signs of the equation.

Table 1: Efforts and flows of different domains of thermodynamics.

Domain	Effort variable	Flow variable
Mechanical translation	Linear speed (m s^{-1})	Force (N)
Mechanical rotation	Rotation frequency (rad s^{-1})	Torque (N m)
Fluidics	Pressure (N m^{-2}) or (J m^{-3})	Volume deformation (m^3 s^{-1})
Electricity	Electric field D (V)	Electric current D (A)
Magnetism	Magnetic field H (A)	Magnetic flow B (Wb s^{-1})
Thermal domain	Temperature (K)	Entropy flow (W K^{-1}) or (J K^{-1} s^{-1})
Chemistry	Chemical potential (J kg^{-1})	Mass flow (kg s^{-1})

In Table 1, it is interesting to note that:

- Chemical energy appears there as stored in matter in the form of a force field between the particles of which it is composed, effort, called chemical potential, being measured in Joules per particle.
- Bond graphs can be used to represent tensor magnitudes such as are found in continuum mechanics [16].

Another great idea of bond graphs is to represent causality, that is, to distinguish permanently the variables that are the causes of the phenomena from those that are its effects, independent variables from dependent variables as it were, Fig. 4.

We can sum up the principles of bond graphs in diagrams showing the relations between the various concepts used to represent dynamic systems, linear or otherwise, since bond graphs can represent non-linear systems. We can start with the circular relation that exists between the variables, Fig. 5.

The diagram above is a modified version of the circular diagram that Thoma has used in his courses, the so-called 'Thoma Carrousel' [4], Fig. 6.

Observing the Thoma Carrousel, we note that there is still a direct relation R both ways between effort and flow. We can represent it by the diagram in Fig. 7.

To be complete it would still be necessary to introduce memristance, an element that connects accumulations of effort and flow (practical applications of which are beginning to emerge in 2009), Fig. 8.

To use bond graphs in thermodynamics, Thoma has introduced a new element, resistive source, called RS, whereby thermal phenomena can be represented by a production of entropy, Fig. 9.

We can show element RS by the diagram in Fig. 10.

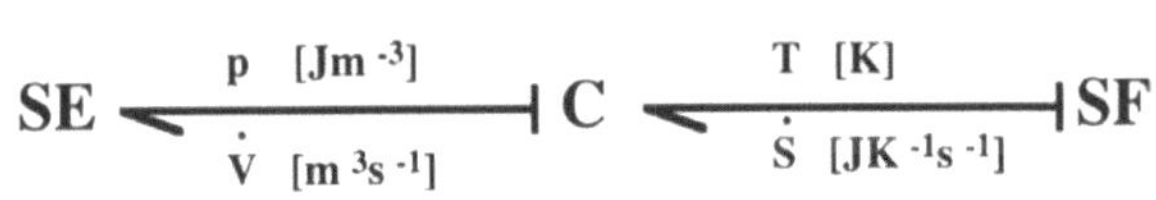

Figure 4: Representation of a particle of matter interacting hydraulically and thermally with its environment. (i) Here, the small vertical line at the right end of the left arrow, representing the hydraulic link, called a causality line, indicates that the matter C considered is subjected to a fixed pressure p, possibly constant, by the source of effort SE (source of pressure in hydraulics). (ii) The small causality line on the right of the right arrow, representing the thermal bond, indicates that the entropy flow is fixed by the source of flow SF (i.e. by the environment).

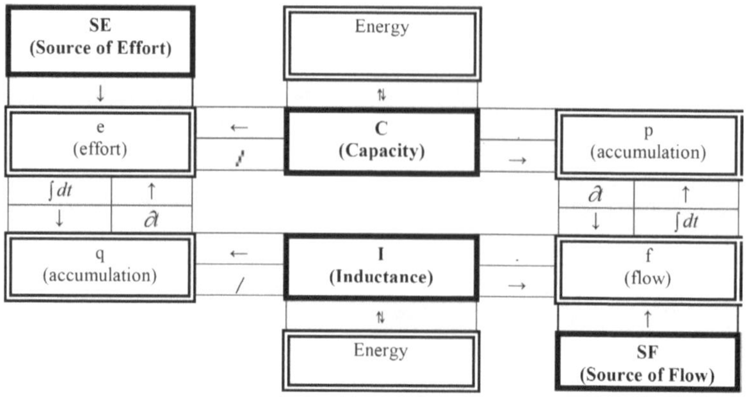

Figure 5: Circular arrangement of variables to represent dynamic systems. (i) Elements representing system parameters are surrounded by a thick black line and those representing variables are surrounded by a double line. (ii) Energy originating from changes in flow *f* is stored in a form *p* in the reality considered (element C in bond graphs). (iii) Energy from changes in effort *e* is stored in a form *q* in the reality considered (element I in bond graphs). (iv) Effort *e* is fixed by a source of effort (element SE in bond graphs) and flow *f* is fixed by a source of flow (element SF in bond graphs). (v) A double arrow appears in the relation between energy and elements C and I because these can accumulate and return energy. (vi) Letters C and I represent relations that are not necessarily linear.

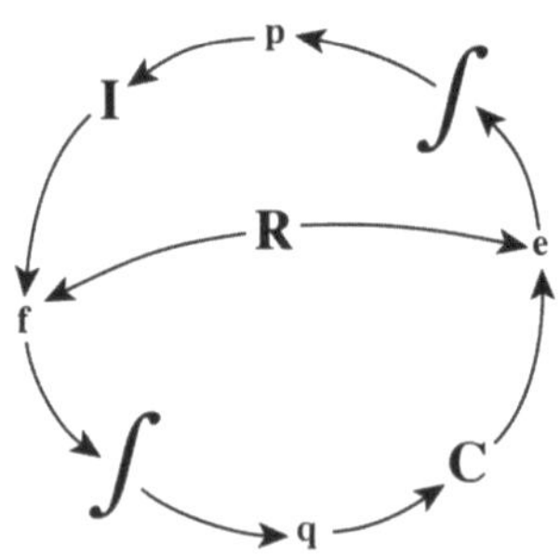

Figure 6: The Thoma Carrousel in its integral causality version. (i) The variables of a dynamic system are organised circularly in bond graphs: if one starts with a flow *f*, one integrates it temporarily in a displacement *q*, which gives an effort *e* at the output of capacitor C which is integrated temporarily in a momentum *p* which returns a flow *f* at the output of inductor I. (ii) The reasoning reciprocal to integration would be valid for derivation if it is turned the other way round.

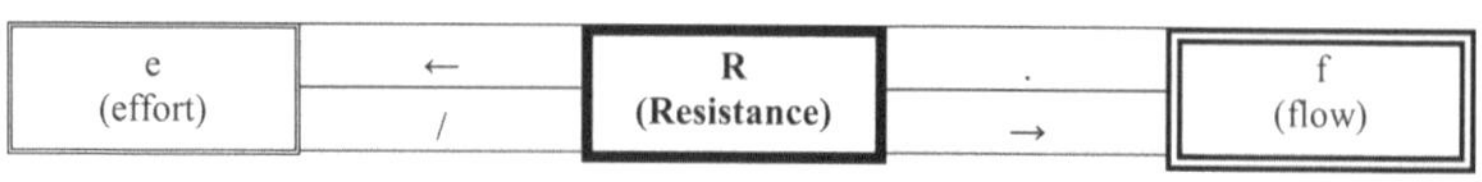

Figure 7: Direct relation between effort and flow (element R in bond graphs): an effort *e* multiplied by a resistance *R* gives a flow *f*, and a flow *f* divided by a resistance R gives an effort *e*.

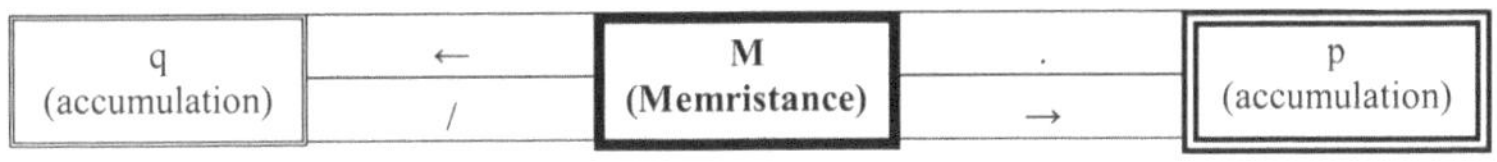

Figure 8: Direct relation between accumulation of effort and accumulation of flow: an accumulation of effort q divided by a memristance M gives an accumulation of flow p, and an accumulation of flow p multiplied by a memristance M gives an accumulation of effort q.

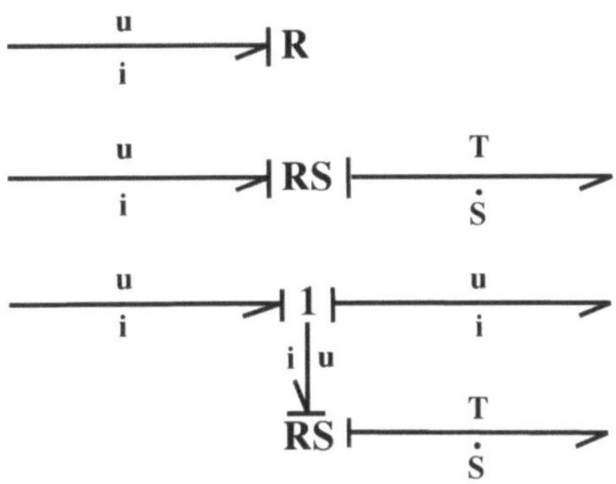

Figure 9: Introduction of the element RS ('an R set by an S' as Thoma calls it). (i) Element R at the top of the figure becomes element RS in the middle because all dissipation, by friction in particular, generates entropy. The irreversibility of this production appears clearly from the fact that the energy arrow producing entropy emerges from element RS, which implies that power always emerges from element RS into the environment, according to the laws of nature. (ii) The causal sign, which distinguishes the dependent variables from the independent ones, is arbitrary on the left bond of element RS. Conversely, the arrow representing thermal power emerging from the right of that RS has a causality line at its left end, which indicates that the input of RS is the temperature of the environment and the output is entropy flow, in the sense of systems theory. (iii) With element RS, it becomes very easy to introduce a thermal part in the study of a complex system: one needs to couple only this thermal part by a link (represented in the figure by the symbol 1) to another energy domain, to an electrical part for example, as shown in the third bond graph at the bottom of the figure.

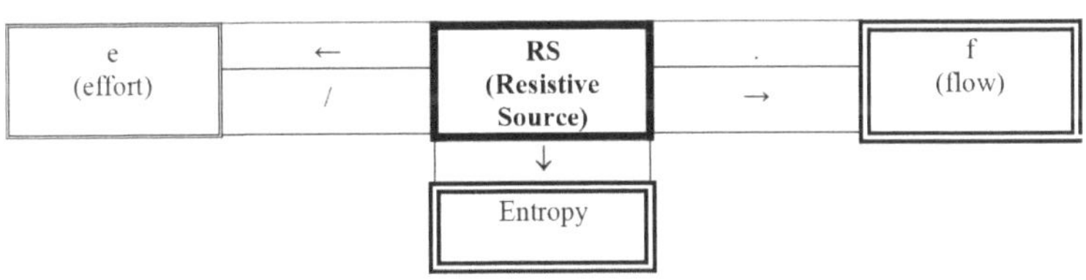

Figure 10: Schematic representation of element RS (resistive source). (i) Here the arrow linking element RS to Entropy only points in one direction, contrary to those making a connection between elements that can store and return energy in the previous diagrams, this being a clear representation of the irreversibility of nature. (ii) After the introduction of element RS, it becomes very easy to represent the production of entropy from the various friction phenomena acting in nature, including chemical friction, Fig. 11.

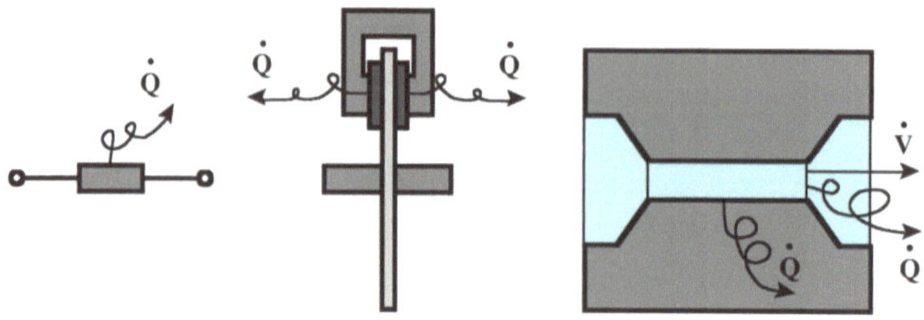

Figure 11: Various modes of entropy-generating friction: electrical on the left, mechanical in the centre and hydraulic on the right.

3 Systemics

In some discontinuous representations of reality, such as bond graphs, for example, reality is considered as made up of a network of elements in interaction. The lines connecting the elements have no other meaning than to show a conceptual relationship between these elements. At first sight, the spatial positions and dimensions of the elements have no effect on the study of the reality considered. Such representations, although approximate, are more than sufficient for some studies.

3.1 Structure

Once the elements and their interactions are defined we still need to be able to describe, in a systematic manner, the structure itself of the interactions occurring in the reality studied, that is to say, to be able to connect multiple elements to each other to constitute a whole in interaction.

To do this, bond graphs arbitrarily use the symbol '0' to represent a point of potential (effort) known in relation to a reference: all energy arrows attached to this link have an identical effort, which amounts to saying that the elements attached to it are topologically in parallel (see Fig. 12).

To be systematically complete, one still needs to be able to the show points of reality at which the flow is common to all energy arrows connected to them. As we have seen in Fig. 13, bond graphs conventionally use the symbol '1' to represent such links in series. This type of link provides, in particular, a very easy way of including the thermal domain in the analysis of electric or other circuits, the R elements, a first approximation of reality, being replaced by RS elements, which include the fact that all types of friction produces entropy conservatively. And often, as experience shows, the thermal part should be included, since power lost by friction appears as heat (or entropy × absolute temperature) and heats the installation, often leading to dangerous levels of overheating that must be taken into account. To repeat, a pure R element does not exist and must be replaced by the RS field as soon as one wants to include thermal effects in the bond graph. In this sense, entropy can be considered as a kind of substance that is always linked to the transmission of heat and thereby the RS field becomes power conserving.

Other productions of entropy arise from mixing chemicals or from chemical friction ([13, p. 23], and with reversible mixing with selective pistons, p. 74). These modes, easily represented by bond graphs, are interesting in principle, but not relevant for this study. Although bond graphs

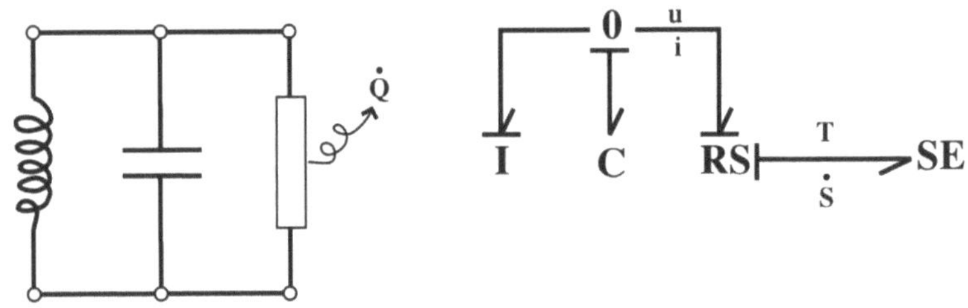

Figure 12: Left: schematic representation of a parallel electric circuit generating entropy. Right: a bond graph representation of the same circuit. In the bond graph, it can be noted that the energy arrow emerging on the right of element RS represents dissipation in the form of entropy production and is orientated positively towards the environment, itself represented by source of effort SE which sets the temperature.

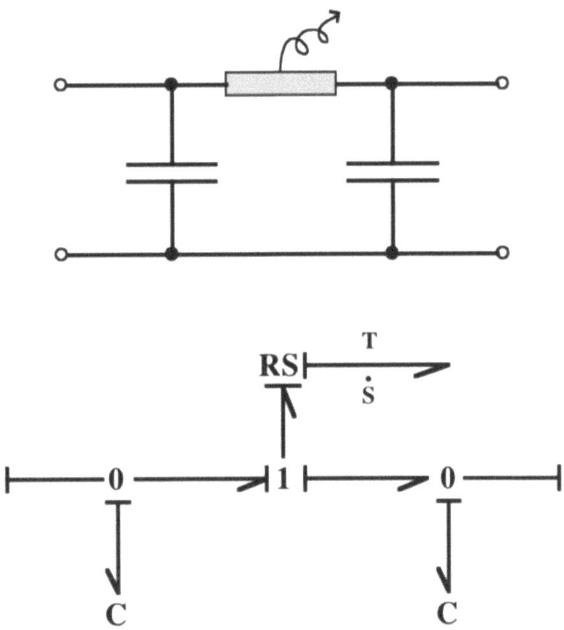

Figure 13: Representation of a link in series by bond graph symbol 1. (i) In the bond graph, junction '1' compels the flow to be identical in the three energy bonds connected to it. Resistive source RS situated between the two capacitors C produces a flow of entropy as long as there is a voltage imbalance between the two capacitors. (ii) The new entropy is indestructible and must be eliminated, for example, in an inverse source on the right of the bond graph. As the causal sign on the RS element shows, the resistive part can depend on temperature.

can account immediately for the thermal aspects of an electric system, it still remains to state where the entropy generated by the resistance is going to go, that is, to show in a calculable manner how the thermal effects propagate in the circuit. For example, if a resistor is surrounded by two capacitors, the one to which the entropy is going to propagate must be determined, and bond graphs can do this.

3.2 Dynamic systems

We show how the problem can be solved graphically and also how a bond graph allows the derivation of the various flows of entropy acting in an electric circuit: just as electric capacitors are said to be electrically charged, thermal capacitors can be said to be thermally charged, the thermal charge being the entropy.

We note that the electric flux in the circuit is

$$i = \frac{(u_1 - u_2)}{R}$$

and the dissipated power:

$$\Phi = (u_1 - u_2) \cdot i = \frac{(u_1 - u_2)^2}{R}$$

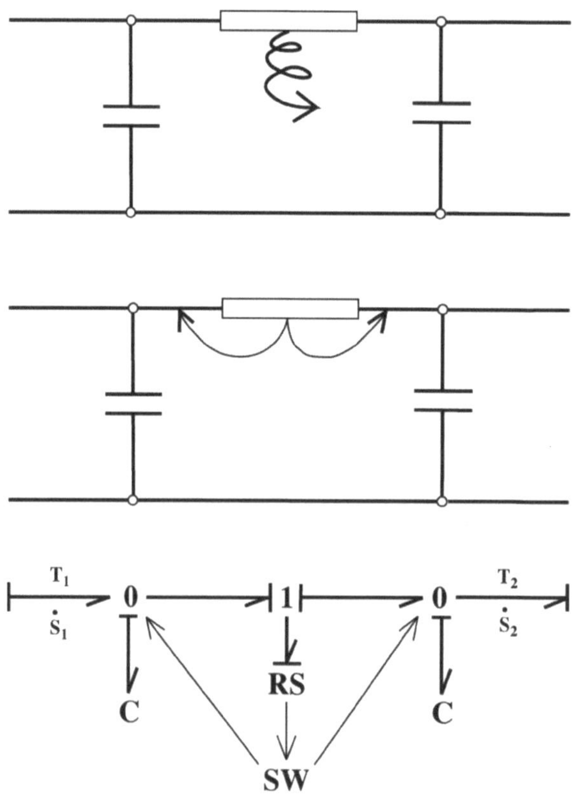

Figure 14: Representation of entropy distribution in an electric circuit. (i) The power dissipated in the resistor is always positive whatever the sign of the electric current. (ii) The entropy generated by the resistor is injected into capacitors C around it. (iii) The introduction of a switch SW allows the entropy created in the centre to be transferred to the left or right and enables us to understand how the entropy is injected into the capacitor C that is at the lowest temperature.

We see in particular that the dissipated power is always positive whatever the direction of flow, as indicated by the formula.

3.2.1 *Multiport C elements*

We can continue to explore the electric domain by considering networks of capacitors, known as multiports C or, in the bond graph community, simply C-fields. An example of a multiport is given in Fig. 15 in which two capacitors, one on the left and one on the right, are connected by a coupling capacitor, in the centre; this has a radical effect on the behaviour of the circuit. The capacitive reactance that can be measured on energy bond 2 depends on what arrives from side 1. If the sources of flow are replaced by sources of effort, a change in side 2 sends a small amount of power to side 1, which affects the capacitance of side 2.

Multiports are very important for representing standard thermodynamics. They can easily show situations in which hydraulic and thermal phenomena appear simultaneously. For example, Fig. 16 shows a bond graph that enables us to calculate the change in entropy at constant volume and constant pressure and, after multiplication by the absolute temperature, the change in specific heat values, since that change simply comes from the cross terms of the multiport C [13].

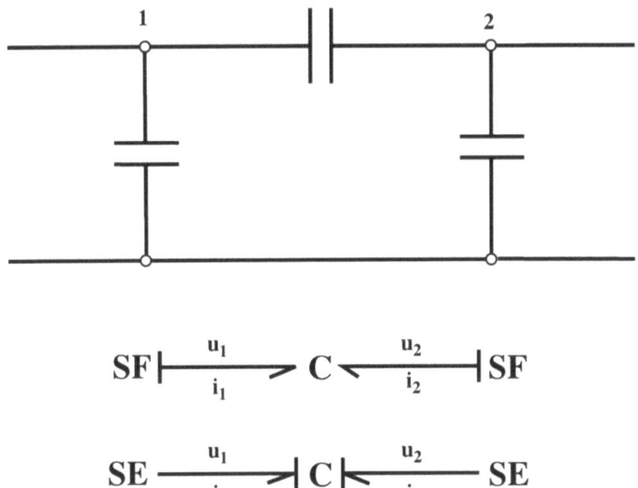

Figure 15: Example of an electric circuit, here a network of capacitors, represented at the top in schematic form and at the bottom in bond graph form by a multiport C. (i) In the top circuit, the capacitor in the centre couples the two left and right parts. (ii) At the bottom are two examples of a multiport C, one with flow sources and the other with effort sources, a circumstance which imposes different causalities on the multiport C.

Figure 16: Hydraulic and thermal multiport C with sources of effort.

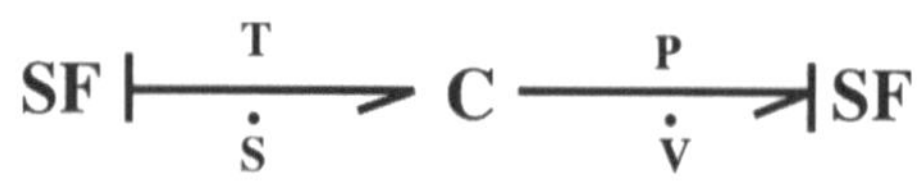

Figure 17: Multiport C between two sources of flow. If the sources of flow are set to zero, the power circulating in the two energy bonds disappears and multiport C conserves the energy.

Accumulation equation: Let us consider a particle of matter that presents coupled thermal and hydraulic phenomena. If both entropy and volume are injected into this particle, it accumulates both thermal power and hydraulic power. We can show such a phenomenon by the bond graph in Fig. 17, which presents two sources of flow SF, one thermal and the other hydraulic, and a multiport C representing matter:

The power circulating in the bonds is:

$$\dot{E} = T \cdot \dot{S}$$

$$\dot{E} = -p \cdot \dot{V}$$

Multiplying by an interval of time and integrating for a short time, one obtains the familiar equation:

$$dE = T \cdot dS - V \cdot dP$$

which represents the accumulation, in the form of energy, of the power supplied to multiport C by the two energy bonds.

Power in hydraulic circuits: Bond graph representation reveals its full power when we try to understand what happens in pipe circuits carrying fluids. The total power transported in a pipe is

$$\dot{E} = \dot{U} + p \cdot \dot{V}$$

This power is made up of the sum of the variation in internal energy and in hydraulic power, itself represented by the product of the pressure times the volume flow of the fluid considered, that pressure being the only measurement that really matters in practice in hydraulic circuits.

3.2.2 Heat conduction and entropy conduction

The power of representation of bond graphs can also be demonstrated by taking as examples systems in which we only seek to understand the thermal phenomena that develop in them.

Heat conduction through a wall is the most typical example of this. We have seen how bond graphs have drawn our attention to the fact that all resistive phenomena generate entropy, as shown in the top circuit in Fig. 18. If this production of entropy is not accounted for, we have what we call a pseudo-bond graph, with variables T and Q as conjugate variables instead of and for a true bond graph. The representation of the wall then becomes a simple RC network as in the lower circuit in Fig. 18 because the heat flux is conserved in conduction.

If we make a bond graph representation of the wall, we obtain a pseudo-bond graph as shown at the top of Fig. 19 and a true bond graph as shown in the lower diagram in Fig. 19:

We see that for problems of heat conduction, the true bond graph is more complex than the pseudo-bond graph. This explains why pseudo-bond graphs are used almost universally in this domain: many people even try to avoid the notion of entropy altogether.

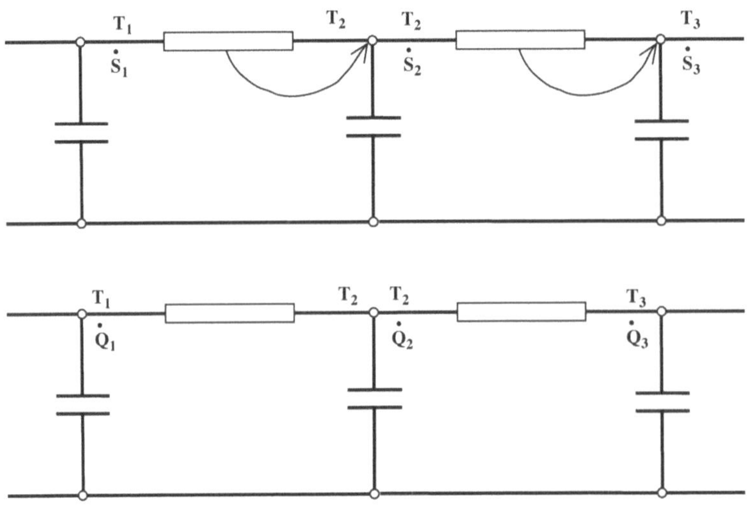

Figure 18: Thermal circuit of a house wall. The thermal circuit at the bottom is simple because it does not show the entropy flows generated by the resistors as shown in the top circuit: the heat flux is conserved.

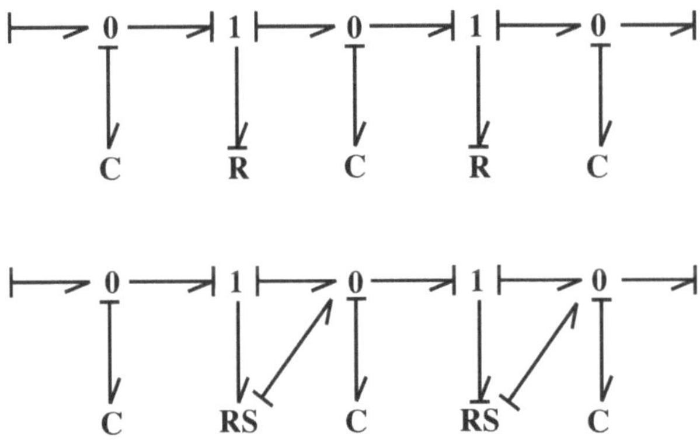

Figure 19: Bond graph representation of a wall with a pseudo bond graph (top) and a true bond graph (below). (i) The true bond graph at the bottom is more complex because the new entropy generated by the RS element, which has replaced the R element, is injected into the multiport C on its right. (ii) Note that since in a house the entropy and heat flows always from the interior to the exterior, we do not need the switch SW of Fig. 10.

To sum up, in true bond graphs the product of effort times flow gives a power, that is, a flow of energy, while in pseudo-bond graphs this is not the case: the product of effort times flow does not have the meaning of a flow of energy. The classic example of this is the transmission of heat in which one considers heat flow and temperature.

3.2.3 *Perfect gas equation*

A perfect gas is a very good approximation of the behaviour of fluids at high temperatures and low densities, as well as matter in general in a very diluted form.

Its behaviour equation is as follows:

$$p = \frac{m}{V} \cdot R \cdot T$$

For pseudo-bond graphs, two others, usually called calorific equations, are added to it:

$$\dot{U} = c_v \cdot T \cdot \dot{m}$$

and

$$\dot{S} = c_v \cdot \dot{m}$$

3.2.4 *Origin of heat and enthalpy*

Normally, heat and enthalpy are used in preference to entropy. With water and steam, the origin of temperatures is defined as 0°C for the liquid state and all values of energy variables are taken from there. Thus, one writes:

$$U = c_v \cdot m \cdot T$$

Entropy is treated differently and one writes

$$dS = \frac{dQ}{T}$$

$$dQ = c_v \cdot dT$$

and

$$S = S_1 - S_2 = \int c_v \cdot \frac{dT}{T}$$

such that one obtains, with the usual approximation $c_v = Cte$:

$$S_1 - S_2 = c_v \cdot \ln \frac{T_1}{T_2}$$

According to Nernst's theorem, entropy is zero at null absolute temperature, a purely theoretical value as the entropic content at low temperatures is not known. Moreover, one can also have freezing at low temperatures [13, p. 78], which is why one always starts with intermediate values.

3.2.5 Conduction and convection

So far, we have spoken of the step-by-step conduction of entropy in matter. It still remains to represent the convection of entropy, which is the transport of entropy by the movement of the fluid itself: in this case, the power transmitted is given by the flow of enthalpy.

To represent the phenomenon of convection, we can use vector bond graphs in which the energy bonds no longer represent a single pair of energy variables (effort and flow), but a vector in the algebraic sense, that is, an ordered set of pairs of these variables (efforts and flows). In fact, for convection, we use a pseudo vector bond graph containing two pairs of variables, that is, two efforts, pressure and temperature, and two flows, mass flow and enthalpy flow. The advantage of using enthalpy flow is that the change it undergoes is equal to the power injected in the fluid by the solid walls. We can now construct new thermodynamic elements and add them to the basic elements of bond graphs.

3.3 Bond graphs of pipe circuits with hot gases

It is often necessary to use bond graphs to represent installations including fluid pipe circuits. This can be done immediately using pseudo vector bond graphs with temperature and enthalpy flow as the first pair of variables and pressure and mass flow as the second pair of variables. One can thus develop progressively all of the important components for understanding such circuits, starting with the simplest, resistance to convection which we call RECO, Fig. 20.

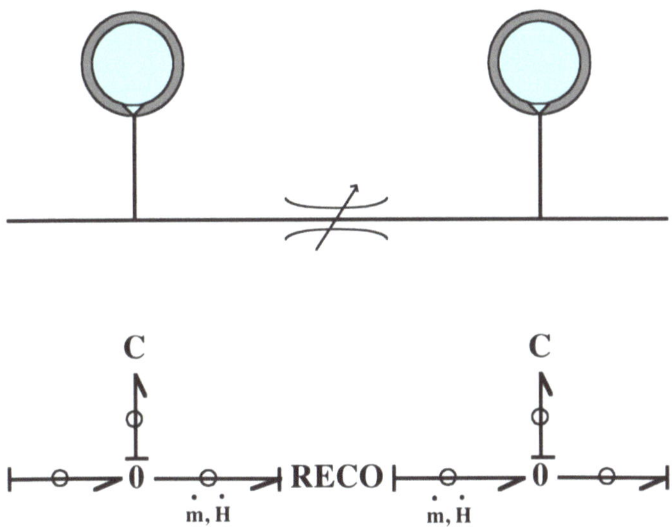

Figure 20: Bond graph of a resistance to convection (RECO). The fact that the bonds are vectors (in the algebraic sense) is indicated by the small circles in the middle of the bonds.

In the RECO of Fig. 20, enthalpy flow is conserved as is mass flow, and we have the equations:

$$m = K_d \cdot A \cdot \frac{\sqrt{(p_1 - p_2)p_2}}{T_1} if\ p_2 > 0.5 \cdot p_1$$

$$m = \frac{K_d}{2} \cdot \frac{A \cdot p_1}{\sqrt{T_1}} if\ p_2 < 0.5 \cdot p_1$$

Remarkably, the temperature at the bottom of the flow does not influence the flow. The next component to be introduced is the heat exchanger, which we call HEXA, in which a flow of thermal power is simply added to the flow of enthalpy:

$$H_{Sortie} = H_{Entrée} + \dot{Q}$$

We can show such a heat exchanger by Fig. 21:

A resistor R can be added to this bond graph to adjust the causalities, Fig. 22:

The next element is a thermofluid machine, named TEFMA, which allows us to show a mechanical energy bond in which the flowing power is , where is the torque and is the angular frequency of rotation of the shaft. In Fig. 23, in addition to the bond supplying power to the environment, an inertia I has been added in series (by link '1' on the right) to make the causalities consistent.

The equation for such a thermo-fluid machine becomes

$$\dot{H}_{Sortie} = \dot{H}_{Entree} + \dot{Q} + M \cdot \dot{\varphi}$$

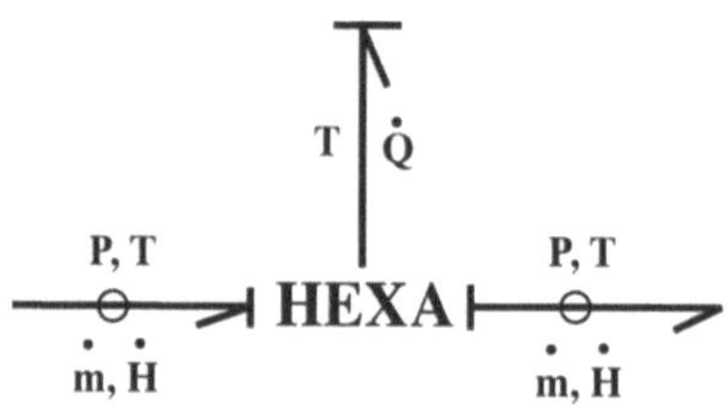

Figure 21: Bond graph of a heat exchanger HEXA, in which we have a balance of power.

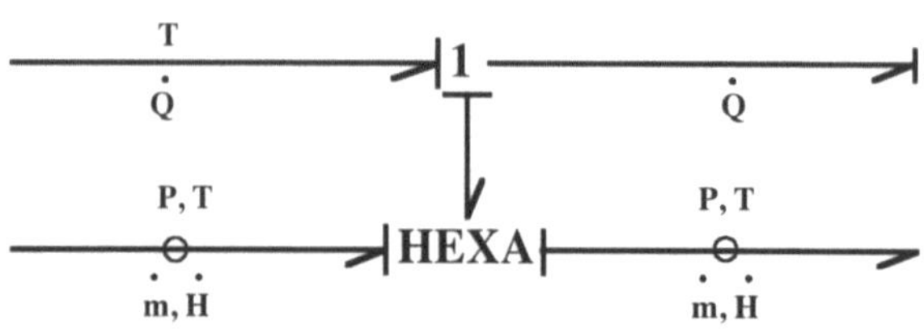

Figure 22: Bond graph of a HEXA with a resistor added to adjust the causalities.

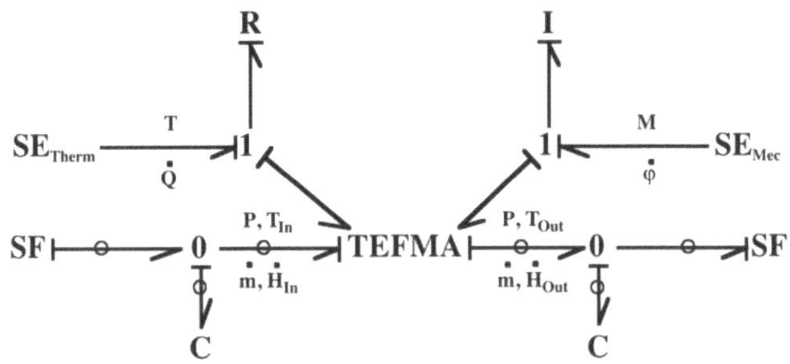

Figure 23: Bond graph of a thermofluid machine TEFMA in which a mechanical branch with inertia I is added at the top right to adapt the causalities. (i) At the top right, we have the mechanical bond that generates the rotation frequency of the shaft in response to the torque of the turbine and the torque of the flywheel which are both connected in series. This is a true bond graph. (ii) The thermal branch, at the top left, receives its temperature at the thermofluid machine TEFMA and goes into a thermal element R: this part of the bond graph is a pseudo-bond graph.

In the elements above, the causality lines are all important: we see that the horizontal vector bonds, representing flows of gas, have as efforts p and T and as flows $\dot{m}$and $\dot{H}$. These are pseudo vector bonds.

3.4 Bond graph of a thermofluid machine

One can assemble all of the elements described above at will to represent any heat engine, a thermofluid one in particular.

Figure 24 brings together the three basic elements we have prepared to construct bond graphs of a hot gas heat engine. At the bottom, we have added two coupling capacitors C to ensure correct causalities. Apart from the coupling capacitors, the flywheel on the turbine shaft also provides consistent causalities. Two small precautions are to be taken when modelling any turbomachine:

- reticulate the pipes well by parallel junctions followed by a leak resistor to control the losses in the pipes, and
- ensure that the velocities at the coupling capacitors are quite low, of a magnitude such that their kinetic power is negligible compared with the static power of the enthalpy flow.

The difference between reversible and irreversible thermodynamics is not brought out clearly in many texts. Using the bond graph method it becomes very simple: if the R elements impose only negligible dissipation, the system is reversible. If not, we have an irreversible system which produces entropy in RS fields. All the frictions reside there, including entropy conduction under finite temperature drop, mixing and chemical effects.

4 Icons for thermofluid machines

Our bond graph representation of thermofluid machines is perfectly adequate, but some people need a schematic representation of the RECO, HEXA and TEFMA machine elements, if only for illustration purposes.

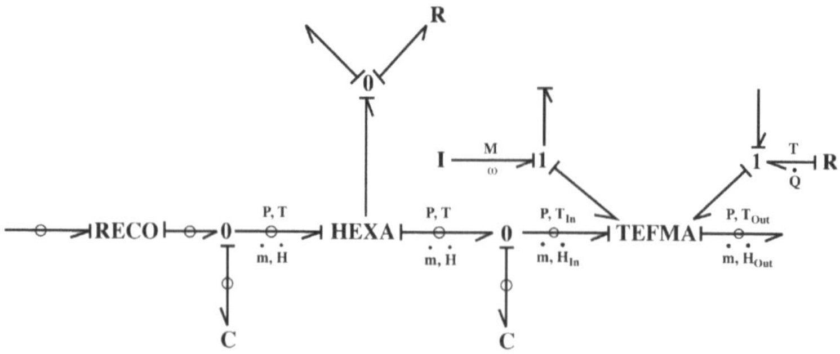

Figure 24: Bond graph containing the three elements RECO, HEXA and TEFMA, separated by coupling capacitors.

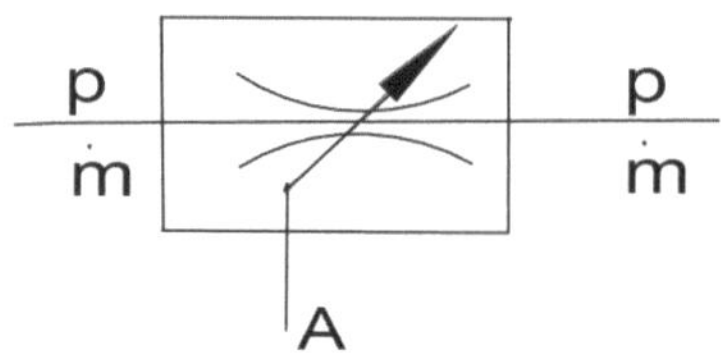

Figure 25: Icon to represent a RECO. The icon represents a small passage in a resistor, the letter A representing the smallest section of the passage.

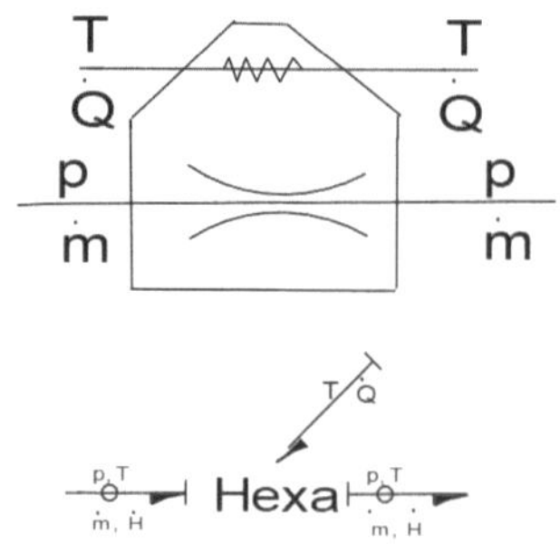

Figure 26: Icon for a HEXA. An additional thermal line in a HEXA (with bond graph below). This figure is an extension of the heat exchanger, in which the hot fluid is depicted by a zig-zag line within the casing.

For the RECO, we can make the proposition as in Fig. 25:

With regard to the heat exchanger HEXA, the icon is a little more complex because it has to contain two lines of power, Fig. 26:

The icon for the TEFMA, for its part, must contain a third line of power, Fig. 27:

Note that the coupling capacitors do not appear in the RECO icon. Therefore, Fig. 28 proposes an icon for a turbomachine with coupling capacitors and a flywheel on the shaft:

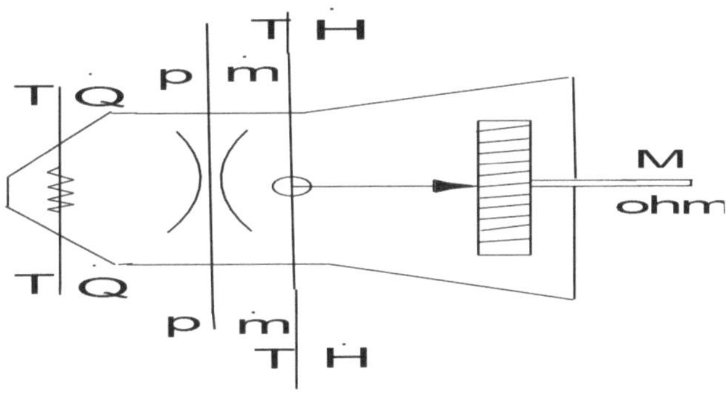

Figure 27: Icon of a thermofluid machine TEFMA with a turbine.

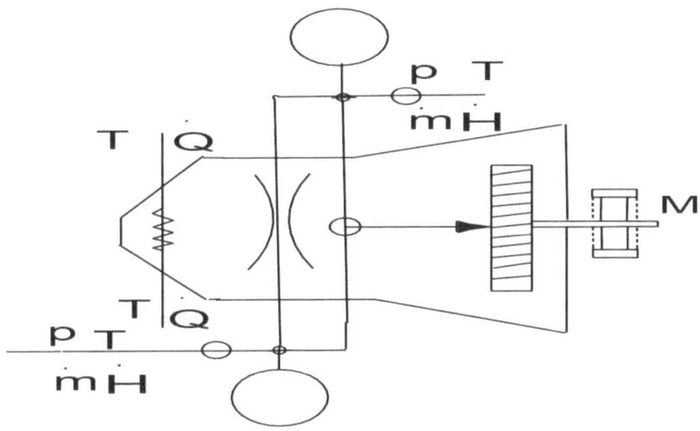

Figure 28: Complete icon of a TEFMA, with icons for the coupling capacitors and for the inertia of the shaft.

5 The Carnot cycle

Carnot's original heat engine was a steam engine that consisted of a cylinder, a piston, a connecting rod and a transmission shaft, together with a heat source and a cold source, alternatively entering into thermal contact with the cylinder to exchange entropy.

Carnot's idea was to introduce two partial cycles, in which the working medium is completely isolated thermally, adiabatics in modern terminology. At the end of heating, for example, the medium is disconnected from the heat source, the piston continues its expansion and consequently the medium cools, until it reaches the low temperature at which it can be connected to the cold source without thermal shock (without thermal short circuit, or 'jolt' one might say).

Next, the steam used is expelled by movement of the piston. At the end of the expulsion, the medium is again disconnected from the cold source, but the piston continues to move. It compresses the medium and raises its temperature until it reaches the high temperature. Then, the cycle starts over again.

By this decomposition of the process, Carnot avoids the two thermal short circuits, or, as he put it, the production of 'calorique' (entropy) by conduction between a finite temperature difference. The instants at which switching occurs between the isotherms and the adiabatics must be well synchronised with the inversion of piston movement: errors in these instants diminish the power of the machine. Carnot, moreover, also understood that a heat engine can be used inversely, by passing the 'calorique' from the cold source to the heat source. The power then outputs the 'calorique' from the cold source, causing its temperature to fall further by injection of mechanical power.

5.1 Bond graph of the Carnot cycle

We can show an elementary Carnot heat engine schematically, Fig. 29:

Such a heat engine can be represented very easily using bond graphs (Fig. 30).

The working cycle of this machine is the following: to start with, the piston is at the extreme right and begins to move to the left. The switch connects the heat source and the hot vapour enters the cylinder. At a certain point, the switch goes off, the piston continues to travel and the medium to expand by cooling until it has reached the temperature of the cold source. Then, the piston reverses its movement and discharges the cold medium, cooled by expansion. Towards the end of expulsion, the switch disconnects the cold source, the piston reverses its travel and compresses the remaining medium until it again reaches the temperature of the heat source. And the cycle can begin again with new 'calorique' or steam. By this process, Carnot avoids thermal contact by the steam with different temperatures, that is, thermal short circuits.

Sometimes it is said that the Carnot cycle must take place infinitely slowly to avoid entropy being generated by the irreversibilities. This is not the case because, apart from the fact that such a machine would have no useful power output, thermal short circuits are avoided by the adiabatics whereby the temperatures adjust their values. On the contrary, it must be executed at a suitable intermediate speed at which the resistors in series and the resistor in parallel towards the environment produce only a tolerable amount of entropy.

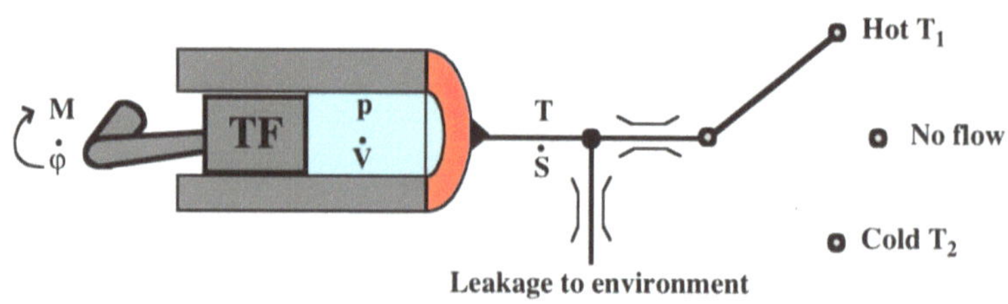

Figure 29: A thermal machine producing the Carnot cycle. It is made up of a horizontal cylinder, a piston, a link mechanism and thermal connections with the heat and cold sources and the environment. The three-way switch on the right is important in that it allows the machine to be disconnected from the sources to produce the adiabatic parts of Carnot's cycle.

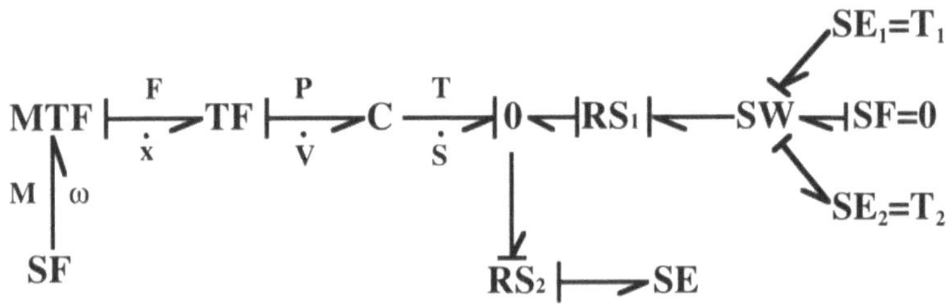

Figure 30: Bond graph of a Carnot heat engine. (i) At the bottom left, we have a mechanical load represented by the source of flow SF. The power conveyed by the mechanical energy bond that goes upwards from SF is characterised by moment M, which is its extensive flow variable, its angular momentum flow in other words, and by angular frequency ω which is the intensive variable, or effort. (ii) The mechanical energy rotating the shaft is transformed into mechanical energy for translatory motion of the piston by the crankshaft represented by a modulated transformer MTF. (iii) The linear mechanical energy obtained, whose extensive variable is displacement x and intensive variable force F (inverse mechanical analogy here), is transformed into hydraulic energy by the piston itself represented by transformer TF. At the output of this transformer, the intensive variable is therefore pressure P and the extensive variable is volume V. (iv) The medium contained in the cylinder is represented by the multiport C since it accumulates both hydraulic energy on the left, whose intensive variable is pressure P and extensive variable is volume V, and on the right thermal energy whose intensive variable is temperature T and extensive variable is entropy S. (v) The part of the machine that allows entropy (but not matter) to pass easily is represented by thermal resistors RS.

5.2 Properties of 'calorique' (entropy)

We come now to the properties of entropy that we can consider as a thermal charge, since it is to thermal energy transmission what electric charge is to electric energy transmission.

It is generated by two principal mechanisms:

- the transfer of thermal power by a finite temperature difference, such a process conserving thermal power.
- friction, whether produced in electric wires or mechanically, for example, such a process also conserving power.

Two additional processes produce entropy:

- chemical friction, in chemical reactions;
- the mixing of chemicals, which must be avoided in good practice in steam engines, and which Carnot calls the mixing of cold air and hot air (or steam), which produces useless calorique, that is to say entropy in modern terms. He also said in a roundabout way: temperature must change only by compression and expansion and not by conduction.

We shall not deal here with the latter two phenomena, that is, chemical phenomena, as they have been dealt with elsewhere [13, p. 74]. We shall note simply that a reversible mixture can be produced by using, in a cylinder, selective pistons which do not produce entropy, and that the combustion part of a heat engine can be represented by bond graphs in which the intensive variable is the chemical potential and the extensive variable the quantity of matter, as mentioned in Table 1.

5.3 Removing entropy

Entropy, this sort of thermal charge, is always necessary for transmitting energy in the form of thermal power, just as an electric charge is necessary for transmitting energy in the form of electric power. After its use, the thermal charge must be eliminated in a cold source, the inverse power necessary at the cold source thus limiting the power output of thermal engines.

When introducing bond graphs, we said that entropy is a simple concept resulting directly from the notion of heat as conceived by Carnot in his equation:

$$\dot{Q} = \dot{S} \cdot T$$

Although the concept of entropy is *a priori* simple, two subtleties regarding it are worth mentioning here:

- Zero entropy is a difficult notion because entropy is not well defined: in general, one considers entropy from a reference value, usually 0° or 273 K. It follows from this that the practice of postulating null entropy at absolute zero temperature must be relativised since entropy can be set in some bodies by very rapid cooling to extremely low temperatures.
- All irreversible processes occurring on earth are negligible due to the earth's ejection back into space of the immense flow of entropy that it receives from solar radiation. It follows that a calculation according to Boltzmann's formula, stipulating an equality of powers, gives a rather low temperature at the surface and a difference between surface and clouds, an essential cause of the greenhouse effect occurring in the atmospheric layer.

5.4 Carnot's efficiency formula

A reversible heat engine is an engine that does not produce new entropy in other words, that conserves entropy. No entropy leaves from the mechanical side and only the entropy that arrives must be eliminated in the cold source, which requires some power:

$$\dot{Q}_2 = T_2 \cdot \dot{S}$$

in such a way that only the difference with the power that arrives is available for conversion, which limits efficiency to

$$\eta_{Carnot} = \frac{(T_1 - T_2) \cdot S}{T_1 \cdot S} = \frac{T_1 - T_2}{T_1}$$

A point to note in passing is that all temperatures must be taken as absolute. If we take the temperatures as excesses of temperature, then we need to add a offset $T_{Offset} = 273\ K$, which gives the following formula:

$$\eta_{Carnot} = \frac{(T_1 - T_2)\cdot S}{(T_1 + T_{offset})\cdot S} = \frac{(T_1 - T_2)}{(T_1 + T_{offset})}$$

for the Carnot efficiency, expressed with temperatures in degrees centigrade (Celsius).

For example, for a heat source at 200°C and a cold source at 100°C, we can obtain a Carnot efficiency of 21%. These temperatures are typical of simple machines without a condenser in which the steam used is ejected to the environment, as in steam locomotives. This value is an upper limit, obtained in an ideal engine without losses. It shows, as Carnot pointed out, that the high temperature should be made the highest possible and the low temperature the lowest possible to obtain good levels of efficiency.

5.5 Conclusion on the Carnot cycle

Our incursion into the nascent practice of steam engines shows that Carnot's ideas were truly astounding for understanding these thermal engines. It is a theory of thermal engines for generating mechanical power that today still informs the design of modern internal combustion engines. It can be summed up by the following main points:

- Repeated or cyclic operation to obtain a machine with a useful life.
- Discharge of the used steam to the environment.
- The use of inverted sources for discharging entropy.

All theory on heat engines is deduced from the above ideas. As Carnot said: 'avoid increases in temperature by conduction and raise temperature by reducing volume. '

Appendix

2005 Edition of Carnot's *Reflections on the Motive Power of Fire*

In 2005, Dover Publications republished Carnot's work as edited by E. Mendoza [2].

We have reproduced, with the permission of the Publishers, the book's front cover. This shows Sadi Carnot at work with an oil lamp of the time.

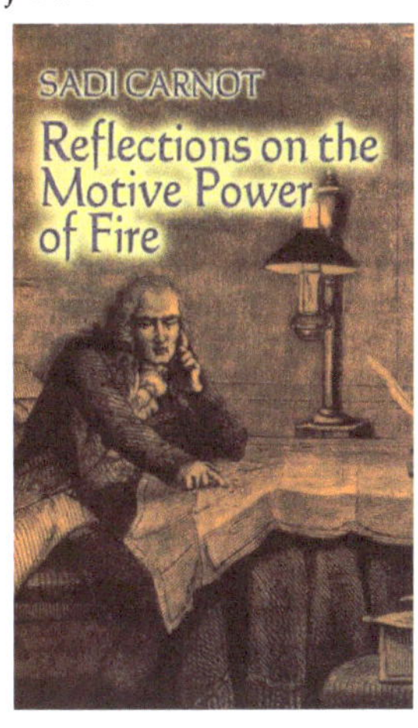

References

[1] Carnot, S., *Réflexions sur la puissance motrice du feu et sur les machines propres à développer cette puissance*, Bibliothèque du Conservatoire National des Arts et Métiers: Paris, 1824.

[2] Clapeyron, E., & Clausius, R. *Reflections on the Motive Power of Fire by Sadi Carnot and other Papers on the Second Law of Thermodynamics*, translated by E. Mendoza, Dover: New York, 1960.

[3] Callendar, H.L., *The Theory of Heat and Carnot's Principle*, Presidential Address, *Proceedings of the Physical Society, London*, **23**, pp. 153–189, 1911. (First use of entropy as thermal charge after Carnot.)

[4] Thoma, J., *Introduction to Bond Graphs and Their Applications*, Pergamon Press: Oxford, 1975.

[5] Job, G., *Neudarstellung der Wärmelehre: die Entropie als Wärme*, Academische Verlagsgesellschaft: Berlin, 1972. (Early publication with entropy as thermal charge.)

[6] Falk, G., *Theoretische Physik II – Allgemeine Dynamik*, Springer: Heidelberg, 1976. (Detailed standard text of thermodynamics with entropy as thermal charge.)

[7] Falk. G., Entropy, a resurrection of caloric – a look at the history of thermodynamics. *European Journal of Physics*, **6**, pp. 108–115, 1985. (History of thermodynamics explained in English.)

[8] Fuchs, H., *The Dynamics of Heat*, Springer: Heidelberg, 1996. (Extensive book with entropy as thermal charge, written without Bondgraphs.)

[9] Paynter, H. *Analysis and Design of Engineering Systems*, Class Notes, MIT Course 2.751, MIT Press: Cambridge, MA, 1960.

[10] Hogan, *Modeling and Simulation of Dynamic Systems*, Class Notes, MIT Course 2.141, MIT Press: Cambridge, MA, 2009. (Excellent course with deep insights.)

[11] Thoma, J., *Simulation by Bond graphs*, Springer: Heidelberg, 1990. (Thoma's course at the University of Waterloo, Canada.)

[12] Thoma, J. & Bouamana, O. *Modeling and Simulation in Thermal and Chemical Engineering*, Springer: Heidelberg, 2000. (Chemical reactions and entropy stripping.)

[13] Thoma, J. & Mocellin, G., *Simulation With Entropy in Engineering Thermodynamics – Understanding Matter and Systems With Bondgraphs*, Springer: Heidelberg, 2006. (Thermodynamics using bondgraphs and entropy as thermal charge. New standard work of the authors.)

[14] Karnopp, D., Margolis, D. & Rosenberg, R., *System Dynamics – A Unified Approach*, Wiley: New York, 1990. (Standard work on Bondgraphs by three evangelists of the method.)

[15] Dauphin-Tanguy, G. (ed.), *Les Bondgraphes*, Hermes: Paris, 2000. (Volume with articles from different fields, including electrical engineering and the use of small resistors to resolve causal conflicts.)

[16] Mocellin, G., The bond graphs method applied to social and life science, *International Conference on Bond Graph Modeling and Simulation*, Phoenix, USA, 2001.

Chapter 12

Thermodynamic entropy and temperature rigorously defined without heuristic use of the concepts of heat and empirical temperature

E. Zanchini[1] & G.P. Beretta[2]
[1]*Dipartimento di Ingegneria Industriale, Università di Bologna, Italy.*
[2]*Dipartimento di Ingegneria Meccanica e Industriale, Università di Brescia, Italy.*

1 Introduction

The physical foundations of a variety of emerging technologies — ranging from the applications of quantum entanglement in quantum information to the applications of non-equilibrium bulk and interface phenomena in microfluidics, biology, materials science, energy engineering, etc. — require understanding thermodynamic entropy beyond the equilibrium realm of its traditional definition. This article presents a rigorous logical scheme that provides a generalized definition of entropy free of the usual unnecessary assumptions, which constrain the traditional treatments to the equilibrium domain.

The dawning of thermodynamics can be traced back to the experiments of Galileo to measure the degrees of hot and cold, followed by the invention of the sealed-stem alcohol thermometer, by the Grand Duke Ferdinand II of Tuscany (1640). During the eighteenth century and the first half of the nineteenth century, the changes in temperature of physical bodies were thought to be caused by the flow of a fluid called *caloric*, which could neither be created nor destroyed.

An important step in the development of thermodynamics was the publication, in 1824, of the celebrated booklet by Carnot "*Réflections sur la puissance motrice du feu et sur les machines propres à développer cette puissance*" [1]. Carnot introduced the concept of thermodynamic cycle and stated that the highest efficiency of a cyclic heat engine is obtained when the engine is reversible. Carnot still believed in the conservation of caloric; he thought that motive power was the result of caloric "falling down" from a hot to a cold body.

The convertibility of work into heat was proved, between 1840 and 1848, by the experiments of Mayer [2] and Joule [3]. In particular, Joule showed experimentally that heat and work could produce the same effect on bodies when used in a fixed proportion. He concluded [4] that both heat and work can result in a change of something *stored* in the bodies, which is conserved. Thus, the conservation of caloric was disproved and the experimental foundations of the First Law became available.

In 1848, William Thomson, later known as Lord Kelvin, *introduced* the thermodynamic temperature scale [5]; then, in 1849 [6], he pointed out the conflict between the caloric basis of Carnot's argument, in which heat (or caloric) is conserved, and the conclusion reached by Joule, in which the sum of heat and work is conserved. In 1850, Clausius [7] reconciled Carnot's principle with Joule's results by introducing the concept that bodies possess a property, which he finally called entropy in 1865 [8], having the following characteristics: in the absence of heat exchange with other bodies, it either remains constant if the body undergoes a reversible process, or increases; during a heat exchange, entropy is transferred to or from a body in proportion to the heat transferred. The latter characteristic limits the efficiency of any work-producing cyclical engine, as required by Carnot's principle. Clausius concluded that (Clausius statement of the Second Law): *a transfer of heat from a body at any temperature to a body at higher temperature, without other external effects, is impossible.*

In 1872, Maxwell stated the transitivity of mutual thermal equilibrium (zeroth law of thermodynamics). Between 1873 and 1878, J. Willard Gibbs gave important contributions to thermodynamics: he introduced, for instance, the temperature versus entropy diagram, the concept of free energy, and the phase rule [9]. A few years later, in 1901, Gibbs presented a rigorous and general treatment of statistical mechanics [9].

In 1897, Planck [10] stated the Second Law in the form that is still used in most textbooks (and is called Kelvin–Planck's statement of the Second Law): *it is impossible to construct an engine which, working in a cycle, produces no effect except the raising of a weight and the cooling of a heat reservoir.* In 1908, Poincaré [11] presented a complete structure of classical thermodynamics, which we call *Poincaré–Planck thermodynamics*.

The basic approach of *Poincaré–Planck thermodynamics* is still used in several university textbooks, with very small changes. In this approach, the First Law is stated as follows: in a cycle, the work done by a system is proportional to the heat received by the system. In symbols, for a cycle:

$$Q = JW, \tag{1}$$

where J is a universal constant, which depends only on the system of units. With the advent of the International System of Units (SI), eqn (1) is rewritten as

$$Q = W. \tag{2}$$

Equation (2) is used to deduce that, in a process of a system A from the initial state A_1 to the final state A_2, the quantity $Q - W$ depends only on the states A_1 and A_2

Then, one defines the energy difference between A_2 and A_1 as the value of $Q - W$ for A in the process, i.e.

$$E_2^A - E_1^A = (Q - W)_{12}^A. \tag{3}$$

However, eqn (3) is clearly vitiated by a logical circularity, because it is impossible to define *heat* without a previous definition of energy.

The circularity of the logic leading to the definition of energy through eqn (3) was understood and resolved in 1909 by Carathéodory [12], who defined an *adiabatic* process without employing the concept of heat and stated the First Law as follows: *the work performed by a system in any adiabatic process depends only on the end states of the system*. However, Carathéodory's definition of adiabatic process applies only to stable equilibrium states; therefore, to extend the treatment to non-equilibrium states one needs a new definition of adiabatic process. Carathéodory also proposed a new statement of the Second Law (in terms of adiabatic accessibility), which has been used only in a few axiomatic treatments.

In 1937, Fermi [13] presented a well-known treatment of classical thermodynamics. In this treatment, Carathéodory's statement of the First Law is employed and rigorous theorems are used to define the thermodynamic temperature of a *heat source* and the entropy of a system. However, some unsatisfactory aspects still remain: the unnecessary concept of *empirical temperature* is used; the concepts of *heat* and of *heat source* are not defined rigorously; a *reversible* process is defined as a sequence of stable equilibrium states, i.e. as a *quasistatic* process. Moreover, an incompleteness in the deductive scheme is still present: it is not proved that the thermodynamic temperature of a heat source is independent of the initial state of the reference heat source used to define it. Indeed, to define the thermodynamic temperature of a heat source, Fermi considers a reversible cyclic engine which absorbs a quantity of heat Q_2 from a source at (empirical) temperature T_2 and supplies a quantity of heat Q_1 to a source at (empirical) temperature T_1. He states that if the engine performs n cycles, the quantity of heat subtracted from the first source is $n\ Q_2$ and the quantity of heat supplied to the second source is $n\ Q_1$. Thus, Fermi assumes implicitly that the quantity of heat exchanged in a cycle between a source and a reversible cyclic engine is independent of the initial state of the source. This incompleteness in the deductive scheme of Fermi's thermodynamics is resolved only in the treatment presented here.

Other well-known presentations of thermodynamics based on Carathéodory's statement of the First Law are e.g. those by Pippard [14] and by Zemansky [15]. In the latter, a definition of reversibility conceptually independent of quasistaticity is introduced.

A few decades after Fermi's contribution, two schools of thermodynamics produced relevant further developments. The *Prigogine school* [16] studied the extension of the theory to non-equilibrium states and developed the *thermodynamics of irreversible processes*, pioneered in 1931 by Onsager [17]. The *Keenan school* deepened the conceptual foundations of Carathéodory–Fermi thermodynamics and strengthened the bridge between quantum mechanics and thermodynamics. Some improvements of the logical foundations of thermodynamics due to the *Keenan school* are as follows.

Hatsopoulos and Keenan [18] analyzed deeply the meaning of Kelvin–Planck's statement of the Second Law. They pointed out that, with the term *reservoir*, Planck did not mean a system in either metastable or unstable equilibrium, but a system in *stable equilibrium*; otherwise, the statement of the Second Law would be false. However, when stable equilibrium states are defined rigorously, Kelvin–Planck's statement of the Second Law becomes a corollary of the definition. They called *stable equilibrium state* any state for which *no finite change of state of the system can occur without a corresponding finite permanent change of the state of the system's environment*; then, they stated the Second Law as follows (Hatsopoulos–Keenan statement of the Second Law):

A system having specified allowed states and an upper bound in volume can reach from any given state a stable state and leave no net effect on the environment [18, p. 34, 373]. They presented the definition of entropy in two ways: the first through the concept of heat (which they defined rigorously)and the second without the concept of heat. The latter, however, is incomplete, because according to it the entropy difference between two states of a system can be measured only by means of a *standard thermal reservoir*, chosen once and for all.

Gyftopoulos and Beretta [19] completed the definition of entropy outlined by Hatsopoulos and Keenan; they presented a treatment of thermodynamics in which the definition of entropy is not based on the concepts of heat and of quasistatic process, so that the definition applies, potentially, also to local non-equilibrium states. Recently, Beretta and Zanchini improved the treatment presented in Ref. [19] and developed a logical scheme for the definition of entropy which is outlined in Ref. [20] and is presented here in more detail.

In the present article, Ref. [19] is assumed as a starting point, but the basic definitions of system, state, isolated system, environment, process, separable system, and parameters of a system are deepened, by developing a logical scheme outlined in Ref. [21]. The operative and general definitions of these concepts as presented here are valid also in the presence of internal semipermeable walls and reaction mechanisms. The treatment of Ref. [19] is simplified by identifying the minimal set of definitions, assumptions, and theorems, which yield the definition of entropy and the principle of entropy non-decrease in the most direct way. Moreover, the definition of a reversible process is given with reference to the concept of *scenario*; the latter is the largest isolated system whose subsystems are available for interaction, for the class of processes under examination. Thus, the operativity of the definition is improved and the treatment becomes compatible also with old [22] and recent [23] interpretations of entropy and irreversibility in the quantum theoretical framework.

2 Aims and structure of the present treatment

In this article, we present a treatment of the foundations of thermodynamics, focused on generalizing the definition of thermodynamic entropy, which is free of conceptual loops, as well as of undefined or unnecessary concepts.

First of all, we state operative definitions of all the basic concepts employed in the treatment, such as those of system, state of a system, isolated system, environment of a system, system separable from its environment, system uncorrelated with its environment, and reversible process. Our definitions are completely general: they apply also to systems with movable internal semipermeable walls and allowed chemical reactions, as well as to systems contained in electric, magnetic, and/or gravitational force fields. They hold both for many particle systems and for few particle systems.

To simplify the treatment, after stating the basic definitions, we consider here only closed systems and states of a closed system A in which A is separable and uncorrelated from its environment. This restriction is implicit in all traditional treatments of thermodynamics. We then define a *weight process*, i.e. a process such that the only net effect in the system's environment is the change in the level of a weight in a gravity field. We use the weight process, instead of the commonly employed adiabatic process, because it is less restrictive (it concerns only the end states of system and environment) and has a simple and rigorous definition.[1]

In Section 4, by employing the concept of weight process, we state the First Law and present the definition of energy. As in Ref. [19], we also prove that the First Law entails, as a consequence, the impossibility of a perpetual motion machine of the first kind (PMM1). Since our statement of the First Law is completely independent of the concept of heat (which we never use, either explicitly or implicitly), we break the first conceptual loop present in Poincaré's treatment of thermodynamics: the concept of heat employed to define that of energy. Moreover, we extend the definition of energy given by Carathéodory to non-equilibrium states.

In Section 5, we present our definition of thermodynamic entropy, which employs the auxiliary concepts of thermal reservoir and of temperature of a thermal reservoir. We do not use the Zeroth Law and the empirical temperature, for the following reasons:

- as we prove in Section 7, the Zeroth Law is not an independent law in our exposition of thermodynamics, but a consequence of the First Law, the Second Law, and some auxiliary assumptions (these assumptions are used, almost always implicitly, in every treatment of thermodynamics: in our treatment, for the sake of logical clarity, we make them stand out explicitly);
- as is also shown in Refs. [18] and [19], empirical temperature is unnecessary for the definition of temperature of a thermal reservoir;
- in traditional treatments of thermodynamics, based on the Zeroth Law and on the empirical temperature, first *thermometers are used to define* several different empirical temperature scales, then the absolute scale of an ideal gas thermometer is selected, then the thermodynamic temperature is defined, then it is proved that the absolute scale of an ideal gas thermometer coincides with the thermodynamic temperature; finally, all the empirical temperature scales are rejected and the thermodynamic temperature is chosen, *because it is independent of the properties of thermometers*; this long circuit is useless and confusing.

In our treatment, it is clear that thermometers are not involved in the definition of temperature of a thermal reservoir, nor in the definition of temperature of a system in a stable equilibrium state given in Section 6; they emerge only later as *instruments for practical indirect measurements of the temperature.*

Following Refs. [18] and [19], we base our definitions of thermodynamic entropy and temperature on the concept of *stable equilibrium state*. For an isolated system A, a state A_e is called an *equilibrium state* if it is time invariant. For a non-isolated system A, a state A_e is called an equilibrium

[1] By the same method, we could define an *adiabatic process* as a process such that, at every time instant, the only effect of the process in the system's environment is the change in the level of a weight in a gravity field. This definition is not employed in the present paper, but we note that it is valid even if A evolves through nonequilibrium states, like the definition of weight process that we do employ. On the contrary, Carathe´odory's definition of an adiabatic process holds only for a quasistatic process; the whole treatment of Carathe´odory's thermodynamics holds only for stable equilibrium states.

state if upon isolation of the system it is possible to reproduce the same state A_e and such state is time invariant. An equilibrium state of A is called a *stable equilibrium state* if it cannot be changed without either a permanent change of the region of space occupied by A or a permanent change of the state of the environment of A.

We state the Second Law as follows: among all the states of a system A such that the constituents of A are contained in a given set of regions of space $\boldsymbol{R}^A$, there is a stable equilibrium state for every value of the energy E^A. Then, we prove by an easy Lemma, that for fixed $\boldsymbol{R}^A$ and E^A, the stable equilibrium state of A is unique. By means of the definition of stable equilibrium state and of the Second Law, we obtain, through a simple theorem, a broad extension of the Kelvin–Planck statement of the Second Law: for every set of regions of space $\boldsymbol{R}^A$ which contain the matter of A and for every value of the energy of A, there exists a unique state of A such that it is impossible to lower the energy of A and raise a weight without either a permanent change of the regions of space $\boldsymbol{R}^A$ or other effects external to A (impossibility of a perpetual motion machine of the second kind, PMM2).

At this stage, we define a thermal reservoir as a system R contained in a fixed region of space, such that any pair of identical copies of the reservoir, R and R^d, is in mutual stable equilibrium when R and R^d are in stable equilibrium states. Then, by means of two basic theorems, we define the temperature of a thermal reservoir and the entropy difference between any pair of states (A_1, A_2) of any system A. We also prove the additivity of entropy differences, the principle of entropy non-decrease, and the highest entropy principle.

In Section 6, we prove the existence of the fundamental relation for the stable equilibrium states of any closed system A, and finally define temperature (as well as pressure, and other generalized forces). In Section 7, we prove that two closed systems A and B, with given regions of space $\boldsymbol{R}^A$ and $\boldsymbol{R}^B$ occupied by the constituents of A and of B, are in mutual stable equilibrium if and only if they have the same temperature. A proof of the Zeroth Law follows as a straightforward corollary.

As a result, we obtain a complete, logically sound, rigorous treatment of the foundations of thermodynamics, without undefined concepts, without conceptual loops (such as, heat in the definition of energy), without restrictions to some kind of system (such as systems without internal constraints, or systems without external force fields, or systems with a very large numbers of particles), without restrictions to stable equilibrium states (such as those which appear in Carathéodory's treatment), and without the use of unnecessary concepts (such as those of heat and of empirical temperature).

3 Basic definitions

3.1 Constituents and amounts of constituents

We call *constituents* the material particles chosen to describe the matter contained in any region of space R, at a time instant t. Examples of constituents are: atoms, molecules, ions, protons, neutrons, and electrons. Constituents may combine and/or transform into other constituents according to a set of model-specific *reaction mechanisms*. We call *amount of constituent i* in any region of space R, at a time instant t, the number of particles of constituent i contained in R, at time t.

3.2 Region of space which contains particles of the *i*th constituent

We call *region of space which contains particles of the* ith *constituent* a connected region R_i of physical space (the three-dimensional Euclidean space) in which particles of the *i*th constituent are contained. The boundary surface of R_i may be a patchwork of *walls*, i.e. surfaces impermeable to particles of the *i*th constituent, and ideal surfaces (permeable to particles of the *i*-th constituent). The geometry of the boundary surface of R_i and the permeability features of its component walls and ideal surfaces can vary in time, and so can the number of particles contained in R_i.

3.3 Collection of matter and composition

We call *collection of matter*, denoted by C^A, a set of particles of one or more constituents which is described by specifying the allowed reaction mechanisms between different constituents and, at any time instant t, the set of r connected regions of space, R^A = \textsl = $R_1^A,\ldots, R_i^A,\ldots,R_r^A$, each of which contains n_i^A particles of a single kind of constituent. The regions of space $\boldsymbol{R}^A$ can vary in time and overlap. Two regions of space may contain the same kind of constituent provided that they do not overlap. Thus, the *i*-th constituent could be identical to the *j*-th constituent, provided that R_i^A and R_j^A are disjoint. If, due to changes with time, two regions of space which contain the same kind of constituent begin to overlap, then starting from that instant a new collection of matter must be considered. Similarly, the collection must be redefined when one region splits into two or more disjoint regions.

3.3.1 *Comment*

This method of description allows us to consider from the outset the presence of internal walls and/or internal *semipermeable* membranes, i.e. surfaces that which can be crossed only by some kinds of constituents and not others. An example of the method is illustrated in Fig. 1: a collection of matter C^A with constituents O_2 and N_2, with a movable external wall and with two movable internal membranes, permeable to O_2 and to N_2, respectively, is represented by two overlapping regions of space, R_1^A and R_2^A, each bounded by a movable wall: R_1^A contains O_2, while R_2^A contains N_2.

In the simplest case of a collection of matter without internal partitions, the regions of space $\boldsymbol{R}^A$ coincide at every time instant.

The amount n_i of the constituent in the *i*th region of space can vary in time for two reasons:

- *Matter exchange*: during a time interval in which the boundary surface of R_i is not entirely a wall, particles may be transferred into or out of R_i; we denote by $\dot{\boldsymbol{n}}^{A\leftarrow}$ the set of time rates at which particles are transferred in or out of each region, assumed positive if inward, negative if outward.
- *Reaction mechanisms*: in a portion of space where two or more regions overlap, the allowed reaction mechanisms may combine particles of one or more regions and then transform into particles of one or more other regions, according to well-specified proportions (e.g. stoichiometry).

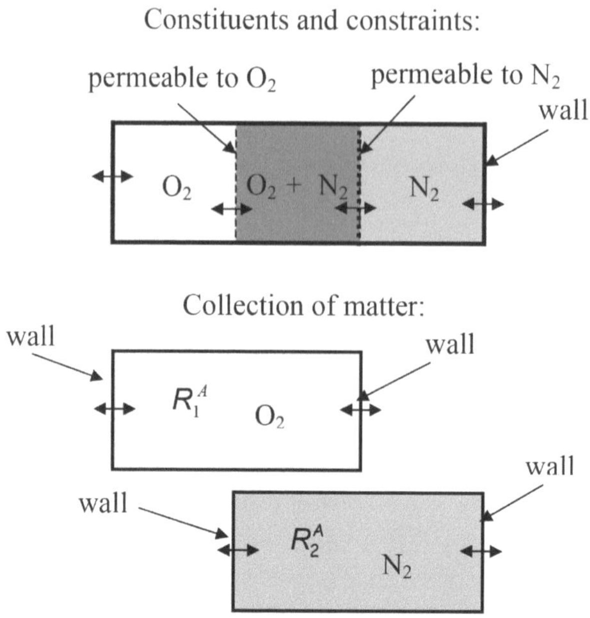

Figure 1: Collection of matter with constituents O_2 and N_2, with two movable internal membranes: the overlapping regions of space R_1^A and R_2^A are split for clarity.

3.4 Compatible compositions: set of compatible compositions

We say that two compositions, $\boldsymbol{n}^{1A}$ and $\boldsymbol{n}^{2A}$, of a given collection of matter C^A are *compatible*, if the change between $\boldsymbol{n}^{1A}$ and $\boldsymbol{n}^{2A}$ or vice versa can take place as a consequence of the allowed reaction mechanisms without matter exchange. We call *set of compatible compositions* for a system A the set of all the compositions of A, which are compatible with a given one. We denote a set of compatible compositions for A by the symbol $(\boldsymbol{n}^{0A}, \nu^A)$. By this we mean that the set of τ allowed reaction mechanisms is defined like for chemical reactions by a matrix of stoichiometric coefficients $\nu^A = [\nu_k^{(\ell)}]$, with $\nu_k^{(\ell)}$ representing the stoichiometric coefficient of the k-th constituent in the l-th reaction. The set of compatible compositions is a τ-parameter set defined by the reaction coordinates $\varepsilon^A = \varepsilon_1^A, \ldots, \varepsilon_\ell^A, \ldots, \varepsilon_\tau^A$ through the proportionality relations

$$\boldsymbol{n}^A = \boldsymbol{n}^{0A} + \nu^A \cdot \varepsilon^A, \tag{4}$$

where $\boldsymbol{n}^{0A}$ denotes the composition corresponding to the value zero of all the reaction coordinates ε^A. To fix ideas and for convenience, we select $\varepsilon^A = 0$ at time $t = 0$ so that $\boldsymbol{n}^{0A}$ is the composition at time $t = 0$ and we may call it the *initial composition*.

3.5 External force field

Let us denote by **F** a force field given by the superposition of the gravitational field **G**, the electric field **E**, and the magnetic field **H**. Let us denote by Σ_t^A the union of all the regions of space $\boldsymbol{R}_t^A$ in which the constituents of C^A are contained, at a time instant t, which we also call region of space occupied by C^A at time t. Let us denote by Σ^A the union of the regions of space Σ_t^A, i.e. the union of all the regions of space occupied by C^A during its time evolution.

We call *external force field for* C^A *at time* t, denoted by $\mathbf{F}_{e,t}^A$ the spatial distribution of $\mathbf{F}$ which is measured at time t in Σ_t^A if all the constituents and the walls of C^A are removed and placed far away from Σ_t^A.

We call *external force field for* C^A, denoted by $\mathbf{F}_e^A$, the spatial and time distribution of $\mathbf{F}$ which is measured in Σ^A if all the constituents and the walls of C^A are removed and placed far away from Σ^A.

3.6 System: properties of a system

We call *system* A a collection of matter C^A defined by the initial composition $\boldsymbol{n}^{0A}$, the stoichiometric coefficients ν^A of the allowed reaction mechanisms, and the possibly time-dependent specification, *over the entire time interval of interest*, of:

- the geometrical variables and the nature of the boundary surfaces that define the regions of space $\boldsymbol{R}_t^A$,
- the rates $\dot{\boldsymbol{n}}_t^{A\leftarrow}$ at which particles are transferred in or out of the regions of space, and
- the external force field $\mathbf{F}_e^A$ for C^A,

provided that the following conditions apply:

1. an ensemble of identically prepared replicas of C^A can be obtained at any instant of time t, according to a specified set of instructions or preparation scheme;
2. a set of measurement procedures, $P_1^A, \ldots, P_n^A$, exists, such that when each P_i^A is applied on replicas of C^A at any given instant of time t, the arithmetic mean $\langle P_i^A \rangle_t$ of the numerical outcomes of repeated applications of P_i^A is a value which is the same for every subensemble of replicas of C^A (the latter condition guarantees the so-called statistical *homogeneity* of the ensemble); $\langle P_i^A \rangle_t$ is called the *value of* P_i^A for C^A at time t;
3. the set of measurement procedures, $P_1^A, \ldots, P_n^A$, is *complete* in the sense that the set of values $\{\langle P_1^A \rangle_t, \ldots, \langle P_n^A \rangle_t\}$ allows to predict the value of any other measurement procedure satisfying conditions 2 and 3.

Then, each measurement procedure satisfying conditions 2 and 3 is called a *property* of system A, and the set $P_1^A, \ldots, P_n^A$ a *complete set of properties* of system A.

The amounts of constituents, n_t^A, are properties according to the above definition, but, for clarity, in the following definition of state of a system we list them separately and explicitly.

3.7 State of a system

Given a system A as just defined, we call *state of system A at time t*, denoted by A_t, the set of the values *at time t* of

- all the properties of the system or, equivalently, of a complete set of properties, $\{\langle P_1\rangle_t,\ldots,\langle P_n\rangle_t\}$;
- the amounts of constituents, n_t^A;
- the geometrical variables and the nature of the boundary surfaces of the regions of space, $\mathbf{R}_t^A$;
- the rates $\dot{\boldsymbol{n}}_t^{A\leftarrow}$ of particle transfer in or out of the regions of space; and
- the external force field distribution for A at time t, $\mathbf{F}_{e,t}^A$.

With respect to the chosen complete set of properties, we can write

$$A_t \equiv \left\{\langle P_1\rangle_t,\ldots,\langle P_n\rangle_t; n_t^A; \mathbf{R}_t^A; \dot{n}_t^{A\leftarrow}{}_t; \mathbf{F}_{e,t}^A\right\}. \tag{5}$$

For shorthand, states $A_{t_1}, A_{t_2},\ldots,$ are denoted by $A_1, A_2, \ldots$. Also, when the context allows it, the value $\langle P^A\rangle_{t_1}$ of property P^A of system A at time t_1 is denoted depending on convenience by the symbol P_1^A, or simply P_1.

3.8 Closed system and open system

A system A is called a *closed system* if, at every time instant t, the boundary surface of every region of space R_{it}^A is a wall. Otherwise, A is called an *open system*.

3.8.1 Comment

For a closed system, in each region of space R_i^A, the number of particles of the i-th constituent can change only as a consequence of allowed reaction mechanisms.

3.9 Composite system and subsystems

Given a system C in the external force field $\mathbf{F}_e^C$, we say that C is the *composite* of systems A and B, denoted AB, if (a) there exists a pair of systems A and B such that the external force field which obtains when both A and B are removed and placed far away coincides with $\mathbf{F}_e^C$; (b) no region of space R_i^A overlaps with any region of space R_j^B; and (c) the $r_C = r_A + r_B$ regions of space of C are $R^C = R_1^A,\ldots, R_i^A,\ldots,R_{r_A}^A, R_1^B,\ldots, R_j^B,\ldots,R_{r_B}^B$. Then, we say that A and B are *subsystems* of the *composite system* C, and we write $C = AB$ and denote its state at time t by $C_t = (AB)_t$.

3.10 Isolated system

We say that a closed system I is an isolated system in the stationary external force field $\mathbf{F}_e^I$, or simply an *isolated system*, if, during the whole time evolution of I: (a) only the particles of I are present in Σ^I; (b) the external force field for I, $\mathbf{F}_e^I$, is stationary, i.e. time independent.

3.10.1 Comment

In simpler words, a system I is isolated if, at every time instant: no other material particle is present in the whole region of space Σ^I which will be crossed by system I during its time evolution; if system I is removed, only a stationary (vanishing or non-vanishing) force field is present in Σ^I.

3.11 Separable closed systems

Consider a composite system AB, with A and B closed subsystems. We say that systems A and B are *separable* at time t if:

- the external force field for A at time t coincides (where defined) with the external force field for AB at time t, i.e. $\mathbf{F}_{e,t}^{A} = \mathbf{F}_{e,t}^{AB}$;
- the external force field for B at time t coincides (where defined) with the external force field for AB at time t, i.e. $\mathbf{F}_{e,t}^{B} = \mathbf{F}_{e,t}^{AB}$

3.11.1 Comment

In simpler words, system A is separable from B at time t, if at that instant the force field produced by B is vanishing in the region of space occupied by A and vice versa. During the subsequent time evolution of AB, A and B need not remain separable at all times.

3.12 Subsystems uncorrelated from each other

Consider a composite system AB such that at time t the states A_t and B_t of the two subsystems fully determine the state $(AB)_t$, i.e. the values of all the properties of AB can be determined by *local* measurements of properties of systems A and B. Then, at time t, we say that the states of subsystems A and B are *uncorrelated from each other*, and we write the state of AB as $(AB)_t = A_tB_t$. We also say, for brevity, that A and B are *systems uncorrelated from each other* at time t.

3.13 Subsystems correlated with each other

If at time t the states A_t and B_t do not fully determine the state $(AB)_t$ of the composite system AB, we say that A_t and B_t are *states correlated with each other*. We also say that A and B are *systems correlated with each other* at time t.

Comment. Two systems A and B that are uncorrelated from each other at time t_1 can undergo an interaction such that they are correlated with each other at time $t_2 > t_1$.

Comment. Correlations between isolated systems. Let us consider an isolated system $I = AB$ such that, at time t, system A is separable and uncorrelated from B. This circumstance does not exclude that, at time t, A and/or B (or both) may be correlated with a system C, even if the latter is isolated, e.g. it is far away from the region of space occupied by AB. Indeed, our definitions of separability and correlation are general enough to be fully compatible with the notion of quantum correlations, i.e. *entanglement*, which plays an important role in modern physics. In other words, assume that an isolated system U is made of three subsystems A, B and C, i.e. $U = ABC$, with C isolated and AB isolated. The fact that A is uncorrelated from B, so that according to our notation we may write $(AB)t = A_tB_t$, does not exclude that A and C may be entangled, in such a way that the states A_t and C_t do not determine the state of AC, i.e. $(AC)_t \neq A_tC_t$, *nor* we can write $U_t = (A)_t(BC)_t$.

3.14 Environment of a system and scenario

If for the time span of interest, a system A is a subsystem of an isolated system $I = AB$, we can choose AB as the isolated system to be studied. Then, we call B the *environment* of A, and we call AB the *scenario* under which A is studied.

Comment. The chosen scenario AB contains as subsystems all and only the systems that are allowed to interact with A; thus, all the remaining systems in the universe, even if correlated with AB, are considered as not available for interaction.

3.15 Process and cycle

We call *process* for a system A from state A_1 to state A_2 in the scenario AB, denoted by $(AB)_1 \rightarrow (AB)_2$, the time evolution from $(AB)_1$ to $(AB)_2$ of the isolated system AB which defines the scenario. We call *cycle* for a system A a process whereby the final state A_2 coincides with the initial state A_1.

3.16 Process between uncorrelated states and external effects

A process in the scenario AB in which the end states of system A are both uncorrelated from its environment B is called *process between uncorrelated states* and denoted by $\Pi_{12}^{A,B} \equiv (A_1 \rightarrow A_2)_{B_1 \rightarrow B_2}$. In such a process, the change of state of the environment B from B_1 to B_2 is called *effect external to A*. Traditional expositions of thermodynamics consider only this kind of process.

3.17 Reversible process and reverse of a reversible process

A process for A in the scenario AB, $(AB)_1 \rightarrow (AB)_2$, is called a *reversible process* if there exists a process $(AB)_2 \rightarrow (AB)_1$ which restores the initial state of the isolated system AB. The process $(AB)_2 \rightarrow (AB)_1$ is called *reverse* of process $(AB)_1 \rightarrow (AB)_2$. In other words, a process of an isolated system $I = AB$ is reversible if it can be reproduced as a part of a cycle of the isolated system I. For a reversible process between uncorrelated states, $\Pi_{12}^{A,B} \equiv (A_1 \rightarrow A_2)_{B_1 \rightarrow B_2}$, the *reverse* will be denoted by $-\Pi_{12}^{A,B} \equiv (A_2 \rightarrow A_1)_{B_2 \rightarrow B_1}$.

Comment. The reverse process may be achieved in more than one way (in particular, not necessarily by retracing the sequence of states $(AB)_t$, with $t_1 \leq t \leq t_2$, followed by the isolated system AB during the forward process).

Comment. The reversibility in one scenario does not grant the reversibility in another. If the smallest isolated system which contains A is AB and another isolated system C exists in a different region of space, one can choose as environment of A either B or BC. Thus, the time evolution of A can be described by the process $(AB)_1 \to (AB)_2$ in the scenario AB or by the process $(ABC)_1 \to (ABC)_2$ in the scenario ABC. For instance, the process $(AB)_1 \to (AB)_2$ could be ireversible; however, by broadening the scenario so that interactions between AB and C become available, a reverse process $(ABC)_2 \to (ABC)_1$ may become possible. On the other hand, a process $(ABC)_1 \to (ABC)_2$ could be irreversible on account of an irreversible evolution $C_1 \to C_2$ of C, even if the process $(AB)_1 \to (AB)_2$ is reversible in the scenario AB.

Comment. A reversible process need not be slow. In the general framework, we are setting up, it is noteworthy that nowhere we state nor do we need the concept that a process to be reversible needs to be *slow* in some sense. Actually, as well represented in Ref. [19] and clearly understood within dynamical systems models based on linear or non-linear master equations, the time evolution of the state of a system is the result of a competition between (hamiltonian) mechanisms, which are reversible and (dissipative) mechanisms which are not. Therefore, to design a reversible process in the non-equilibrium domain, we most likely need a *fast* process, whereby the state is changed quickly by a fast hamiltonian dynamics, leaving negligible time for the dissipative mechanisms to produce irreversible effects.

3.18 Restriction to closed systems, separable and uncorrelated from their environments

In the following, to simplify our treatment, we will consider only *closed systems* and only states of a closed system A in which A is separable and uncorrelated from its environment. Moreover, for a composite system AB, we will consider only states such that the subsystems A and B are separable and uncorrelated from each other.

Comment. This restriction is an implicit assumption in all traditional treatments of thermodynamics. Here, rather than an assumption, we state it to delimit the scope of the paper. Elsewhere, we will show how the treatment can be extended to open systems, and to non-separable and correlated systems.

3.19 Weight

We call *weight* a system M always separable and uncorrelated from its environment, such that:

- M is closed, it has a single constituent contained in a single region of space whose shape and volume are fixed.
- It has a constant mass m.
- In any process, the difference between the initial and the final state of M is determined uniquely by the change in the position z of the center of mass of M, which can move only along a straight line whose direction is identified by the unit vector $\mathbf{k} = \nabla z$.
- Along the straight line, there is a uniform stationary external gravitational force field $\mathbf{G}_e = -g\mathbf{k}$, where g is a constant gravitational acceleration.

As a consequence, the difference in potential energy between any initial and final states of M is given by $mg(z_2 - z_1)$.

3.20 Weight process and work in a weight process

A process of a system A is called a *weight process*, denoted by $(A_1 \to A_2)^W$, if the only effect external to A is the displacement of the center of mass of a weight M between two positions z_1 and z_2. We call *work performed by A in the weight process*, denoted by the symbol $W_{12}^{A\to}$, the quantity

$$W_{12}^{A\to} = mg(z_2 - z_1) \ . \tag{6}$$

We call work *received by A in the weight process* the opposite of this work; two equivalent notations for this work are $-W_{12}^{A\to} = W_{12}^{A\leftarrow}$.

4 Definition of energy for a closed system

4.1 First Law

Every pair of states (A_1, A_2) of a closed system A can be interconnected by means of a weight process for A. The works performed by the system in any two weight processes between the same initial and final states are identical.

4.2 Definition of energy for a closed system. Proof that it is a property

Let (A_1, A_2) be any pair of states of a closed system A. We call *energy difference* between states A_2 and A_1 either the work $W_{12}^{A\leftarrow}$ received by A in any weight process from A_1 to A_2 or the work $W_{21}^{A\leftarrow}$ done by A in any weight process from A_2 to A_1; in symbols:

$$E_2^A - E_1^A = W_{12}^{A\leftarrow} \quad \text{or} \quad E_2^A - E_1^A = W_{21}^{A\to} . \tag{7}$$

The First Law guarantees that at least one of the weight processes considered in eqn (7) exists. Moreover, it entails the following consequences:

(a) If both weight processes $(A_1 \to A_2)^W$ and $(A_2 \to A_1)^W$ exist, the two forms of eqn (7) yield the same result, i.e. $W_{12}^{A\leftarrow} = W_{21}^{A\to}$;

(b) The energy difference between two states A_2 and A_1 depends only on the states A_1 and A_2.

(c) (*Additivity of energy differences*) consider a pair of states A_1B_1 and A_2B_2 of the composite system AB; then

$$E_2^{AB} - E_1^{AB} = E_2^A - E_1^A + E_2^B - E_1^B \ ; \tag{8}$$

(d) (*Energy is a property*) let A_0 be a reference state of a closed system A, to which we assign an arbitrarily chosen value of energy E_0^A; the value of the energy of A in any other state A_1 is determined uniquely by the equation

$$E_1^A = E_0^A + W_{01}^{A\leftarrow} \quad \text{or} \quad E_1^A = E_0^A + W_{10}^{A\rightarrow}, \tag{9}$$

where $W_{01}^{A\leftarrow}$ or $W_{10}^{A\rightarrow}$ is the work in any weight process for A either from A_0 to A_1 or from A_1 to A_0; therefore, energy is a property of A.

Rigorous proofs of these consequences can be found in Refs. [19,21,24], and will not be repeated here.

Theorem 1. Impossibility of PMM1. The work performed by a system in any cyclic weight process is zero.

Proof. If the final state A_2 of a system A coincides with the initial state A_1, then A_2 and A_1 can be interconnected by a zero-work weight process for A, in which nothing happens. On account of the First Law, the work is zero in any other weight process for A from A_1 to A_2 or from A_2 to A_1.

5 Definition of thermodynamic entropy for a closed system

Equilibrium state of a closed system. A state A_t of a closed system A, with environment B, is called an *equilibrium state* if:

- state A_t is stationary, i.e. the time derivatives of the properties of A, at time t, are vanishing;
- state A_t can be reproduced, as a stationary state, while A is an isolated system in the force field external to AB, $\mathbf{F}_e^{AB}$.

Stable equilibrium state of a closed system. An equilibrium state of a closed system A is called a *stable equilibrium state* if it cannot be modified in any process in which neither the geometrical configuration of the walls, which bind the regions of space $\boldsymbol{R}^A$ where the constituents of A are contained, nor the state of the environment B of A have net changes.

Assumption 1: restriction to normal systems. We call *normal system* any system A that, starting from every state, can be changed to a non-equilibrium state with higher energy by means of a weight process for A in which the regions of space $\boldsymbol{R}^A$ occupied by the constituents of A have no net change.

From here on, we consider only normal systems; even when we say only *system* we mean a *normal system*.

Comment. For a normal system, the energy is unbounded from above; the system can accommodate an indefinite amount of energy, such as when its constituents have translational, rotational, or vibrational degrees of freedom. In traditional treatments of thermodynamics, Assumption 1 is *not stated explicitly, but it is used*, for example, when one states that any amount of work can be transferred to a thermal reservoir by a stirrer. Notable exceptions to this assumption are important quantum theoretical model systems, such as spins, qubits, qudits, etc., whose energy is bounded from above. The extension of our treatment to such so-called *special systems* is straightforward, but we omit it here for simplicity.

Second Law. Among all the states of a system A such that the constituents of A are contained in a given set of regions of space $\boldsymbol{R}^A$, there is a stable equilibrium state for every value of the energy E^A.

Lemma 1. Uniqueness of the stable equilibrium states. There can be no pair of different stable equilibrium states of a given closed system A with identical regions of space $\boldsymbol{R}^A$ and the same value of the energy E^A.

Proof. If two such states existed, by the First Law and the definition of energy they could be interconnected by means of a zero-work weight process. Therefore, at least one of them could be changed to a different state with no external effect, and, hence, would not satisfy the definition of stable equilibrium state.

Lemma 2. State Principle. The stable equilibrium states of a given closed system A are uniquely identified by the regions of space $\boldsymbol{R}^A$ occupied by the constituents of A and by the value of the energy E^A, i.e. there exists a single-valued function

$$A_{se} = A_{se}\,(E^A, \boldsymbol{R}^A), \tag{10}$$

where A_{se} denotes the state of A in the sense of eqn (5).

Proof. The thesis follows immediately from the Second Law and Lemma 1.

Comment. Recall that the composition n^A belongs to the set of compatible compositions $(\boldsymbol{n}^{0A}, \nu^A)$ fixed once and for all by the definition of the system.

Theorem 2. Impossibility of a PMM2. If a system A is in a stable equilibrium state, it is impossible to lower its energy by means of a weight process for A in which the regions of space $\boldsymbol{R}^A$ occupied by the constituents of A have no net change.

Proof. Suppose that, starting from a stable equilibrium state A_{se} of A, by means of a weight process $\prod_1$ with positive work $W^{A\rightarrow} = W > 0$, the energy of A is lowered and the regions of space $\boldsymbol{R}^A$ occupied by the constituents of A have no net change. On account of Assumption 1, it would be possible to perform a weight process $\prod_2$ for A in which the regions of space $\boldsymbol{R}^A$ occupied by the constituents of A have no net change, the weight M is restored to its initial state so that the positive amount of energy $W^{A\leftarrow} = W > 0$ is supplied back to A, and the final state of A is a non-equilibrium state, namely, a state clearly different from A_{se}. Thus, the zero-work sequence of weight processes $(\prod_1, \prod_2)$ would violate the definition of stable equilibrium state.

Comment. Extension of the Kelvin–Planck statement of the Second Law. Theorem 2 proves that the impossibility of a PMM2 (perpetual motion of the second kind) is a corollary of the definition of stable equilibrium state and of Assumption 1 (used implicitly in all traditional treatments), as noted in Refs [18,19, p. 64]. The definition of stable equilibrium state, Assumption 1 and our statement of the Second Law yield an important extension of the Kelvin–Planck statement. They prove that, for any normal system A, for every value of the energy of A and every position of the walls (or semipermeable membranes) which constrain the matter of A, there exists a state of A such that it is impossible to lower the energy of A and raise a weight without other effects.

Lemma 3. Any stable equilibrium state A_{se} of a system A is accessible via an irreversible zero-work weight process from any other state A_1 in which A has the same regions of space $\boldsymbol{R}^A$ and the same value of the energy E^A.

Proof. By the First Law and the definition of energy, A_{se} and A_1 can be interconnected by a zero-work weight process for A. However, a zero-work weight process from A_{se} to A_1 would violate the definition of stable equilibrium state. Therefore, the process must be in the direction from A_1 to A_{se}. The absence of a zero-work weight process in the opposite direction, implies that any zero-work weight process from A_1 to A_{se} is irreversible.

Corollary 1. Any state of a system A can be changed to a unique stable equilibrium state by means of a zero-work weight process for A in which the regions of space $\boldsymbol{R}^A$ have no net change.

Proof. The thesis follows immediately from the Second Law, Lemma 1, and Lemma 3.

Systems in mutual stable equilibrium. We say that two systems A and B, each in a stable equilibrium state, are in mutual stable equilibrium if the composite system AB is in a stable equilibrium state.

Identical copy of a system. We say that a system A^d, always separable from A and uncorrelated with A, is an *identical copy* of system A (or, a *duplicate* of A) if, at every time instant:

- the difference between the set of regions of space $\boldsymbol{R}^{A^d}$ occupied by the matter of A^d and that $\boldsymbol{R}^A$ occupied by the matter of A is only a rigid translation $\Delta\mathbf{r}$ with respect to the reference frame considered, and the composition of A^d is compatible with that of A;
- the external force field for A^d at any position $\mathbf{r} + \Delta\mathbf{r}$ coincides with the external force field for A at the position $\mathbf{r}$.

Identical states of a system A and of an identical copy of A. We will say that the states A_1 of a system A and A_1^d of an identical copy of A are *identical states* if the value of any property of A in state A_1 coincides with the value of the same property for A^d in state A_1^d, except for a rigid translation $\Delta\mathbf{r}$ of the regions of space $\boldsymbol{R}^{A^d}$ with respect to $\boldsymbol{R}^A$.

Thermal reservoir. We call thermal reservoir a system R with a single constituent, contained in a fixed region of space, with a vanishing external force field, with energy values restricted to a finite range such that in any of its stable equilibrium states, R is in mutual stable equilibrium with an identical copy of R, R^d, in any of its stable equilibrium states.

Comment. Every single-constituent system without internal boundaries and applied external fields, and with a number of particles of the order of 1 mol (so that the *simple system* approximation as defined in Ref. [19, p. 263] applies), when restricted to a fixed region of space of appropriate volume and to the range of energy values corresponding to the so-called *solid–liquid–vapor triple-point* stable equilibrium states, is a an excellent approximation of a thermal reservoir.

Reference thermal reservoir. A thermal reservoir chosen once and for all will be called a *reference thermal reservoir*. To fix ideas, we will see below that in the International System of Units, we choose as a practical and easily reproducible approximation of a reference thermal reservoir, one having water as constituent, with a volume, an amount and a range of energy values which correspond to the so-called *solid–liquid–vapor triple-point* stable equilibrium states.

Standard weight process. Given a pair of states (A_1, A_2) of a system A and a thermal reservoir R, we call *standard weight process* for AR from A_1 to A_2 a weight process for the composite system AR in which the end states of R are stable equilibrium states. We denote by $(A_1R_1 \to A_2R_2)^{\mathrm{sw}}$ a standard weight process for AR in which A goes from A_1 to A_2 and by $\left(\Delta E^R\right)^{\mathrm{SW}}_{A_1A_2}$ the corresponding energy change of the thermal reservoir R.

Assumption 2. Every pair of states (A_1, A_2) of a system A can be interconnected by a reversible standard weight process for AR, where R is an arbitrarily chosen thermal reservoir.

Comment. Statements of the Second Law. The combination of Assumption 2 with the statement of the Second Law and Lemma 1 given above forms our re-statement of the *Gyftopoulos–Beretta statement of the Second Law* [19, p. 62, 63], which, in turn, is a restatement of that introduced by Hatsopoulos and Keenan [18, p. 34, 373]. Our motivation for separating the statement proposed in Ref. [19] in three parts is twofold: on one hand, it allows us to emphasize that the uniqueness of the stable equilibrium states (which in Ref. [19] is a part of the postulate) can be proved; on the other hand, it allows us to separate *logically independent* assumptions, i.e. assumptions such that a violation of the first would not imply a violation of the second, and *vice versa*.

In addition to the Kelvin–Planck statement discussed above, also the well-known historical statements due to Clausius and to Carathéodory unfold as rigorous theorems in our logical scheme. Proofs can be found in Ref. [19, p. 64, 121, 133], as well as in Section 7 of this paper.

Theorem 3. For a given system A and a given reservoir R, among all the standard weight processes for AR between a given pair of states (A_1, A_2), the energy change $\left(\Delta E^R\right)^{\mathrm{SW}}_{A_1A_2}$ of the thermal reservoir R has a lower bound which is reached if and only if the process is reversible.

Proof. Let $\prod_{AR}$ denote a standard weight process for AR from A_1 to A_2, and $\prod_{AR\mathrm{rev}}$ a reversible one; the energy changes of R in processes $\prod_{AR}$ and $\prod_{AR\mathrm{rev}}$ are, respectively, $\left(\Delta E^R\right)^{\mathrm{SW}}_{A_1A_2}$ and $\left(\Delta E^R\right)^{\mathrm{swrev}}_{A_1A_2}$. With the help of Fig. 2, we will prove that, regardless of the initial state of R:

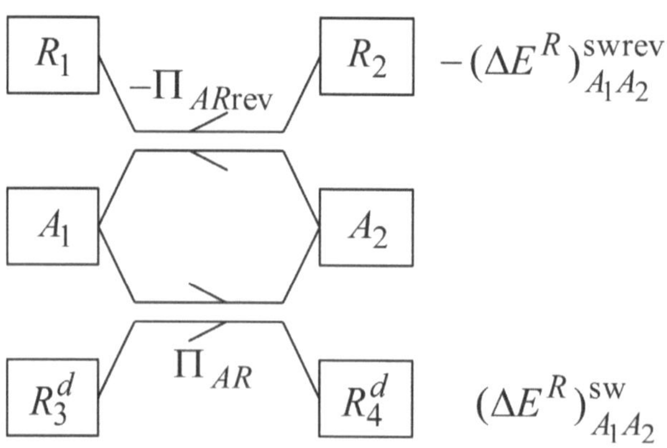

Figure 2: Illustration of the proof of Theorem 3: standard weight processes $\prod_{AR\mathrm{rev}}$ (reversible) and $\prod_{AR}$; R^d is a duplicate of R; see text.

(a) $\left(\Delta E^R\right)_{A_1A_2}^{\text{swrev}} \le \left(\Delta E^R\right)_{A_1A_2}^{\text{sw}}$;

(b) if also $\prod_{AR}$ is reversible, then $\left(\Delta E^R\right)_{A_1A_2}^{\text{swrev}} = \left(\Delta E^R\right)_{A_1A_2}^{\text{sw}}$;

(c) if $\left(\Delta E^R\right)_{A_1A_2}^{\text{swrev}} = \left(\Delta E^R\right)_{A_1A_2}^{\text{sw}}$, then also $\prod_{AR}$ is reversible.

Proof of (a). Let us denote by R_1 and by R_2 the initial and the final state of R in process $\prod_{ARrev}$. Let us denote by R^d the duplicate of R which is employed in process $\prod_{AR}$, by $^{\Pi_{AR}}$ and by R_3^d the initial and the final state of R^d in this process. Let us suppose, *ab absurdo*, that $\left(\Delta E^R\right)_{A_1A_2}^{\text{swrev}} > \left(\Delta E^R\right)_{A_1A_2}^{\text{sw}}$. Then, the sequence of processes $(-\prod_{ARrev}, \prod_{AR})$ would be a weight process for RR_d in which, starting from the stable equilibrium state $R_2R_3^d$, the energy of RR^d is lowered and the regions of space occupied by the constituents of RR^d have no net change, in contrast with Theorem 2. Therefore, $\left(\Delta E^R\right)_{A_1A_2}^{\text{swrev}} = \left(\Delta E^R\right)_{A_1A_2}^{\text{sw}}$.

Proof of (b). If $\prod_{AR}$ is reversible too, then, in addition $E$to $\left(\Delta E^R\right)_{A_1A_2}^{\text{swrev}} \le \left(\Delta E^R\right)_{A_1A_2}^{\text{sw}}$ the relation $\left(\Delta E^R\right)_{A_1A_2}^{\text{sw}} \le \left(\Delta E^R\right)_{A_1A_2}^{\text{swrev}}$ must hold too. Otherwise, the sequence of processes $(\prod_{ARrev}, -\prod_{AR})$ would be a weight process for RR^d in which, starting from the stable equilibrium state $R_1R_4^d$ the energy of RR^d is lowered and the regions of space occupied by the constituents of RR^d have no net change, in contrast with Theorem 2. Therefore, $\left(\Delta E^R\right)_{A_1A_2}^{\text{swrev}} = \left(\Delta E^R\right)_{A_1A_2}^{\text{sw}}$.

Proof of (c). Let $\prod_{AR}$ be a standard weight process for AR, from A_1 to A_2, such that $\left(\Delta E^R\right)_{A_1A_2}^{\text{sw}} = \left(\Delta E^R\right)_{A_1A_2}^{\text{swrev}}$ and let R_1 be the initial state of R in this process. Let $\prod_{ARrev}$ be a reversible standard weight process for AR, from A_1 to A_2, with the same initial state R_1 of R. Thus, R_3^d coincides with R_1 and R_4^d coincides with R_2. The sequence of processes $(\prod_{AR}, -\prod_{ARrev})$ is a cycle for the isolated system ARB, where B is the environment of AR. As a consequence, $\prod_{AR}$ is reversible, because it is a part of a cycle of the isolated system ARB.

Theorem 4. Let R' and R'' be any two thermal reservoirs and consider the energy changes, $\left(\Delta E^{R'}\right)_{A_1A_2}^{\text{swrev}}$ and $\left(\Delta E^{R''}\right)_{A_1A_2}^{\text{swrev}}$, respectively, in the reversible standard weight processes and , where (A_1, A_2) is an arbitrarily chosen pair of states of any system A. Then, the ratio $\left(\Delta E^{R'}\right)_{A_1A_2}^{\text{swrev}} / \left(\Delta E^{R''}\right)_{A_1A_2}^{\text{swrev}}$:

(a) is positive;

(b) depends only on R' and R'', i.e. it is independent of (i) the initial stable equilibrium states f R' and R'', (ii) the choice of system A and (iii) the choice of states A_1 and A_2.

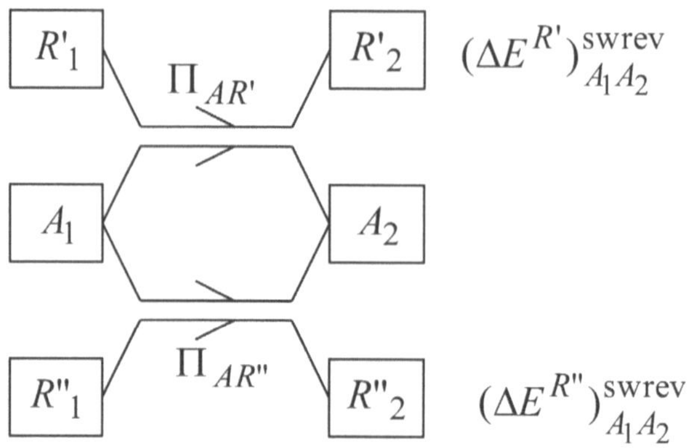

Figure 3: Illustration of the proof of Theorem 4, part (a): reversible standard weight processes $\prod_{AR}'$ and $\prod_{AR}''$, see text.

Proof of (a). With the help of Fig. 3, let us suppose that $\left(\Delta E^{R'}\right)^{\text{swrev}}_{A_1A_2} < 0$. Then $\left(\Delta E^{R''}\right)^{\text{swrev}}_{A_1A_2}$ cannot be zero. In fact, in that case the sequence of processes ($\prod_{AR}'$, $-\prod_{AR}''$), which is a cycle for A, would be a weight process for R' in which, starting from the stable equilibrium state R'_1, the energy of R' is lowered and the regions of space occupied by the constituents of R' have no net change, in contrast with Theorem 2. Moreover, $\left(\Delta E^{R''}\right)^{\text{swrev}}_{A_1A_2}$ cannot be positive. In fact, if it were positive, the work performed by $R'R''$ as a result of the overall weight process ($\prod_{AR}'$, $-\prod_{AR}''$) for $R'R''$ would be

$$W^{R'R''\rightarrow} = -\left(\Delta E^{R'}\right)^{\text{swrev}}_{A_1A_2} + \left(\Delta E^{R''}\right)^{\text{swrev}}_{A_1A_2}, \tag{11}$$

where both terms are positive. On account of Assumption 1 and Corollary 1, after the process ($\prod_{AR}'$, $-\prod_{AR}''$), one could perform a weight process $\prod_R''$ for R'' in which a positive amount of energy equal to $\left(\Delta E^{R''}\right)^{\text{swrev}}_{A_1A_2}$ is given back to R'' and the latter is restored to its initial stable equilibrium state. As a result, the sequence ($\prod_{AR}'$, $-\prod_{AR}''$, $\prod_R''$) would be a weight process for R' in which, starting from the stable equilibrium state R'_1 , the energy of R' is lowered and the regions of space occupied by the constituents of R' have no net change, in contrast with Theorem 2. Therefore, the assumption $\left(\Delta E^{R'}\right)^{\text{swrev}}_{A_1A_2} < 0$ implies $\left(\Delta E^{R''}\right)^{\text{swrev}}_{A_1A_2} < 0$.

Let us suppose that $\left(\Delta E^{R'}\right)^{\text{swrev}}_{A_1A_2} > 0$. Then, for process $-\prod_{AR}'$ one has $\left(\Delta E^{R'}\right)^{\text{swrev}}_{A_1A_2} < 0$. By repeating the previous argument, one proves that for process $-\prod_{AR}''$ one has $\left(\Delta E^{R''}\right)^{\text{swrev}}_{A_1A_2} < 0$. Therefore, for process $\prod_{AR}''$ one has $\left(\Delta E^{R''}\right)^{\text{swrev}}_{A_1A_2} > 0$.

Proof of (b). Given a pair of states (A_1, A_2) of a system A, consider the reversible standard weight process $\Pi_{AR'} = (A_1R'_1 \to A_2R'_2)^{\text{swrev}}$ for AR', with R' initially in state R'_1, and the reversible standard weight process $\Pi_{AR''} = (A_1R''_1 \to A_2R''_2)^{\text{swrev}}$ for AR″, with R″ initially in state R'_1. Moreover, given a pair of states A'_1, A'_2 of another system A', consider the reversible standard weight process $\Pi_{A'R'} = (A'_1R'_1 \to A'_2R'_3)^{\text{swrev}}$ for $A'R'$, with R' initially in state R'_1, and the reversible standard weight process $\Pi_{A'R''} = (A'_1R''_1 \to A'_2R''_3)^{\text{swrev}}$ for $A'R''$, with R'' initially in state R''_1.

With the help of Fig. 4, we will prove that the changes in energy of the reservoirs in these processes obey the relation

$$\frac{\left(\Delta E^{R'}\right)^{\text{swrev}}_{A_1A_2}}{\left(\Delta E^{R''}\right)^{\text{swrev}}_{A_1A_2}} = \frac{\left(\Delta E^{R'}\right)^{\text{swrev}}_{A'_1A'_2}}{\left(\Delta E^{R''}\right)^{\text{swrev}}_{A'_1A'_2}} \; . \tag{12}$$

Let us assume: $\left(\Delta E^{R'}\right)^{\text{swrev}}_{A_1A_2} > 0$ and $\left(\Delta E^{R'}\right)^{\text{swrev}}_{A'_1A'_2} > 0$, which implies, on account of part (a) of the proof, $\left(\Delta E^{R''}\right)^{\text{swrev}}_{A_1A_2} > 0$ and $\left(\Delta E^{R''}\right)^{\text{swrev}}_{A'_1A'_2} > 0$. This is not a restriction, because it is possible to reverse the processes under examination. Now, as is well known, any real number can be approximated with arbitrarily high accuracy by a rational number. Therefore, we will assume that the energy changes $\left(\Delta E^{R'}\right)^{\text{swrev}}_{A_1A_2}$ and $\left(\Delta E^{R'}\right)^{\text{swrev}}_{A'_1A'_2}$ are rational numbers, so that whatever is the value of their ratio, there exist two positive integers m and n such that $\left(\Delta E^{R'}\right)^{\text{swrev}}_{A_1A_2} / \left(\Delta E^{R'}\right)^{\text{swrev}}_{A'_1A'_2} = n/m$, i.e.

$$m \left(\Delta E^{R'}\right)^{\text{swrev}}_{A_1A_2} = n \left(\Delta E^{R'}\right)^{\text{swrev}}_{A'_1A'_2} \; . \tag{13}$$

Therefore, as sketched in Fig. 4, let us consider the sequences $\prod_A$ and Π_A defined as follows. $\prod_A$ is the following sequence of weight processes for the composite system $AR'R''$: starting from the

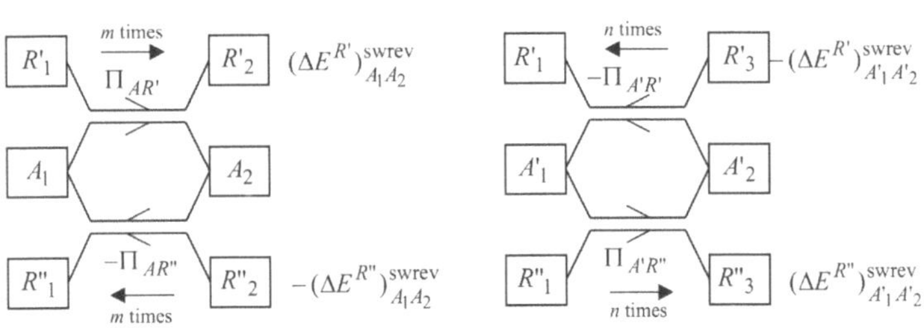

Figure 4: Illustration of the proof of Theorem 4, part (b): sequence of processes ($\prod_A$, $\prod_A'$), see text.

initial state R'_1 of R' and R''_2 of R'', system A is brought from A_1 to A_2 by a reversible standard weight process for AR', then from A_2 to A_1 by a reversible standard weight process for AR''; whatever the new states of R' and R'' are, again system A is brought from A_1 to A_2 by a reversible standard weight process for AR' and back to A_1 by a reversible standard weight process for AR'', until the cycle for A is repeated m times. Similarly, $\prod_A'$ is a sequence of weight processes for the composite system $A'R'R''$ whereby starting from the end states of R' and R'' reached by sequence $\prod_A$, system A' is brought from A_1 to A_2 by a reversible standard weight process for $A'R''$, then from A_2 to A_1 by a reversible standard weight process for A'R'; and so on until the cycle for A' is repeated n times.

Clearly, the composite sequence ($\prod_A$, $\prod_A'$) is a cycle for AA'. Moreover, it is a cycle also for R'. In fact, on account of Theorem 3, the energy change of R' in each process $\prod_{AR'}$ is equal to $\left(\Delta E^{R'}\right)^{\text{swrev}}_{A_1A_2}$ regardless of its initial state and in each process $-\prod_{A'R'}$ is equal to $-\left(\Delta E^{R'}\right)^{\text{swrev}}_{A'_1A'_2}$. Therefore, the energy change of R' in the sequence (Π_A, Π'_A) is $m\left(\Delta E^{R'}\right)^{\text{swrev}}_{A_1A_2} - n\left(\Delta E^{R'}\right)^{\text{swrev}}_{A'_1A'_2}$ and equals zero on account of eqn (13). As a result, after $(\Pi_A \Pi'_A)$, reservoir R' has been restored to its initial state, so that $(\Pi_A \Pi'_A)$ is a reversible weight process for R''.

Again on account of Theorem 3, the overall energy change of R'' in the sequence is $-m\left(\Delta E^{R''}\right)^{\text{swrev}}_{A_1A_2} + n\left(\Delta E^{R''}\right)^{\text{swrev}}_{A_1A_2}$. If this quantity were negative, Theorem 2 would be violated. If this quantity were positive Theorem 2 would also be violated by the reverse of the process $(-\Pi'_A - \Pi_A)$. Therefore, the only possibility is that $-m\left(\Delta E^{R''}\right)^{\text{swrev}}_{A_1A_2} + n\left(\Delta E^{R''}\right)^{\text{swrev}}_{A_1A_2} = 0$, i.e.

$$m\left(\Delta E^{R''}\right)^{\text{swrev}}_{A_1A_2} = n\left(\Delta E^{R''}\right)^{\text{swrev}}_{A'_1A'_2} \tag{14}$$

Finally, taking the ratio of eqns (13) and (14), we obtain eqn (12) which is our thesis.

Temperature of a thermal reservoir. Let R be a given thermal reservoir and R^0 a reference thermal reservoir. Select an arbitrary pair of states (A_1, A_2) of any system A, and consider the energy changes $\left(\Delta E^{R}\right)^{\text{swrev}}_{A_1A_2}$ and $\left(\Delta E^{R^o}\right)^{\text{swrev}}_{A_1A_2}$ in two reversible standard weight processes from A_1 to A_2, one for AR and the other for AR^0, respectively. We call *temperature* of R the positive quantity

$$T_R = T_{R^o} \frac{\left(\Delta E^{R}\right)^{\text{swrev}}_{A_1A_2}}{\left(\Delta E^{R^o}\right)^{\text{swrev}}_{A_1A_2}}, \tag{15}$$

where T_{R^0} is a positive constant associated arbitrarily with the reference thermal reservoir R^0. Clearly, the temperature T_R of R is defined only up to an arbitrary multiplicative constant. If for R^0 we select a thermal reservoir having water as constituent, with energy restricted to the solid–liquid–vapor triple-point range, and we set $TR^0 = 273.16$ K, we obtain the unit kelvin (K) for the thermodynamic temperature, which is adopted in the International System of Units (SI).

Corollary 2. The ratio of the temperatures of two thermal reservoirs, R' and R'', is independent of the choice of the reference thermal reservoir and can be measured directly as

$$\frac{T_{R'}}{T_{R''}} = \frac{\left(\Delta E^{R'}\right)^{\text{swrev}}_{A_1A_2}}{\left(\Delta E^{R''}\right)^{\text{swrev}}_{A_1A_2}}, \tag{16}$$

where $\left(\Delta E^{R'}\right)^{\text{swrev}}_{A_1A_2}$ and $\left(\Delta E^{R''}\right)^{\text{swrev}}_{A_1A_2}$ are the energy changes of R' and R'' in two reversible standard weight processes, one for AR' and the other for AR'', which interconnect the same but otherwise arbitrary pair of states (A_1, A_2) of any system A.

Proof. Let $\left(\Delta E^{R^0}\right)^{\text{swrev}}_{A_1A_2}$ be the energy change of the reference thermal reservoir R^0 in any reversible standard weight process for AR^0 which interconnects the same states (A_1, A_2) of A. From eqn (15) we have

$$T_{R'} = T_{R^\circ} \frac{\left(\Delta E^{R'}\right)^{\text{swrev}}_{A_1A_2}}{\left(\Delta E^{R^\circ}\right)^{\text{swrev}}_{A_1A_2}}, \tag{17}$$

$$T_{R''} = T_{R^\circ} \frac{\left(\Delta E^{R''}\right)^{\text{swrev}}_{A_1A_2}}{\left(\Delta E^{R^\circ}\right)^{\text{swrev}}_{A_1A_2}}, \tag{18}$$

therefore, the ratio of eqns (17) and (18) yields eqn (16).

Corollary 3. Let (A_1, A_2) be any pair of states of a system A, and let $\left(\Delta E^{R}\right)^{\text{swrev}}_{A_1A_2}$ be the energy change of a thermal reservoir R with temperature T_R, in any reversible standard weight process for AR from A_1 to A_2. Then, for the given system A, the ratio $\left(\Delta E^{R}\right)^{\text{swrev}}_{A_1A_2} \backslash T_R$ depends only on the pair of states (A_1, A_2), i.e. it is independent of the choice of reservoir R and of its initial stable equilibrium state R_1.

Proof. Let us consider two reversible standard weight processes from A_1 to A_2, one for AR' and the other for AR'', where R' is a thermal reservoir with temperature T_R' and R'' is a thermal reservoir with temperature T_R''. Then, eqn (16) yields

$$\frac{\left(\Delta E^{R'}\right)^{\text{swrev}}_{A_1A_2}}{T_{R'}} = \frac{\left(\Delta E^{R''}\right)^{\text{swrev}}_{A_1A_2}}{T_{R''}} \; . \tag{19}$$

Definition of entropy for a closed system. Proof that it is a property. Let (A_1, A_2) be any pair of states of a system A, and let R be an arbitrarily chosen thermal reservoir placed in the environment B of A. We call *entropy difference* between A_2 and A_1 the quantity

$$S_2^A - S_1^A = -\frac{\left(\Delta E^{R}\right)^{\text{swrev}}_{A_1A_2}}{T_R} \, , \tag{20}$$

where $\left(\Delta E^{R}\right)^{\text{swrev}}_{A_1A_2}$ is the energy change of R in any reversible standard weight process for AR from A_1 to A_2, and T_R is the temperature of R. On account of Corollary 3, the right-hand side of eqn (20) is determined uniquely by states A_1 and A_2.

Let A_0 be a reference state of A, to which we assign an arbitrarily chosen value of entropy S_0^A. Then, the value of the entropy of A in any other state A_1 of A is determined uniquely by the equation

$$S_1^A = S_0^A - \frac{\left(\Delta E^{R}\right)^{\text{swrev}}_{A_0A_1}}{T_R} \, , \tag{21}$$

where $\left(\Delta E^{R}\right)^{\text{swrev}}_{A_0A_1}$ is the energy change of R in any reversible standard weight process for AR from A_0 to A_1, and T_R is the temperature of R. Such a process exists for every state A_1, on account of Assumption 2. Therefore, entropy is a property of A, defined for every state A_1 of A.

Theorem 5. Additivity of entropy differences. Consider the pairs of states ($C_1 = A_1B_1$, $C_2 = A_2B_2$) of the composite system $C = AB$. Then,

$$S^{AB}_{A_2B_2} - S^{AB}_{A_1B_1} = S_2^A - S_1^A + S_2^B - S_1^B \; . \tag{22}$$

Proof. Let us choose a thermal reservoir R, with temperature T_R, and consider the sequence $(\prod_{AR}, \prod_{BR})$ where $\prod_{AR}$ is a reversible standard weight process for AR from A_1 to A_2, while

$\prod_{BR}$ is a reversible standard weight process for BR from B_1 to B_2. The sequence ($\prod_{AR}$, $\prod_{BR}$) is a reversible standard weight process for CR from C_1 to C_2 , in which the energy change of R is the sum of the energy changes in the constituent processes $\prod_{AR}$ and $\prod_{BR}$, i.e. $\left(\Delta E^R\right)_{C_1C_2}^{\text{swrev}} = \left(\Delta E^R\right)_{A_1A_2}^{\text{swrev}} + \left(\Delta E^R\right)_{B_1B_2}^{\text{swrev}}$. Therefore,

$$\frac{\left(\Delta E^R\right)_{C_1C_2}^{\text{swrev}}}{T_R} = \frac{\left(\Delta E^R\right)_{A_1A_2}^{\text{swrev}}}{T_R} + \frac{\left(\Delta E^R\right)_{B_1B_2}^{\text{swrev}}}{T_R} \quad . \tag{23}$$

Equation (23) and the definition of entropy (20) yield eqn (22).

Comment. As a consequence of Theorem 5, if the values of entropy are chosen so that they are additive in the reference states, entropy results as an additive property.

Theorem 6. Let (A_1, A_2) be any pair of states of a system A and let R be a thermal reservoir with temperature T_R. Let $\prod_{ARirr}$ be any irreversible standard weight process for AR from A_1 to A_2 and let $\left(\Delta E^R\right)_{A_1A_2}^{\text{swirr}}$ be the energy change of R in this process. Then

$$-\frac{\left(\Delta E^R\right)_{A_1A_2}^{\text{swirr}}}{T_R} < S_2^A - S_1^A \quad . \tag{24}$$

Proof. Let $\prod_{ARrev}$ be any reversible standard weight process for AR from A_1 to A_2 and let $\left(\Delta E^R\right)_{A_1A_2}^{\text{swirr}}$ be the energy change of R in this process. On account of Theorem 3,

$$\left(\Delta E^R\right)_{A_1A_2}^{\text{swrev}} < \left(\Delta E^R\right)_{A_1A_2}^{\text{swirr}} \quad . \tag{25}$$

Since T_R is positive, from eqns (25) and (20) one obtains

$$-\frac{\left(\Delta E^R\right)_{A_1A_2}^{\text{swirr}}}{T_R} < -\frac{\left(\Delta E^R\right)_{A_1A_2}^{\text{swrev}}}{T_R} = S_2^A - S_1^A \quad . \tag{26}$$

Theorem 7. Principle of entropy non-decrease. Let (A_1, A_2) be a pair of states of a system A and let $(A_1 \rightarrow A_2)^W$ be any weight process for A from A_1 to A_2. Then, the entropy difference $S_2^A - S_1^A$ is equal to zero if and only if the weight process is reversible; it is strictly positive if and only if the weight process is irreversible.

Proof. If $(A_1 \rightarrow A_2)^W$ is reversible, then it is a special case of a reversible standard weight process for AR in which the initial stable equilibrium state of R does not change. Therefore,

$\left(\Delta E^R\right)^{\text{swrev}}_{A_1 A_2} = 0$ and by applying the definition of entropy, eqn (20), one obtains

$$S_2^A - S_1^A = -\frac{\left(\Delta E^R\right)^{\text{swrev}}_{A_1 A_2}}{T_R} = 0 \ . \tag{27}$$

If $(A_1 \to A_2)^{\text{W}}$ is irreversible, then it is a special case of an irreversible standard weight process for AR in which the initial stable equilibrium state of R does not change. Therefore, $\left(\Delta E^R\right)^{\text{swirr}}_{A_1 A_2} = 0$ and eqn (24) yields

$$S_2^A - S_1^A > -\frac{\left(\Delta E^R\right)^{\text{swirr}}_{A_1 A_2}}{T_R} = 0 \ . \tag{28}$$

Moreover, if a weight process $(A_1 \to A_2)^{\text{W}}$ for A is such that $S_2^A - S_1^A = 0$ then the process must be reversible, because we just proved that for any irreversible weight process $S_2^A - S_1^A > 0$; if a weight process $(A_1 \to A_2)^{\text{W}}$ for A is such that $S_2^A - S_1^A > 0$, then the process must be irreversible, because we just proved that for any reversible weight process $S_2^A - S_1^A = 0$.

Corollary 4. If states A_1 and A_2 can be interconnected by means of a reversible weight process for A, they have the same entropy. If states A_1 and A_2 can be interconnected by means of a zero-work reversible weight process for A, they have the same energy and the same entropy.

Proof. These are straightforward consequences of Theorem 7 together with the definition of energy.

Theorem 8. Highest entropy principle. Among all the states of a system A such that the constituents of A are contained in a given set of regions of space $\boldsymbol{R}^A$, and the value of the energy E^A of A is fixed, the entropy of A has the highest value only in the unique stable equilibrium state A_{se} determined (Lemma 1) by $\boldsymbol{R}^A$ and E^A.

Proof. Let A_g be any other state of A in the set of states considered here. On account of the First Law and of the definition of energy, A_g and A_{se} can be interconnected by a zero-work weight process for A, either $(A_g \to A_{se})^W$ or $(A_{se} \to A_g)^W$. However, the existence of a zero-work weight process $(A_{se} \to A_g)^W$ would violate the definition of stable equilibrium state. Therefore, a zero-work weight process $(A_g \to A_{se})^W$ exists and is irreversible, so that Theorem 7 implies $S_{se}^A > S_g^A$.

6 Fundamental relation, temperature, and Gibbs relation (for a closed system)

Let us recall that, for all the stable equilibrium states of a (closed) system A in a scenario AB: the external force field for A coincides (where defined) with the external force field for AB, which is the same for all the stable equilibrium states of A; moreover, all the compositions of A belong to the same set of compatible compositions (n^{0A}, ν^A).

Set of equivalent stable equilibrium states. We call *set of equivalent stable equilibrium states* of a system A, denoted ESE^A, a subset of its stable equilibrium states such that any pair of states in the set:

- differ from one another by some geometrical features of the regions of space $\boldsymbol{R}^A$;
- have the same composition;
- can be interconnected by a zero-work reversible weight process for A and, hence, by Corollary 4, have the same energy and the same entropy.

Parameters. We call *parameters* of a system A, denoted by $\beta^A = \beta_1^A,\dots,\beta_s^A$ a minimal set of real variables sufficient to fully and uniquely parameterize all the different sets of equivalent stable equilibrium states ESE^A of A. In the following, we will consider systems with a finite number s of parameters.

Examples. Consider a system A consisting of a single particle confined in spherical region of space of volume V; the box is centered at position r which can move in a larger region where there are no external fields. Then, it is clear that any rotation or translation of the spherical box within the larger region can be effected in a zero-work weight process that does not alter the rest of the state. Therefore, the position of the center of the box is *not* a parameter of the system. The volume instead is a parameter. The same holds if the box is cubic. If it is a parallelepiped, instead, the parameters are the sides l_1, l_2, l_3 but not its position and orientation. For a more complex geometry of the box, the parameters are any minimal set of geometrical features sufficient to fully describe its shape, regardless of its position and orientation. The same if instead of one, the box contains many particles. Suppose now we have a spherical box, with one or many particles, that can be moved in a larger region where there are k subregions, each much larger than the box and each with an external electric field everywhere parallel to the x axis and with a uniform but different magnitude E_{ek}. As part of the definition of the system, let us restrict it only to the states such that the box is fully contained in one of these regions. For this system, the magnitude of E_e can be changed in a weight process by moving A from one uniform field subregion to another, but this in general will vary the energy. Therefore, in addition to the volume of the sphere, this system will have k as a parameter identifying the subregion where the box is located. Equivalently, the subregion can be identified by the parameter E_e taking values in the set $\{E_{ek}\}$. For each value of the energy E, system A has a set ESE^A for every pair of values of the parameters (V, E_e), with E_e in $\{E_{ek}\}$.

Corollary 5. Fundamental relation (for a closed system). On the set of all the stable equilibrium states of a system A, the entropy is given by a single valued function

$$S^A = S_{se}^A(E^A, \beta^A) \; , \tag{29}$$

which is called *fundamental relation* for the stable equilibrium states of A. Moreover, also the reaction coordinates are given at stable equilibrium by a single-valued function

$$\varepsilon^A = \varepsilon^A_{se}(E^A, \beta^A) , \tag{30}$$

which specifies the unique composition compatible with the initial composition $\boldsymbol{n}^{0A}$, called the *chemical equilibrium composition*.

Proof. On account of the Second Law and Lemma 1, among all the states of a system A with energy E^A, the regions of space $\boldsymbol{R}^A$ identify a unique stable equilibrium state. This implies (Lemma 2) the existence of a single-valued function $A_{se} = A_{se}(E^A, \boldsymbol{R}^A)$, where A_{se} denotes the state, in the sense of eqn (5). By definition, for each value of the energy E^A, the values of the parameters β^A fully identify all the regions of space $\boldsymbol{R}^A$ that correspond to a set of equivalent stable equilibrium states ESE^A, which have the same value of the entropy and the same composition. Therefore, the values of E^A and β^A fix uniquely the values of S^A_{se} and of ε^A_{se}. This implies the existence of the single-valued functions written in eqns (29) and (30).

Comment. Clearly, for a non-reactive system, the composition is fixed and equal to the initial, i.e. $\varepsilon^A_{se}(E^A, \beta^A) = 0$.

Usually [18, 19], in view of the equivalence that defines them, each set ESE^A is thought of as a single state called "a stable equilibrium state" of A. Thus, for a given closed system A (and, hence, given initial amounts of constituents), it is commonly stated that the energy and the parameters of A determine "a unique stable equilibrium state" of A, which is called "the chemical equilibrium state" of A if the system is reactive according to a given set of stoichiometric coefficients. For a discussion of the implications of eqn (30) and its reduction to more familiar chemical equilibrium criteria in terms of chemical potentials, see, e.g. Ref. [25].

Theorem 9. For any (normal) system, for fixed values of the parameters the fundamental relation eqn (29) is a strictly increasing function of the energy.

Proof. Consider two stable equilibrium states A_{se1} and A_{se2} of a system A, with energies E^A_1 and E^A_2, entropies S^A_{se1} and S^A_{se2}, and with the same regions of space occupied by the constituents of A (and therefore the same values of the parameters). Assume $E^A_2 > E^A_1$. By Assumption 1, we can start from state A_{se1} and, by a weight process for A in which the regions of space occupied by the constituents of A have no net changes, add work so that the system ends in a non-equilibrium state A_2 with energy E^A_2. By Theorem 7, we must have $S^A_2 \geq S^A_{se1}$. Now, on account of Lemma 3, we can go from state A_2 to A_{se2} with a zero-work irreversible weight process for A. By Theorem 7, we must have $S^A_{se2} > S^A_2$. Combining the two inequalities, we find that $E^A_2 > E^A_1$ implies $S^A_{se2} > S^A_{se1}$.

Corollary 6. The fundamental relation for any (normal) system A is invertible with respect to E^A and can be rewritten in the form

$$E^A = E^A_{\text{se}}(S^A, \beta^A) . \tag{31}$$

Proof. By Theorem 9, for fixed β^A, eqn (29) is a strictly increasing function of E^A. Therefore, it is invertible with respect to E^A and, as a consequence, can be written in the form (31).

Restriction, for simplicity: exclusion of ground states. Hereafter, for simplicity, we will consider all the stable equilibrium states of any system A whose energy E^A is higher than the lowest allowed energy of A for the given set of values of the parameters β^A, namely $E^A > E^A_{\text{min}}(\beta^A)$. In fact, some properties of the lowest energy stable equilibrium states, also called *ground states*, are not defined through the same statements employed for the other stable equilibrium states, but by taking the limit for $E^A \to E^A_{\text{min}}(\beta^A)$ of the properties of the stable equilibrium states with energy $E^A > E^A_{\text{min}}(\beta^A)$. Moreover, a complete description of the properties of ground states requires the use of the Third Law, which is not discussed in this paper.

Assumption 3. The fundamental relation eqn (29) is continuous and twice differentiable with respect to each of the variables E^A and β^A. Moreover, the relation $A_{se} = A_{se}(E^A, \mathbf{R}^A)$ implied by Lemma 2 is continuous in E^A.

Temperature of a system in a stable equilibrium state. Consider a stable equilibrium state A_{se} of a system A identified by the values of E^A and β^A. The partial derivative of the fundamental relation eqn (31) with respect to S^A is denoted by

$$T^A = \left(\frac{\partial E^A_{\text{se}}}{\partial S^A}\right)_{\beta^A} . \tag{32}$$

Such derivative is always defined on account of Assumption 3. When evaluated at the values of E^A and β^A that identify state A_{se}, it yields a value that we call the *temperature* of A in state A_{se}. Clearly, eqn (32) implies

$$\frac{1}{T^A} = \left(\frac{\partial S^A_{\text{se}}}{\partial E^A}\right)_{\beta^A} . \tag{33}$$

Corollary 7. For any stable equilibrium state of any (normal) system, the temperature is non-negative.

Proof. The thesis follows immediately from the definition of temperature, eqn (32), and Theorem 9.

Theorem 10. When applied to a thermal reservoir R, eqn (32) yields that all the stable equilibrium states of a thermal reservoir have the same temperature which is equal to the temperature T_R of R defined by eqn (15).

Proof. Let R be any thermal reservoir and let (A_1, A_2) be any pair of states of a system A. Let

Π_{ARrev} be any reversible standard weight process for AR from A_1 to A_2. On account of the definition of standard weight process, the end states R_1 and R_2 of R are stable equilibrium states; moreover, the definition of R ensures that the region of space occupied by R is fixed, i.e. the parameters of R never change. On account of the definition of entropy difference between A_2 and A_1, of the principle of entropy non-decrease, and of the additivity of entropy differences (Theorem 5), one has

$$S_2^A - S_1^A = -\frac{\left(\Delta E^R\right)_{A_1A_2}^{\text{swrev}}}{T_R} = -\left(S_2^R - S_1^R\right) . \tag{34}$$

Equation (34) yields the following relation between the end states of R and T_R,

$$T_R = \left(\frac{\Delta E_{\text{se}}^R}{\Delta S^R}\right)_{\beta^R} . \tag{35}$$

In the limit of infinitely small energy and entropy changes of R, eqn (35) yields

$$T_R = \left(\frac{\partial E_{\text{se}}^R}{\partial S^R}\right)_{\beta^R} = T_1^R . \tag{36}$$

Since R_1 is an arbitrary initial state of R, it follows that all the stable equilibrium states of R have temperature equal to T_R.

Gibbs relation (for a closed system). By differentiating eqn (31), one obtains (omitting the superscript "A" and the subscript "se" for simplicity)

$$dE = T\,dS + \sum_{j=1}^{s} F_j\,d\beta_j , \tag{37}$$

where F_j is called generalized force conjugated to the jth parameter of A, $F_j = (\partial E_{se}/\partial\beta_j)_{s,\beta'}$. If all the regions of space $\boldsymbol{R}^A$ coincide and the volume V of any of them is a parameter, the negative of the conjugated generalized force is called *pressure*, denoted by p, $p = -(\partial E_{se}/\partial V)_{s,\beta'}$.

Comment. Quantum formalism for the fundamental relation. Let us recall that the measurement procedures that define energy and entropy must be applied, in general, to a (homogeneous) ensemble of identically prepared replicas of the system of interest. Because the numerical outcomes may vary (fluctuate) from replica to replica, the values of the energy and the entropy defined by these procedures are arithmetic means. Therefore, what we have denoted so far, for simplicity, by the symbols E^A and S^A should be understood as $\langle E^A\rangle$ and $\langle S^A\rangle$. Where appropriate, like in the quantum formalism implementation, this more precise notation should be preferred. Then, written in full notation, the fundamental relation eqn (29) is

$$\langle S^A \rangle_{se} = S^A_{se}(\langle E^A \rangle, \beta^A) \ , \tag{38}$$

and the corresponding Gibbs relation

$$d\langle E \rangle = T d\langle S \rangle + \sum_{j=1}^{s} F_j d\beta_j \ . \tag{39}$$

Comment. Validity for macroscopic as well as microscopic systems. As already emphasized in Ref. [19], it is important to note that none of the results derived up to this point requires any restriction on whether the amounts of constituents of the system should be large or small. The only exception is the comment on the practical existence of systems which satisfy, at least approximately, the definition of a thermal reservoir. However, thermal reservoirs play only an auxiliary role in our logical scheme: namely, they allow a simplification of the treatment. We will publish elsewhere an equivalent and equally general definition of entropy which does not require the concept of a thermal reservoir. Hence, all the results obtained in this paper are valid for all systems, microscopic and macroscopic.

Comment. Validity for equilibrium as well as non-equilibrium states. Moreover, our definition of entropy holds for all states, equilibrium and non-equilibrium, at least insofar as the crucial Assumption 2 is indeed a general law of physics.

Comment. Further stable-equilibrium-state results require the simple system model, and hence hold only for macroscopic or mesoscopic systems. For systems with very large amounts of constituents, rarefaction effects near walls and internal partitions are negligible and, therefore, the *simple system* model [19, p. 263] becomes an excellent approximation. The validity of the simple system model entails a number of further standard and important results about the thermodynamics of stable equilibrium states, such as the Euler and the Gibbs–Duhem equations, and their well-known and innumerable implications.

7 Proofs of Clausius and Carathéodory statements of the Second Law and of the Zeroth Law

In this section, we prove Carathéodory statement of the Second Law as a straightforward consequence of the impossibility of a PMM2 (Theorem 2). Then, with reference to closed systems with fixed regions of space occupied by their constituents, we prove that the temperature of any system is a strictly increasing function of its energy and, for a pair of systems A and B, the temperature equality is a necessary and sufficient condition for A and B to be in mutual stable equilibrium. Finally, by employing these results, we prove the Zeroth Law and Clausius statement of the Second Law.

Theorem 11. Carathéodory statement of the Second Law. In the (arbitrarily small) neighborhood of any stable equilibrium state A_{se1} of a system A, there exist stable equilibrium states of

A which are inaccessible from A_{se1} by means of a weight process[2] for A.

Proof. Let A_{se1} be a stable equilibrium state of A, with regions of space $\boldsymbol{R}_1^A$ occupied by the constituents of A and let $E_{\min}^A(\boldsymbol{R}_1^A)$ be the energy of the ground state of A which corresponds to the given regions of space. Let ε be an arbitrarily small positive energy value such that $0 < \varepsilon < E_1^A - E_{\min}^A(\boldsymbol{R}_1^A)$ and consider the energy values E^A in the interval $E_1^A - \varepsilon < E^A < E_1^A$. On account of the Second Law, for each value E^A in such interval and the given regions $\boldsymbol{R}_1^A$ there exists a stable equilibrium state $A_{se}(E^A, \boldsymbol{R}_1^A)$. By the assumed continuity of $A_{se} = A_{se}(E^A, \boldsymbol{R}^A)$ (Assumption 3) it follows that $E^A \to E_1^A$ implies $A_{se}(E^A, \boldsymbol{R}_1^A) \to A_{se1}$ and, therefore, these stable equilibrium states are in the neighborhood of state A_{se1}. On account of Theorem 2, none of these states is accessible from A_{se1} by means of a weight process for A.

Theorem 12. Consider two closed systems A and B, with fixed regions of space $\boldsymbol{R}^A$ and $\boldsymbol{R}^B$ occupied by their constituents, with corresponding parameters β^A and β^B. Then, the following are necessary conditions for A and B to be in mutual stable equilibrium:

- their states A_1 and B_1, with energy values E_1^A and E_1^B, respectively, are stable equilibrium states;
- the temperatures of A and B are equal, i.e.

$$T_1^A = T_1^B \quad ; \tag{40}$$

- there exists a positive amount of energy δ such that, if $-\delta < \varepsilon < \delta$,

$$T^A(E_1^A + \varepsilon, \beta^A) < T^B(E_1^B - \varepsilon, \beta^B) \text{ if } \varepsilon < 0 \ , \tag{41}$$

$$T^A(E_1^A + \varepsilon, \beta^A) > T^B(E_1^B - \varepsilon, \beta^B) \text{ if } \varepsilon > 0 \ , \tag{42}$$

where, of course, $T(E, \beta)$ denotes the inverse of $\partial S_{se}(E, \beta)/\partial E$.

Proof. If either A_1 or B_1 were not a stable equilibrium state, then by Lemma 3 it could be changed to a different state in a zero-work weight process. Therefore, also $C_1 = A_1B_1$ could be changed to a different state with no external effects; thus, it could not be a stable equilibrium state.

Let us denote by $\Gamma^C(E_1^C)$ the set of all the states of $C = AB$ such that: A and B are in stable equilibrium states; the constituents of A and B are contained in the sets of regions of space $\boldsymbol{R}^A$ and $\boldsymbol{R}^B$; the energy of C has the value $E_1^C = E_1^A + E_1^B$. On account of Theorem 8, a necessary condition for C_1 to be a stable equilibrium state is that C_1 be the unique highest entropy state in the set $\Gamma^C(E_1^C)$. By the additivity of entropy, we have

---

[2] We have replaced the concept of adiabatic process, employed by Carathéodory, with the less restrictive concept of weight process, employed in our treatment.

$$S^C = S^A + S^B. \tag{43}$$

Because in the set $\Gamma^C(E_1^C)$ the states of A and B are stable equilibrium, by eqn (29) we can write $S^A = S_{se}^A(E^A, \beta^A)$ and $S^B = S_{se}^B(E^B, \beta^B)$, where β^A and β^B are the values of the parameters of A and B which correspond to the regions of space $\boldsymbol{R}^A$ and $\boldsymbol{R}^B$. Moreover, since $E^A + E^B = E_1^C$, and $E_1^C = E_1^A + E_1^B$ is fixed, $E^A = E_1^A + \varepsilon, E^B = E_1^B - \varepsilon$. Therefore, we may write S^C as

$$S^C = S_{se}^A(E_1^A + \varepsilon, \beta^A) + S_{se}^B(E_1^B - \varepsilon, \beta^B) \ , \tag{44}$$

and, by differentiation with respect to ε, we readily obtain

$$\frac{\partial S^C}{\partial \varepsilon} = \left(\frac{\partial S_{se}^A}{\partial E^A}\right)_{\beta^A} - \left(\frac{\partial S_{se}^B}{\partial E^B}\right)_{\beta^B} = \frac{1}{T^A(E_1^A + \varepsilon, \beta^A)} - \frac{1}{T^B(E_1^B - \varepsilon, \ \beta^B)} \ . \tag{45}$$

Necessary conditions for C_1 (corresponding to $\varepsilon = 0$) to be the unique state which maximizes the entropy S^C in the set $\Gamma^C(E_1^C)$ are

$$\left.\frac{\partial S^C}{\partial \varepsilon}\right|_{\varepsilon=0} = 0 \ , \tag{46}$$

and,

$$\left.\frac{\partial S^C}{\partial \varepsilon}\right|_{\varepsilon<0} > 0 \ , \quad \left.\frac{\partial S^C}{\partial \varepsilon}\right|_{\varepsilon>0} < 0 \ . \tag{47}$$

Equations (45) and (46) prove eqn (40). Equations (45) and (47) prove eqn (41) and (42).

Assumption 4. Any system A, in any stable equilibrium state A_{se1}, is in mutual stable equilibrium with an identical copy A_d of A, in the same state.

Corollary 8. For the set of stable equilibrium states of a system A which correspond to a fixed set of values of the parameters β^A, the temperature of A is a strictly increasing function of the energy of A.

Proof. Let A_{se1} be any stable equilibrium state of a closed system A, with regions of space $\boldsymbol{R}^A$ occupied by the constituents of A, which correspond to the values β^A of the parameters, and with an energy value E_1^A Let A^d be an identical copy of A and let A_{se1}^d be the stable equilibrium state of A^d which is identical with A_{se1}. On account of Assumption 4, $C_1 = A_{se1}A_{se1}^d$ is a stable equilibrium state of $C = AA^d$. Therefore, by Theorem 12 and the fact that A and A^d, being identical, have identical fundamental relations, there exists a positive amount of energy δ such that, for any $0 < \varepsilon < \delta$,

$$T^A(E_1^A+\varepsilon,\beta^A) > T^{A^d}(E_1^{A^d}-\varepsilon,\beta^A) = T^A(E_1^A-\varepsilon,\beta^A)\ . \tag{48}$$

Therefore, in the neighborhood of A_{se1}, for fixed values of the parameters β^A the temperature of A is a strictly increasing function of the energy of A. Since A_{se1} has been chosen arbitrarily, the thesis of Corollary 8 is proved.

Comment. Corollary 8 proves that the second necessary condition for the mutual stable equilibrium of two closed systems with fixed regions of space occupied by their constituents, given by eqns (41) and (42), is automatically fulfilled (once Assumption 4 is made, or implied). For instance, for $\varepsilon > 0$, using eqn (40) and Corollary 8, we have

$$T^A(E_1^A+\varepsilon,\beta^A) > T_1^A = T_1^B > T^B(E_1^B-\varepsilon,\beta^B)\ . \tag{49}$$

Corollary 9. The fundamental relation $S_{\text{se}}^A(E^A,\beta^A)$ is a concave function of the energy E^A. Therefore, for any pair of values $E_1^A\ and\ E_2^A$ within the range allowed by the values of the parameters β^A, we have

$$S_{\text{se}}^A(E_2^A,\beta^A) \le S_{\text{se}}^A(E_1^A,\beta^A) + \frac{E_2^A - E_1^A}{T_1^A}\ . \tag{50}$$

Proof. A function $f(x)$ is called concave if, for any pair of points (x_1, x_2) in the domain of f, the segment with extremes $(x_1, f(x_1))$ and $(x_2, f(x_2))$ has no point above the graph of f. If f is differentiable, then a necessary and sufficient condition for f to be concave is that the derivative of f is a decreasing function of x. Therefore, the fundamental relation $S_{\text{se}}^A(E^A,\beta^A)$ is a concave function of the energy E^A, because its partial derivative with respect to E^A, $1/T^A$, is a decreasing function of E^A (Corollary 8). For a concave differentiable function f, for any pair of points (x_2, x_1) in the domain of f, $f(x_2) \le f(x_1) + f'(x_1)(x_2 - x_1)$. Relation eqn (50) is a direct consequence of this property.

Corollary 10. If two closed systems A and B have fixed regions of space $\boldsymbol{R}^A$ and $\boldsymbol{R}^B$ occupied by their constituents, then a necessary and sufficient condition for A and B to be in mutual stable equilibrium is that their states A_1 and B_1 be stable equilibrium states with the same temperature, namely $T_1^A = T_1^B$.

Proof. By Theorem 12, the condition is necessary. We will now prove that it is sufficient. Let A_1 and B_1 be stable equilibrium states of A and B with the same temperature $T_1^A = T_1^B$, and let $E_1^C = E_1^A + E_1^B$ be the the energy of the composite system C in state $C_1 = A_1\ B_1$. On account of the Second Law and Lemma 1, for the given regions of space $\boldsymbol{R}^A$ and $\boldsymbol{R}^B$ and the given energy value E_1^C, there exists a unique stable equilibrium state of C. The state $C_1 = A_1\ B_1$ fulfills the necessary condition for the mutual stable equilibrium of A and B given by eqn (40). Moreover, by Corollary 8, no other state of C with energy E_1^C fulfills this condition. If fact, if the energy of A is lower than E_1^B and the energy of B is higher than E_1^B, then $T^A < T^B$; if the energy of A is higher than E_1^A and the energy of B is lower than E_1^B, then $T^A > T^B$.

Corollary 11. Zeroth Law. Let *A*, *B*, and *C* be closed systems, with fixed regions of space occupied by their constituents, each in a stable equilibrium state. If system *A* is in mutual stable equilibrium with system *B*, and system *B* is in mutual stable equilibrium with system *C*, then *A* is in mutual stable equilibrium with *C*.

Proof. On account of Theorem 12, if *A* is in mutual stable equilibrium with *B*, and *B* is in mutual stable equilibrium with *C*, then $T^A = T^C$. Hence, on account of Corollary 10, *A* is in mutual stable equilibrium with *C*.

Comment. This corollary proves that the Zeroth Law is a consequence of the First and of the Second Law, and of Assumptions 1–4; these assumptions, or equivalent ones, are used, either explicitly or implicitly, in all treatments of thermodynamics.

Comment. Practical measurements of temperature. In practice, the temperature of a system *A* in a stable equilibrium state is measured indirectly, through a *thermometer*. The latter is a system *B* such that the temperature of *B* is directly related (through a calibration procedure) to another easily measurable property of *B*. The thermometer can then be brought in mutual stable equilibrium with *A* in such a way as not to modify appreciably the state of *A*. The reading of the temperature of the thermometer *B* yields, by Theorem 12, an indirect reading of the temperature of system *A*.

Theorem 13. Clausius statement of the Second Law. Given a pair of systems *A* and *B*, initially in stable equilibrium states A_{se1} and B_{se1} with different temperatures, T_1^A and $T_1^B < T_1^A$, it is impossible that a process of an isolated system *I* which contains *A* and *B* results in no changes of state for all the other subsystems of *I* while a positive amount of energy $E^{B\to A}$ is transferred from *B* to *A*.

Proof. Suppose, *ab absurdo*, that such a process, denoted by $\prod$, occurred, and denote by A_2 and B_2 the final states of *A* and *B* in this process. After $\prod$, it is possible to perform a process $\prod'$ for *I*, composed of a zero-work weight process $A_2 \to A_{2se}$ for *A* and a zero-work weight process $B_2 \to B_{2se}$ for *B*, such that A_{2se} and B_{2se} are stable equilibrium states and the regions of space occupied by the constituents of *A* and of *B* did not change. By Theorem 7, the entropy change of *I* in process $\prod'$ is positive, or zero if $\prod'$ is a process where nothing happens, i.e. if both A_2 and B_2 are stable equilibrium states. Thus, the entropy change of *I* in the sequence ($\prod$, $\prod'$) is greater than or equal to that in process $\prod$, i.e.

$$(\Delta S^I)_\Pi \le (\Delta S^I)_{(\Pi,\Pi')} = (S^A_{se2} - S^A_{se1}) + (S^B_{se2} - S^B_{se1}) \ , \tag{51}$$

where Theorem 5 has been applied. The right-hand side of eqn (51) can be evaluated using (twice) Relation eqn (50), to obtain

$$(S^A_{se2} - S^A_{se1}) + (S^B_{se2} - S^B_{se1}) \le \frac{E_2^A - E_1^A}{T_1^A} + \frac{E_2^B - E_1^B}{T_1^B} = \left[\frac{1}{T_1^A} - \frac{1}{T_1^B}\right] E^{B\to A} \ , \tag{52}$$

where in the last step we used the energy balance equations

$$E_2^A - E_1^A = E^{B\to A} \text{ and } E_2^B - E_1^B = -E^{B\to A} \quad . \tag{53}$$

By the condition $T_1^B < T_1^A$ and the assumption *ab absurdo* that $E^{B\to A} > 0$, the last term in Relations eqn (52) is strictly negative and, therefore, $(S_{\text{se2}}^A - S_{\text{se1}}^A) + (S_{\text{se2}}^B - S_{\text{se1}}^B) < 0$. When this is inserted in Relations eqn (51), we find $(\Delta S^I)_{\Pi} < 0$, which contradicts the principle of entropy non-decrease (Theorem 7).

8 Conclusions

This paper presents a rigorous and general treatment of the foundations of thermodynamics, based on operative definitions of all the concepts employed, such as those of system, state, isolated system, environment of a system, process, separable system, system uncorrelated from its environment, and parameters of a system. The treatment holds for any system, even in the presence of internal semipermeable walls and reaction mechanisms, as well as external force fields. The concept of empirical temperature is not used, and the Zeroth Law is proved to be a consequence of the First and of the Second Law, together with some important auxiliary assumptions that are here stated explicitly, while in most other treatments are used implicitly. The concepts of heat and of quasistatic process are not even mentioned, and the whole logical scheme is built so that the definition of entropy is valid also for non-equilibrium states, both for macroscopic systems and those involving only a few particles.

A definition of thermal reservoir less restrictive than in previous treatments is adopted: it is operationally very well approximated by a sufficiently large amount of any single-constituent simple system contained in a fixed region of space, provided that the energy values are restricted to the finite range corresponding to a triple point.

The proof that entropy is a property of the system is completed by a new explicit proof that the entropy difference between two states of a system is independent of the initial state of the thermal reservoir chosen to measure it.

The definition of a reversible process is given with reference to a given *scenario*, i.e. the largest isolated system whose subsystems are available for interaction; thus, the operativity of the definition is improved and the treatment is compatible also with recent interpretations of irreversibility in the quantum mechanical framework which we will discuss elsewhere.

Our contribution yields a deeper understanding of the foundations of thermodynamics, and can be useful as a starting point for researches on non-equilibrium thermodynamics and on the relations between quantum mechanics and thermodynamics. It also provides an effective alternative approach to teaching thermodynamics which improves, by making it even more rigorous and general, the method developed first by Hatsopoulos and Keenan and then by Gyftopoulos and Beretta. We are aware that, compared to traditional treatments of thermodynamics, this approach requires a bigger effort to be fully understood, especially from the teachers side, to clear out some of the common misconception and logical loops that affect the traditional treatments. Once this is achieved, the great strength of the method is in the initial part of the treatment, which contains rigorous definitions of all the concepts employed. It is precisely the logical rigor of the method, which yields a deeper insight into thermodynamics and also an increased confidence and capability in problem solving.

We have been using this approach for several years, in courses of thermodynamics both for undergraduate and for graduate students, and our teaching was clearly understood and well appreciated by a vast majority of the students.

Acknowledgments

G.P. Beretta gratefully acknowledges the Cariplo–UniBS–MIT-MechE faculty exchange program co-sponsored by UniBS and the CARIPLO Foundation, Italy, under grant 2008-2290.

References

[1] Carnot, N.L.S., *Réflections sur la puissance motrice du feu et sur les machines propres á dévelop-per cette puissance*, Bachelier: Paris, 1824.
[2] *von Mayer*, R., *Die Mechanik der Warme in gesammelten Schriften*, Cotta: Stuttgart, 1893.
[3] Joule, J.P., *The Scientific Papers of James Prescott Joule*, Physical Society of London: London, 1884.
[4] Joule, J.P., On the mechanical equivalent of heat. *B.A.A.S. Report*, **II**, pp. 21–22, 1848; *Philosophical Transactions of the Royal Society of London*, **140**, pp. 61–82, 1850.
[5] Thomson, W., On an absolute thermometric scale founded on Carnot's theory of the motive power of heat, and calculated from Regnault's observations, Read at the Cambridge Philosophical Proceedings for June 5, 1848 and published in the *Philosophical Magazine* in October 1848; reprinted in *Mathematical and Physical Papers of William Thomson*, **I**, pp. 100–106, 1882.
[6] Thomson, W., An account of Carnot's theory of the motive power of heat, with numerical results deduced from Regnault's experiments on steam. *Trans. Edinburgh R. Soc.* XVI, reprinted in *Mathematical and Physical Papers of William Thomson*, **I**, pp. 116–155, 1882.
[7] Clausius, R.J.E., On the motive power of heat, and on the laws which can be deduced from it for the theory of heat. *Annalen der Physik und Chemie* (ed. J.C. Poggendorff), **79**, pp. 368–397, pp. 500–524, 1850.
[8] Clausius, R.J.E., U¨ er Verschiedene fu¨ r die Anwendungen Bequeme Formen der Hauptgleichungen der Mechanischen Warmetheorie. Memoir read at the Philos. Soc. Zürich on April 24; *Pogg. Ann.*, **125**, pp. 353–400, 1865. English translation in: *The Second Law of Thermodynamics*; Kestin, J. (ed); Dowden, Hutchinson and Ros: Stroudsburg, pp. 162–193, 1976.
[9] Gibbs, J.W., *The Scientific Papers of J.W. Gibbs*, Longmans, Green, and Co.: New York, 1906.
[10] Planck, M., *Treatise on Thermodynamics*, translated by A. Oggs from the 7th German edition, Longmans, Green, and Co.: London, 1927 (the first German edition appeared in 1897).
[11] Poincaré, H., *Thermodynamique*, Gautier-Villars: Paris, 1908.

[12] Carathéodory, C., Untersuchungen ueber die Grundlagen der Thermodynamik. *Math. Ann.*, **67**, pp. 355–386, 1909.

[13] Fermi, E., *Thermodynamics*, Prentice-Hall: New York, 1937.

[14] Pippard, A.B., *Elements of Classical Thermodynamics*, Cambridge University Press: Cambridge, 1957.

[15] Zemansky, M.W., *Heat and Thermodynamics*, McGraw-Hill: New York, 1968.

[16] Prigogine, I., *Introduction to Thermodynamics of Irreversible Processes*, Interscience: New York, 1961.

[17] Onsager, L., Reciprocal relations in irreversible processes. *Phys. Rev.*, **38**, p. 405; **38**, p. 2265, 1931.

[18] Hatsopoulos, G.N. & Keenan, J.H., *Principles of General Thermodynamics*, Wiley: New York, 1965.

[19] Gyftopoulos, E.P. & Beretta, G.P., *Thermodynamics. Foundations and Applications*, Dover: Mineola, NY, 2005 (1st edn, Macmillan, 1991).

[20] Zanchini, E. & Beretta, G.P., Rigorous axiomatic definition of entropy valid also for non-equilibrium states, *AIP Conf. Proc. Series on Meeting the Entropy Challenge*, **1033**, pp. 296–301, 2008.

[21] Zanchini, E., Thermodynamics: energy of closed and open systems, *Il Nuovo Cimento*, **101B**, pp. 453–465, 1988. Zanchini, E., Thermodynamics: energy of nonsimple systems and second postulate, *Il Nuovo Cimento B***107A**, pp. 123–139, 1992.

[22] Hatsopoulos, G.N. & Beretta, G.P., Where is the entropy challenge? *AIP Conf. Proc. Series on Meeting the Entropy Challenge*, **1033**, 2008, pp. 34–54; Beretta, G.P., et al., Quantum thermodynamics: a new equation of motion for a single constituent of matter. *Nuovo Cimento B*, **82**, pp. 169–191, 1984; Beretta, G.P., Nonlinear quantum evolution equations to model irreversible adiabatic relaxation with maximal entropy production and other nonunitary processes. *Rep. Math. Phys.*, **64**, 139–168, 2009.

[23] Bennett, C.H., The second law and quantum physics, pp. 66–79, and Lloyd, S., The once and future second law of thermodynamics, pp. 143–157, *AIP Conf. Proc. Series on Meeting the Entropy Challenge*, **1033**, 2008; Goldstein, S. et al., Canonical typicality. *Phys. Rev. Lett.*, **96**, 050403, pp. 1–3, 2006; Maccone, L., Quantum solution to the arrow-of-time dilemma. *Phys. Rev. Lett.*, **103**, 080401, pp. 1–4, 2009.

[24] Zanchini, E., On the definition of extensive property energy by the first postulate of thermodynamics. *Found. Phys.* **16**, pp. 923–935, 1986.

[25] Beretta, G.P. & Gyftopoulos, E.P., Thermodynamic derivations of conditions for chemical equilibrium and of Onsager reciprocal relations for chemical reactors. *J. Chem. Phys.*, **121**, pp. 2718–2728, 2004.

Thermodynamics in the wider context of science

Hot and cold taps. (Marianne Greated 2014, pencil and coloured pencil)

Chapter 13

William Thomson (Lord Kelvin) and thermoelectricity

L. Anatychuk
Institute of Thermoelectricity, National Academy of Sciences, Ukraine
Ministry of Education and Science of Ukraine, Ukraine

Abstract

Some information on the results of Thomson's scientific research in the field of thermoelectricity is given below. The experiments that preceded this research and served, together with his thermodynamic approach, to Thomson as a basis for discovery of the fundamental relations in thermoelectricity named after him are characterized. Discoveries of the Thomson thermoelectric effect and transversal thermo-EMF in anisotropic crystals are also considered. Practical applications of the regularities and phenomena in thermoelectricity discovered by Thomson are described together with the prospects of their further applications.

1 Introduction

William Thomson (Fig. 1) belonged to that rare type of a scientist who combined the qualities of a brilliant theoretician capable of applying mathematical instruments to physical phenomena description, a perspicacious physicist who deeply understood the essence of the surrounding world phenomena, a careful experimenter possessing accurate skills in laboratory investigations and a talented engineer who aptly applied physical laws to development and design of various apparatus useful for humanity and speeding up scientific and technical progress.

His knowledge and skills Thomson achieved from the brilliant minds of his time. It was his father who cultivated his love of mathematics. Later Laplace, Lagrange, Fourier, Liouville, Charles and Sturm became his teachers of mathematics. Renier was the one to teach him the secrets of the experiment. This marvelous experimenter performed measurements so accurate that their results have not lost significance up to now. Having used his experimental approaches together with development of those of his own, Thomson created the first ammeters and voltmeters, absolute electrometers and other numerous devices. Faraday, the great experimenter, influenced Thomson greatly. Thomson performed a series of brilliant applied works based on his discoveries, knowledge and experience.

Thermoelectricity did not belong to his main scientific priorities. Nevertheless, it was due to Thomson's research work that very important fundamental results for this interesting and constantly developing scientific trend were also obtained. The description of Thomson's fundamental contribution to thermoelectricity is given below.

Figure 1: William Thomson (1824–1907).

2 On the events that preceded Thomson's discoveries in thermoelectricity

The equilibrium thermodynamics of the reversible processes created by Thomson served as a basis for his discoveries in thermoelectricity. Its application to the description of thermoelectric phenomena was one of the vivid results of the said theory [1–9].

Two thermoelectric effects were well known to Thomson.

2.1 The first thermoelectric effect: the Seebeck effect

Seebeck's discovery was really dramatic [10]. It was not the aim Seebeck set to find the way of direct heat conversion into electricity. He was enchanted by Oersted's experiments (Fig. 2) on the magnetic field induction under the influence of the electric current [11]. While varying them differently, Seebeck noticed that if the electric circuit consisted of two dissimilar materials (e.g. bismuth and copper) and if one of the combinations of these materials was subjected to heat, a magnetic field appeared inside such circuit that was registered by the deflection of the magnetic needle (Fig. 3). Seebeck called this effect **thermomagnetism**, as he really observed the transformation of heat into the magnetic field.

Seebeck reported on his discovery of thermomagnetism to Berlin Academy of Sciences in 1821. It was received with great enthusiasm. Seebeck's reports on thermomagnetism were printed in the Proceedings of the Royal Academy of Sciences in Berlin which were issued as late as 1825 [12].

The information on the Seebeck effect of thermomagnetism, though, was soon spread among scientists. It reached Oersted by the end of 1822.

Having repeated the Seebeck experiment, Oersted described it as the most wonderful discovery among the discoveries he had ever faced. And, as the electric current was the reason for

Figure 2: Occurrence of the magnetic field at the electric current flow in Oersted's experiment.

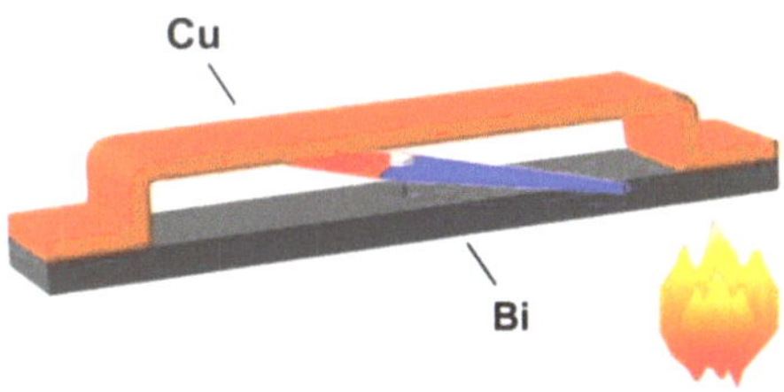

Figure 3: Schematic of the Seebeck's discovery of thermomagnetism.

magnetic field induction occurrence in Oersted's discovery, it was but natural to assume that the magnetic field in the Seebeck's experiment was excited by current as well (Fig. 4).

To prove his assumption, Oersted broke the circuit in Seebeck's experiment (Fig. 5) and connected a galvanometer which registered the electric current to the break.

In 1823, Oersted published the results obtained in the *Annales de Chimie et de Physique*. In the same year, he reported on his discoveries to the French Academy of Sciences [13].

He started his report with words: 'I have the honour to present to the Assembly the amazing experiments that helped Seebeck prove the possibility of obtaining electric current in the circuit consisting of solid conductors exceptionally by distorting temperature balance in them'. He suggested that this effect should be called *Thermoelectricity*.

It is evident that Oersted had not forgotten about Seebeck. Furthermore, he granted Seebeck with the priority in the discovery of thermoelectricity.

Such a name for the Seebeck effect was correct and so it found total support among scientists.

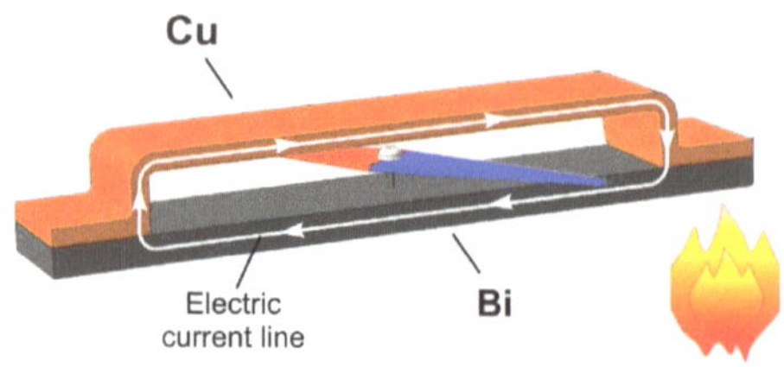

Figure 4: Idea of the current excitation in the Seebeck's experiment.

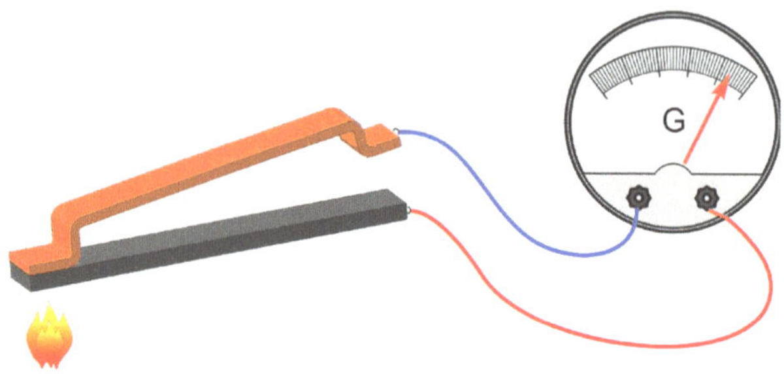

Figure 5: Oersted's experiment that verified the electric nature of the Seebeck effect.

Despite this, Seebeck did not acknowledge the evident results of Oersted's experiments. In his book *Semiconductor thermoelements*' Ioffe writes the following: 'Seebeck not only refused to accept such natural explanation of his discovery but fought vigorously against it within several years accusing representatives of the theory of thermoelectric currents of being fascinated with the 'fashion' due to Oersted's discovery' [14].

Ioffe suggested that such Seebeck's behavior was caused by his wish to explain the nature of the magnetic field of the Earth with a help of his effect of thermomagnetism.

All the same, it became generally accepted to define the Seebeck effect as thermoelectromotive force that appears at the free ends of the broken electric circuit that consists of two dissimilar materials A and B, provided the place of their connection is heated and free ends are kept under the steady temperature (Fig. 6).

Oersted had a good understanding of the possibility of practical application of the Seebeck effect for the creation of sources of electricity. In 1824, together with Fourier, he fabricated the first thermoelectric generator which, due to its reliability and stability, was further used in a great number of important experimental researches.

Such thermoelectric source of electricity was employed by Ohm. He obtained rather precise results on the connection between the electric voltage, current and resistance of the electric circuit due to the exceptional stability of the source (Fig. 7). These experimental results served as a base for discovery by Ohm of his law in 1826 [15].

Peltier also used a thermoelectric power source for his experiments. The said experiments led to the discovery of still another thermoelectric effect named after Peltier.

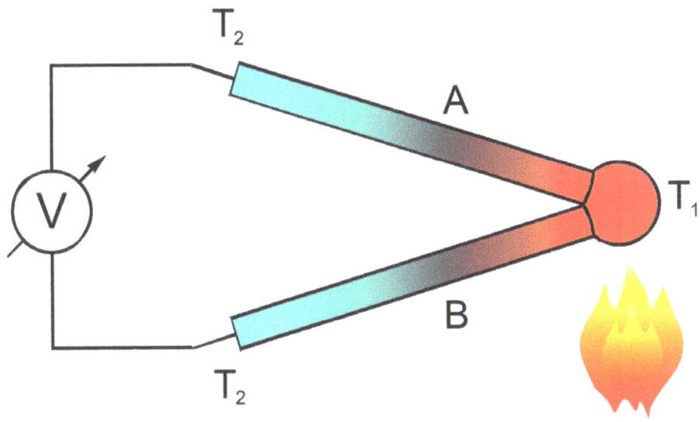

Figure 6: Seebeck effect.

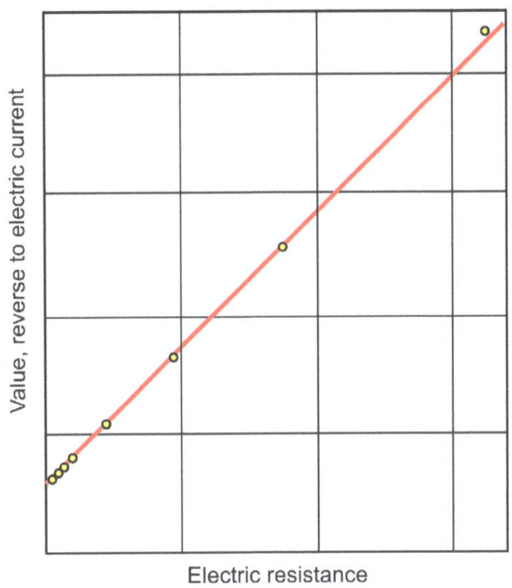

Figure 7: Results of Ohm's experiments on the linear dependence between the electric resistance and current. Experimental results are marked with circles.

2.2 Second thermoelectric effect: the Peltier effect

Peltier, a watchmaker from Paris, started self-education at the age of 30 and became attracted by electric measurements that were in trend in those days. He started with the measurements of bismuth and antimony conductivities for which purpose he cast the 0.5 mm diameter rods, 45 mm long. These rods played an important role in the discovery of his famous effect by Peltier.

Peltier also tried to check one of the theories of emergence of heat while the electric current passes through the conductor. For this purpose, he used rods sending current through them and measuring the level of their heating with the help of a thermocouple (Fig. 8). It was a French physicist Becquerel who suggested temperature measuring with the help of thermocouple in 1826. Peltier found out that in the central part of the rod, as it was expected, the temperature was nearly the same and higher than the ambient temperature. But, to his surprise, he discovered that the temperature on one end of the rod turned out to be higher than that on the opposite.

Temperature abnormalities were observed at the place of dissimilar materials junctions. That was nothing but the Peltier effect [16].

At first, the Peltier's discovery was treated with suspicion because the temperature abnormalities were quite small and Peltier's reputation as a researcher was not high enough.

More explicit data were given by the so called Peltier thermal cross (Fig. 9) that consisted of the said two bars (bismuth and antimony) connected with each other in the middle. Electric current was sent through two nearest ends of the bars while two other parts of the bars formed a thermocouple for measurement of the temperature at the place of their connection. When the current was sent one way, the junction became heated; at the reverse direction, it became cooled. Even this experiment could not persuade skeptics in the existence of the Peltier effect. It was Lentz (1838) who put the end to these arguments by having successfully frozen water with the help of bismuth–antimony alloy.

The events enumerated above demonstrate a very uncertain situation regarding thermoelectric effects that existed by the time Thomson started his research into thermoelectricity. The experimental data, especially those of the Peltier effect, were, in general, of the qualitative character, unlike, for instance, rather precise Ohm's researches, which led him to the discovery of his law.

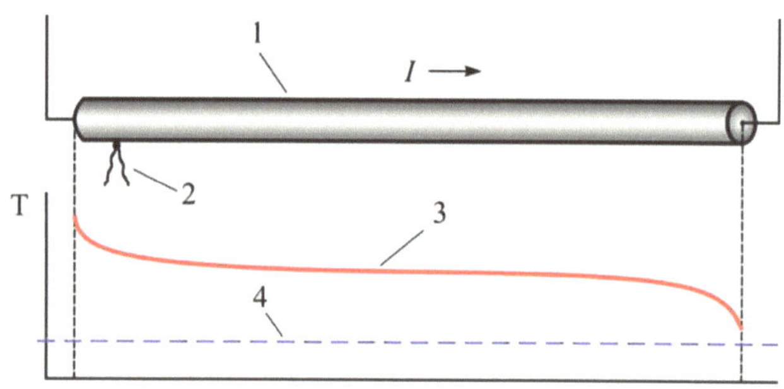

Figure 8: Peltier's experiment schematic. 1 – Bi or Sb rod, 2 – thermocouple for temperature measurement, 3 – temperature along the rod, 4 – ambient temperature.

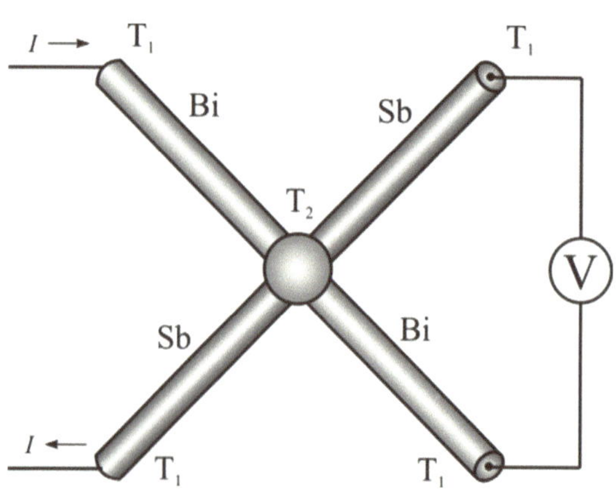

Figure 9: Peltier thermal cross.

There existed absolutely no qualitative or quantitative connections between the Seebeck and Peltier effects. A theory was absent that might have been able to describe them.

That is why the foundation of the thermoelectric effect theory on such a shaky basis demanded from Thomson not only thorough understanding of the essence of thermoelectric phenomena but also enigmatic intuition of the scientist, facing the choice of major factors and neglecting of minor ones in order to describe processes which occur in a thermoelectric circuit.

3 Fundamental thermoelectric Thomson relations

Thermoelectricity was one of the first areas of physics where Thomson applied thermodynamics of equilibrium processes developed by him. Thomson's main guiding idea was the assumption that thermoelectric phenomena are reversible and so might be described with the help of methods of equilibrium thermodynamics of reversible processes.

3.1 First thermoelectric Thomson relation

Thomson used a physical model of a thermoelectric circuit (Fig. 10) which consisted of two conductors whose junctions where kept under absolute temperatures T and $T + dT$. Thermo-EMF αdT appearing in the circuit carries the electric charge q through the conductors. Here, α is the Seebeck coefficient for the pair of metals. When thermoelectric current passes through this circuit, Peltier heat πq is emitted at one junction and absorbed at the cold one $(\pi + d\pi)q$, π being the Peltier coefficient. According to the second law of thermodynamics, Peltier heat πq is converted into electric energy $q\alpha dT$ with the maximum possible within the Thomson thermodynamic theory efficiency of the Carnot cycle $\frac{dT}{T}$

$$q\alpha dT = \frac{dT}{T}\pi\alpha. \tag{1}$$

Hence, the first Thomson thermoelectric relation appears [5] that combines the coefficients of two thermoelectric phenomena, that is,those of Seebeck and Peltier:

$$\pi = \alpha T. \tag{2}$$

3.2 Second Thomson thermoelectric relation

Thomson also considered that to preserve heat balance in the thermoelectric circuit, there should be another distributed thermoelectric effect that leads to either heat emission or heat absorption in the bulk of semiconductor with the current provided the temperature balance is violated there (Fig. 11). Therefore, in each of conductors A and B that form a thermoelectric circuit heat $\tau_a qdT$ and $\tau_b qdT$ should be either emitted or absorbed; where τ_a, τ_b here are the Thomson coefficients of materials. The law of energy conservation for a thermoelectric circuit has then the form of

$$\alpha qdT - (\pi + d\pi)q + \pi q + \tau_a qdT - \tau_b qdT = 0, \tag{3}$$

where the first component is electric energy, the second and third are Peltier heat at junctions, the fourth and the fifth are Thomson heat in conductors A and B.

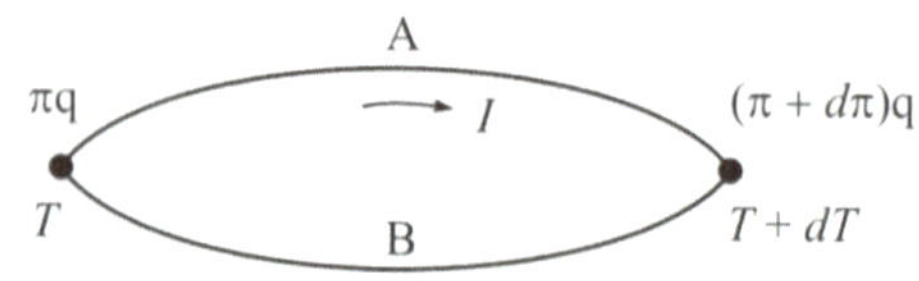

Figure 10: Thomson's thermoelectric circuit.

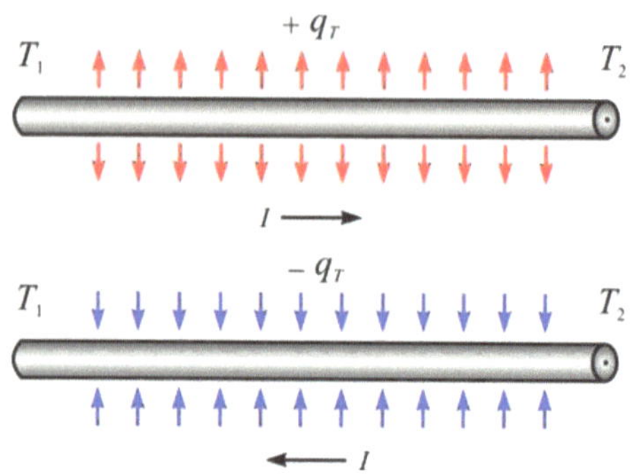

Figure 11: Thermoelectric Thomson effect.

Reducing the last equation by q and termwise dividing it by dT, as well as using from expression (1)

$$\frac{d\pi}{dT} = \frac{d\alpha}{dT} T + \alpha$$

it was obtained that

$$\tau_a - \tau_b = T \frac{d\alpha}{dT}. \tag{4}$$

The formula (4) is the second Thomson relation and the effect of Thomson heat occurrence is called the thermoelectric Thomson effect [1,5].

These results were published by Thomson in his work 'On a Mechanical Theory of Thermoelectric Currents' in December 1851. They gave start to the creation of the phenomenological theory of thermoelectricity.

This was the first case of theoretically predicted thermoelectric effect unlike the Seebeck and Peltier effects which were, by common knowledge, discovered by chance.

4 Experimental verification of Thomson thermoelectric effect

Having received the first and the second relations for thermoelectricity, Thomson was well aware of the fact that they were based on the hypothesis which he formulated as follows: 'The electromotive forces produced by inequalities of temperature in a circuit of different metals, and the thermal effects of electric currents circulating in it, are subject to the laws which would follow from the general principles of the dynamical theory of heat if there were no conduction of heat from one part of the circuit to another' [5].

Thomson saw no ways to prove his hypothesis by theoretical means so the crucial role here he gave to the experiment: 'In adopting this hypothesis, it must be distinctly understood that it is only a hypothesis, and that, however probable it may appear, experimental evidence in the special phenomena of thermoelectricity is quite necessary to prove it' [5].

Such experimental verification would be well grounded on the base of the reliably discovered and measured Thomson effect.

Therefore, Thomson started active search for his effect. First experiments gave no positive results as well as the following ones with no positive effect.

It was only at the end of November 1853 that Thomson reached the desirable target [6]. Considering the reasons for previous failures, an experimental setup was created where, to improve heat exchange, the layered samples of metal stripes were used instead of bars. They went through three containers with water (Fig. 12). The middle container was filled with hot water, and the outside ones - with cold. Water temperature was registered with thermometers. Such a setup enabled creation of temperature gradients of different direction along the conductor. When electric current was sent through it, such conditions led to emission and, correspondingly, absorption of the Thomson heat which was registered with the thermometers. When the current was reversed, the change in temperatures exactly corresponded to the predicted properties of the Thomson effect. This meant success. The prediction of Thomson thermodynamic theory of thermoelectricity was verified.

For a long time, the researchers failed to repeat Thomson's experiment. It was only in 1867 that the French physicist F. Le Roux [17] got indisputable proofs of the existence of the Thomson effect. It is interesting to notice that the main advantage in his experiment was the use of thermocouples instead of gas thermometers for temperature measurements.

5 On the effect of thermodynamic reversibility on Thomson relation

As stated above, Thomson was well aware that he had obtained his discoveries in thermoelectricity with the use of insufficiently reliable assumptions of the absence of the influence of irreversible effects connected with the Joule heat emission and heat transfer with the thermocouple thermal conductivity. Heat exchange with the environment was totally neglected [5].

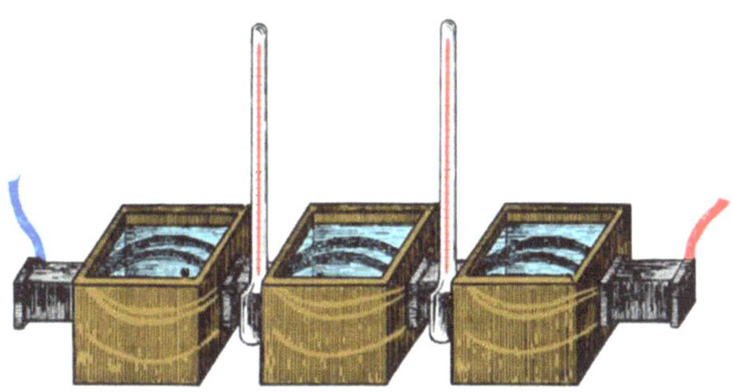

Figure 12: Experimental setup where Thomson observed his thermoelectric effect.

Such assumptions might have been approximately valid for the Joule heat only provided the currents were small. Actually, due to the quadratic dependence of the Joule heat on the current, its value at small currents might have appeared to be considerably smaller than that of the Peltier heat.

Thomson's theory was severely criticized by Boltzmann. He pointed out that the Joule heat irreversibility may really be neglected when currents are small. Though, irreversibility caused by heat transfer due to the thermal conductivity through the thermocouple from one junction to another should be considered because both this heat and the Thomson effect are proportional to temperature gradients in the first power.

But more correct descriptions of thermoelectric effects based on the irreversible processes thermodynamics [18–25, etc.] gave results not differing from those discovered by Thomson earlier. Therefore, **The first Thomson relation is always true**.

6 Generalization of Thomson relations pursuant to thermodynamics of irreversible processes

Relations (2) and (4) were obtained in a more general form for the cases of anisotropic medium in the magnetic field. In accordance with the thermodynamic theory of irreversible processes, formulae (2) and (4) were obtained in the form of

$$T\alpha_{km}\left(\vec{B}\right)=\Pi_{mk}\left(-\vec{B}\right), \tag{5}$$

$$\tau_{T,ik}(\vec{B})=T\frac{\partial\alpha_{ki}(-\vec{B})}{\partial T}, \tag{6}$$

whereas a particular case for anisotropic medium, the relations obtained by Thomson follow from.

From the first and second Thomson relations for anisotropic medium there appears the interrelation between three tensors of thermoelectric coefficients

$$\hat{\tau}+\hat{\alpha}-\frac{\partial\hat{\Pi}}{\partial T}=0. \tag{7}$$

These fundamental relations allow, for example, easy calculation of Peltier coefficients, provided the Seebeck coefficients are known for the given temperature, or determination of the Thomson coefficients provided the Seebeck coefficients and their temperature dependencies are known.

Moreover, they ensure a deeper understanding of the physical nature of thermoelectric effects. Thus, by combining (5) and (6) for an anisotropic medium we have

$$\tau_{T,ik}=\frac{\partial\Pi_{ik}}{\partial T}-\alpha_{ik}. \tag{8}$$

As seen from the formula (8), two mechanisms participate in the formation of the Thomson heat. First, the existence of the temperature gradient leads to appearing of the thermoelectric field proportional to α_{ik}. Charges, while moving in this field, perform either positive or negative (depending on the current direction) work, thus leading to either emission or absorption of the heat. Second, changes in the Peltier coefficient $\partial\Pi_{ik}/\partial T$ in the non-isothermal medium stipulate the inhomogeneity of the media (even homogeneous in the isothermic conditions) thus leading to the Peltier effect along this homogeneity.

With the use of the relation (5) it may be shown that the isothermic Nernst-Ettingshausen coefficient is connected with the adiabatic Ettingshausen coefficient through the relation

$$TQ_i^{\perp} = P_E\kappa_{11}, \tag{9}$$

which is called the Bridgman relation and is a corollary of the application of the first Thomson relation (5) to gyrotropic medium.

7 First Thomson relation novel form

It is well known that the connection between electric field intensity vector **E**, current density **j** and resistivity ρ is defined from the Ohm law

$$\mathrm{E} = \rho\mathbf{j}, \tag{10}$$

whereas the relation between temperature gradient ∇T, heat flux vector **q** and thermal conductivity coefficient κ is given by the Fourier law

$$\mathbf{q} = -\kappa\nabla T. \tag{11}$$

These equations are valid when either electric or thermal processes occur in the medium. Therefore, these equations are not suitable for simultaneous thermoelectric processes description.

Generalization of eqns (10) and (11) for a thermoelectric medium were found [22,23,26] on the basis of the irreversible processes thermodynamics, in the form of

$$j_i = \sigma_{ik}E_k - \sigma_{ik}\alpha_{km}\frac{\partial T}{\partial x_m}, \tag{12}$$

$$q_i = \Pi_{lk}j_k - \kappa_{lm}\frac{\partial T}{\partial x_m}. \tag{13}$$

Here E_k is the electric field intensity component, j_k is the electric current density vector component, q_i is the heat flux density vector component, σ_{ik} is the electric conductivity tensor component, α_{im} is the thermo-EMF tensor component, κ_{im} is the thermal conductivity tensor component, Π_{ik} is the Peltier coefficient tensor component, indices i, k, l, m pass the meanings 1, 2, 3. Equations (12) and (13) are called generalized Ohm and Fourier laws, respectively. They are well known and applied to solution of various thermoelectric problems.

In [27], the condition of the correctness of the system of eqns (12) and (13) was determined by Luste. The system contains six non-uniform linear equations related to six quantities, namely three components of the electric field E_i and three components of the temperature gradient $\partial T/\partial x$. Such a system is known to have a unique solution provided its determinant does not equal to zero:

$$\begin{vmatrix} \rho_{11} & \rho_{12} & \rho_{13} & \alpha_{11} & \alpha_{12} & \alpha_{13} \\ \rho_{21} & \rho_{22} & \rho_{23} & \alpha_{21} & \alpha_{22} & \alpha_{23} \\ \rho_{31} & \rho_{32} & \rho_{33} & \alpha_{31} & \alpha_{32} & \alpha_{33} \\ \Pi_{11} & \Pi_{12} & \Pi_{13} & -\kappa_{11} & -\kappa_{12} & -\kappa_{13} \\ \Pi_{21} & \Pi_{22} & \Pi_{23} & -\kappa_{21} & -\kappa_{22} & -\kappa_{23} \\ \Pi_{31} & \Pi_{32} & \Pi_{33} & -\kappa_{31} & -\kappa_{32} & -\kappa_{33} \end{vmatrix} \neq 0 \qquad (14)$$

Condition (14) may be presented in the dimensionless form

$$Z^* T \neq -1, \qquad (15)$$

where

$$Z^* T = \frac{\Pi^* \alpha^*}{\rho^* \kappa^*}, \qquad (16)$$

$$\Pi^* = \vec{\tau}_0 (\breve{\Pi} \vec{j}_0), \qquad (17)$$

$$\alpha^* = \vec{\tau}_0 \left(\breve{\alpha} \vec{j} \right), \qquad (18)$$

$$\rho * = \vec{j}_0 (\breve{\rho} \vec{j}_0), \qquad (19)$$

$$\kappa^* = \vec{\tau}_0 (\breve{\kappa} \vec{\tau}_0), \qquad (20)$$

where $\vec{\tau}_0, \vec{j}_0$ are unit temperature gradient and electric current density vectors.

In expressions (15) and (16), the product $Z^* T$ is of special interest. This product usually appears when thermoelectric energy conversion is described in the expressions for the efficiency, where

$$ZT = \frac{\alpha^2}{\kappa \rho} T \qquad (21)$$

is the most significant parameter of the thermoelectric material quality on which thermoelectric energy conversion efficiency depends.

Actually, in an isotropic medium kinetic coefficients $\breve{\rho}, \breve{\sigma}, \breve{\alpha}, \breve{\Pi}, \breve{\kappa}$ are scalars and hence, if the first Thomson relation is used, we obtain from (21) for an isotropic medium $Z^*T = ZT$.

But ZT and Z^*T were not obtained from the energy description of thermoelectric energy conversion. Therefore, this parameter has a wider meaning. It determines the compatibility of the Ohm and Fourier generalized laws at which the said laws describe thermoelectric phenomena simultaneously. Hence, when Z^*T is used, the conditions under which thermoelectric phenomena occur can be defined.

Let us give a more detailed attention to the said conditions.

Suppose we generalize the first Thomson thermoelectric relation for the case of free anisotropic media in the magnetic field. For this purpose, we need to compare the generalized coefficient Π^* with the product α^*T:

$$\Pi^* = \tau_0 \left(\hat{\Pi} j_0\right),$$

$$\alpha^* T = T\tau_0 (\hat{\alpha} j_0) = \Pi^* \delta, \tag{22}$$

where

$$\delta = \frac{\alpha^*\left(\vec{B}\right)}{\alpha^*\left(-\vec{B}\right)}. \tag{23}$$

Having taken into consideration (4), the dimensionless parameter δ may be presented in the form

$$\delta\left(j_0, \tau_0, B\right) = \frac{j_0 \hat{\alpha}(B) \tau_0}{j_0 \hat{\alpha}\left(-B\right) \tau_0}, \tag{24}$$

From which it is evident that this parameter in the absence of the magnetic field becomes

$$\delta\left(\vec{j}_0, \vec{\tau}_0, 0\right) = 1. \tag{25}$$

In the presence of the magnetic field, (23) may become the value different from (Fig. 1) for that case only, if

$$\hat{\alpha}\left(B\right) \neq \hat{\alpha}\left(-B\right). \tag{26}$$

The said becomes possible provided a change in the Seebeck coefficient takes place when the direction of the magnetic field becomes reverse. It does not occur in the majority of materials, though, hence

$$\hat{\alpha}\left(B\right)=\hat{\alpha}\left(-B\right) \text{ and } \delta = 1 \tag{27}$$

But, for some low symmetry crystals, a so-called commutation effect [14] was observed when the change in the magnetic field direction at the absolute value of the field intensity preservation leads to a significant change of the Seebeck coefficient. As it was shown in reference [28], on the basis of the thermodynamic theory, the parameter δ should satisfy the condition

$$-1 < \delta \leq 1, \tag{28}$$

and generalized thermal efficiency $Z^{*}T$ may lay within the limits of

$$-1 < Z^{*}T < \infty. \tag{29}$$

The last equation also answers the question that was a subject of a lively discussion within many years, namely on the so-called 'theoretical limit' of $Z^{*}T$ thermal efficiency of thermoelectric materials. It follows from the said equation that the upper positive limit for $Z^{*}T$ does not exist and from the standpoint of thermodynamics generalized thermal efficiency may be arbitrary large. In the same time, in the crystals where commutation effect is possible, that generalized thermal efficiency $Z^{*}T$ might be even negative, not exceeding the figure of one in absolute value, though.

From eqns (22) and (23), considering thermodynamic limits for δ, it is evident that

$$\Pi^{*} \geq \alpha^{*}T\,. \tag{30}$$

The above expression is the generalization of the first Thomson relation. It was first formulated by Luste.

For an isotropic medium only the relation (30) has the form of a strict equality (2), but for a general case, when the magnetic field is present in the crystals with low symmetry, it has a form of inequality.

Consequently, the application of irreversible processes thermodynamics to description of Thomson relations for more complicated anisotropic and gyrotropic media results in the development of the thermoelectricity thermodynamic theory as well as to the disclosure of new regularities whose physical meaning should be still understood.

8 On practical applications of Thomson relations

Apart from theoretical value, Thomson relations have practical applications. In energy applications (thermoelectric generators or coolers), though, the Thomson effect is rather small as compared to the effects of electric power generation, the Peltier effect, heat transfer along the legs of thermocouples and the Joule heat, and equals some percent of their value only. That is why the Thomson effect is usually neglected at energy devices design.

There are places, however, where it should not be neglected and those are measuring technique in the first place.

The Thomson effect should be considered at precise measurements of the thermoelectric energy converters, generator modules parameters (Fig. 13).

Properties of modules are determined at the application of various operating modes: open electric circuit, maximum efficiency, maximum power, etc. Here, the Thomson effect $\pm Q_T$ is considered in the heat and energy balances equation:

$$Q_0 = Q_\kappa - \frac{1}{2}Q_I + Q_\Pi \pm Q_T + Q_E + Q_A \tag{31}$$

where Q_0 is the heat power supplied to the module, Q_κ is the heat transferred due to the legs thermal conductivity, Q_I is the Joule heat, Q_Π is the Peltier heat, Q_T is the Thomson heat, $+Q_E$ is the external circuit electric power, and Q_A are the heat losses due to deviations from adiabaticity. The error in the definition of modules parameters where the Thomson effect is considered is lessened down to 1.5–3%.

Application of the first Thomson relation at the precise measuring of thermoelectric materials properties is also of great help [29] (Fig. 14). To measure the thermo-EMF coefficient, electric current is sent through the sample under measurements in the direction at which the junction

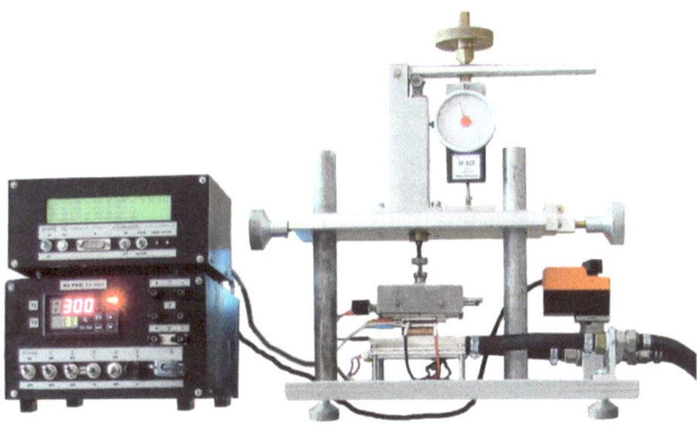

Figure 13: Equipment Altec-10002 for precise measurements of the generator module parameters with the Thomson effect considered [30].

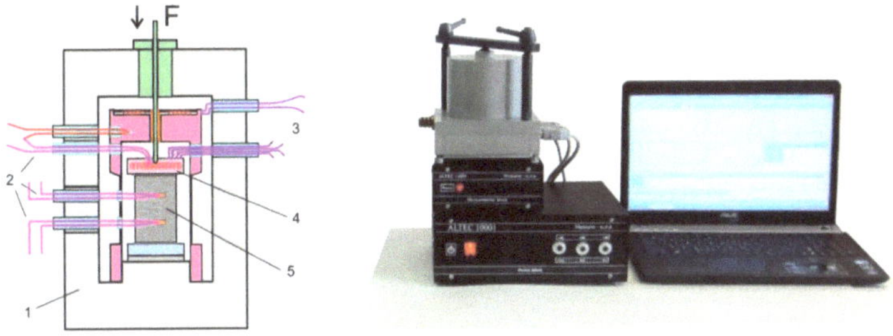

Figure 14: Schematic and appearance of the automated measuring setup for the material thermoelectric properties determination. 1 – thermostat, 2 – thermocouples, 3 – current leads, 4 –reference heater, 5 – sample.

adjoining the reference heater is cooled. After that current I is sent through the reference heater with resistance R, heat release here compensating for the Peltier effect. Such a compensation is more precisely defined by the absence of thermo-EMF in the sample.

In such a case

$$Q_{\Pi} = I^2 R, \tag{32}$$

and in accordance with the first Thomson relation

$$\alpha = \frac{I^2 R}{T}. \tag{33}$$

Of still more importance is consideration of the Thomson effect at the development of standards of alternating current and voltage. For this purpose, thermoelectric converters of the alternating current into direct one are used, which consist of an electric heater 1 and a thermal couple 2 [31].

Primarily, the required AC current $I_{\approx}$ is sent through the heater, thermo-EMF $E_{\approx}$ being fixed (Fig. 15). Then, direct current with a value providing EMF of the thermocouple equal to $E_{\approx} = E_{=}$ is sent through the heater. For this case, $I_{=}$ should be equal to the required value of the alternating current $I_{\approx}$. Since $I_{=}$ can be determined with the error of $10^{-4} - 10^{-6}$ percentage, the precision of reproduction of the alternating current unit will be totally dependent on the precision of a thermoelectric converter. The Thomson effect becomes here the main error that distorts the identity of temperature distribution in the heater when both direct and alternating current are sent through it. For this reason, a thermocouple is heated to different temperatures $I_{=} - I_{=} = \Delta T$ at equal functioning values of alternating and direct currents that results in the error of their comparison up to 5–8 percentage.

To reduce the said error, a thermopile is installed along the heater (Fig. 16) which helps minimize the Thomson effect operation error [31]. Application of such thermoelectric converters implements reproduction of the alternating current and voltage units with the accuracy up to 10^{-4} percentage. They are used for national standards and facilities for alternating currents measurements (Fig. 17). Film thermoelectric AC converters were developed in USA (NIST), Germany (PTB/IPHT), Korea (KRISS) [33,34], etc.

9. Thomson transversal EMF: anisotropic thermoelements

In his paper [35], Thomson scrutinized thermoelectric phenomena in crystalline media. For the uniaxial crystals case, he introduced the Seebeck thermo-EMF coefficients α_1 and α_2 along

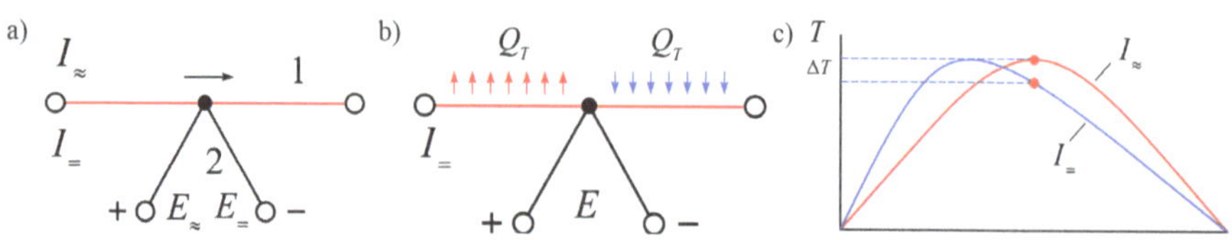

Figure 15: Schematic of the elemental thermoelectric AC converter (a), the Thomson effect Q_T (b) and symmetry distortion in temperature distribution in the heater (c) due to the Thomson effect.

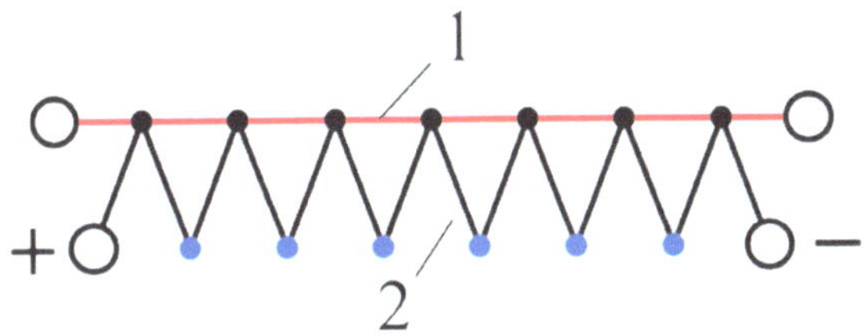

Figure 16: Schematic of the multiple-unit thermoelectric AC converter.

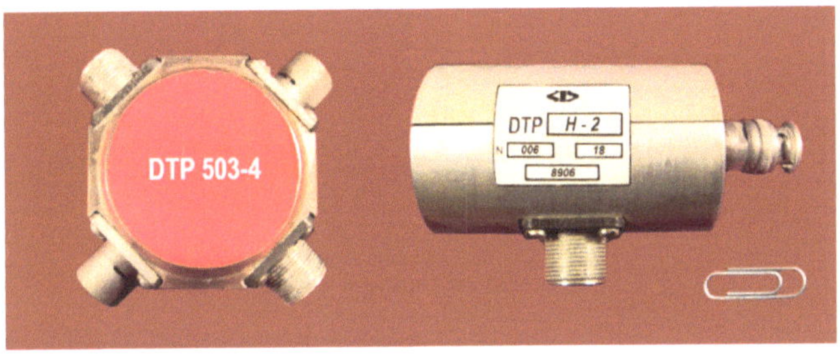

Figure 17: Thermoelectric converters of current (a) and voltage (b) for AC metrology [32].

crystallographic axes x_1 and x_2, which could have been different, $\alpha_{11} \neq \alpha_{22}$. Therefore, such crystals possess thermo-EMF anisotropy. If a rectangular sample is cut from this material (Fig. 18) at the angle φ to crystallographic axis x_1 and temperature gradient is established along the sample ∇T, thermo-EMF coefficient along this direction will be described by the formula

$$\alpha_{\parallel} = \alpha_{11}\cos^2\varphi + \alpha_{22}\sin^2\varphi. \tag{34}$$

Of principal importance here is the discovery of the Seebeck transversal effect by Thomson; the said effect occurring perpendicularly to the temperature gradient in an anisotropic material. Its value was calculated by Thomson in the form of

$$\alpha_{\perp} = \left(\alpha_{11} - \alpha_{22}\right)\sin\varphi\cos\varphi. \tag{35}$$

Somewhat later, Borelius [36], guided by the Thomson thermo-EMF transversal coefficient, wrote the value of the transversal thermo-EMF in the obvious form

$$E_{\perp} = \left(\alpha_{11} - \alpha_{22}\right)\sin\varphi\cos\varphi\frac{a}{b}\left(T_1 - T_2\right). \tag{36}$$

The fact that the value of thermoelectromotive forces transversal to the temperature gradient becomes dependent on the samples a and b geometry, is noteworthy.

In fact, it meant that, relying on the Thomson transversal thermo-EMF, it is possible to develop a thermoelectric energy converter (Fig. 19), different in essence from conventional thermocouples. It is traditionally called an anisotropic thermoelement.

Such a thermoelement has a set of attractive properties.

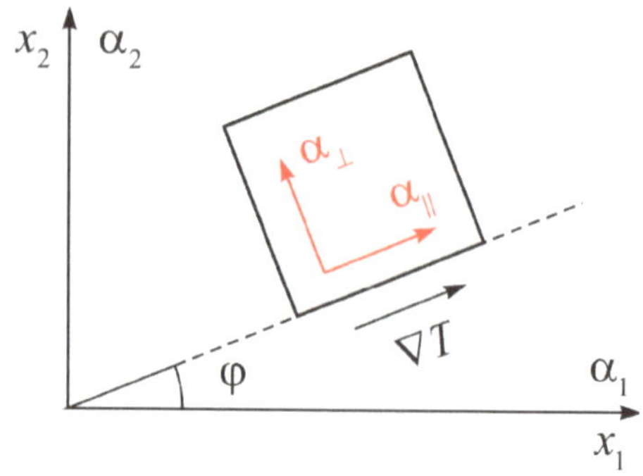

Figure 18: Conditions for the longitudal $a_\parallel$ and the transversal $a_\perp$ thermo-EMF coefficients occurrence.

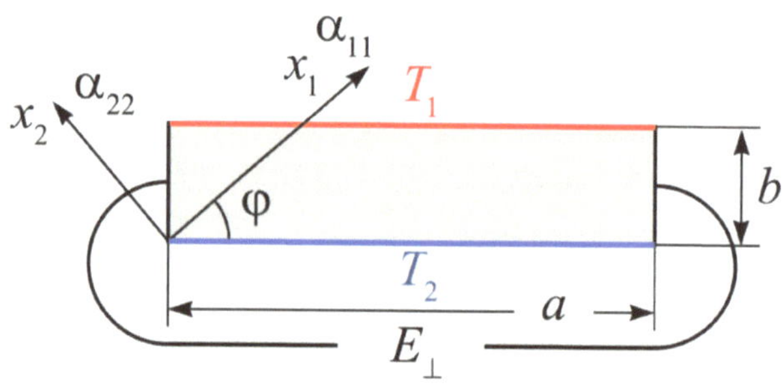

Figure 19: Schematic of an anisotropic thermoelement based on the Thomson transversal thermo-EMF.

First, unlike thermocouples, it may be fabricated from one material, thus excluding commutation problems, accompanying thermopiles.

Second, different from thermocouples, the value of the developed thermo-EMF depends on the sample geometry. The bigger the relation a/b, the higher the developed voltage.

Third, such a thermoelement is extremely manufacturable: once a rectangular sample with the anisotropic thermo-EMF is cut from the crystalline material, the electricity generator is ready.

The patent for the thermoelectric electricity source based on the Thomson transversal EMF was received in 1964 [37,38].

The efficiency of such a thermoelement is determined from the expression [39].

$$\eta = \frac{1}{4} Z_A \eta_K \bar{T}, \tag{37}$$

where

$$Z_A = \frac{\left(\alpha_{11} - \alpha_{22}\right)^2 \sigma}{\kappa}. \tag{38}$$

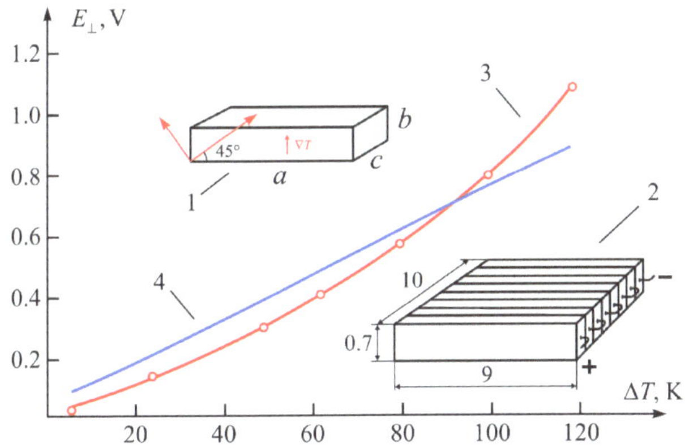

Figure 20: Results of testing of a thermoelectric generator of anisotropic elements. 1 – basic thermoelement, 2 – thermopile of anisotropic thermoelements, 3 – thermopile electric voltage $E_{\perp}$, 4 – calculated values $E_{\perp}$, ΔT – temperature difference on upper and lower surfaces of the battery.

In these expressions η_K is the Carnot efficiency, $\bar{T}$ is an average temperature, σ and κ are electric and thermal conductivities of the material, respectively. The expression Z_A characterizes thermoelectric material figure of merit and bears the same meaning as the Ioffe criterion for thermocouple elements.

The first thermoelectric generator based on the Thomson transversal EMF was fabricated and examined by the author of the present work in 1964 [40].

For experiments single crystals of CdSb with the thermo-EMF anisotropy $\alpha_{22} - \alpha_{33} = 150\ \mu\text{V/K}$ were used. Eight anisotropic thermoelements with the dimensions $a = 9$ mm, $b = 7$ mm, $c = 7$ mm , whose edge a was oriented along the crystallographic axis $< 100 >$ at the angle of 45° to the direction of the temperature gradient were cut from such material (Fig. 20). With the said orientation (see Formula 36) the value $E_{\perp}$ reaches its maximum. Thermoelements were connected in series into a thermopile and placed between the flat heater and the cooler with a help of thin electrical insulation. CdSb low thermal conductivity ($\approx 2{\cdot}10^{-2}$ W/cm·K) allowed easy obtaining of sufficiently big temperature differences. At $T_1 = 23$°C and $T_2 = 140$°C, a transversal thermo-EMF 1.1 V was obtained.

The above results gave a convincing corroboration of the existence of the Thomson transversal thermo-EMF, as well as the correctness of the Borelius's formula together with the possibility of developing of thermoelectric electricity sources based on the transversal thermo-EMF.

As follows from (38), thermo-EMF anisotropy has the most significant effect on a thermoelectric material figure of merit. Unlike electric conductivity σ and thermal conductivity κ that always occur in electroconductive materials, thermo-EMF anisotropy $\Delta\alpha$ is quite a rare phenomenon. The reason for this is that for $\Delta\alpha$ to occur specific conditions are required. The following mechanisms should be present for thermo-EMF anisotropy occurrence:

(1) Existing of several scattering mechanisms in the presence of one type of carriers (either electrons or holes).

(2) Photon-electron drag effect.

(3) Simultaneous existence of several carrier types in the presence of a single anisotropic scattering mechanism.

The analysis has shown that for the first case the thermo-EMF anisotropy is insignificant and, thus, of no practical interest. For the second case, though, the growth of the thermo-EMF anisotropy is accompanied by the growth of thermal conductivity, neither of the said resulting in the figure of merit Z_A of the material but insignificant improvement. Of most prospective character is the third mechanism. In particular, when two types of carriers, electrons and holes, are present, maximum values of the thermo-EMF anisotropy are obtained if at least one of the conditions below is satisfied:

$$\begin{aligned} K_p / K_n \gg 1, \\ K_n / K_p \gg 1, \end{aligned} \tag{39}$$

where $K_n = \dfrac{\sigma_{22}^n}{\sigma_{11}^n}$; $K_p = \dfrac{\sigma_{22}^p}{\sigma_{11}^p}$; σ_{11}^n and σ_{22}^n are electronic constituents of the biaxial semiconductor electric conductivity tensor components, whereas σ_{11}^p and σ_{22}^p are their hole constituents.

Maximum thermo-EMF anisotropy values are obtained in the mixed conductivity area by introducing a donor impurity provided hole mobility u_p is larger than electron mobility u_n, or that of acceptor impurity, if $u_p < u_n$.

The above results were confirmed empirically for the CdSb anisotropic material, doped with impurity (Fig. 21). The highest known values for the thermo-EMF anisotropy $\Delta\alpha$ were obtained together with the corresponding increase in Z_A.

And still, figure of merit Z_A absolute values remain low enough. This explains the limited application of anisotropic thermoelements as generators. Search for the high-efficient anisotropic materials belongs to the future.

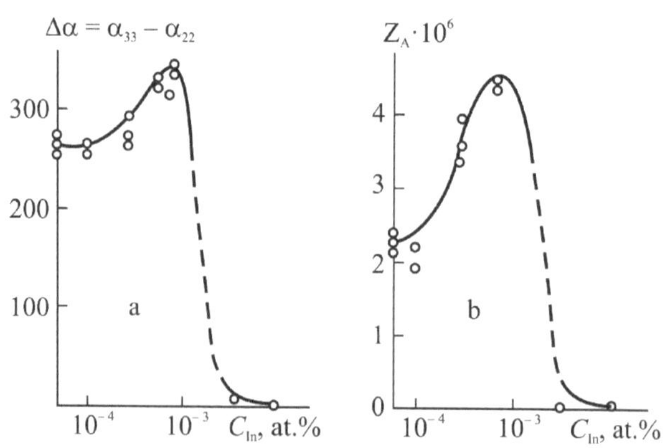

Figure 21: Dependencies of thermo-EMF anisotropy $\Delta\alpha$ (a) and thermoelectric figure of merit (b) of CdSb single crystals in concentration at 330 K.

The existing thermo-EMF anisotropy high values together with the possibility of increasing the anisotropic thermoelements electromotive forces by simple increase in a/b ratio makes them more and more attractive for the application of susceptible sensors. The increase in anisotropic thermoelements response time by means of their thickness lessening also serves their further promotion.

Notwithstanding the attractive features of anisotropic thermoelements, they have not yet enjoyed wide practical application as thermogenerators because the materials with high values of the anisotropic figure of merit Z_A have not been found yet. The solution belongs to the future.

However, anisotropic thermoelements are finding a wider application in measuring technology. Potentiality of increase in thermoelectric sensors sensitivity by simple increase of the relation a/b makes them more and more attractive. Moreover, the speed of such devices rises with the decrease of the anisotropic thermoelement thickness.

That is why such sensitive and high-performance elements have found application as laser radiation receivers for power energy and energy distribution determination, especially in the high-power technological lasers (Fig. 22). Anisotropic thermoelements are of great use when applied in various heat-flux sensors for remote temperature measurements in different industrial processes.

Of special interest are thermoelements based on the Thomson anisotropic thermo-EMF made in the form of flat rectangular spirals cut from one single crystal [41,42]. Each turn of the spiral (Fig. 23) is nothing but four anisotropic elements connected in series. Electric voltage on the ends of the spiral occurs when the inner cavity of the spiral is heated whereas its outer surfaces are kept under the set temperature. Electric voltage here increases in proportion to the number of spiral turns. For instance, a spiral made of CdSb with the volume of 1 cm^3 at the temperature difference 50 K develops electric voltage as high as about 1000 V!

Such spiral structures possess exceptionally high sensitivity and register heat fluxes that approach the level of inner thermal noises. As commutation junctions are practically absent in such spirals, it excludes the influence of contact noises, thus favoring obtaining of ultimate sensitivity.

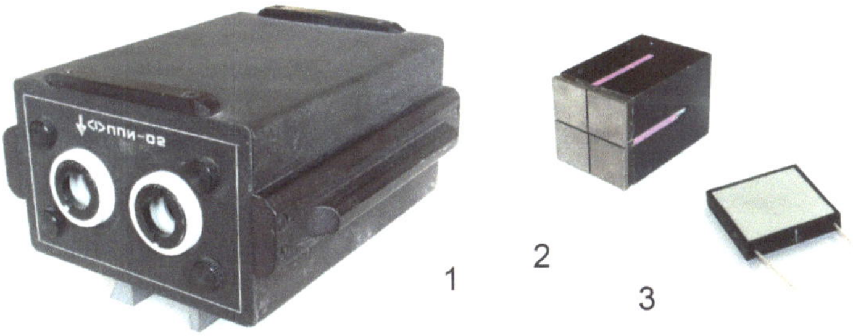

Figure 22: Devices with anisotropic thermoelements for radiation registering designed and developed in the Institute of Thermoelectricity: 1 – absolute differential receiver for continuous laser radiation power measurements, 2 – basic matrix for power laser CO_2 measurements, 3 – laser emission sensor with the sensitivity of 3 V/W.

The study of the properties of heat flux sensors made from such spirals was given special attention to. For this purpose, the information and energy theory of thermoelectric measuring systems [41] was applied, where the specific informativity equaling to the number of information bits, produced by a sensor per unit of time, was used as a sensor universal parameter.

Such an approach enabled creation of the methods of computer simulation of measuring devices with the said sensors to reach ultimate sensitivity. For example, with such approaches employed, superinformative microcalorimeters with ultimate values of sensitivity and high performance were developed (Fig. 24). Their application allows solution of both practical and scientific problems, which could not be realized with the help of conventional measuring equipment.

Thus, they help in determining of the expected lifetime of lithium chemical electricity sources, those sources being used in the devices of high importance, where their premature failure is inadmissible. The information on heat products of the microorganisms for the determination of the optimal dosage of antibiotics was obtained with such microcalorimeters. Characteristic dependencies of microorganisms' heat generation were found that could help in their identifi-

Figure 23: Anisotropic thermoelement of spiral shape.

Figure 24: Ultrasensitive fast-acting differential microcalorimeter (10^{-8} W), containing spiral anisotropic thermoelements, using the Thomson transversal thermo-EMF (1986) [43].

cation. Oscillating processes of their vital heat production important for search of mechanisms of microorganisms' informational exchange were discovered.

An important fact should be noted here that the increase in the energy needs of a human being is about 4% per year. The need in information at the same time rises as high as 100% every 2 years. Therefore, the increase in information becomes more important for humanity progress than energetics (Fig. 25).

The significance of informational systems becomes vital in such conditions. The role of measuring devices with thermoelectric sensors also increases.

It would be appropriate to remind here that it was Mendeleev, a metrologist by profession, who affirmed that, if the resolution of measuring devices is improved by one order, it would be a good premise for new discoveries in natural sciences.

Information and energy analysis shows that application of thermoelectric sensors, the ones with the Thomson's transversal thermo-EMF included will enable improvement of informativity of a set of important measuring devices ten- to hundredfold (Fig. 26). That is why, a transversal thermoelectric effect discovered by Thomson remains highly in demand in the present and will remain as such in the future.

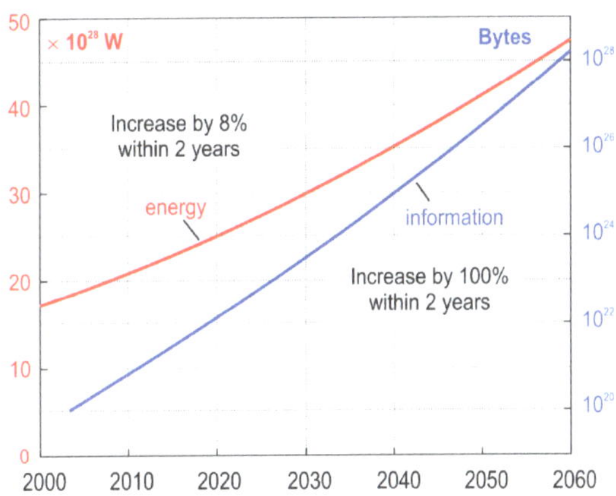

Figure 25: Comparison of prospects of energy and information consumption.

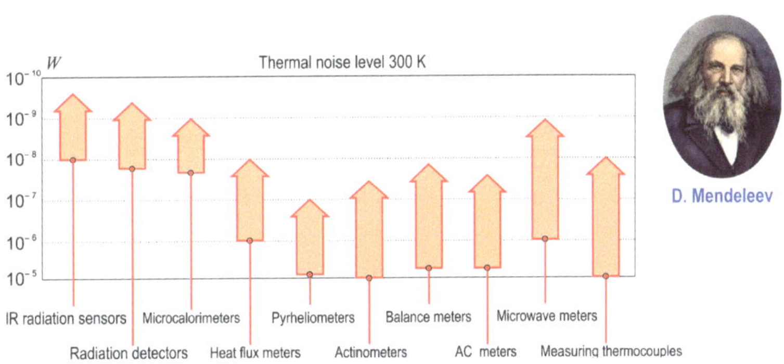

Figure 26: Prospects of development of measuring technology based on thermoelectric sensors application.

References

[1] Thomson, W., On a mechanical theory of thermoelectric currents. *Proceedings of Royal Society of Edinburgh*, **III**, pp. 91–98, 1857.

[2] Thomson, W., Account of researches in thermo-electricity. *Proceedings of the Royal Society of London*, **VII**, pp. 49–58, 1855.

[3] Thomson, W., Account of experimental investigations to answer questions originating in the mechanical theory of thermo-electric currents. *Proceedings of the Royal Society of Edinburgh*, **VII**, pp. 255, 1857.

[4] Thomson, W., Dynamical theory of heat (part VI), A mechanical theory of thermo-electric currents in crystalline solids. *Proceedings of the Royal Society of Edinburgh*, **III**, pp. 255–256, 1857.

[5] Thomson, W., On the dynamical theory of heat (part V), Thermo-electric currents. *Transactions of the Royal Society of Edinburgh*, **XXI**, pp. 123–171, 1854.

[6] Thomson, W., The Bakerian lecture – on the electro-dynamic qualities of metals. *Philosophical Transactions of the Royal Society of London*, **146(3)**, pp. 649–751, 1856.

[7] Thomson, W., On the thermoelectric position of aluminium (part 2). *Mathematical and Physical Papers*, **II**, pp. 181, 1884.

[8] Thomson, W., On the effects of mechanical strain on the thermo-electric qualities of metals (Part 2). *Mathematical and Physical Papers*, **II**, pp. 173–174, 1884.

[9] Thomson, W., Apercu sur des recherches relatives aux effets des courants electriques dans des conducteurs inegalement echauffes, et a d'autres points de la thermo-electricite. *Mathematical and Physical Papers*, **I**, pp. 460–463, 1882.

[10] Velmre, E., Thomas Johann Seebeck (1770-1831), *Proc. Estonian Acad, Sci. Eng.* **13**, pp. 276-282, 2007.

[11] Oersted, H.C., Experimenta circa effectum conflictus electrici in acum magneticam. *Journal für Chemie und Physik*, **29**, pp. 275–281, 1820.

[12] Seebeck, T.J., Magnetische Polarisation der Metalle und Erze durch Temperatur-Differenz, Abhandlungen der Königlichen Akademie der Wissenschaften zu Berlin, Königliche Akademie der Wissenschaften, pp. 289–346, 1825.

[13] Fourier J.-B.J. & Oersted, H.C., Sur quelques nouvelles Experiences thermoelectriques faites. *Annales de Chimie et de Physique*, **22**, pp. 375–389, 1823.

[14] Ioffe, A.F., *Semiconductor Elements*, Publishing House of the Academy of Science of the USSR: Moscow-Leningrad, 1969.

[15] Ohm, G.S., Bestimmung des Gesetzes, nach welchem Metalle die Contaktelektricit, nebst einem Entwurfe zu einer Theorie des Voltaischen Apparates und des Schweiggerschen Multiplicators. *Jahrbuch der Chemie und Physik*, **16**, pp. 137–166, 1826.

[16] Peltier, M., Nouvelles Experiences sur la Caloricite des courans electriques. *Annales de Chimie et de Physique*, **56**, pp. 371–386, 1834.

[17] Le Roux, F.R., Recherches sur les courants thermo-elecriques. *Annales de Chimie et de Physique*, **10**, pp. 201–291, 1867.

[18] Tolman, R.C. & Fine, P.C., On the irreversible production of entropy. *Reviews of Modern Physics*, **20**, pp. 51–77, 1948.

[19] Onsager, L., Reciprocal relations in irreversible processes. *Physical Review*, **37**, pp. 405–426, 1931.

[20] Callen, H.B., The application of Onsager's reciprocal relations to thermoelectric, thermomagnetic, and galvanomagnetic effects. *Physical Review*, **73**, pp. 1349–1358, 1948.

[21] Fieshi, R., De Groot, S.R. & Mazur, P., Thermodynamical theory of galvanomagnetic phenomena. *Physica*, **20**, pp. 67–76, 1954.
[22] Domenicali, C.A., Irreversible thermodynamics of thermoelectricity. *Reviews of Modern Physics*, **26(2)**, pp. 237–275, 1954.
[23] Samoilovich, A.G. & Korenblit, L.L., Recent state of the theory of thermoelectric and thermomagnetic phenomena in semiconductors. *Uspekhy Fizicheskikh Nauk*, **49(2)**, pp. 243–272, 1953.
[24] Nye, J.E., *Physical Properties of Crystals*, Oxford University Press: New York, 1957.
[25] Harman, T.C. & Honig, J.M., *Thermoelectric and Thermomagnetic Effects and Applications*, McGraw-Hill Book Co Press: New York, San Francisco, Toronto, London, Sydney, p. 377, 1967.
[26] Prigogine, I., *Introduction to Thermodynamics of Irreversible Processes*, Interscience Publishers Press: New York, London, p. 132, 1962.
[27] Luste, O.J., Thermodynamic limits of the thermoelectric figure of merit. *Journal of Thermoelectricity*, **1**, pp. 21–23, 1993.
[28] Tuz, Y.M., Dobrolyubova, M.V., Artyukhova, Y.V., Strunina, Y.A., Bodnaruk, V.I. & Taschuk, D.D., AC voltage unit standard based on thermoelectricconverters. *Journal of Thermoelectricity*, **3**, pp. 76–86, 2012.
[29] Anatychuk, L.I. & Lysko, V.V., On improvement of the accuracy and speed in the process of measuring characteristics of thermoelectric materials, *11th European Conference on Thermoelectrics*, November 18–20, Noordwijk, The Netherlands, p. 64, 2013.
[30] Anatychuk, L.I. & Havrylyuk, M.V., Procedure and equipment for measuring parameters of thermoelectric generator modules. *Journal of Electronic Materials*, **40(5)**, pp. 1292–1297, 2011.
[31] Bodnaruk, V.I. & Taschuk, D.D., Status and trends in development of AC measuring thermoelectric converters. *Journal of Thermoelectricity*, **1**, pp. 72–81, 2009.
[32] Anatychuk, L.I. & Bodnaruk, V.I., Thermoelectric AC converters for measuring equipment. *Journal of Thermoelectricity*, **1**, pp. 68–82, 2007.
[33] Klonz, M., Laiz, H. & Kessler, E., Development of thin-film multijunction thermal converters at PTB/IPHT. *IEEE Transactions on Instrumentation and Measurement*, **50(6)**, pp. 1490-1498, December 2001.
[34] Lee, Y.H., Kwon, S.W., Kim, K.J., Park, S.I. & Ihm, Y.E., Performance improving of KRISS micromachined multijunction thermal converter. *Digest of Conference on Precision Electromagnetic Measurements*, London, p. 591, June 2004.
[35] Thomson, W., On thermoelectric currents in linear conductors of crystalline substance. *Mathematical and Physical Papers*, **I**, pp. 266–273, 1882.
[36] Borelius, G., Grundlagen des metallischen Zustandes. *Handbuch der Metallphysik*, Vol. 1, Akademische Verlagbuchhandlung: Leipzig, pp. 185–485, 1935.
[37] Pilat, I.M., Samoilovich, A.G. & Anatychuk, L.I., Thermoelement, USSR Author's Certificate SU230915, 1963.
[38] Anatychuk, L.I., Pilat, I.M. & Samoilovich, A.G., Thermoelement, Patent JP652882.
[39] Samoilovich, A.G. & Slipchenko, V.N., Research on the efficiency of anisotropic thermoelements. *Fizika i Tekhnika Poluprovodnikov*, **9**, pp. 1897–1901, 1975.
[40] Anatychuk, L.I., Investigation of electric, thermoelectric and galvanomagnetic properties in cadmium–antimony, Dissertation for a degree of candidate of physical and mathematical sciences, Chernivtsi, p. 254, 1964.
[41] Anatychuk., L.I., *Thermoelectricity*, Vol. 1, Institute of Thermoelectricity Press: Kyiv, Chernivtsi, pp. 321–323, 1998.

[42] Anatychuk, L.I. & Luste, O.J., Thermoelement, Patent UK1335303, 1971.
[43] Anatychuk, L.I. & Luste, O.J., *Microcalorimetry*, Vyscha Shkola Press: Lviv, p.180, 1981.

Chapter 14

Kelvin and the age of the earth

J.A.T. Halls[1] & D. Wilkie[2]
[1]*Brunel Institute for Bioengineering, Brunel University London, United Kingdom.*
[2]*Formerly of British Nuclear Fuels, United Kingdom.*

Abstract

'For having, in the natural history of this earth, seen a succession of worlds, we may from this conclude that there is a system in nature; in like manner as, from seeing revolutions of the planets, it is concluded, that there is a system by which they are intended to continue those revolutions. But if the succession of worlds is established in the system of nature, it is in vain to look for anything higher in the origin of the earth. The result, therefore, of our present enquiry is, that we find no vestige of a beginning, – no prospect of an end.' James Hutton, Theory of the Earth, 1788.

'I think we may with much probability say that the consolidation cannot have taken place less than 20,000,000 years ago, or we should have more underground heat than we actually have, nor more than 400,000,000 years ago, or we should not have so much as the least observed underground increment of temperature.' Sir William Thomson (Lord Kelvin), On the Secular Cooling of the Earth, 1862.

'I do not presume to throw the slightest doubt upon the accuracy of any of the calculations made by such distinguished mathematicians as those who have made the suggestions I have cited. On the contrary, it is necessary to my argument to assume that they are all correct. But I desire to point out that this seems to be one of the many cases in which the admitted accuracy of mathematical processes is allowed to throw a wholly inadmissible appearance of authority over the results obtained by them. Mathematics may be compared to a mill of exquisite workmanship, which grinds you stuff of any degree of fineness; but, nevertheless, what you get out depends on what you put in; and as the grandest mill in the world will not extract wheat-flour from peascods, so pages of formulæ will not get a definite result out of loose data.' Thomas Henry Huxley, Address to the Geological Society, 1869.

1 Early estimates

Speculation on the age of the earth has existed for millennia. Many, like Aristotle have suggested that it is infinitely old, but most in early years have taken it to be finite and essentially unchanging albeit with variations in weather and with occasional volcanic eruptions and earthquakes superimposed.

The development of Judaic religions with a single deity who created the world and all things in it imposed finite limits on the existence of the Earth. The future limit lay in the hands of the deity and was therefore inherently unpredictable, but the creation was a matter of biblical record and could therefore be deduced by reference to the books of scripture.

It was only natural that someone would attempt this calculation, at least to define the date of the creation. Several attempts were made including that of Theophilus, 7th Bishop of Antioch (see Albritton [1]) who carefully traced the biblical records to arrive at a date for the creation of 5529 BC. Others came to similar conclusions including Julius Africanus who held that Christ had come during the sixth 'day' of the cosmic week, where each day lasted for a 1000 years.

Several quantitative estimates of the earth's age were made in the 17th century, the most famous and oft quoted being that of the Irish protestant Archbishop of Armagh and Primate of all Ireland James Ussher (1581–1656). With considerable care and outstanding scholarship, he used historical records of events such as the deaths of Julius Caesar in 44 BC and of Alexander in 323 BC to establish a definitive chronology [10]. Beyond the reign of Nebuchadnezzar around 600 BC, he relied on biblical texts and chose the version of the Hebrew bible prepared by the Masaretes, which is the basis of the modern Hebrew bible. From all this, he deduced that creation took place at nightfall preceding 23 October 4004 BC, that is, 4000 years before the birth of Christ. His chronology was added to the Authorised Version of the bible at one stage and is still included in the marginal annotations of Gideon Bibles. Many, including Charles Darwin (see Fig. 2), thought it was part of the authorised bible and that may be why some Christian commentators have come to accept it as precise, despite information from geology and evolution and the fossil record.

Possibly the first attempt to describe a non-biblical origin of the Earth came from Benoit de Maillet [11]. His account was not published until 1748, 10 years after his death, having been circulated amongst his contemporaries as a handwritten manuscript, since de Maillet was well aware of the power and influence of the church and the views that they would take of his work. His theory was based on the assumption that at some time in the past the Earth had been completely covered in water, from which man and the animals had developed and emerged onto dry land as the level of the oceans dropped over time. From observations of the rate at which sea levels were declining, and also evidence of ancient cities and settlements which were originally built at sea level but were now up to 6000 feet above the ocean, he determined that the age of the oldest of these settlements must be around 2.4 Ma.

A more practical approach to the problem was taken by George-Louis Leclerc, Comte de Buffon (1707–1788). Newton, in his *Principia Mathematica*, had established that the Earth must once have been molten and that it had cooled in the shape of an oblate spheroid. On the basis of the time, it took a 1 inch sphere of iron to cool from red heat he calculated that a mass of iron the size of the Earth would take 50,000 years to cool from red heat, although he felt that this figure was likely to be an underestimate and suggested that some practical experiments would be advisable. Buffon was attempting to compile an encyclopaedia of all knowledge of nature and natural history. This *Histoire Naturelle, Generale et Particuliere* was to occupy 50 volumes with numerous additions, although only 35 volumes were completed in Buffon's lifetime. The twentieth volume, Epochs of Nature (1778) [12] was concerned with a history of the development of the Earth from a molten globe to the present day. He based his calculations on experimental data, using 10 spheres of iron up to 5 inches in diameter in ½ inch increments, specially manufactured at his own foundry so that he could be sure of their quality and uniformity. He heated the spheres to white heat and timed the cooling to red heat, absence of glow and to be holdable in a bare hand. From this he calculated that a mass the size of the earth would cool to below incandescence in 42,964 years and to its present temperature in 96,670 years. After making adjustments for the delaying effect of the sun's heat and also some historical data he produced a revised figure of 74,832 years, although he never felt that this really allowed sufficient time for observed geological formations such as the tremendous thickness of sedimentary rocks exposed by the Alps. Unpublished manuscripts, however, show several longer chronologies, even up to 3 Ma [1].

A more philosophical approach was taken by Immanuel Kant (1724–1804) who supposed that divine creation lay in the marvellous creation of laws of nature which by their simple and consequent operation would result in the formation and development of the universe according to God's divine plan. Where Newton had declared that his laws could not explain the development of the planetary system and that it had been given its present structure by God, Kant in 1755 [13], developing on the work of the English astronomer Thomas Wright (1711–1786), used Newtonian mechanics to describe not just the development of the solar system, but the development of the entire universe. He considered that while the past history of the Universe must be finite, the future development was infinite in its extent. His ideas are not totally irreconcilable with modern theories of the development of the universe.

2 The development of uniformitarianism

Two theories, the Neptunian and the Plutonic, were propounded based on the Earth that evolved through a series of catastrophes. Georges Cuvier (1769–1832) and Jean-Baptiste Lamarck (1744–1829) espoused and developed the Neptunian school of geological development, although they differed on the interpretation of the Noachian flood. The Neptunian school in general preferred to favour periodic catastrophic flooding as the major cause of geological change. By invoking these major upheavals, it became possible to account for the observed geological development without the need for extensive periods of time.

The opposing theory came from the Plutonist school who favoured Buffon's fiery beginning for the Earth and assumed that the Earth was either molten at its centre or at least had a molten shell around a solid core. This molten region was then the cause of volcanoes and earthquakes in shaping the planet.

It was not until the second half of the eighteenth century that a serious geological examination took place. This was due to Edinburgh born James Hutton (1726–1797, see Fig. 1), who after studying medicine at Edinburgh, Paris and Leyden took up agriculture in Norfolk and afterwards on his estate in Berwickshire. He returned to Edinburgh in 1768 where he joined up with other intellectuals including Joseph Black the founder of quantitative chemistry and discoverer of latent heat, specific heat and carbon dioxide; philosopher Adam Ferguson, a member of the 'common sense' school of philosophy established by John Reid; John Playfair, mathematician and physicist; and Sir James Hall, geologist. These and others including David Hume and Adam Smith formed what was designated in the late nineteenth century as the Scottish Enlightenment.

Hutton lived in a house on St. John's Hill, 50 yards south of the Royal Mile in the Old Town of Edinburgh where the intellectuals would meet in each others' homes to discuss their new

Figure 1: James Hutton (left) and Charles Lyell (right).

ideas. From his house, he would be able to see Arthur's Seat and Salisbury Crags, volcanic outcrops where he would make his initial discoveries. After a number of years visiting many sites in Scotland and England during which he collected samples and developed his theories, he presented his ideas to the Royal Society of Edinburgh in 1788 in *A Theory of the Earth* [7] which was later expanded to a two volume book in 1795 [14]. (A third volume, was published much later by Sir Archibald Geikie in 1899.) In these two volumes he demonstrated that the internal heat of the earth caused intrusions into the crust, the molten rock then solidifying and forming granite. His theory can be summarised as follows:

- There occurs gradual degradation of the land by erosion, particularly through river action.
- The eroded matter is transported out to sea and deposited as sediments.
- The sediments consolidate at the bottom of the sea. In due course these sediments are elevated by land upheavals which form new land surfaces which in turn are subject to erosion producing repetition of the cycle.

Hutton concluded that he could see 'no vestige of a beginning and no prospect of an end'. 'Thus the present is a key to the past.' As a result, the theory became known as uniformitarianism. His ideas were made accessible to a much wider audience by the publication in 1802 of John Playfair's book *Illustrations of the Huttonian Theory of the Earth* [15].

Hutton's theory did not necessarily rule out a creation but indicated an equilibrium state extending into the distant past. His ideas were, of course, at variance with the biblical chronology expounded by Ussher and were opposed by the church and by geologists throughout Britain and Europe who generally held theological views in these times. However, the ideas survived their criticisms, forming the basis of modern geology of which Hutton is generally accepted as the founder.

In 1824 John Coleridge, newly appointed editor of the *Quarterly Review*, asked the young barrister, Charles Lyell, (see Fig. 1), to prepare articles for the journal. Lyell was much more interested in geology than in law but had seen no way to make a good living from the subject so the possibility of writing about geology and being paid for it was a godsend. Over the next 2 years, Lyell produced five articles including a lengthy and extensive review of the Geological Society's Transactions; he was familiar with these having been elected a secretary of the society in 1823. Late in 1827 he began writing his *Principles of Geology* [16], which was to become one of the standard texts in the subject and which would influence an entire generation of geologists. The first volume was issued in 1830 and the entire work was continuously revised and extended throughout his life. He accepted, with some reservations, the principles of Huttonian theory as propounded by Playfair and gently but firmly repudiated the claims of the catastrophists. His belief was that the geological evidence firmly proclaimed that geological development had taken place over periods of time which were vast in comparison to historical time. He further believed that the processes at work in the past were not essentially different from the processes which can be observed today. He did not make any attempt to quantify the age of the Earth but showed from descriptions of geological processes now in action that these were sufficient to account for the geological record given sufficient time in which to operate.

Lyell was not strictly an evolutionist but his model of the development of the Earth was one of gradual change over long periods of time. This suggestion of gradual change may have influenced the thinking of Charles Darwin who took a copy of volume 1 of Lyell's *Principles* with him on his voyage on the Beagle. Although Darwin had had only a relatively superficial training in geology, he made extensive geological observations on his travels. His ideas on the origin and development of species developed over a long period of time and was outlined in a pair of unpublished essays written in 1842 and 1844. The main treatise *On the Origin of Species* was not published until 1859 [17] and even then was considered to be the abstract of a much more extensive work. The similar theories of Wallace, however, led his friends, especially Charles Lyell and Joseph Hooker, to push him to a joint presentation of his ideas and an earlier publication than he would have preferred [18].

The theory of the origin of species through a process of natural selection relied in large part on the availability of extensive periods of time in which to operate and Darwin cites Lyell to

Figure 2: Charles Darwin.

support the thesis that there would have been plenty of time available. In the first edition of Origin, he also provided a brief illustration to support this contention. He estimated that the denudation of the Weald of Kent must have taken something around 300 Ma and that this represented but a part of post-Cambrian time. As a brief aside we must throughout bear in mind the distinction between estimates of times *since* the Cambrian as evidenced by sedimentary deposits or rates of erosion – the geological stratification is unreliable for pre-Cambrian times – and estimates of the time since the Earth cooled from a molten globe which must include all of the pre-Cambrian epoch in addition.

Darwin's estimate was provided, essentially, as an illustration. He was clearly in no doubt that sufficient periods of time were available. However, in the controversy generated by Origin this estimate was pounced upon and extensively criticised to such an extent that Darwin removed the reference from the second and subsequent editions, Although he still cited Lyell as support for the claim that the availability of time was not an issue.

Darwin's health had suffered badly as a result of his voyage and he found visits to London to be very debilitating. He was therefore not well placed to defend his very controversial theories. The role of defender was therefore taken by his friend and adviser Thomas Henry Huxley.

Huxley was already becoming a very influential figure and was to become even more so. The son of a schoolmaster in Ealing he had trained as a surgeon and got an appointment as a surgeon's mate in the Royal Navy. He served originally at Haslar and was then appointed to be Surgeon on a voyage to the South Seas. On this voyage, he made a detailed study of marine invertebrates, sending reports back to London. On his return, after some disputes with the navy about being paid to develop his observations, he was awarded a medal by the Royal Society and later was appointed Professor of Natural History at the School of Mines. This was the start of a long and influential career during which he was instrumental in developing the principle that science should be a recognised profession. He formed the X-Club in association with a number of equally erudite friends and this group held great power over the scientific community in London.

3 Lord Kelvin

William Thomson, later to be Lord Kelvin, had a very different background, having a classical, if precocious, education leading to a degree from Cambridge, and eventually with the aid of some political manoeuvrings, the chair of Natural Philosophy at Glasgow [19]. A condition of his appointment was that he should, within a month, write an essay in Latin on the movement of heat through the body of the Earth. He was obviously dissatisfied with this essay since he burnt it immediately after its acceptance. The essay developed a theme which he had originally enunciated in the *Cambridge Mathematical Journal*. In a paper of 1842 'On the linear motion of heat' [20] he considered the motion of heat in an infinite solid, developing an equation from which it would be possible to find the temperature at any point a given distance from the zero-plane at a given time. This paper developed in some detail the work of Fourier and was the basis for much of his later work on determining the Age of the Earth. It was followed in 1844 by a 'Note on some Points in the Theory of heat' [21]. This second paper considered the problem of the motion of heat in solid bodies given an arbitrary initial distribution. Thomson also considered the case of negative values for time, that is, given the present conditions, can we deduce the conditions in the past? He considered three classes of cases depending on the degree of convergence of the terms in the controlling expression. He concluded that in one case it would be possible to assign a finite age given the initial distribution; in another it would not be possible and in the third the distribution at time zero must be an initial distribution which could not have been the result of any previous distribution of heat.

The importance of these two papers and the development in his inaugural essay can be seen from the note which he attached to the paper when it was included in his volume of reprints:-

'An application to terrestrial temperature of the principles set forth in the first part of this paper relating to the age of thermal distributions, was made the subject of the author's Inaugural Dissertation on the occasion of his induction to the University of Glasgow, in October 1846, De Motus Caloris per Terrae Corpus; which, more fully developed afterwards, gave a very decisive limitation to the possible age of the earth as a habitation for living creatures, and proved the untenability of the enormous claims for TIME which, uncurbed by any physical science, geologists and biologists had begun to make and to regard as unchallengeable'.

Only one page of draft of that inaugural essay remains, heavily corrected by his father:-

'De Distributione Caloris per Corpus Terrae.

In caloris theoria mathematica, problemata saepe pro-ponuntur in quibus data temperaturae distributione in initio, distributio in quolibet temporis sequentis puncto invenienda est In plurimis horum problematum quaestio oritur, an possibile sit, ad tempus aliquid datum, invenire distributionem antecedentem ex qua distributio data produceretur per motum liberum caloris. Si hoc fieri potest, inveniendum est maximum temporis intervallum inter distributionem datam et distributionem antecedentem ex qua data derivari potest. Veruntamen distributio data talis esse potest ut intervallum valorem maximum non habere possit. Haec considerationes suggerunt omnes distributiones quae possunt existere in tria genera redigi posse ; sc.

1. Distributiones quae e distributione antecedente produci nequeunt.
2. Distributiones quae produci possunt e distributionibus determinatis existentibus in temporibus datis ante tempus datarum distributionum, intervallis limites certos haud excedentibus...'

In 1852 he published his paper 'On a universal tendency in nature to the dissipation of mechanical energy' [22] in which he formulated the second law of thermodynamics which states that in any conversion of energy from one form to another a proportion of the energy is dissipated in the form of heat.

Kelvin was first and foremost a mathematician and secondly an experimental physicist and practising engineer. He considered mathematics to be the fundamental tool for both disciplines and the essential path to the truth. Reason told him that if the second law of thermodynamics were true, then it must inevitably mean that the Universe had at some time had a beginning and at some time would die in heat death with energy equally distributed everywhere and no energy potential available to do any work. If the universe had a beginning and an end then so must the Earth and the Solar system and it should be possible to find some limiting case conditions that could place bounds on their ages.

In particular, Kelvin was intensely dissatisfied with the principles of uniformitarianism either in the form suggested by Hutton and Playfair or as accepted by Lyell. Thermodynamics proved beyond doubt that the earth could not be ageless and must have a beginning and an end. Life itself could only have been brought into being by a creator, although it was conceivable that life had been created elsewhere and brought to this planet in an advanced form of development [23]. The future of mankind was also in the hands of the creator even though the potential lifetime of the Earth was limited.

Kelvin was extremely tenacious with his theories. Once he had established the principles he continued to develop the ideas, refining the details to accommodate criticisms, but always maintaining the integrity of the central principles. Even his undergraduate essay 'On the Figure of the Earth' was revised and corrected in later years. In large part, it was this re-iteration of his ideas that kept them alive and maintained his influence over the years.

4 Geological time

After leaving Cambridge Kelvin went to Paris to study under Henri Victor Regnualt. There he read Fourier's *Analytic Theory of Heat* which contained the equation for diffusion of heat through a solid body. Kelvin had already come across this equation when he was an undergraduate at Glasgow University in 1839 under Professor Nichol in the Natural Philosophy Classes. In 1841, Kelvin responded anonymously to a book by Kelland at Cambridge criticising Fourier's mathematics [24]. Despite Playfair's influence Cambridge had not yet fully accepted the advances on the Continent. Kelvin decided to use the equation to calculate the age of the earth from when it became a solid body, which was the general view of the state of the earth in Kelvin's time. He also specifically stated that he assumed there was no internal heat generation which was also the general belief at the time. Assuming that the earth had a uniform temperature to begin with and now possessed a temperature gradient as a result of cooling he could calculate, using the diffusion equation, the time taken to achieve this gradient.

In 1868, 10 years after the publication of *On the Origin of Species*, Kelvin addressed the Geological Society of Glasgow on the subject of Geological Time [25]. He had addressed the issue several times discussing the possible age of the sun or of the Earth and attacking the principles of Uniformitarianism, but with little reaction from the geological community.

On this occasion, he spoke more forcibly to the geologists, announcing that 'a great reform in geological speculation seems now to have become necessary'. He explained how the geologists of the previous century had tried to bring geology within the physical sciences and to move away from theologically based chronologies. At the same time, the astronomers had been making great steps in the theory of the motions of planetary bodies. However, the theory of the stability of the planetary movements had been misinterpreted and where the original founders of the theories had acknowledged the simplifications that had been made, these caveats were overlooked by the geologists who adopted the theories. The geologists had tried to use the stability of the heavens to suppose that the universe, including the earth, was essentially unchanging, thus leading to Playfair's statement 'we can discern neither a beginning or the end'.

Nothing, proclaims Kelvin, could be further from the truth. The theorem of the French mathematicians regarding the motions of heavenly bodies, he explains, is a theorem of approximate application and neglects frictional resistance of every kind.

Having established the foundations of his arguments, Kelvin then proceeded to take a quotation from Playfair as his text and to counter the arguments contained within it from two different approaches. He considered firstly the motions of the heavenly bodies including the Earth and secondly the phenomena presented by the earth's crust.

The arguments which he presents are based on a number of earlier papers and lectures extending back to an essay of 85 pages written when he was 16 'On the Figure of the Earth'. That essay was in four parts, Physical Theory, Disturbance in the Moon's Action, Geodetic Measures and Pendulum Observations. While it drew on the work of Airy, Poisson, Pentecoulant, Pratt and Laplace, it was a remarkable work in its own right extending its arguments to cases not considered by the earlier workers. Kelvin kept the manuscript and added amendments to it in 1844, 1866 and 1907; just 2 months before his death he was still returning to this boyhood study.

The 1866 amendments coincide with Kelvin's delivery of the Rede lecture after receiving an honorary doctorate of Laws at Cambridge. The full lecture is not recorded and witnesses describe how Kelvin would keep leaving the main topic to expose or discourse upon common fallacies, eventually returning to somewhere near where he had left the main topic of discussion which was the dissipation of energy. This made the lecture very difficult for his listeners to follow and appears to be typical of his approach to lecturing at Glasgow where it was not uncommon for less committed undergraduates to slip out of his lectures once their presence had been recorded, leaving a hard core of more gifted students to delight in following his intellectual wanderings.

An abstract of the Rede lecture was published in the Cambridge Chronicle a few days later and is also reprinted in the *Life of William Thomson* [19]. In the lecture, he not only considered the amount of heat being lost from the sun (7000 horse-power per square foot) some small fraction of which falling upon the earth providing the energy of coal, and growing combustibles and fuel, but also the loss of heat from the earth of 180 million horse-power from the whole surface.

Tidal friction was also considered as giving rise to dissipation of the earth's energy of rotation [26], the ultimate tendency of which would be to make the earth, sun and moon turn like parts of a rigid body. The effect might possibly amount to as much as one or two hundred seconds in a century. However, astronomical clock making was not wonderful and astronomical clocks could only be made to perform two or three times as accurately as a good pocket watch. Therefore, the only means of determining retardation of the earth was by comparison with the moon, which itself was subject to great irregularities. Adams and Delaunay had calculated that the earth had lost on the moon by some ten seconds in the previous hundred years, but it could not be concluded that this was entirely due to tidal retardation. For example, melting snows and ice caps that caused in imperceptible increase in ocean depths of one and a half inches would slow the earth by two tenths of a second per year. Also, while the direct effect of the moon on the tides in the Thames from London Bridge to the Nore is to retard the rotation of the earth, from the Nore to Exmouth and Jersey the effect is to accelerate the rotation. However, the fact that there is loss of energy by fluid friction makes it certain that the sum of the retarding and accelerating forces results in a net retardation.

The lecture ended with three conclusions, the third of which stated:

'within a finite period of time past the earth must have been, and within a finite period of time to come, the earth must again be unfit for the habitation of man as at present constituted, unless operations have been or are to be performed, which are impossible under the laws to which the known operations going on at present in the material world are subject.'

4.1. Tidal retardation of the earth

In 'On geological time' [25], Kelvin goes into more detail about tidal processes. The distribution of water over the surface of the earth is not uniform, but it can be represented by a model of the earth with a uniform layer of water which has the same net effect as the actual oceans. This model is referred to as the tidal spheroid. The tides, which occur twice in each day, are caused because the moon attracts the water on the side of the earth nearest to it into a bulge and also attracts the earth itself more than the total mass so that there is a bulge in the water mass on both sides of the planet. If the moon were stationary with respect to the earth then there would be a bulge in the water level directly below the moon and diametrically opposite it. However, since the earth is rotating with respect to the moon and also the moon is rotating about the earth, and since the mass of water cannot respond to the gravitational field of the moon instantaneously, the tidal bulge lags behind the position of the moon and this in itself demonstrates that frictional processes in the movement of the body of water do occur. The lost energy is dissipated in the form of heat as shown by Joule.

Similar arguments apply to solar tidal forces except that the solar tides have a period of 24 hours and the lunar tidal forces have a period slightly longer due to the moon's rotation about the earth. When the lunar and solar tides coincide the tides are higher than average, called spring tides, and when the moon and sun are at right angles the tides are lower, called neap tides. The

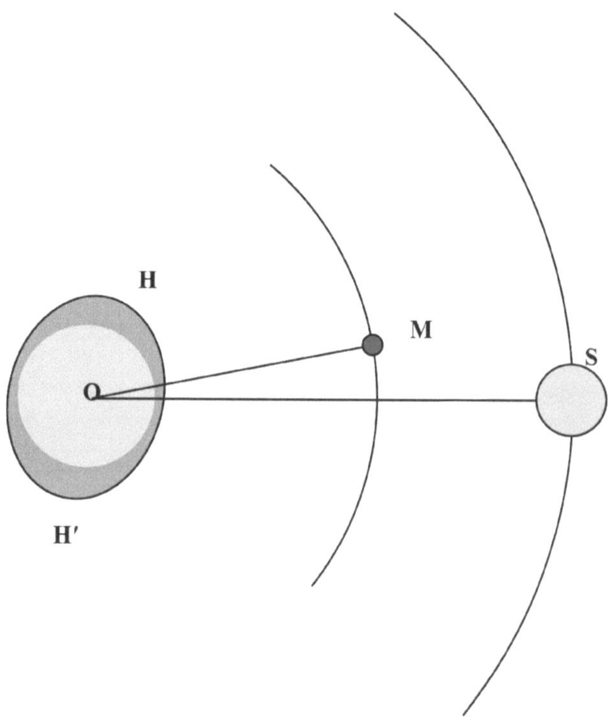

Figure 3: Position of the Sun, Moon and Earth at Spring Tides.

detailed processes are more complex due to tidal streams and in some areas the tides are exceptionally high due to the tidal timing matching a resonance in the local tidal flows.

Since it is clear that energy is being dissipated, and that the primary source of this energy is due to the relative motions of the earth, moon and sun, then the ultimate result of the dissipation of this energy must eventually lead to a static situation in which the relative motions reduce to zero, the moon rotating about the earth in exactly 1 day and with the earth and moon perpetually showing the same sides to one another. Similarly, the Earth–moon complex would show just one face perpetually towards the sun and the three bodies would rotate as a single rigid complex with no tidal forces being exerted on the oceans. The two bulges would stay in the same location forever. To achieve this state, it is clear that the rotation of the earth must be decreasing, and the angular momentum of the moon must be increasing, the orbital distance of the moon also increasing.

To consider this more analytically, consider the situation in Fig. 3 which shows the relative positions of the earth, moon and sun and the tidal bulge at the time of spring tides, viewed from the north. If there were no tidal friction then OMS would be in a straight line and HH´ would be perpendicular to it. But from observation the spring tides are late by between a day and half to three days after the new moon and the full moon. The line OM is therefore inclined forward from OS by the distance that the moon would move in the time of the spring tide lag – for a lag of 12 hours this would be 6°.

Now, the waters of the earth are drawn along with the motion of the earth by friction on the bottom and by friction of water on water and the overall effect is as if the earth and water were rotating within a friction strap. The forces on this friction strap would amount to the weight of two million tons according to the calculations taken from the Rede lecture. This supposes the axis HH´ to be inclined to OM at an angle of 45°, the position which would result in the greatest amount of retardation. If the angle HOM were between 45° and 84° the frictional resistance would be between this amount and one fifth of it. But if the lateness of the spring tides is at least 12 hours, then the angle MOS cannot be less than 6° and the angle HOS is certainly something short of 90°. Therefore, the average spheroidal tide must be less than 1⅛ feet or the amount of resistance to the earth's rotation must exceed one fifth of the estimated superior limit.

The general tendency of the action is therefore to diminish the velocity of the earth's rotation and to lengthen the duration of the day. Similarly, the waters exert a force on the moon forwards in her orbit which has the effect of making the moon's motion slower and to increase her distance from the earth. If the moon were a viscous mass, then she would also feel enormous tidal effects which would tend to slow her rate of rotation until she ended up with one face turned permanently towards the earth, as is indeed the case, and we may therefore suppose it likely that the moon was once a liquid mass.

It is all very well to show that these processes occur in principle, but it is quite another to be able to quantify them especially give the very poor quality of astronomical clocks. There were also arguments to show that the rotation of the earth had not changed by one ten millionth part of twenty four hours since 721BC based on observations of eclipses in 721 BC and 313 BC in Babylon. Dunthorne [27] had used these observations to confirm a suspicion of Halley, that the moon's mean angular motion had been accelerated relative to the earth by 20 seconds of angular velocity/century per century. Laplace, however, while accepting this conclusion, had explained it by showing that the planets indirectly cause an acceleration of the moon's angular velocity through their influence in producing a secular diminution of the eccentricity of the earth's orbit. Calculating out this effect tallied precisely with the supposition that the earth's velocity of rotation had been constant since 721BC.

But, in 1853, Adams – a great English physical astronomer – pointed out an error in Laplace's process, in that he had omitted to take into account the tangential component of the sun's disturbing force on the moon and reworked the problem with this correction. He found that there was an unaccounted for gain of 11.4 seconds of arc per century in the angular velocity of the moon's motion, and Delaunay in France soon confirmed this result, suggesting that the discrepancy could be due to a diminution in the earth's rotational velocity by tidal friction. Adams, Kelvin and Tait then used this hypothesis with certain assumptions as to the proportion of retardations due to the moon and the sun, to show that the earth would get 22 seconds of time behind a thoroughly perfect clock after a century. This in itself discredits Playfair's statement that there is nothing in the motions of the heavenly bodies that tends to their own dissolution or to a permanent alteration of the existing state of things.

If we consider the effect of centrifugal force on the earth a hundred million years ago, then this would be some 3% greater than at present and if the earth had consolidated at this time, under those conditions then it might easily be of the form we observe. However, if we go back in time as much as one thousand million years, then the speed of rotation would have been more than twice as fast as at present and if it solidified then it would be very different in form from that which we observe.

At the very least, whether the earth's lost time is 22 seconds or considerably more or less than 22 seconds, it is evident that there must be a progress of events towards a state infinitely different from the present.

4.2 The sun's heat

Next, Kelvin went on to consider the question of the sun's heat, for if Playfair considered planetary motions to be unchanging then it would be necessary to consider also the source of the sun's heat. This might be considered a miraculous body, created by a Creative Power and so designed as to produce unending, constant supplies of heat in violation of those laws of action between matter and matter which we are allowed to investigate on the earth. But, he says, we also know that the Creative Power has created in our minds a wish to investigate and a capacity for investigating; and there is nothing too rash, there is nothing audacious in questioning human assumptions regarding Creative Power. Indeed the very geologists who support Playfair are strenuous in insisting that we must consider the laws observable in the present state of things as perennial laws.

He then considered all the energy that would be released by the planets falling into the sun. Jupiter alone would provide 32,240 years of the sun's heat and all together would provide some 46,000 year's heat. Given that, it is accepted that the sun has already, in geological periods, emitted maybe a thousand times as much heat as would be produced by all the planets falling together into the sun. But Playfair and his followers take no consideration of this prodigious dissipation of energy.

4.2.1 Meteoric theory

In 1854, Kelvin considered three possible sources of the sun's heat [28]: firstly, that the sun has only primeval heat and is cooling as a result of the dissipation of this heat; secondly that the sun is 'burning' and generating heat through chemical processes; and finally that the sun acquires its heat from the energy provided by meteors falling into it and releasing their kinetic energy in the form of heat. The first of these possibilities he had previously shown to be entirely untenable since, unless the sun's thermal conductivity was enormously greater than any terrestrial matter the sun would become dark in a matter of minutes, or days, or at best a few years at the present rate of emission.

Next, he considered the meteoric theory proposed by Mayer and by Waterston. The fact that there are meteors arriving at the earth shows that there must always be meteors falling into the sun, and since Joule had shown the enormous quantities of heat generated from a meteor's motion relative to the earth in his explanation of falling stars, it follows that some of the sun's heat must be due to meteors.

According to Pouillet, the amount of heat incident per second on a square foot at the earth's surface is 0.06 of a thermal unit (pound-water-degree) centigrade. Since the distance from the earth to the sun is 95,000,000 miles and the sun's radius is 441,000 miles, the rate of emission of heat from the sun must be:

$0.06 \times (95{,}000{,}000/441{,}000)^2 = 2781$ thermal units/second/square foot of his surface, which is equivalent to 3,869,000 ft lb-weight/sec/sq ft.

Neglecting atmospheric resistance and the mile and a quarter per second speed of rotation of the sun's surface, the impact speed of a meteor would be that obtained by falling from an infinite distance, which is that obtained by the action of a constant force equal to its weight at the sun's surface, operating through a space equal to his radius. The force of gravity at the sun's surface being about 28 times that at the earth's surface, the speed is

$$\sqrt{\frac{2\times 28\times 32.2\times 441{,}000}{5280}} = 390 \text{ miles per second}$$

And the mechanical value per pound of meteoric matter is

$$28\times 441{,}000\times 5280 = 65{,}000{,}000{,}000 \text{ ft lb-weight}$$

Hence, the amount of meteoric matter required to account for the sun's heat would be 0.000060 pounds per second, or about one pound per square foot every five hours. If this density of meteors were falling into the sun, whether on linear, parabolic or hyperbolic paths, we would find that the earth was intercepting far more of them than is actually the case. It, therefore, follows that the majority of these meteorites are orbiting the sun inside the orbit of the earth and Kelvin suggests that it is these bodies which form the zodiacal light, the outer edge of which appears to reach nearly to the earth [29]. In this case, the meteors would approach the sun on a gradual spiral path with a speed little more than that of a circular orbiting body. The speed lost by a meteor in this case is $1/\sqrt{2}$ of that due to gravitation from an infinite distance or 276 miles per second and, therefore, the mass of meteorites required is roughly double the previous estimate. The whole surface of the sun would, therefore, be covered to a depth of 60 feet in a year and the diameter of the sun would increase by 1 mile in 88 years. In 40,000 years the [angular] diameter would grow by only 1 second of arc which would be undetectable without the aid of powerful telescopes.

Finally, he considers what chemical actions would be required to produce heat at a rate of 7000 horse-power per square foot. This would require 1500 lbs of coal to be burnt per hour, but it would not be possible to provide oxygen to keep the fire burning at a sufficient rate. Therefore the action must be more like gunpowder which does not require an externally provided accelerant. He assumes that the heat of combustion could not be much more than 4000 thermal units per pound of matter – the greatest thermal reaction known being somewhat less than that. But the sun loses 2781 thermal units per second for each square foot implying a loss of about 0.7 of a pound of matter per square foot per second. At this rate, the sun would lose about 0.5 feet of thickness in a minute or 55 miles in a year. At the same rate, a mass the size of the sun would burn away in 8000 years, so it can be concluded that the sun does not get its energy by chemical action.

4.2.2 The sun as a cooling body

In 1862 Kelvin addressed this subject again [30,31], pointing out that the amount of zodiacal matter would have to amount to at least 1/5000th of the mass of the sun, and that this would be expected to have some noticeable effect on the orbits of the planets. Indeed, some such variations had been found by M. le Verrier in his observations of Mercury [32], but the amount of matter indicated was far too small. The mass of material would need to be very close to the sun indeed and the density would then be such that comets passing this close would be perceptibly slowed. He concluded that the meteoric theory was, therefore, unlikely and since no chemical theory was tenable he re-addressed the problems of treating the sun as a cooling body. The considerations here are complicated by the fact that the interior of the sun must be at very high pressure and density and that it is not reliably known how matter behaves under these conditions, but on the basis that the sun's specific heat is more than ten times and less than 10,000 times that of liquid water, then the temperature must sink by 100°C in sometime between 700 and 7000 years. He then attacks Darwin's estimate of 300 Ma for the denudation of the Weald [17] on the basis of whether it is more credible to assume that the conditions of the sun's matter differ 1000 times more than dynamics compel us to suppose, or that a stormy sea and channel tides of extreme violence should encroach on a chalk cliff 1000 times more rapidly than Darwin's estimate of one inch per century.

Finally, he considers what the source of the sun's heat could have been, if not by the act of a creative power. He concludes that this must have been by meteoric accretion and that this theory would have no difficulty in accounting for 20,000,000 years of heat, and that the age of the sun is probably not more than 100 Ma and almost certainly not more than 500 Ma.

4.3 Cooling of the Earth

Thirdly, Kelvin considered the question of the underground heat of the earth, to see whether this might provide justification for Playfair's assertion that the phenomena presented all through its crust gave no evidence of a beginning and no progress or advance towards an end. The argument is based on the observation that if you bore into the earth it is warm, and the deeper you go the warmer it gets. He admits that some geologists acknowledge that there may be such a temperature gradient in some places, but claim that this is not sufficient to suppose that it is true everywhere. Kelvin had long advocated, and now reiterated his suggestions, that there should be a global survey of underground temperatures in widely varying locations. In fact, at a recent meeting of the British Association (BA) at Dundee, a committee had been established for investigating underground temperature.

The determination of the age of the earth from knowledge of the existing temperature gradient, and the surface temperature is mathematically feasible but relies on a number of assumptions about the condition of the earth which were well recognised by Kelvin. The thermal conductivity of rock had been established by observation at a number of locations and the variations due to daily, annual and semi-annual changes in the surface temperature had been well observed by a number of people including Fourier. Secondly, the nature of the interior of the earth needed consideration and Kelvin worked on the basis that the earth was essentially a homogeneous ball of consolidated rock that was cooling from an initial temperature of around 7000°F. In 'On geological time' he did not go into the details of these calculations as they had been discussed in great depth in his earlier papers 'On the secular cooling of the earth' [8] and 'On the rigidity of the earth' [33].

Secular cooling was a most detailed and reasoned analysis of the earth considered as a cooling body. The exposition is based on Fourier's mathematical theory of the conduction of heat and its essential prerequisite that there was an initial state prior to which the cooling functions are divergent, and which therefore must have been arrived at through some other mechanism. Kelvin had established the criterion for such an initial distribution in the *Cambridge Mathematical Journal* some 19 years previously and had outlined the application of Fourier's theory to the cooling earth in his inaugural dissertation.

An essential factor in applying the cooling solution is to determine the average rate of increase in temperature with depth and, as noted above, Kelvin had called repeatedly for global geothermic surveys to be carried out so that this figure could be determined with a reasonable degree of certainty. For the time being he had to use the average of data from three different locations, in which gave a rate of 1°F for every 50 feet of depth. This temperature gradient can only result from a continual loss of heat from the surface of the earth and must be supplied either from an initial store of heat or from some form of chemical action. In either case, the source of the heat must be limited. Kelvin dismisses the suggestions of Lyell that the chemical sources might be regenerated by electrolytic action as being fundamentally opposed to the second law of thermodynamics.

He next considers the suggestion of Poisson that the present underground heat was the result of the earth having passed through hotter stellar regions at some time in its past. Kelvin calculated, using values of thermal capacity for the earth's crust obtained from Forbes' measurements in Edinburgh, that if such a transit had taken place between 1250 and 5000 years previously the region of space must have been 25–50°F above the present surface temperature of the earth in order to account for the observed temperature gradient. However, no historical record indicates that such an event had occurred within this time scale. But if the heating had occurred as much as 20,000 years ago, then the temperature must have been 100°F more than at present which would have been fatal to life. The further back such an event had happened, the hotter it must have been, until at 200 Ma ago the temperature would have been 10,000°F which is a somewhat high estimate of the temperature of melting rock. The more acceptable figure of 7,000°F would have occurred some 95 Ma ago.

These estimates were based on Fourier's solution for the problem of finding the rate of variation of temperature from point to point, and the actual temperature at any point in an infinite solid given as an initial condition that the temperature had two different values on the two sides of an infinite plane. The solution is

$$\frac{dv}{dx} = \frac{V}{\sqrt{\pi kt}} e^{-\frac{x^2}{4kt}} \tag{1}$$

$$v = v_0 + \frac{2V}{\sqrt{\pi}} \int_0^{\frac{x}{2\sqrt{kt}}} dz e^{-z^2} \tag{2}$$

where

k denotes the thermal conductivity of the solid, measured in terms of the thermal capacity of the unit of bulk, V the half the difference of the two initial temperatures, v_0 their arithmetical mean, t the time,

x the distance of any point from the middle plane,

v the temperature of the point x at time t, and $\frac{dv}{dx}$ the rate of variation of the temperature per unit length perpendicular to the isothermal plane.

Kelvin demonstrates the solution by verifying

1. That the expression for satisfies Fourier's equation for heat:

$$\frac{dv}{dt} = k\frac{d^2v}{dx^2}$$

2. That when $t = 0$, the expression for v becomes $v_0 + V$ for all positive, and $v_0 - V$ for all negative values of x.

3. That the expression for is the differential coefficient of the expression with reference to.

(1) and (3) being proved directly by differentiation, and (2) by showing that when and is positive:

$$v = v_0 + \frac{2V}{\sqrt{\pi}} \int_0^{\infty} dz e^{-z^2} \tag{3}$$

And since the known value of the integral is : $\frac{1}{2}\sqrt{\pi}$

$$v = v_0 - V$$

And, for all values of the second term has equal positive and negative values for equal positive and negative values of, so that when $t = 0$ and x is negative:

$$v = v_0 + V$$

These results are expressed graphically in Fig. 4 reproduced from [8], which shows the temperature as a function of the distance from the surface and the rate of increase in temperature with depth. It can be shown that this is without significant error for a solid sphere, initially heated to a uniform temperature and suddenly having its surface reduced to some other constant temperature, which is maintained thereafter. For a globe with a diameter of 8000 miles, the error will not be significant for more than 1000 Ma since the variation in temperature is insignificant below a depth of 568 miles and therefore the surface may be well considered as a plane surface.

To use the diagram to show the variation, which would now exist if the initial temperature drop had occurred 100 Ma ago, with a temperature reduction at the surface of V °F, then on the scale of depth below the surface OX, length 'a' represents 400,000 ft.

For the rate of increase of temperature with depth, length 'b' on the scale parallel to OY represents 1/354,000 of V per foot, so that if V =7000°F 'b' will represent 1/50°F per foot.

For the excess of temperature above that at the surface, length 'b' on the scale parallel to OY represents or $V / \frac{1}{2}\sqrt{\pi}$ 7900°F when V =7000°F.

It can be seen that while the rate of increase in temperature with depth is around 1/50°F per foot for the first 100,000 feet or so, after that the rate decreases dramatically and at 800,000 feet would be 1/50 of that value. Below a depth of about 100 miles the temperature of the earth, whether liquid or solid, would essentially be the proper melting temperature at that depth.

Kelvin addressed directly two major arguments against this theory (1) that the earth was once all melted, or at least the surface layers were all melted and that it was never a uniform 7000°F throughout its mass; and (2) that no natural action could have instantaneously reduced the surface temperature by some 7000°F and then maintained that condition thereafter. For the latter objection, he argued that whatever the initial condition, the surface would have cooled quite rapidly, over a period of weeks or months or years, to establish a solid crust which with a surface temperature not vastly different from that at present.

With regard to the former argument, he pointed out that it is difficult to establish the precise nature of the body of the earth or the process of its solidification without more detailed knowledge of the thermal conductivity and the heat capacity of the rock. The earth may have started as a completely liquid mass, or as a solid cold core which was then heated by the impact of meteorites resulting in a solid cooler core surrounded by a molten layer. The model behaves differently depending on whether the rock shrinks on cooling, or, as is the case with water, it expands. In the former case the temperature will increase with pressure and depth, and in the latter the temperature at higher pressures and depths will be lower than near the surface. In this case, solidification will occur from the centre outwards to result in a solid mass. In the former case, the surface would first solidify, but would break up due to buckling and shrinkage and the solid portions would sink through the underlying liquid. This process would continue resulting in a honeycomb structure of solid masses interspersed with liquid rock. Kelvin accepted

Bischof's findings that melted granite, slate and trachyte all contracted by about 20% on freezing, justifying the use of this model.

However, many geologists still supported a model of the earth that comprised a liquid core supporting a thin solid crust and Kelvin addressed this directly in 1862 [33], when he considered the effects the rigidity of the earth on the solar semiannual and lunar fortnightly nutations and he was able to demonstrate that the earth must be at least as rigid as glass, and probably much more so. He also returned to the subject in 1872 [34] and 1876 [35].

All of these ideas were brought together in the paper he presented to the Glasgow geological society [25] in order to support his cry that 'A great reform in geological speculation seems now to have become necessary'. Taking all these considerations into account, he concluded that some upper bound must be placed on the age of the earth, whether it be 50, 100 or 300 Ma, but in any case it was clear that the earth could not be older than 1000 Ma. Most likely he said was that the age of the earth was some such period as 100 Ma.

The strength of this attack could not be allowed to go unchallenged. In 1869, Thomas Henry Huxley (Fig. 5) made his presidential address to the Geological Society [9] and made use of the opportunity to refute Kelvin's arguments. He is often cited as rejecting Kelvin's estimates in defence of Darwin's theory of natural selection. However, although he was convinced by the arguments for evolution, and the progressive development of species, he was not totally convinced that natural selection was the process by which this was achieved. He was quite firm in the statement in his address that Biology had to take its time from Geology and adapt its theories to fit with the time scales that were available from geological principles.

Neither did he particularly refute the age estimates that Kelvin had produced – indeed, his own estimates of the age of the Earth based on sedimentation rates agreed on a figure around 100 Ma. He did take grave exception to the suggestions from Kelvin that Geologists were fundamentally wrong in their approaches to the subject.

$$ON = x \qquad a = 2\sqrt{kt}$$

$$NP' = be^{-x^2/a^2} = y' \qquad \frac{dv}{dx} = \frac{V}{a}\cdot\frac{NP}{b\frac{1}{2}\sqrt{\pi}}$$

$$NP = ONP'A \div a = \frac{1}{a}\int_0^x y'dx \qquad v - v_0 = V\cdot\frac{NP}{b\frac{1}{2}\sqrt{\pi}}$$

The curve OPQ shows excess of temperature above that of the surface.

The curve AP'R shows rate of augmentation of temperature downwards.

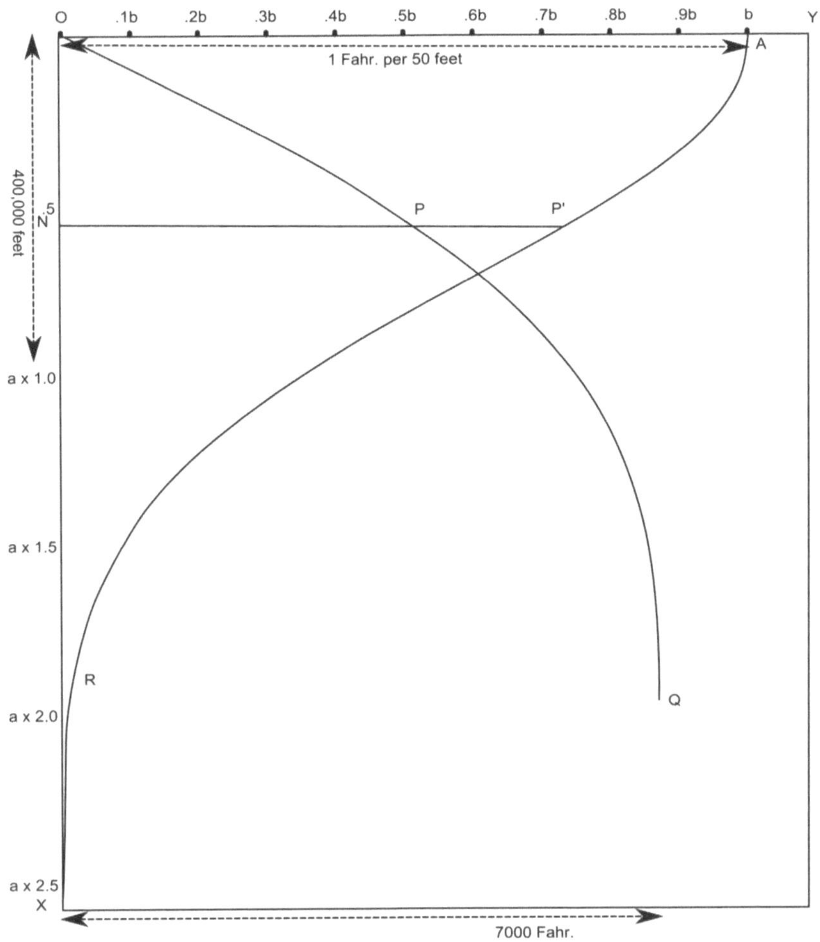

Figure 4: Increase of temperature downwards in the earth (redrawn from [29])

It is obviously Uniformitarianism which the distinguished physicist takes to be the representative of geological speculation in general. And thus a first issue is raised, inasmuch as many persons (and those not the least thoughtful among the younger geologists) do not accept strict Uniformitarianism as the final form of geological speculation. We should say, if Hutton and Playfair declare the course of the world to have been always the same, point out the fallacy by all means, but in so doing do not imagine that you are proving modern geology to be in opposition to natural philosophy. I do not suppose that, at the present day, any geologist would be found to maintain absolute Uniformitarianism, to deny that the rapidity of the rotation of the earth may be diminishing, that the sun may be waxing dim, or that the earth itself may be cooling. Most of us, I suspect, are Gallios, "who care for none of these things," being of opinion that, true or fictitious, they have made no practical difference to the earth, during the period of which a record is preserved in stratified deposits.

The accusation that we have been running counter to the principles of natural philosophy, therefore, is devoid of foundation. The only question which can arise is whether we have, or have not, been tacitly making assumptions which are in opposition to certain conclusions which

Figure 5: Thomas Henry Huxley.

may be drawn from those principles. And this question subdivides itself into two:–the first, are we really contravening such conclusions? the second, if we are, are those conclusions so firmly based that we may not contravene them? I reply in the negative to both these questions, and I will give you my reasons for so doing. Sir William Thomson believes that he is able to prove by physical reasonings, "that the existing state of things on the earth, life on the earth–all geological history showing continuity of life–must be limited within some such period of time as one hundred million years" [sic].

The first inquiry which arises plainly is, has it ever been denied that this period may be enough for the purposes of geology?'

He also examined Kelvin's assumptions and shows that in all three of his approaches there is lack of precise data and that slight changes in the assumed values could lead to wide variations in the estimates of the Earth's age.

'I do not presume to throw the slightest doubt upon the accuracy of any of the calculations made by such distinguished mathematicians as those who have made the suggestions I have cited. On the contrary, it is necessary to my argument to assume that they are all correct. But I desire to point out that this seems to be one of the many cases in which the admitted accuracy of mathematical processes is allowed to throw a wholly inadmissible appearance of authority over the results obtained by them. Mathematics may be compared to a mill of exquisite workmanship, which grinds you stuff of any degree of fineness; but, nevertheless, what you get out depends on what you put in; and as the grandest mill in the world will not extract wheat-flour from peascods, so pages of formulæ will not get a definite result out of loose data.'

At this point, there is little disagreement, and in his second and subsequent editions of *On the Origin of Species* Darwin withdrew his estimates of the age of the Weald – although he did leave in place his generic statements that there was plenty of time available in which evolution could take place.

Huxley's views about the early history and the formation of the Earth were based on the works of Kant, who had described in great detail how the planets and the Sun could coalesce out of clouds of dust, showing through basic Newtonian principles how the planets obtained their spins and their respective positions in the solar system [13]. He also showed how celestial bodies could go on to collapse so far that the immense pressures and temperatures resulting from the collapse would lead to their eventual explosion producing a new cosmic dust cloud from which new planets could then be built. Kant's exposition was extremely prescient, even suggesting that the universe was full of millions of galaxies and that stars were not confined to the Milky Way. However, his arguments were susceptible to many criticisms in the light of classical physics and chemistry, and it is only recently that we have been discovering just how galaxies and planetary systems are formed and how they eventually die. In particular, his cyclic views were quite unacceptable to strict thermodynamicists who could look ahead to the eventual heat death of the universe, and who could not see where the immeasurable quantities of energy required could possibly come from.

Kelvin himself responded critically to Huxley's paper and shortly afterwards P.G. Tait, a colleague and friend of Kelvin, writing anonymously made a vigorous attack, quoting lengthy passages out of context and completely missing the essential points that Huxley was trying to make [36].

Kelvin's response, while no less vigorous, was more closely reasoned in a paper of three parts entitled 'On Geological Dynamics' [37]. The first part was a direct response to Huxley's address, while the second and third parts addressed the questions of the origin and total amount of Plutonic energy, and a note on the meteoric theory of the sun's heat. In the first part, he re-affirms that in his paper of 1868 he was not desiring a reform of all of geology, but explicitly the principle of Uniformitarianism. Where Huxley had suggested that 'Most of us, I suspect, are Gallios, "who care for none of these things," being of the opinion that, true or fictitious, they have made no practical difference to the earth during the period of which a record is preserved in stratified deposits', Kelvin ripostes that precisely for this reason it is important that evidence for opposing opinions should be thoroughly sifted and not merely disposed as matters of opinion, or 'of faith beyond the realm of reason.' He expresses considerable satisfaction that Huxley should find his estimates of the age of the earth acceptable to geologists, and concludes by asserting that experimental and mathematical investigations of underground temperature should be an integral part of geology, and that he himself should not be regarded 'as a mere passer-by but as one constantly interested in their grand subject and anxious in any way, however slight, to assist them in their search for truth.'

Unlike the rather longer rebuttal by Tait, this response by Kelvin shows the regard and friendliness between Kelvin and Huxley which is often overlooked. Indeed, Kelvin invited Huxley to accompany him on a summer cruise on his yacht, the *Lalla Rookh*, along with Tait, Helmholtz and Tyndall, although it is regrettable that such an intellectually stimulating summer holiday failed to materialise.

Despite Kelvin having stated that uniformitarianism violated the second law of thermodynamics, some geologists and biologists continued to support uniformitarianism and other ideas that violated the second law to the end of the nineteenth century and indeed used these unsupportable ideas to reject Kelvin's criticisms, for example, by suggesting continuous creation of energy in the sun to provide the energy for uniformitarianism. Not only did they not understand the thermodynamic concepts involved in Kelvin's work but they did not see the relevance to their work of a law derived from mechanics (a subject looked down upon in those days).

In the middle of the nineteenth century, the geologists started to quantify the dating of rocks more precisely. The most significant attempt was made by James Croll (1821–1890) who did not regard himself as a geologist. He was the Keeper of the modest Andersonian Museum in Glasgow and during his tenure became a self-taught physicist with a particular interest in glaciation. He also studied astronomy in relation to this topic.

Croll supported the physicists', in particular Kelvin's opposition to uniformitarianism and Kelvin's upper limit of 100 Ma for the age of the earth, and he put forward reasons for accepting this value. He could not, therefore, accept 20 Ma for the age of the sun and so he put forward a theory that would give a higher value but could not provide data to quantify it [38]. He also rejected the estimate of the earth's age calculated from tidal retardation [39] as the equatorial bulge at the time of the earth's solidification would have been eroded over time and be quite different from the existing shape. Assuming a constant shape throughout would lead to an erroneous result. He set about developing a quite different method.

Croll hypothesised that the shape of the earth's orbit around the sun varies slowly with time because of the gravitational influence of all bodies in the solar system [40]. This problem is not solvable analytically (the n-body problem) but he assumed that the orbit varies from near circularity as at present to greatest ellipticity in which configuration the relative lengths of the seasons are different. The hemisphere experiencing winter at aphelion, the remotest part of the orbit, would suffer long hard winters and would not fully recover in the short summer. Reinforcing this difference would be the gradual accumulation of snow and ice. By contrast, the other hemisphere would enjoy a moderate climate with long pleasant summers and short mild winters. Other climatic changes would occur such as shifts in the winds and in tropical currents away from the cold regions and increased precipitation in the cold regions. The snow and ice would reflect the sun's heat and dense fogs would shield the cold regions from solar radiation and eventually an ice age would occur. Arising from the precession of the equinoxes the extremes in climate would alternate from one hemisphere to the other every 10,500 years.

By 1867, Croll had begun to gather data to substantiate his theory. He calculated orbital eccentricities for the last million years and found three periods of extremely high eccentricity between 950,000 and 750,000 years ago. He also found several periods of slightly less eccentricity at later times, the most recent being 100,000 and 200,000 years ago. He then had to decide which of these many periods corresponded to the ice age currently being investigated by geologists. He initially looked at the most recent period of extreme eccentricity around 750,000 years ago but concluded that there would be no trace of glaciation by now. In addition, Lyell had calculated that if the last ice age occurred 850,000 years ago the date of the Cambrian period must extend back nearly 240 Ma, much longer than 100 Ma which Croll had come to accept. If, however, Croll chose the most recent period of relatively extreme eccentricity which ended 8,000 years ago and made certain changes to Lyell's calculation, he reduced the Cambrian period to 60 Ma and hence an age of the earth that corresponded approximately to his preferred value, thus getting agreement between geology and physics. He published his findings in 1868. It is obvious, however, that the age of the earth deduced by this means is very dependent on the degree of eccentricity and on what date is chosen for the ice age, remains of which can still be seen. Lyell's choice is closer to the middle of the last ice age and would have been a better choice for Croll if he had not allowed the 100 Ma estimate by Kelvin to influence his choice. Kelvin seems to have ignored Croll's work perhaps because of its several unsupported assumptions.

Numerous estimates of the age of the Earth were produced using various techniques during the late nineteenth and early twentieth centuries and Dalrymple [3] conveniently tabulates these for us. Between 1860 and 1917, he lists 32 estimates based on the rates of sedimentation or erosion, varying between 3 and 15,000 Ma, but with nearly half of them falling in the range of 50–500 Ma for the period of time since the Cambrian era. Estimates on the basis of orbital physics ranged from Tait's <10 Ma to Kelvin's <1000 Ma (although Jeffreys, on the basis of the eccentricity of the orbit of Mercury suggested the higher figure of 3,000 Ma in 1918 [41]). Estimates based on ocean chemistry such as sodium, chloride or sulphate accumulation up to 1931 produced 12 estimates between 25 Ma and an upper bound of 340 Ma. Cooling of the Earth provided 10 estimates ranging from Tait's low figure of 10–15 Ma to Holmes estimate of >1314 Ma and Haughton's >1280 Ma, but mostly in the range of 20–100 Ma. Solar cooling tended to provide lower estimates from 4.4–5.8 Ma from Ritter in 1899 to Kelvin's 10–500 Ma from 1862.

5 Growing opposition

A particularly influential paper with Kelvin was that of Clarence King in 1893, on 'The Age of the Earth' [42], in which he presented temperature distributions calculated for various initial Earth temperatures and cooling times using the same assumptions and mathematics as Kelvin. King's work was based on the assumption of the solidity of the Earth, combined with estimates of the change of the melting point of diabase (a shallow intrusive rock similar in composition to basalt) as a function of depth. These were based on the work of Barus [43,44] who measured changes in volume and latent heat of fusion of diabase as it cooled and solidified from 1500° C. From them, King was able to predict that the change in melting point with temperature amounted to 0.025°C/atmosphere and was linear. The straight line showing this is in Fig. 6.

A consequence was that any reasonable initial temperature for the earth lay on the solid side of the line and therefore the earth was probably never totally molten. Figure 6 shows the variation of temperature with depth, for the earth, for a range of initial assumptions. Those used by Kelvin (curve b) produce a temperature curve which intersects the diabase melting line at two points, implying a molten layer between 42 km and (outside the range of the data plotted) 364 km from the surface, with a solid crust only 42 km thick. Increasing the cooling time to 600 Ma (curve c) avoids the liquid layer but results in a temperature gradient near the surface of only 1°C/68 m – only half the average measured value.

Excluding results that provided near-surface temperature gradients significantly different from the BA's average, and also estimates that required a combination of time and temperature which was more than was absolutely needed, King settled on an a final figure for the age of the earth of 24 Ma (curve e). In 1897, Kelvin presented a paper on 'The Earth as an Abode Fitted for Life' to the Victoria Institute [45] in which he pointed out that changing his initial temperature assumption to only 2000°F – the approximate temperature of lava, would have reduced his 100 Ma estimate to only 10 Ma, and that as a result of refinements in his calculations he would not differ by much from Clarence King's figure of 24 Ma.

While there was still some adherence to the principles of Uniformitarianism among the geological community, most had been content to live with the possibility of an Earth that had been in existence for 100–500Ma. However, the later reductions in the estimates began to conflict with other geological evidence and there was a feeling that the physicists were getting carried away by their theoretical approach with insufficient attention to the fundamental geological principles.

John Perry had been an assistant and occasional collaborator with Kelvin and was well acquainted with Kelvin's approach to the Age of the Earth problem. However, he was not convinced that Kelvin was right in the face of growing discord from the geological community. He had attempted to raise the issue with both Kelvin and Tait but had received no satisfactory response. In 1895 he therefore published his concerns in Nature [46–48] after allowing Kelvin and Tait to see the manuscript beforehand and to provide answers to it in the same edition of Nature.

Perry did not attempt to attack Kelvin's mathematics or procedures, but questioned the basic assumption that the Earth could be modelled as a solid cooling mass. Fisher, in 1881 had shown that the crust could move and changed in height with the advance and retreat of the ice ages [49]. A molten interior to the Earth would allow convection processes as well as conduction, and a honeycomb model would need Kelvin's estimates to be increased 56-fold, bringing them close to

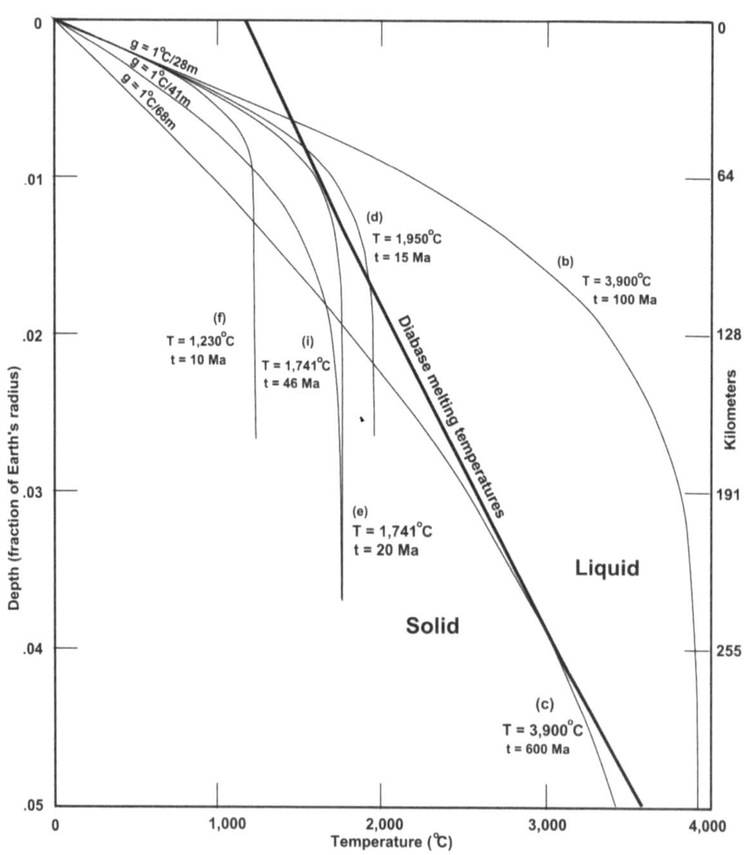

Figure 6: Some of Clarence King's temperature versus depth curves for determining the age of the Earth (adapted from Dalrymple [3, p. 42], after King [42]).

the actual figure of around 4.5 Ga. Perry showed that for a slab of unlimited thickness, initially at uniform temperature – say 4000°C – and with one face thereafter kept at zero temperature, then from the theory of dimensions, if the conductivity is increased m times and all lengths are increased n times, all temperatures remain the same if the times are increased n^2/m times. The gradient at the surface will be altered n^{-1} times, but that can be brought back to an assigned value, with the same value of surface temperature, by altering the conductivity and thickness of an outer surface stratum. In the particular case of a globe, the size of the earth (radius R = 6.38×10^8 cm) and a conductivity k of 0.47 (79 times that of surface rock), and k/c = 0.165, which is 14 times that of surface rock, where c is the thermal capacity, then, with a shell of rock with a thickness of 4×10^5 cm (about 2½ miles) superposed, the time of cooling to the present surface gradient of 1 C in 2740 cm would be as much as 98×10^8 years. With k set to 195 times and k/c to 35 times the value for surface rock, and with a shell 20 miles (3.272×10^6 cm), the time would be 127×10^8 years or 12.7 Ga.

Tait rejected Perry's discussions out of hand, but Kelvin did admit that perhaps he should have placed the upper limit at 4000 Ma instead of 400 Ma. Kelvin accepted the principles of Perry's calculations but did not believe that the model of an earth with a shell of constant thickness and smaller conductivity and thermal capacity was a valid one. Although he welcomed the possibility of determining the change in conductivity and thermal capacity of various rocks at elevated temperatures experimentally, he felt that these assumptions could not be made on the basis of existing data.

Later, on the basis of fresh data on thermal conductivity from Weber and from Barus and Clarence King, and taking into account the augmentation of melting temperature with pressure, Kelvin arrived at a revised estimate of 24 Ma [50] and stated that until there was more knowledge of the augmentation or diminution of thermal conductivity with increasing temperature it would be quite uninteresting to publish any closer estimate.

Perry then wound up the discussion [48] with a full review of the evidence and concluded: 'To sum up, we can find no published record of any lower maximum age of life on the earth as calculated by physicists (I leave out the estimates based upon the assumption of uniform density in the sun, and also those of Mr Clarence King) than 400 million years. From the three physical arguments, Lord Kelvin's higher limits are 1000, 400, and 500 million years. I have shown that we have reasons for believing that the age, from all three, may be very considerably under-estimated. It is to be observed that if we exclude everything but the arguments from mere physics, that the probable age of life on the earth is much less than any of the above estimates; but if the palaeontologists have good reasons for demanding much greater times, I see nothing from the physicist's point of view which denies them four times the greatest of these estimates.'

One reason for Kelvin's obduracy was his over-riding conviction that the possible age of the Sun provided an ultimate limit on the possible age of the Earth. Even this assumption did not go entirely unchallenged, however, and when Thomas Chrowder Chamberlin of the University of Chicago also attacked Kelvin's estimates he made a prophetic statement:

'Is present knowledge relevant to the behaviour of matter under such extraordinary conditions as obtain in the interior of the sun sufficiently exhaustive to warrant the assertion that no unrecognized sources of heat reside there? What the internal composition of the atoms may be is yet an open question. It is not improbable that they are complex organizations and seats of enormous energies. Certainly, no careful chemist would affirm either that the atoms are really elementary or that there may not be locked up in them energies of the first order of magnitude. No cautious chemist would probably venture to assert that the component atomecules, to use a convenient phrase, may not have energies of rotation, revolution, position and be otherwise comparable in kind and proportion to those of a planetary system. Nor would he probably be prepared to affirm or deny that the extraordinary conditions which reside in the centre of the sun may not set free a portion of this energy' [51] [sic].

Darwin had asked his son George, a competent mathematician, to check Kelvin's work but he found no error. The discrepancy must lie in the assumptions, which Kelvin clearly stated as he was accustomed to do, a feature of his work noted by Maxwell. One assumption is the uniformity of the earth in terms of thermal conductivity. It is not an unusual procedure to construct simple models in science in order to obtain an order of magnitude estimate. Taking account of non-uniformity has little effect on the result. The second assumption is that of no internal heat generation. This is 60 years before it was discovered that radioactivity existed and that heat was produced by this fission. Ernest Rutherford (1871–1937) in his famous lecture on radioactivity given at the Royal Institution in 1904 declared that this was the source of the discrepancy. Various attempts were made by Rutherford and his contemporaries to support Rutherford's statement, but these led to conflicting results, mainly because of inadequate data. The omission of internal heat generation is taken by many to be Kelvin's mistake but it was not a mistake because the existence of radioactivity and its heating effect were unknown. However, recent work by England, Molnar and Richter [52], for example, demonstrated that using modern estimates of radioactive heating, whether taken as uniform for simplicity or as it is actually distributed has little effect on Kelvin's answer for the age of the earth, although it has a dominant effect on the age of the sun where the energy arises from nuclear fusion – an effect which was not discovered for another 50 years. When radioactivity was discovered, Kelvin maintained that it could not account for the discrepancy between his estimates of the age of the earth and that required by the geologists and biologists. In this he was right. However, he did not accept the physicists' explanation for the phenomenon and attempted to explain it in mechanical terms just as he and Helmholtz continued to try and find a mechanical theory of electromagnetism.

6 Radioactivity as an additional source of heat

Although radioactive heating has been shown to be insufficient to substantially change the estimate of the earth's age, the phenomenon of radioactivity itself possesses the ability to give a measure of time as a consequence of the time it takes for a radioactive element to decay. This decay time is measured by the half-life, that is, the time it takes for half of the radioactive matter to decay. Frederick Soddy (1877–1956) not only observed the heating effect of the release of alpha-particles but discovered isotopes, elements with the same chemical properties but different atomic weights (and hence atomic number). Many of these isotopes are radioactive, for example, there is the very stable uranium 238 isotope and the relatively unstable uranium 235 one. Soddy measured some of these half-lives and also discovered some of the daughter products. Uranium 235 has a half-life of 704 million years (Table 1), which means that half of it decays in that time, three-quarters in twice that time and seven-eighths in thrice that time. The ultimate decay product is the non-radioactive isotope lead 207. Thus, from measured amounts of uranium 235 and lead 207 in a sample of rock, it is possible to calculate the age of the rock. Various elements can be used depending on the composition and age of the rock. In early investigations, the decay product helium was the choice but helium tended to escape from the rock in situ and from the specimens under test in the laboratory and could only supply a lower estimate of the age of the rock. For example, Robert John Strutt, 4th Baron Lord Rayleigh (1875–1947), son of the Nobel prize winner, arrived at 2000 Ma for the age of the earth, an underestimate of the now accepted value but much longer than earlier geological estimates by non-radioactive means.

The person who put the method on a firm foundation was Arthur Holmes (1890–1965) who, from 1913 onwards [53], made careful measurements of half-lives and daughter products of a large number of radioactive elements. It was this method based on his data that established the dates of the oldest rocks and gave a continuous time history for all geological work including fossil dating. A list of geological periods is given in Table 2.

Conclusions

The conflict among physicists, geologists and evolutionists to reach a consensus on the age of the earth lasted for well over half a century. Two camps were led by towering figures in science. On the one hand, were the physicists led by Lord Kelvin supported by Tait and Helmholtz. All three believed in a mechanical universe and all three were intransigent in their views. On the other side were the geologists and evolutionists led by Sir Charles Lyell aided by Huxley and to a lesser extent Darwin, the first two as equally intransigent as the physicists. Kelvin having conceived, at an early date, methods for calculating the age of the earth did little to develop new methods or to investigate changes to his original assumptions for half a century. Lyell who coined the term uniformitarianism in geology following the work of Hutton and Playfair stuck to the idea for a similar length of time. It was only when other physicists and geologists developed new methods and gathered more information and accepted the truth of the second law of thermodynamics that some semblance of agreement began to emerge. Finally, it was the development of rock dating by radioactive elements that produced general acceptance of the ages of rock and fossils and of the earth itself.

Kelvin and Tait's greatest desire was to overthrow the principle of Uniformitarianism. This theory had originated in the eighteenth century, and in less extreme form had been widely, if often passively, accepted in the geological community by the middle of the nineteenth century. It was largely due to a single text book, and on the basis that reason based on observation seemed to suggest that long periods of time could account most easily for the geological record. But geology was more concerned with sequence than with absolutes and while uniformitarianism was not denounced, absolute time scales were not a necessity. Wherever geologists or astronomers did make attempts at some absolute dating, the results held so many uncertainties that Kelvin's original estimates were never really considered to be unacceptable. Only later, when the estimates started getting lower and lower were they seriously questioned.

Table 1: Half-life value for various elements.

Element	Decay product	Half-life
Uranium 235	Lead 207	704 million years
Uranium 238	Lead 206	4.5 billion years
Carbon 14	Nitrogen 14	5730 rs

Table 2: Modern classification of geological periods.

Period	Millions of years ago	Features
Precambrian	4500–590	Oldest rock found so far, 3800 Fossils rare. Earth's crust formed 4500
Cambrian	590–515	Abundancy of fossils including primitive representatives of modern invertebrates, trilobites, brachiopods, gastropods
Ordovician	515–445	Graptolite fossils abundant in deep water. Ice age 500
Silurian	445–415	First true fish appearedFirst evidence of plant life
Devonian	415–370	Invertebrate fossils such as corals, brachiopods, ammonoids, crinoids in marine deposits, old red sandstone
Carboniferous	370–280	Land plants prolific and led to major coal deposits. Amphibians more prolific. Some reptiles evolve
Permian	280–240	Ice age 250
Triassic	240–200	Dinosaurs appeared
Jurassic	200–135	Dinosaurs flourished and developed
Cretaceous	135–65	Dinosaurs died out
Tertiary	65–1.8	Modern invertebrates and mammals evolved. Flowering plants dominant
Quaternary	1.8–present day	Ice age 1.8 m to 10,000 years. Fossils of horses, pigs, elephants. Man dominant

Uniformitarianism never quite died completely, and even Einstein considered that even if the life of the Earth was finite the life of the universe as a whole was infinite and he applied a correction factor to his theories to allow this to be true. He was never completely happy with this correction, which later was removed. Even today modern theories suggest a cycle of universe birth, expansion, collapse and eventual re-birth.

The entire debate is an interesting example of the way in which science progresses, with theories proposed by influential figures, rising to prominence with them, and never quite fading away until their proponents also passed away. The conflict sadly detracts from the reputations of Lyell and Kelvin. In particular, Kelvin who was revered as the greatest physicist of the century has in retrospect suffered most. This is partly because he was so far out on the age of the earth and partly because of his attachment to mechanical illustrations and explanations of the topics constituting the basic physics of his time. However, his appreciation of physics as a whole can still inspire those who now attempt to do the same for a subject now of massively greater breadth.

References

Note: The first six references have been drawn on extensively throughout and where original sources were not available the references are noted as 'cited in' one of these publications although generally the references will have occurred in more than one of these sources. MPP refers to Mathematical and Physical Papers, 6 vols, Cambridge, 1882–1911; PLA refers to Popular Lectures and Addresses, 3 vols., London, 1891–1894; T&T' refers to [54].

[1] Albritton, C.C., *The Abyss of Time*, Dover Publications, Inc.: Mineola, New York, 2002 (Unabridged republication of the edition published by J.P. Tarcher, Los Angeles, 1986).
[2] Burchfield, J.D., *Lord Kelvin and the Age of the Earth*, University of Chicago Press: Chicago and London, 1990 (First published 1975).
[3] Dalrymple, G.B., *The Age of the Earth*, Stanford University Press: California, 1991.
[4] Jackson, P.W., *The Chronologer's Quest*, Cambridge University Press: Cambridge and New York, 2006.
[5] Smith, C. & Wise, M.N., *Energy and Empire – A Biographical Study of Lord Kelvin*, Cambridge University Press: Cambridge, 1989.
[6] Thompson, S.P., *The Life of William Thomson, Baron Kelvin of Largs*, MacMillan and Co.: London, 1910.
[7] Hutton, J., Theory of the Earth; or an investigation of the laws observable in the composition, dissolution, and restoration of land upon the globe. *Transactions of the Royal Society of Edinburgh*, **1(2)**, pp. 209–304, 1788.
[8] Thomson, W., On the secular cooling of the earth. *Transactions of the Royal Society of Edinburgh*, **23**, pp. 157–170, 1864 (paper read 28 April 1862); *Transactions of the Royal Society of Edinburgh*, **XXIII**, pp. 157–170, 1864; *Philosophical Magazine.*, **XXV**, pp. 1–14, Jan 1863; *The Edinburgh New Philosophical Journal*, **XVI**, pp. 151–152, 1862; MPP, **III art xciv**, pp. 295–311; T&T', Appendix D.
[9] Huxley, T.H., Anniversary address of the president: geological society. *Quarterly Journal of the Geological Society of London*, **25**, pp. xxxvii–liii, 1869.
[10] Ussher, J., *The Annals of the World. Deduced from the Origin of Time and Continued to the Beginning of the Emperour Vespasian's Reign, and the Total Destruction and Abolition of the Temple and Commonwealth of the Jews*, E. Tyler for J. Crook and G. Bedell: London, 1658 (cited in [3]).
[11] Maillet, B.d., *Telliamed, or Conversations Between and Indian Philosopher and a French Missionary on the Diminution of the Sea*, University of Illinois Press: Urbana, 1968 (Originally published in 1748).
[12] Comte de Buffon, G.-L.L., *Des Epoques de la Nature*, vol. 10, Ser C, Museum National d'Histoire Naturelle: Paris, 1962.
[13] Kant, I., *Universal Natural History and Theory of the Heavens or an Essay on the Constitution and the Mechanical Origin of the Entire Structure of the Universe Based on Newtonian Principles*, Vancouver Island University: Vancouver 1998, revised 2008, 1755 (Based on George Reimer's edition of the Complete Works of Immanuelle Kant, 1905).
[14] Hutton, J., *Theory of the Earth With Proofs and Illustrations*, printed for Cadell and Davies: London; printed for William Creech, Edinburgh, 1795 (cited in [1]).
[15] Playfair, J., *Illustrations of the Huttonian Theory of the Earth*, Caldell and Davies: Edinburgh, 1802 (cited in [1]).
[16] Lyell, C., *Principles of Geology*, John Murray: London, 1830–1833 (cited in [1]).
[17] Darwin, C., *On the Origin of Species*, Harvard University Press: Cambridge, MA and London, England, 1859 (facsimile of the first edition published in 2003).
[18] Darwin, F., *The Life and Letters of Charles Darwin*, vol II, John Murray: London, 1887.
[19] Thompson, S.P., *The Life of William Thomson, Baron Kelvin of Largs*, Chapter 4, The Glasgow Chair, MacMillan and Co. Ltd.: London, pp. 160–189, 1910.
[20] Thomson, W., On the linear motion of heat. *Cambridge Mathematical Journal*, **III**, pp. 170–174, 1842; *Cambridge Mathematical Journal*, **III**, pp. 206–211 1843; MPP, **I art iv**, pp. 10–15; MPP, **I art v**, pp. 16–21.
[21] Thomson, W., Note on some points in the theory of heat. *Cambridge Mathematical Journal*, **IV**, pp. 62–64, 1844; MPP, **I art xi**, pp. 36–38 (including a note added in 1881).

[22] Thomson, W., On a universal tendency in nature to the dissipation of mechanical energy. *Philosophical Magazine Series 4*, **4**, pp. 304–306, 1852.

[23] Thomson, W., Address to the British Association at Edinburgh. *British Association Report*, **XLI**, pp. lxxxiv–cv, 1871; *Nature*, **IV**, pp. 262–270, Aug 3 1871; *American Journal of Science*, **II(3)**, pp. 269–294, 1871; MPP, **II art. lxvi**, pp. 25–27; PLA, **II**, pp. 132–205.

[24] Thomson, W., Note on a passage in Fourier's 'heat. *Cambridge Mathematical Journal*, **III**, pp. 25–27, 1841; MPP, **I art ii**, pp. 7–9.

[25] Thomson, W., On geological time. *Transactions of the Geological Society of Glasgow*, **3(1)**, pp. 1–28, 1871 (paper read Feb 27 1868); PLA, **II**, pp. 10–64.

[26] Thomson, W., Extract 'On the observations and calculations required to find the tidal retardation of the earth's rotation', *The Rede Lecture, Cambridge*, 23 May 1866; PLA, **II**, pp. 65–72; MPP, **III art xcv**, pp. 337–341.

[27] Dunthorne, R., On the acceleration of the moon. *Philosophical Transactions*, **46**, pp. 162–172, 1749.

[28] Thomson, W., On the mechanical action of radiant heat or Light. On the power of animated creatures over matter. On the sources available to man for the production of mechanical effect. *Proceedings of the Royal Society of Edinburgh*, **III**, pp. 108–113, 1857 (Read 2 Feb 1852); MPP, **I art lviii**, pp. 505–510.

[29] Thomson, W., On the mechanical energies of the solar system. *Proceedings of the Royal Society of Edinburgh*, **III**, pp. 241–244, 1854, ; Appendix on the age of the sun added Aug 15 1854; *Proceedings of the Royal Society of Edinburgh*, **XXI**, pp. 63–80, 1857; *Comptes Rendu*, **XXXIX**, pp. 682–687, Oct. 1854; *Philosophical Magazine*. **VIII**, pp. 409–430, Dec 1854; MPP, **II art lxvi**, pp. 1–25.

[30] Thomson, W., On the age of the sun's heat. *Macmillan's Magazine*, pp. 288–393, March 5 1862; PLA, **I**, pp. 349–368.

[31] Thomson, W., Physical considerations regarding the possible age of the sun's heat. *British Association Report*, **pt ii**, pp. 27–28, 1861, *Philosophical Magazine*, **XXIII**, pp. 158–160, Feb 1862; PLA, **I**, pp. 349–368.

[32] Thomson, W., Recent investigations of M. Le Verrier on the motion of mercury. *Proceedings of the Philosophical Society of Glasgow*, **IV**, pp. 263–266, 1859 (Read Dec 14 1859); MPP, **V**, pp. 134–137.

[33] Thomson, W., On the rigidity of the Earth. *Proceedings of the Philosophical Society of Glasgow*, **V**, pp. 169–170, 1862 (Read Mar 26 1862); *Proceeding of the Royal Society London*, **XII**, pp. 103–104, 1862 (Read May 15 1862 – abstract); *Phil. Trans.*, **CLIII**, pp. 573–582, 1863 (appendix added Jan 2 1864); MPP, **III art xcv**, pp. 312–336; *Phil. Mag.*, **XXV**, pp. 149–151, Feb 1863; T&T', pp. 689–704.

[34] Thomson, W., The rigidity of the earth. *Nature*, **V**, pp. 223–224, 18 January 1872.

[35] Thomson, W., Review of the evidence regarding the physical condition of the Earth. *British Association Report*, **pt ii**, pp. 1–12, 1876, Presidential Address to the Mathematical and Physical Science Section of the British Association at Glasgow, 7 September 1876; *Nature*, **XIV**, pp. 138–161, 1876; *American Journal of Science*, **XII**, pp. 336–354, 1876; MPP, **III art xcv**, pp. 320–335; PLA, **II**, pp. 238–272.

[36] Tait, P.G., Geological time. *North British Revolution*, **50**, pp. 215–233, 1869.

[37] Thomson, W., Geological Dynamics. *Transactions of the Geological Society of Glasgow*, **III**, pp. 215–240, 1871 (Read April 5 1869); *Geological Magazine*, **6**, pp. 472–476, 1869; PLA, **II**, pp. 73–127.

[38] Croll, J., On geological time and the probable date of the glacial and upper Miocene period. *Philosophical Magazine*, **Ser 4**, **35**, pp. 363–384, 1868; *Philosophical Magazine*, **Ser 4**, **36**, pp. 362–386, 1868.

[39] Croll, J., On the age of the Earth determined from tidal retardations. *Nature*, **4**, pp. 323–324, Aug 24 1871.
[40] Croll, J., On the eccentricity of the Earth's orbit and its physical relations to the glacial epoch. *Philosophical Magazine*, **Ser 4**, **33**, pp. 119–131, 1867.
[41] Jeffreys, H., On the early history of the solar system. *Monthly Notice of the Royal Astronomical Society*, **78**, pp. 424–441, 1918 (cited in [3]).
[42] King, C., The age of the Earth. *American Journal of Science*, **Ser 3**, **45**, pp. 1–20, 1893, (cited in[3]).
[43] Barus, C., The contraction of molten rock. *American Journal of Science*, **142**, pp. 498–499, 1891.
[44] Barus, C., Criticism of Mr Fisher's remarks on rock fusion. *American Journal of Science*, **146**, pp. 140–141, 1893.
[45] Thomson, W., The age of the Earth as an abode fitted for life. *Philosophical Magazine Series 5*, **47**, pp. 69–90, 1899 (1897 address before the Victoria Institute, with numerous additions written at various times from June 1897 to May 1898).
[46] Perry, J., On the age of the Earth. *Nature*, **51(1314)**, pp. 224–227, 1895.
[47] Perry, J., On the age of the Earth. *Nature*, **51(1319)**, p. 341, 1895.
[48] Perry, J., On the age of the Earth. *Nature*, **51(1329)**, pp. 582–585, 1895.
[49] Fisher, O., On the depression of ice-loaded lands. *Geological Magazine Series 2*,, **9**, p. 526, 1882.
[50] Thomson, W., The age of the Earth. *Nature*, **51**, pp. 438–440, 1895.
[51] Chamberlain, T.C., Lord Kelvin's address on the age of the Earth as an abode fitted for life. *Science*, **9**, pp. 11–18, 1899.
[52] England, P.C., Molnar, P., and Richter, R.F.M., Kelvin, Perry and the age of the Earth. *American scientist*, **95**, pp. 342–349, July-August 2007.
[53] Holmes, A., *The Age of the Earth*, Nelson: London, 1937.
[54] Thomson, W. and Tait, P.G., *Elements of Natural Philosophy*, 2nd edn, Oxford: Cambridge 1879. x

Postscript

Light bulb. (Marianne Greated 2014, pencil on paper)

Chapter 15

Kelvin in the twenty-first century

D.H. Saxon
Kelvin Professor of Physics (1990–2008), University of Glasgow, United Kingdom.

Abstract

Aspects of Kelvin's work that are a subject of study today are outlined focusing on five topics: The Kelvin (foam) problem, Kelvin Waves and their role in the El Nino phenomenon, the application of Stirling engines and the possibility of reliable earthquake prediction,

1 'The Kelvin Problem': space-filling foam

Kelvin's interest in everything has given him an influence that far outlasted his time. One could mention the first and second Thomson relations in thermoelectricity, or the Kelvin angle, which describes the wake of anything (e.g. duck, boat, aircraft carrier) moving through deep water. The tidal predictor (the 'great brass brain') was used for almost a century. Two particular points of interest are the Kelvin foam and Kelvin waves, and there is the renewed interest in Stirling engines for generating electricity.

In 1887, Kelvin was much occupied in trying to understand the nature of the æther, the medium which he saw as supporting electromagnetic waves and which pervades all space and all matter. Because of the extraordinarily high value of the speed of light it needed to be very low density but very strong. This led him to consider a foam rather than a continuous medium, and thus to formulate the "Kelvin problem": How can one partition space into equal-sized cells with a minimum surface area?

Kelvin's proposed solution (the so-called Kelvin Structure) was a tetrakaidekahedron (14-faced solid,) structure, constructed from a regular octahedron by cutting off each of its points, leaving a cell with six regular hexagonal faces and eight square faces [1]. Figure 1 (left) shows the stacking of these cells to make a three-dimensional structure.

For such a structure to be stable it must have minimal energy associated with it. If the energy is proportional to the area (as in surface tension), it obeys Plateau's rules for stable (minimum energy) soap films [2]. Plateau's rules dictate the following. Films meet at edges: always three films meeting at an angle of 120°. Sets of four films meet at corners at the tetrahedral angle

Figure 1: Left, building up the Kelvin foam from identical tetrakaidekahedra. The curvature of the faces is not shown. (Image courtesy of the author, A. Kepert.) Right, the Weaire–Phelan structure. The two cells (12-face and 14-face) have the same volume. (Public domain image by T. Ruen.)

(109.4°). There is a pressure difference across any face proportional to the curvature of the face – strictly the sum of the two principal curvatures. In the Kelvin-foam case the cells are identical and so the pressure difference is zero. The faces in fact turn out to be not totally flat but with the principal curvatures adding to zero.

Is the Kelvin tetrakaidekahedron the answer to the Kelvin problem? Is it the way of partitioning space using equal-sized units with the minimum surface area? For 106 years, it was unchallenged, until Denis Weaire and Robert Phelan (W-P) found a solution with 0.3% less area [3]. Since then, a large effort has produced other structures which do better than Kelvin but not Weaire and Phelan.

The W–P solution uses two different cells (but with the same volume). One is an irregular dodecahedron with pentagonal faces. The second is a 14-faced structure with two hexagonal and 12 pentagonal faces. Again, the faces are slightly curved. Figure 1 (right) illustrates how these stack in three dimensions. Experiments have shown that actual foams with equal volume cells can be made to form a W–P structure.

When the Beijing Olympics' organisers were looking for a structure to symbolise water, they chose to make the 'water cube' in which the swimming and diving events took place, as a Weaire–Phelan structure. This lightness of structure and strength of this structure were remarkable [4], see Fig. 2.

Today the Kelvin problem remains unsolved. The W–P structure has not been bettered, but neither has it been proved to be the best of all possible solutions. As for the aether, in the same year as Kelvin was perfecting his foam, the Michelson-Morley experiment showed that the motion of the earth, or any other body, through the aether is undetectable [5]. This led to the theory of Relativity and the demise of the æther. Yet, today after the discovery of a likely Higgs Boson a more challenging æther is hinted at [6]. But now we seem to have three fields pervading all space: the Higgs field which gives mass; dark matter which may be things we would recognise

Figure 2: The 2008 Beijing Water cube has the Weaire–Phelan structure. (Photograph courtesy of Tristram Carfrae, Arup.)

or may be something else; but dark energy, which makes the universe expand at an accelerating rate, is not yet understood.

2 Kelvin Waves and the El Nino effect

In 1879 Kelvin published a paper 'on the gravitational oscillations of rotating water' [7]. In this work he predicted that the earth's rotation constructs a waveguide that transports (hot) water from the East Indies in an Easterly direction along the equator to South America, forming the basis of the el Nino climate phenomenon. We briefly outline the mechanism here.

An inertial frame is a frame of reference (the 'fixed stars' as we used to say) in which Newton's first law of motion applies. In the absence of an external force, a body moves at a uniform speed in a straight line. If we move to a rotating frame of reference, things are more tricky. The most important rotating frame of reference is the earth itself, which rotates on its axis once in every 23 h and 56 min. For everyday purposes, it is most sensible to describe things securely attached to the earth as at rest in a frame of reference that rotates with the earth.

Such a body of mass m fixed at a point on the surface of the earth experiences a downward force, directed at the centre of the earth of mg, where g is the magnitude of the acceleration due to gravity, and a much smaller centripetal force of magnitude $mrw^2 \cos \lambda$ directed towards the axis of rotation of the earth, where r is the radius of the earth (= 6400 km) approximately), ω is its rate of rotation (1 turn per 24 hours = 1.1×10^{-5} rad s^{-1}) and λ is the latitude.

There is an additional force that comes into play when a body is moving relative to the earth. This is known as the Coriolis force and was of much concern in the nineteenth century to artillery officers, who found that the trajectories of shells were deflected to left or to right as they travelled towards distant targets. The most celebrated use of Coriolis force was the Foucault pendulum, which shows (without needing any reference to the fixed stars) that the earth is rotating. If a pendulum is allowed to swing freely, the plane in which it moves rotates around a vertical axis in time t, relative to local buildings and so on, by $wt \sin \lambda$. The Coriolis force which achieves this is directed at right angles to the motion of the body and also at right angles to the earth's rotation.

Consider a small volume of water near the surface of the ocean and being heated by the sun. It will expand as it is heated and this will cause the surface of the ocean to rise a little: this height difference gives rise to a pressure difference ρgh (where ρ is the water density, and h is the difference in height) that will make the water travel across the surface of the ocean away from its starting point. As it moves, it therefore experiences a Coriolis force deflecting its motion to the right in the Northern hemisphere and to the left in the Southern hemisphere. Let the velocity have Easterly and Northerly components of (v_e, v_n). Then, the Coriolis acceleration has components $(fv_n, -fv_e)$ where the Coriolis factor, $f = 2\omega \sin \lambda$, is zero at the equator (where $\lambda = 0$) and has opposite signs in the two hemispheres.

For Eastbound motion, $v_e > 0$. If this motion is deflected Northward ($v_n > 0$, $\lambda > 0$), it experiences a Southward Coriolis acceleration of magnitude $2wv_e \sin \lambda$, which reverses the Northward velocity. Similarly, if it is deflected Southward, its Southward velocity is reversed. Thus, an Eastbound-moving object at the equator is trapped in a waveguide that directs it to move around the equator.

Conversely, for Westbound motion v_e is negative. The Coriolis acceleration serves to increase any Northward motion in the Northern hemisphere and Southward motion in the Southern hemisphere. Thus Westbound motion around the equator is lost as moving water is transported away from the equator to larger latitudes.

The water in the ocean obeys the wave equation, and the Eastbound motion of water is known is a 'Kelvin Equatorial Wave' [7,8]. Recall, this was driven by heating water via the sun. This warming extends to a depth of water (H) called the thermocline, with much colder water below. The Kelvin Wave speed (c) is given by $c^2 = gH$. The phase and group velocities are identical so the wave travels right across the Pacific Ocean without changing its shape. The speed c is about 3 ms^{-1}, so that it may take a couple of months for the hot water to be transferred to the American face of the Pacific. This is the mechanism that creates the el Nino phenomenon which has such drastic effect on the weather, see Fig. 3 (left).

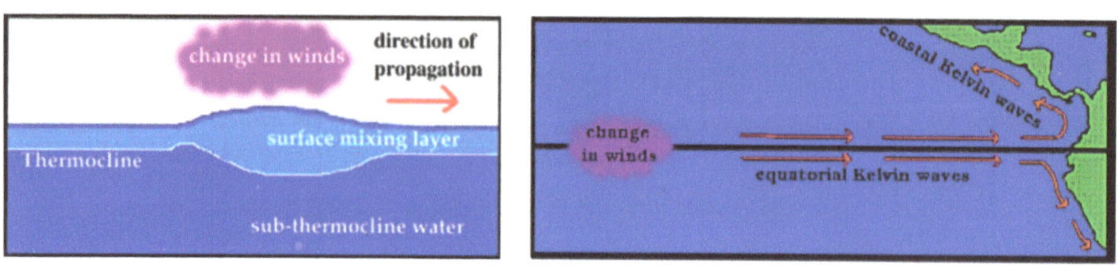

Figure 3: Left: schematic side view of Kelvin equatorial wave and Kelvin coastal wave production. Right: schematic top view of Kelvin equatorial wave and Kelvin coastal wave production. (Both images from Department of Oceanography, Naval Postgraduate School, US Navy.)

When the Kelvin equatorial wave hits the Americas, it will be deflected to run parallel to the coast, Northbound in the Northern hemisphere and Southbound in the Southern. In this case the Coriolis force traps the water in a waveguide, one side being given by the land, and the other by the Coriolis force. This is called a 'Kelvin Coastal Wave', see Fig. 3 (right).

3 The Stirling engine

The Stirling engine has a vital position in history, as it was by making studies on this from 1847 that Kelvin formulated the first and second laws of thermodynamics [9]. The story of this, and of the essential contributions of his brother James, is told in considerable detail in Chapter 3 of this volume: *James and William Thomson: The Creation of Thermodynamics*. The Stirling engine is an external-combustion heat engine with a permanently enclosed gaseous working fluid, see Fig. 4. It has a hot (red) and a cold (blue) reservoir, joined by a pipe that allows the gas to flow between them [10].

The Stirling thermodynamic cycle has four elements, from left to right in Fig. 4:

1. Gas is heated in the hot cylinder, forcing pressure build up and doing useful work pushing the hot piston to the left.
2. The hot piston moves to the right (making the hot cylinder smaller) while the cold cylinder gets larger. Gas is forced down the pipe into the cold cylinder where it is cooled.
3. The piston in the cold cylinder compresses the gas, generating heat which is removed by the cooling source.
4. Gas is forced from the cold cylinder into the hot cylinder where it heats up rapidly and the cycle repeats.

The transfer pipe greatly improves the thermal efficiency. Its thermal capacity acts as a 'regenerator', heating the gas as it flows toward the hot cylinder, and cooling it as it goes the other way. The efficiency approaches that of the Carnot cycle, the most efficient possible. But in real-life systems it comes out to be less efficient than the diesel (Otto) engine. Interest in the Stirling engine languished until recently. There is now renewed interest for a number of reasons.

Domestic heating, hot water and electricity can be produced more efficiently than in a conventional gas-burning heating system and separate mains electricity. This helps in achieving low carbon emissions. Typical systems produce of order 5 kWe of electrical power.

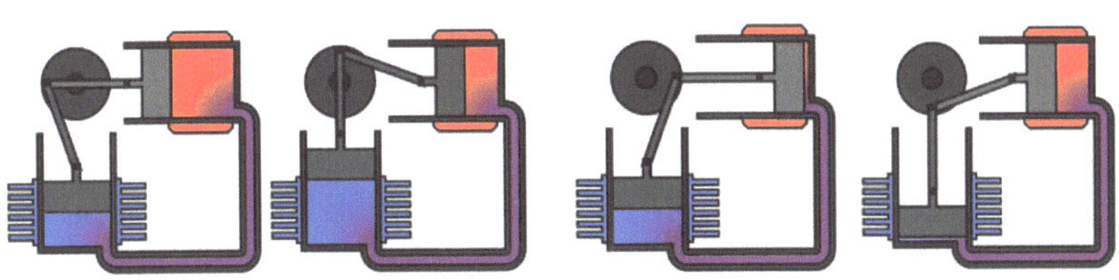

Figure 4: Four steps through the thermodynamic cycle of the Stirling engine. (Images courtesy of the author, Zephyris (Richard Wheeler).)

Because the Stirling engine is an external-combustion engine it is easy to change the fuel. Systems are under development to burn waste previously destined for landfill and to produce combined heat and power.

In sunny place (such as Arizona) large steerable parabolic mirrors are used to focus the sun's rays on to a receiver, which transmits the heat to a Stirling engine. The engine is a sealed system filled with hydrogen. As the gas heats and cools, its pressure rises and falls. The change in pressure drives the piston inside the engine, producing mechanical power, which in turn drives a generator and makes electricity. Systems produce up to 25 kWe each during daylight hours.

Energy-conscious engineering is now revisiting the thermodynamics of heat engines, giving the Natural Philosophy of 1850 renewed life in the 2010s.

4 The Stirling engine in outer space

Space-craft instrumentation is usually powered by solar cells, laid out in large 'sails' to capture the sun's radiation. As one passes into the outer solar system the intensity of solar radiation falls and a new approach is needed. The Advanced Stirling Radioactive Generator (ASRG) uses the heat from ^{238}Pu decay to drive a Stirling engine to generate electricity. ^{238}Pu is very scarce, and so the maximum efficiency is needed. The half-life (14 years) is well matched to the target of 15 years operation. Each generator unit is designed to produce over 100 We from less than 1 kg of Plutonium.

Systems built around the Advanced Stirling Converter (ASC) are currently under development to support future space missions on the Martian surface or in the vacuum of space [11]. The device must be ruggedised to withstand the vibration of the launch and subsequent operation. The device consists of a free-piston Stirling engine and an integral linear alternator that converts the piston reciprocating motion to electrical output. The ASC weighs only about 1.3 kg, has demonstrated a convertor efficiency of 38% at 850°C hot-end temperature and 90°C cold-end temperature. The internal moving components are supported by hydrostatic gas bearings, which allow movement without contact or rubbing. Two ASCs are used in each ASRG, mounted opposite each other and electrically synchronised so that their pistons move in opposite directions, eliminating most of the vibration, see Fig. 5.

The variety of innovative applications of Stirling engines shows that the intellectual vigour Kelvin brought to them in 1847 is by no means exhausted. The search for efficiency, for carbon-neutral electricity generation and waste disposal leaves us knocking at the door of opportunity at the 'final frontier'.

5 Atmospheric electricity

The formulation of the first and second laws of thermodynamics around 1851 enabled one to re-cast physics in terms of energy, and motivated the effort to study (electric) potentials in all possible situations. During the period 1859–1861, Kelvin studied the electrostatic field near the surface of the earth, which is of order 100 V/m, varying according to the weather and other factors.

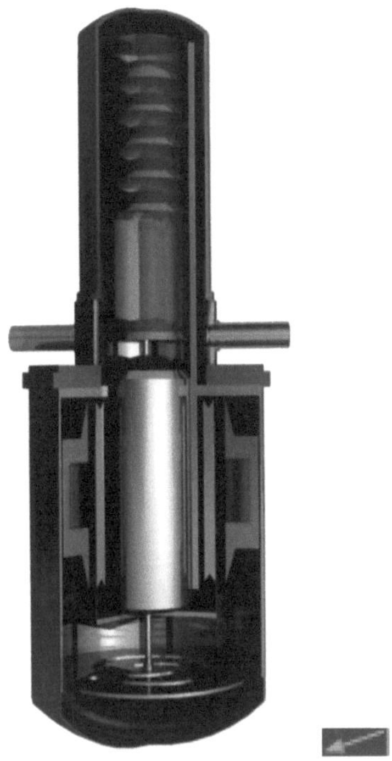

Figure 5: Advanced Stirling Converter. (Image from USA's National Aeronautics and Space Administration)

Kelvin was responsible for two devices for measuring this field, notably the water dropper equaliser sensor. A narrow column of water is slowly poured. The electric field is found from the change in height at the point where the column breaks up into drops. He added photographic recording of the result. He expressed the view that at some date in the future, electrical measurements would form important additions to our 'means for prognosticating the weather' [12]. It is interesting to note that Kelvin was recycling a technology he had just (1858) invented - the syphon recorder (the ancestor of the ink-jet printer.) Both rely on an understanding of the behaviour of ionised fluid droplets [13].

Various observatories were set up around the world, including Kew in 1908 – UK measurements continued to the 1980s. Kakioka in Japan continues to make measurements after over a century [14]. There the matter may perhaps rest, but two events have unexpectedly renewed interest. It appears that earthquakes are preceded over a few days by anomalously high electric fields. Many lives could be saved if this could form the basis of accurate predictions. The answer to the question 'should I stay out of doors through the coming night, or is it safe to sleep indoors', really matters [15]. The heavy prison sentences handed out to scientists who did not predict the 2009 earthquake at L'Aquila in Italy shocked many people for different reasons. The Fukujima earthquake in Japan would have avoided much loss of life if such predictions had been available. A satellite proposal wishes to make a step-change in our knowledge. Over a 3-year period, some 400 earthquakes of magnitude 6–6.9 are expected, and these will form the training data set for quake predictability tests.

6 From the nineteenth to the twenty-first century

In the cover text to the very recent *Kelvin: Life, Labours and Legacy* [16] we read that 'the waxing and waning of the twentieth century and the supplanting of classical physics have eroded his reputation, so that for many scientists he is remembered as little more than a unit of temperature'. This concise study demonstrates that Kelvin's 'sure touch' has resonated in five fields which epitomise our current century: the Olympic movement, the apparent need for fields that fill all space, climate change and environmental disaster and space travel.

References

[1] Thomson, W.S. (Lord Kelvin), On the division of space with minimum partitional area, *Philosophical Magazine*, **24**(151), p. 503, 1887.

[2] Ball, P., Shapes. *Nature's Patterns: A Tapestry in Three Parts*, Oxford University Press: Oxford, pp. 66–71, 97–98, 291–292, 2009.

[3] Weaire, D. & Phelan, R. A counter-example to Kelvin's conjecture on minimal surfaces. *Philosophical Magazine Letters*, **69**, pp. 107–110, 1994.

[4] Fountain, H., A Problem of Bubbles Frames an Olympic Design, *New York Times*, 5 August 2008.

[5] Michelson, A.A. & Morley, E.W., On the relative motion of the earth and the luminiferous ether, *American Journal of Science*, **34**, pp. 333–345, 1887.

[6] Close, F. *The Infinity Puzzle*, Oxford University Press: Oxford, 2011.

[7] Thomson, W.S. (Lord Kelvin), On gravitational oscillations of rotating water, *Proceedings of The Royal Society of Edinburgh*, **10**, pp. 92–100, 1879.

[8] Gill, A.E., *Atmosphere–Ocean Dynamics*, International Geophysics Series 30, Academic Press: Waltham, pp. 378–380, 1982.

[9] *Some Pioneers in Air Engine Design: The Rev. Dr. Robert Stirling*. http://www.stirlingengines.org.uk/pioneers/pion2.html

[10] Walker, G., "A Stirling engine is a mechanical device which operates on a closed regenerative thermodynamic cycle, with cyclic compression and expansion of the working fluid at different temperature levels". *Stirling Engines*, Clarendon Press: Oxford, p. 1, 1980.

[11] NASA, Advanced Stirling Radioisotope Generator. http://microgravity.grc.nasa.gov/SSPO/ASRG/documents/ASRGfacts2_10rev2.pdf, February 2010.

[12] Aplin, K.L. & Harrison A.G., Lord Kelvin's atmospheric electricity measurements. *History of Geo- and Space Sciences*, **4**, pp. 83–95, 2013.

[13] Lindley, D., *Degrees Kelvin: A Tale of Genius, Invention and Tragedy*. Joseph Henry Press: Washington, p. 191, 2004.

[14] Kakioka Magnetic Obeservatory. http://www.kakioka-jma.go.jp/en/index.html

[15] Chmyrev, V., Smith, A., Kataria, D., Nesterov, B., Owen, C., Sammonds, P., Sorokin, V., & Vallianatos, F., Detection and monitoring of earthquake precursors: TwinSat, a Russia-UK satellite project. *Advances in Space Research*, **52**, pp. 1135–1145, 2013.

[16] Flood, R., McCartney M. & Whitaker, A., *Kelvin: Life, Labours and Legacy*, Oxford University Press: Oxford, 2008.

Conclusion

Cable ship. (Marianne Greated 2014, gouache and pencil on paper)

Chapter 16

Honoured by banknotes

I.S. Ruddock[1], A. Whitaker[2] & M.W. Collins[3]
[1]*Department of Physics, University of Strathclyde,United Kingdom.*
[2]*Department of Physics, Queen's University of Belfast, United Kingdom.*
[3]*College of Engineering, Design and Physical Sciences,Brunel University London, United Kingdom.*

Abstract

A frequent theme of a nation's paper currency is the honouring of its past distinguished citizens. Usually, the paper money also features the areas of their achievements. The UK, unique in having no less than eight banks issuing paper money, is no exception. Kelvin has been so honoured over a period of some 30 years by Scotland's Clydesdale Bank and currently the Bank of England commemorates Boulton and Watt. Another theme of banknotes is significant architecture, and the University of Glasgow, Queen's University, Belfast and King's College, Aberdeen (now the University of Aberdeen) have all featured over long periods of time. In this concluding chapter all the above are discussed, the dominant emphasis being on Kelvin himself.

1 Introduction: the UK, its political history and its banknotes

1.1 Paper money – an introduction

One of the key functions of the nationstate is to provide its citizens with legal currency in the form of coinage and paper (sometimes plastic) money. The latter forms the subject of this chapter. Historically, they were often and still are issued by banks – hence, the widespread use of the term banknotes. Besides their essential purpose,notes, in common with postage stamps and coinage, have added value as collectors' items. In fact, the bulk of the entire background material for both the UK notes themselves, and the political history which gave rise to them, may be found in a tripartite set of catalogues of *World Paper Money* [1–3]. These are frequently updated, forming a recent set at the time of writing.

Banknotes have no inherent value and rest for their effectiveness on mutual economic confidence. Briefly to digress, this differs from the original concept of coinage whose value reflected its metal content, for example, gold and silver. These days coinage, too, has virtually no inherent value.Returning to the notes, two key attributes can be identified. Firstly, above all else, notes have to be designed and manufactured to be proof against forgery. This has resulted in complex products containing many security features.What is more significant for us, however, is that the very designscan become a *political statement* of the nationstate.

We take three major world examples:the United States, the People's Republic of China and the European Single Currency. Notes are, respectively, issued by the United States Treasury

Table 1: Designs on Paper Money for three major world currencies.

Country	Currency	Obverse designs	Reverse designs
USA (2006 Series)	US$	US Presidents + independence notables	Federal buildings(various)
PR China(2005 Series) (various)	Yuan	Mao Tse-tung	Natural features + buildings (various)
European Union (single currency area) (2002 Series)	Euro	Architecture(various)	Map of Europe, EU flag + bridges (various)

(Federal Reserve Notes), the People's Bank of China and the European Central Bank. Hence, the description *banknotes* still has meaning. Table 1 (information from [3]) summarises the features of the obverse (front) and reverse (back) sides.

Table 1 shows that the US notes emphasise the US's historical independence and the Republic, the Chinese notes the Communist State, and the European notes the ethos of the trans-national European Union. Prior to the inception of the Euro currency, taking three of the larger constituent nations as examples, Germany, France and Italy all featured past eminent citizens and their accomplishments.

1.2. Political history of the UK

In addition to the significance of current paper money, past issues of currency notes reflect the political history of a nation and the UK is unique in having eight banks issuing legal paper money. The UK's full title is the *United Kingdom of Great Britain and Northern Ireland.* Four political entities are involved: England, Wales, Scotland and Northern Ireland: Table 2 givesa timeline,which summarises the history (information from a 'popular' encyclopedia [4]). Much warfare was involved! *Items* refer to banknote history.

Important comments:

1. In England, the issue of banknotes by private and commercial banks is 'a very complex story' [1, p. 643]. Only the Bank of England now issues legal currency.

2. In Scotland, there were a considerable number of commercial and private banks. Various amalgamations of the former occurred and today three note-issuing banks exist – Royal Bank of Scotland, Bank of Scotland and Clydesdale Bank (all details in 'Scotland' sections in [1–3]).

3. In Ireland, similar to Scotland, a number of commercial banks existed and issued notes. In 1921,this was continued by the banks in Northern Ireland. Today, four note-issuing banks exist – Bank of Ireland, First Trust Bank, Northern Bank (now Danske Bank) and Ulster Bank(details in 'Ireland-Republic' section of [1], 'Ireland' and 'Ireland-Northern' of [2] and 'Northern Ireland' of [3]).

4. Since 1960, a portrait of Queen Elizabeth IIhas always appeared on Bank of England notes. Only four previous monarchs have sofeatured: King George IV (1820–1830), King William IV (1830–1837) and Queen Victoria (1837–1901) on:

Table 2: UK political history and first banknote issues.

England	
829	King Egbert of Wessex accepted as overlord of all England [4, p. 288]
1066	Norman Conquest. William the Conqueror becomes king of England
1603	James VI of Scotland becomes James I of England
1649	Setting up of Commonwealth under Oliver Cromwell
1660	Restoration of monarchy (Charles II)
*1694	Bank of England founded, nationalised in 1946 [2, p. 574; 4, p. 76]*
Wales(termed a Principality)	
1277	Edward 1st of England accepted as overlord of Wales[4, p. 872]
1536–1543	Acts of Union between England and Wales: Welsh representation in the English Parliament
*1694	Bank of England notes cover Wales also*
Scotland	
9th century Scotland united by Kenneth McAlpin [4, p. 739]	
1603	Personal Union of Scottish and English kings (James VI of Scotland becomes James I of England)
*1695	Bank of Scotland issues first notes[2, p. 1062]*
1707	Act of Union with England: new state termed *Great Britain*
Ireland	
Henry II of England (1133–1189) accepted as overlord[4, pp. 388/429, 430]	
1782	Independence granted to Irish Parliament
1801	Act of Union—Irish representation in British parliament, new state termed *United Kingdom of Great Britain and Ireland*
*1797	Bank of Ireland issues first notes [2, p. 721]*
1921	Anglo-Irish Treaty: Irish Free State in South, 6 Northern Counties becoming Northern Ireland, new state termed *United Kingdom of Great Britain and Northern Ireland*

Commercial Bank of Scotland notes from 1827 to 1853 [1,p. 1019]
Provincial Bank of Ireland notesfrom 1825 to 1901 [2, p. 730]

King George V (1910–1936) on

UK treasury notes from 1914 to 1928 [2, pp. 541–542]. This sequence displays the UK title changefrom Ireland to Northern Ireland.

5. Devolved Government exists, currently as follows:
   For Scotland: the *Scottish Parliament*with the *Scottish Government*
   For Wales: the *National Assembly for Wales* with the *Welsh Government*
   For Northern Ireland: the *Northern Ireland Assembly* with the *Northern Ireland Executive.*

1.3 Notes of Scotland and Northern Ireland

Scotland and Northern Ireland, then, are constituent political units within the United Kingdom. Despite its name, the Bank of England issues legal tender notes for the entire UK. What of the Scotland and Northern Ireland notes? They mainly circulate within their respective areas, although from about the 1970s the notes had the word *Sterling* added to *Pounds* (e.g. Royal Bank of Scotland [3, p. 875] and Bank of Ireland [3, p. 750] emphasising their legal currency status within the whole of the UK.

This has significance for the banknote designs. Whereas the Bank of England notes represent the UK as a whole, the Scotland and Northern Ireland notes represent their respective local political units. This does *not* mean their prominent citizens so honoured have only *local* status. The Scotsman Adam Smith epitomises this.

Adam Smith is regarded as the father of economics. He was a peer of James Watt, and this is discussed in Chapter 7, in Section Watt and the Age of Enlightenment. As shown in Table 3, he featured not only on the Clydesdale notes from 1981 to 2008 but also on the current £20 Bank of England notes.

2 Kelvin and the University of Glasgow

2.1 Banknote details

These are given in Table 4.

Table 3: Banknote details: Adam Smith.

Bank	Banknote series	Obverse Design	Reverse design	References
A. Clydesdale	£50: Various 1981–2008	Adam Smith: public address	Blacksmith's & farm implements/ sailing ships	[3, pp. 869–871]
	£50: 2001, Univ. Glasgow, 550th			[3, p. 872]
B. B. of England	£20 2006	Queen E II	Adam Smith portrait: pin factory & quote*	[3, p. 425]

*Quote: *The division of labour in pin manufacturing: (and the great increase inthe quantity of work that results).*

Table 4: Banknote details: Kelvin &and the University of Glasgow.

Bank	Banknote series	Obverse design	Reverse design	References
Clydesdale	A. £10: 1963–1964	Arms of bank	Glasgow Univ.	[3, pp. 868]
	B. £20 and £100: 1971–1981, 1982–1989	Kelvin & compass	Lecture theatre, Glasgow Univ.	[3, pp. 869–870]
	C. £100: 1989–1996	Kelvin & compass	Glasgow Univ.	[3, p. 871]
	D. £100: 2001 Glasgow Univ., 550th	Kelvin & compass	Glasgow Univ.	[3, p. 872]

2.2 Portrait of Lord Kelvin and the compass

Lord Kelvin is shown posing with a compass binnacle on the obverse of Clydesdale Bank £20 and £100 banknotes (Fig. 1).

The engraving was taken from the portrait of 1902, reproduced as Fig. 18 in Chapter 1, *Lord Kelvin*, where he is shown with a Kelvin and Hughes instrument. A similar full size binnacle is on display in the Kelvin Gallery of the University of Glasgow's Hunterian Museum and has the same indicator gauge as can just be seen on the edge of the image on the banknote (Fig. 2). Behind Kelvin on the banknotes is a map purporting to show an Atlantic telegraph cable connecting Bristol with somewhere on the east coast of North America considerably south of Nova Scotia. But, as described in Chapter 1, the early cables were laid on the shortest Atlantic route linking Valentia, Ireland, with Trinity Bay, Newfoundland.

Together, the compass binnacle and the Atlantic cable represent two important elements in the career of Kelvin: the entrepreneur and the industrialist, but only peripherally that of the pre-eminent Victorian scientist. His contributions to the successful laying of the cable in 1866 were rewarded with a knighthood and his continued work on the associated telegraphic equipment provided the substantial royalty income, which formed the basis of his personal wealth.

2.3 Natural Philosophy lecture theatre

Kelvin's academic life is illustrated on the reverses of Clydesdale Bank £20 and £100 banknotes,which show a northern view of his lecture theatre at the University of Glasgow's Gilmourhill campus (Fig. 3). It was located on the first floor of the main building and was occupied in 1870 by Natural Philosophy when the University moved from its original medieval buildings on High Street. The lecturer's bench was at the west end with the raked seating towards the east, despite the original plans showing the opposite arrangement. Because there was also an apparatus room to the east of the lecture theatre, the proposed layout had its merits since demonstration items could have been easily taken back and forth between the bench and the preparation area.

Figure 1: The obverses of the Clydesdale Bank £20 and £100 banknotes. (Reproduced withthe permission of Clydesdale Bank PLC.)

Figure 2: The Kevin & Hughes compass binnacle on view in the Kelvin Gallery of the Hunterian Museum, Glasgow. (Photograph by I.S. Ruddock courtesy of the Hunterian Museum, Glasgow.)

Figure 3: The reverses of the Clydesdale Bank £20 and £100 banknotes. (Reproduced with the permission of Clydesdale Bank PLC.)

Various readily identifiable items of apparatus are set out in both photographs as for a lecture. On the bench in the view from the southeast is a set of tuning forks used to generate beats and demonstrate the addition of waves.At the windows in the same view and on the reverse of the banknotes is a mechanical wave machine (Fig. 4) for creating travelling transverse waves. By turning the handle on the lower right-hand side, the rods attached to the eccentrics cause the red balls to oscillate up and down in relative phase thus giving the illusion of a wave travelling from the right to the left or viceversa depending on the sense of rotation. In the view from the southwest, also at the windows but not shown on the banknotes, is a large flywheel now in the custody of the Hunterian Museum (Fig. 5). Its purpose is not clear but it could have been used to demonstrate angular momentum or mechanical work.

Natural Philosophy vacated the main building of the University and moved into its new accommodation, now known as the Kelvin Building, in 1906. The vacant former Natural

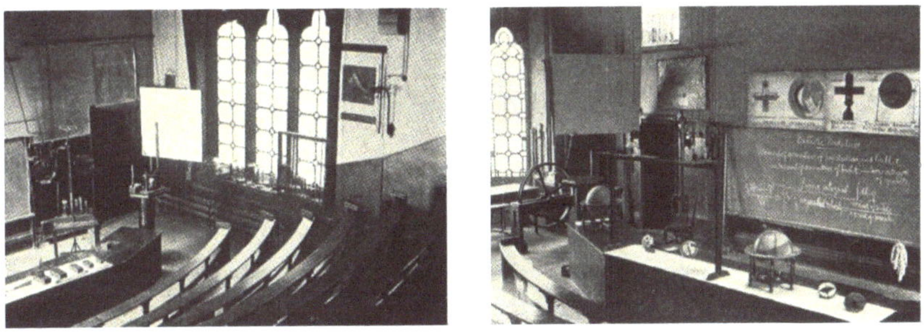

Figure 4: Views from the southeast (left) and northeast (right) of the front of the Natural Philosophy lecture theatre. The southeastern view is taken from a higher position than in the image on the banknotes but shows the same wave machine at the windows on the northern side. (Photographs from Thompson [5].)

Figure 5: Left: a wave machine similar to that in the nineteenth century view of the Natural Philosophy lecture theatre on the banknotes.Right: the flywheel visible in the view from the northeast. (Photographs by I.S. Ruddock, courtesy of the University of Strathclyde and the Hunterian Museum, Glasgow respectively.)

Philosophy lecture theatre was occupied first by Naval Architecture and then History until the mid-1960s when it was converted along with part of the apparatus room into the University's Senate Room [6]. The large lecturer's bench had already been replaced by a smaller one and additional rows of seats added at the front (Fig. 6).The present interior is the result of an upgrade and modernisation in 2009 (Fig. 7) in which the partition between the theatre and the apparatus room was moved further to the east increasing the size of the Senate Room. Due to the historic nature of the Gilbert Scott building, the external view of the former lecture theatre is as built in 1870 (Fig. 8). Behind a door at the bay window at the southwestern corner of the room is probably the world's longest running scientific experiment. Begun by Kelvin in 1872 to investigate the migration of liquids, it consists of two glass tubes of approximately 5 m in length, each containing pairs of liquids.

Figure 6: The view towards the east in the former Natural Philosophy lecture theatre showing at the front the rows of seats added in the twentieth century and a smaller lecturer's bench. (Photograph from University of Glasgow Archive Services, GB0248 PHU2/10.)

Figure 7: The view to the west in 2012 of the University of Glasgow's Senate Room, set out for an examination, looking towards the site of the lecturer's bench. The windows on the right are those visible in the image on the reverse of the banknotes. (Photograph by I.S. Ruddock courtesy of the University of Glasgow.)

Figure 8: The external elevations of the former Natural Philosophy lecture theatre, located above the archway leading from the south side of the West Quadrangle. Left: the northern side showing the three windows visible beside the lecturer's bench on the banknotes. Right: the southern side showing the bay window just visible in the right hand image in Fig. 4. (Photographs by I.S. Ruddock courtesy of the University of Glasgow.)

Figure 9: Left: view from the southwest of the main building of University of Glasgow as used for the reverse of Clydesdale Bank £10 and £100 banknotes. (Reproduced with the permission of Clydesdale Bank PLC.) Right: the same view in 2012 looking across the River Kelvin from Argyle Street. (Photograph by I.S. Ruddock.)

2.4 University of Glasgow

The University of Glasgow relocated to Gilmourhill in the west of the city in 1870 due to the need for expansion and the increasing industrialisation of the High Street area. The move was facilitated by the sale of the old campus to the City of Glasgow Union Railway and the purchase of a cheap green-field site in the newly fashionable West End. The architect of the magnificent main building was (Sir) George Gilbert Scott (1811–1878), responsible for other famous Gothic revival buildings including the Midland Grand Hotel at St Pancras Station, the Albert Memorial in South Kensington and the Italianate styled Foreign and Commonwealth Office in Whitehall (all in London), and St Mary's Episcopal Cathedrals in both Glasgow and Edinburgh. Completed by his son Oldrid Scott, the University is the second largest example in the Gothic revival style in Britain after the Houses of Parliament (Fig. 9).

The decision by the University's Senate not to accept designs from a Glasgow architect resulted in some resentment, especially from Alexander 'Greek' Thomson (1817–1875) who provided Glasgow with iconic buildings in a Greek revival style. In a speech to the Glasgow Architectural Society in 1866, he is recorded as saying '.....there is not a modern Gothic revival building of more than ten years standing that anyone cares a straw about' [7]. It was also felt that this commission was not at the top of Scott's priorities in the 1860s; Thomson continued with

> ...granting that Mr. Scott is all that his most enthusiastic admirers would have him to be, everybody knows that his establishment being the most fashionable in the metropolis, his business is so enormous that, to expect him to bestow more than the most casual consideration upon the work which passes through his office, is altogether unreasonable. Nor could it be supposed that the magnitude of his work was sufficient to have secured more than an ordinary amount of attention at his hands, for it was nothing unusual to him.

Further evidence to support Thomson's view can be found in Scott's own recollections where he refers only once to the University of Glasgow, commenting

> I was commissioned to erect the new University buildings at Glasgow, a very large work, for which I adopted a style which I may call my own invention, having already initiated it in the Albert Institute at Dundee (the McManus Galleries since the 1980s). It is simply a thirteenth or fourteenth century secular style with the addition of certain Scottish features, peculiar in that country to the sixteenth century, though in reality derived from the French style of the thirteenth and fourteenth centuries [8].

Notwithstanding Scott being selected to the chagrin of the locals, the end result has been an asset to Glasgow, with its tower andspire a landmark on the skyline.In the economically vibrant nineteenth century, there were enough opportunities available for creating Glasgow's world renowned Victorian and Edwardian cityscape of buildings of the first rank in a variety of styles including the pioneering 'Glasgow Style' of Art Nouveau, the Greek and Egyptian revival as well as the more familiar Gothic revival and neoclassical.

3 Queen's University of Belfast

3.1 Banknote details

These are given in Table 5.

Table 5: Banknote details: Queen's University, Belfast.

Bank	Banknote series	Obverse design	Reverse design	References
B. of Ireland	£5 to £100: various series 1990–2003	Hibernia + six counties shields	QUB historic building	[3, pp. 751–752]

3.2 Origins of Queen's University of Belfast

The image on the banknotes is of the west front of what was originally Queen's College Belfast. The College was one of three Queen's Colleges, which date from 1845, the others being founded at Cork and Galway. The three Colleges constituted the Queen's University, and it was designed to balance the religious perspective of Trinity College Dublin, which had been established as early as 1591 and which gave preference to students and staff from the Church of Ireland (a component of the Anglican Church). While the Queen's Colleges were not tied to any denomination – indeed, they were criticised as 'the godless colleges' because they did not have chapels – geography dictated that the College at Belfast would have a student body mainly Presbyterian, while at Cork and Galway most students would be Roman Catholic. Teaching at the Colleges began in 1849.

3.3 Queen's and the Thomson family

One of the main candidates for the Presidency of Queen's College Belfast was James Thomson, father of (the later) Lord Kelvin [9]. Thomson had risen from farming stock to be Professor of Mathematics at the Royal Belfast Academical Institution (always known as Inst) from 1815 to 1832. In fact, Inst had school and collegiate sections and Thomson taught in both. It was the product of the liberal Protestantism prevalent in Belfast at the end of the eighteenth and the beginning of the nineteenth centuries. Free from sectarianism and class prejudice, it aimed to provide useful education for all – provided, of course they were male!

Thomson's views were precisely along these lines and he became a figure of considerable influence in Inst. However, after his wife died he moved with his children to take up the Chair of Mathematics at Glasgow University at the end of 1831, where again he became one of the most capable and valued members of staff. Sir Robert Peel, the main inspiration behind the Queen's College scheme, even asked for his advice on how the scheme should be carried out – Thomson emphasised that Queen's College Belfast should definitely *not* be controlled by the Presbyterian Church.

When the Queen's Colleges were set up, it was thus clear that Thomson would be an important candidate for the Presidency. It should be mentioned that there was a suggestion that the collegiate section of Inst could be expanded and strengthened so as to constitute the new college. Yet, by this time, Presbyterianism in Belfast had become split. There was still the liberal or radical range of opinion, often coupled with Unitarianism, or so-called non-subscribing Presbyterianism which did not accept the Westminster Confession. However, there was also now a less tolerant section of Presbyterianism based on strict adherence to the Bible, evangelical fanaticism, hatred of Catholicism, and extreme loyalty to the Union of Ireland with Britain. It might be said that this brand of Presbyterianism became more predominant as the century moved on, and it played a part in stimulating the trouble and violence that has bedevilled the region for much of the decades since.

Inst had attempted to remain neutral in the dispute between the different wings of Presbyterianism, but in so doing had enraged the more extreme element, and there could be no unanimity in support of Inst being upgraded. Rather, an entirely new College was set up; several Inst professors moved over to the new College, but those who did not wish to do so, or who were not selected, were pensioned off. The collegiate section of Inst was closed down, and it retained merely the school section, which remains today.

The greatest advocate of the more orthodox section of Presbyterianism was the firebrand preacher Dr Henry Cooke, a celebrated pulpit orator – exuberant, pugnacious and intolerant. There were many on that wing of the Church who considered him ideal to be the President of the new College. As well as Thomson, a third candidate was another Presbyterian minister but of very different stamp to Cooke. Dr Pooley Henry was a mild man, on good terms with the Catholic Archbishops. He had little leadership ability, but was capable and had already been on several government committees and boards.

Of the candidates, neither of the ministers had research or scholarship credentials. Thomson was not a research worker, but he did have enormous experience in scholarly writing and educational decision-making. Peel, though, initially favoured Cooke, but on studying his writings realised that a College under his leadership would scarcely be attractive to Catholics. He then chose Henry, who was appointed and was to be President of Queen's College Belfast from 1845 to 1879.

Why was Thomson not chosen? It could, of course, just be that Henry was a 'safe pair of hands' and the authorities already had good experience of working with him. On the other hand, it could be that it was part of an elaborate balancing-act. While it was accepted that Presidents of the Colleges at Cork and Galway would both be Catholics, it was felt undesirable that both should be priests, so one priest and one layman were chosen. It may have been felt that Presbyterians would wish to see a minister of their Church as the President of the new College.

Radical opinion, though, disliked the idea of a clerical appointment and was sure that Thomson was the best qualified candidate. The newspaper supported by this section of opinion, the *Northern Whig*, described Henry's appointment as 'infamous' and Thomson's treatment as 'a gratuitous insult'. The rival newspaper, the *Belfast Newsletter*, was equally indignant over the treatment of Cooke.

Thomson was offered the VicePresidency. The salary for this position was £500, as compared to £800 for the Presidency, and Thomson would have lost money by his taking up the post. However, it was certainly the insult that was most hurtful. His family expressed their outrage over the course of events. However, it was not very long before his older son, also James, was appointed to the Chair of Engineering – temporarily from 1853, permanently from 1857. His other son, William, also was to maintain friendly relationships with several members of staff at Queen's College Belfast.

When Thomson turned down the Vice Presidency, it was offered to another outstanding candidate, Thomas Andrews, medically qualified and Professor of Chemistry both at Inst until 1845 and at Queen's College afterwards. He was a researcher of great distinction, being most famous for the so-called Andrews' experiment, which demonstrated the difference between a gas and a vapour.

3.4 The banknotes' view of Queen's

Let us now turn to the view of Queen's as on the banknotes (Fig. 10). It shows the west front of what is called the Lanyon building, named after its architect, Sir Charles Lanyon (1813–1889). By the 1840s, Lanyon was already responsible for a number of important civic buildings and

Figure 10: Left: the reverse of the Bank of Ireland £5 banknote showing the main building of Queen's University of Belfast. (Reproduced with the permission of Clydesdale Bank PLC.) Right: the same view in 2012. (Photograph by A. Whitaker.)

could be spoken of as the most prominent architect in Ulster. He had also been the engineer of the famous Antrim coast road and the Queen's Bridge in Belfast. By the time of his retirement in 1872, the list of his architectural achievements was enormous; he also became successively Mayor of Belfast and MP for Belfast, and he was knighted in 1868. It has been said [9] that he lacked originality of a creative kind, but had a keen eye for the composition of a building and was able to blend features from different architectural sources.

For the design of academic and civic buildings, it was usual to enact an elaborate game of symbol and reference [10]. Quotations of historic precedent were used to demonstrate the authority and pedigree of the building. For academic buildings, the late Gothic and Tudor style of Oxbridge colleges was ideal. Thus, the central tower is closely derived from the Founder's Tower of Magdalen College Oxford, though Lanyon's tower is taller and narrower in proportion than that at Magdalen, and built of brick rather than stone.

Evans and Larmour [10] wrote that: 'Lanyon's west front is the enduring image of Queen's. It has all the virtues of the best Victorian architecture, bold massing, a memorial silhouette and plenty of architectural detail. It is eminently long too in striking an appropriate note of academic pomp and civic circumstance. The long façade is a masterly interplay of progressions and recessions, and the balance between vertical and horizontal lines'.

For everybody in Belfast, the Lanyon building symbolises Queen's, and it is one of the most well-known and well-loved images in Ulster.The building has fortunately survived the years with minimal alteration, none to the west front. We might not have been as lucky as this – in the 1960s, there was a serious proposal to build a multi-storey car-park on the front lawn!

3.5 The early history of Queen's

As for the Queen's Colleges, the early history was far from straightforward, like so much in Ireland. In 1882, in fact, the Queen's University itself was disbanded and was replaced by a new body, the Royal University of Ireland. The Queen's Colleges still existed but their relationship with the Royal University was merely to submit students for examination, along with any other colleges who also wished to submit students. The latter included the new University College

Dublin, which was the successor, financially supported by the state, to the previous independent Catholic University. The Catholic University had been set up in 1851 with Cardinal John Henry Newman as Rector but had achieved little success until this point.

Across the religious divide, there was general great dissatisfaction with the Royal University, and in 1908, the important Irish Universities Act set up two new Universities. Queen's College Belfast became Queen's University Belfast. The Colleges at Galway and Cork joined with University College Dublin to become the National University of Ireland. In general, the result has been popular and successful, and broadly that is the situation today, though, of course, with the foundation of several newer Universities.

4 Kelvin and the Giant's Causeway Tramway

In the section *Electricity Generation* of Chapter 1, *Lord Kelvin*, the story is told of Kelvin's involvement with the pioneering Giant's Causeway, Portrush & Bush Valley Railway and Tramway Co. This was the first hydroelectric railway in the world and the project to build it came about mainly through the efforts of Colonel William Atcheson Traill and his brother Dr. Anthony Traill. W.A. Traill and the Tramway are commemorated on a Northern Ireland banknote of the Northern Bank Ltd, the details being given in Table 6 and illustrated in Fig. 11. A photograph of the tram at Dunluce Castle is included in Chapter 1.

Table 6: Banknote details: Giant's Causeway, Portrush & Bush Valley Railway and Tramway Co. Ltd.

Bank	Banknoteseries	Obverse design	Reverse design	References
Northern Bank	£5:1988–1990 Series	W.A. Traill & Tramway features	Modern technology & bank logo	[3, p. 755]

Figure 11: The obverse of the 1988–1990 series Northern Bank £5 banknote celebrating W.A. Traill and the Giant's Causeway, Portrush & Bush Valley Railway and Tramway showing a tram and trailer at Portrush's railway station. (Reproduced with the permission of Danske Bank PLC.)

5 King's College, Aberdeen (University of Aberdeen)

The iconic building of King's College, Aberdeen, was used in banknote design for well over a century, spanning the takeover of the old North of Scotland Bank by Clydesdale Bank in 1950 (Table 7 and Fig. 12). It enters the storey of this volume by default. In 1856, James Clerk Maxwell became a professor of Natural Philosophy not at King's College, but its sister, Marischal College. In fact, he won this in competition with P.G.Tait [11, p. 175]. When the two colleges united in 1860 to form the University of Aberdeen, Maxwell's appointment was abolished. Moreover, he lost to P.G. Tait when they both competed that same year to become the professor of Natural Philosophy at Edinburgh: this was in succession to Forbes, the founding professor of the post. However, Scotland's loss was England's gain. Maxwell became the professor of Natural Philosophy at King's College, London, in 1860, retiring to his estate in Scotland in 1865. In 1871, he accepted the offer to become the first Cavendish Professor of Physics at Cambridgebefore dying prematurely of cancer in 1879 at the age of 48 [12].

Therefore, King's College Aberdeen is a sort of anti-feature for Maxwell and this volume!

Table 7: Banknote details: King's College Aberdeen (University of Aberdeen).

Bank	Banknote series	Obverse design	Reverse design	References
North of Scotland	Entire output: 1836–1949	King's College	Nil (uniface) or arms of bank	[1, pp. 1022–1025
Clydesdale & North of Scotland	A. £5–£100: 1950–1951	King's College + Glasgow Cathedral	Arms of bank	[3, pp. 867–868]
	B. £5: 1961, 1967 issue	Arms of bank	King's College	[3, pp. 868–869]

Figure 12: The reverse of the 1961 series Clydesdale and North of Scotland Bank £5 banknote showing King's College, Aberdeen. (Reproduced with the permission ofClydesdale Bank PLC.)

6 Boulton and Watt

Sir Mervyn King, Governor of the Bank of England, commented in a News Release in 2011 on '… the invaluable contribution that Boulton and Watt made to the advancement of engineering…' [13]. The current Bank of England £50 note features Matthew Boulton and James Watt (Table 8 and Fig. 13).

Given that the Bank of England series has only four denominations – £5, £10, £20, and £50 notes – it is a very pleasing coincidence that half of the current notes feature the three characters Adam Smith, James Watt, and Matthew Boulton referred to in the above-mentioned section *Watt and the Age of Enlightenment* in Chapter 7.

With reference to the new £50 note (Fig. 13) the Bank of England News release of 2 November 2011 [13] quotes King, as saying additionally to the heading quote: 'The partnership of an innovator and an entrepreneur created exactly the kind of commercial success that we will need in this country as we rebalance our economy over the years ahead.' For our purposes, the intriguing fact is that the partnership is consistently referred to as *Boulton and Watt*, and never the other way round. We refer to Samuel Smiles's *Lives of the Engineers*, most recently published in 2006 [15]. Smiles developed his work over a number of years, culminating in his 1874 version on which [15] is based. Whether in 1874 (Smiles himself), or in 2006 (Shepherd as Editor [16] and Asa Briggs giving an historical Introduction [17]) it is always Boulton who comes first.

Table 8: Banknote details: Matthew Boulton and James Watt.

Bank	Banknote series	Obverse design	Reverse design	References
B. of England	£50: 2011	Queen E II	Portraits of Boulton & Watt, Soho Factory, steam engine & quotes*	[3, p. 425; 14]

*Quotes: Boulton: *I sell here, sir, what all the world desires to have-POWER.*
Watt: *I can think of nothing else but this machine.*

Figure 13: The reverse of the 2011 series Bank of England £50 banknote showing Matthew Boulton (left) and James Watt (right) and the quotations noted above. (Reproduced with the permission of the Bank of England.)

Why is this? Essentially it was because Boulton was the driving force of the partnership. Smiles contrasts the two men in a manner which (for readers well over a century later) is attractive, convincing, and warmly personal. With his 'intense headaches', Watt's 'working power … was limited'. He was greatly annoyed by the 'shortcomings of his workmen' and didn't 'make allowances' for them. He 'never looked on the bright side of things'. In fact, 'Watt's invariable practice was to call upon Boulton for help … Though Watt might be the brain, Boulton was the heart of the concern' (all quotes from [15, p. 106]). Moreover, Boulton's 'open-handed generosity' meant that *'at Watt's request'* (our italics) he changed the agreed '1/3 Watt, 2/3 Boulton' split of profits to 50% each (p. 127). When the boot was on the other foot, however, things were different between them. At the stage immediately before his next 'greatest enterprise' of steam coinage, Boulton himself suffered considerable financial problems and 'failing health' (pp. 126, 127 and Chapter VIII, p. 128 onwards). Watt by then had prospered and 'had already invested his profits elsewhere, *and could not help him*' (p. 126, again our italics).This contrast is repeated in Smiles's overview (pp. 139, 140). Boulton 'was a first-rate man of business'. 'Not the least difficulty … was the timidity and faintheartedness of his partner. For years Watt was on the brink of despair, but Boulton never lost hearthe was Watt's 'very backbone'.

Finally, in Smiles's view this contrast only served to highlight Watt's achievement – 'it is … no exaggeration to say that the steam-engine of Watt, is without exception, the greatest invention of modern times' [15, p. 148].

7 Conclusion

One of the key issues addressed in this volume is the question of Kelvin's public reputation. Connected with this is the recognition of his greatest engineering achievement – his contribution to the trans-Atlantic cable-laying project. While Kelvin's reputation may have suffered in, for example, histories of the Industrial Revolution, in one publicity milieu he has been handsomely recognised: namely within the world of UK banknotes.

In this chapter, the origin, significance and format of these bank notes are explained, together with the complex political history of the UK that they reflect. In fact, not only is Kelvin's cable-laying so honoured, but also his compass design and his contribution to the Traill tramway in Northern Ireland.More generally, within the overall story of this volume, the University of Glasgowand Queens University of Belfast figure prominently; they also appear on UK banknotes.

References

[1] Cuhaj, G.S. ed., *Standard Catalog of World Paper Money: Vol 1: Specialized Issues*, 11th edn, kp books, F + W Publications: Iola, 2009.

[2] Cuhaj, G.S. ed., *Standard Catalog of World Paper Money: Vol 2: General Issues, 1368–1960*, 14th edn, kp books, F + W Publications: Iola, 2012.

[3] Cuhaj, G.S. ed., *Standard Catalog of World Paper Money: Vol 3: Modern Issues 1961–present*, 18th edn, kp books, F + W Publications: Iola, 2012.

[4] Fraser Stewart Book Wholesale Ltd, *The Complete Family Encyclopaedia*, Helicon Publishing: London, 1992.

[5] Thompson, S.P., *Life of Lord Kelvin, Baron Kelvin of Largs*, Macmillan and Co. Limited: London, 1910.

[6] *University of Glasgow: A Short Account of the History of the Alterations Made to the Spaces Currently Housing the Senate Room and Carnegie Room*, University of Glasgow Archives: Glasgow, 2010.

[7] Thomson, A., *Criticism of Sir George Scott's Design for the Building for the University of Glasgow*, manuscript, University of Glasgow Library, 1866.

[8] Scott, G.G., *Personal and Professional Recollections*, Sampson Low, Marston, Searle & Rivington: London, 1879.

[9] Moody, T.W. & Beckett, J.C., *Queen's, Belfast 1845–1949: The History of a University*, Faber and Faber: London, 1958.

[10] Evans, D. & Larmour, P., *Queen's: An Architectural Legacy*, Institute of Irish Studies, Queen's University Belfast: Belfast, 1995.

[11] Smith, C., *The Science of Energy*, Athlone Press: London, 1998

[12] Glazebrook, R.T., *James Clerk Maxwell and Modern Physics*, Cassell and Company: London, 1896.

[13] Bank of England News Release. New-style £50 banknote issued today, 2 November 2011.

[14] www.bankofengland.co.uk/banknotes/Pages/current/current_50.aspx (as at December 2014).

[15] Smiles, S., Matthew Boulton and James Watt, *Lives of the Engineers*, ed. C.J. Shepherd, Abridged Edition, pp. 71–148, Folio Society: London, 2006.

[16] Shepherd, C.J., Editorial Note, *Lives of the Engineers*, ed. C.J. Shepherd, Abridged Edition, pp. 71–148, Folio Society: London, 2006.

[17] Briggs, A., Introduction, *Lives of the Engineers*, ed. C.J. Shepherd, Abridged Edition, pp. 71–148, Folio Society: London, 2006.

Appendix A

Thomson, Brunel and the Atlantic cables of 1865 and 1866

I.S. Ruddock
Department of Physics, University of Strathclyde, United Kingdom.

1 Introduction

When the attempt to lay a working trans-Atlantic telegraph cable was renewed in 1865 after the failure of the 1858 cable, the Atlantic Telegraph Company and its contractor, the Telegraph Construction and Maintenance Company, decided to use only one ship instead of starting mid-ocean with two and laying the cable in opposite directions. William Thomson's connection with the enterprise was as a member of the five-strong Scientific Consulting Committee, which also included Charles Wheatstone and Joseph Whitworth.

By 1865, more than 50 submarine cables had been laid worldwide since 1851 with the longest being that of over 1500 miles connecting Malta and Alexandria. But the challenges now faced were greater on account of the ocean's depth and the resulting strain on the cable during laying, as well as the natural navigation hazards encountered by any ship crossing an ocean. The new Atlantic cable consisted of a core of a single strand of copper wire with six more wound around it within four layers of gutta-percha insulation, each in turn separated by a layer of Chatterton's Compound – a waterproof adhesive. The insulated core was next covered with jute padding before being protected by a spiral sheath of 10 iron wires wound covered in manila yarn saturated with a preservative. Although the planned route from the west of Ireland to Newfoundland was around 1600 nautical miles, the finished cable had a total length of over 2400 nautical miles to allow for the profile of the seabed.

With a mass of over 4000 tons and a correspondingly large volume, the cable demanded the use of a very large vessel, a requirement that could only be satisfied by Isambard Kingdom Brunel's SS Great Eastern which had been launched in 1858. It was not only the largest ship at the time, but it would remain so beyond its scrapping in 1890 until the White Star liners RMS Oceanic and RMS Celtic exceeded its length and gross tonnage in 1899 and 1901, respectively. Great Eastern's career as a passenger ship was short lived because there was insufficient traffic to justify its use on the intended Far East trade, whereas on Atlantic service, it suffered a series of accidents and was the victim of a price war with rival shipping companies. Brunel died in 1859 as preparations for the first revenue earning voyage were being made, but despite Great Eastern being a commercial failure, the ship was an engineering success and its seaworthiness during cable laying vindicated its designer's vision that ships would steadily increase in size.

At the time of its charter to the Telegraph Construction and Maintenance Company, Great Eastern had been purchased by the Great Eastern Steamship Company from the Great Ship Company. Sir Daniel Gooch, a director of the Great Western Railway, and previously Brunel's prodigy as locomotive superintendent, was the company's chairman and as such sailed in 1865.

2 The images

Most of images in this *Story in Pictures* are watercolours painted during the 1865 voyage by Robert Charles Dudley (1826–1900), a popular Victorian artist. They were bequeathed to the Metropolitan Museum of Art, New York by Cyrus Field (American businessman, financier, telegraph entrepreneur and director of the Atlantic Telegraph Company) on his death in 1892. Lithographs based on the watercolours were used to illustrate William Howard Russell's *The Atlantic Cable* published in 1865 [1]. Russell (1820–1907), a journalist who had recently reported from Crimea for The Times, was on board Great Eastern under contract by the Telegraph Construction and Maintenance Company to write an account of the voyage. Detailed accounts of the 1865 and 1866 cabling attempts can be found in [2-4].

3 Preparations

Figures 1–5 and their captions show and describe Great Eastern being prepared for the voyage and then leaving Ireland.

Figure 1: The cable being wound into one of the storage tanks in the works at Greenwich. It was completed at a rate of about 100 miles per week by the Telegraph Construction and Maintenance Company, an organisation recently formed by the merger of Glass, Elliott & Co. and the Gutta Percha Company; the iron wire for the armour was supplied by Webster & Horsfall, Birmingham. From its manufacture to eventual laying, the cable was kept underwater in tanks, both at the factory and while being stored and transported in the hulks. (Metropolitan Museum of Art, New York: Robert Charles Dudley collection.)

Figure 2: Left: the cable being transferred into one of two Royal Navy hulks at Greenwich. (Metropolitan Museum of Art, New York: Robert Charles Dudley collection.)

Figure 3: Right: the loading of the cable from a hulk to Great Eastern at Sheerness. (Metropolitan Museum of Art, New York: Robert Charles Dudley collection.)

Figure 4: The Prince of Wales being shown the cable entering Great Eastern's after tank (one of three) on 24 May 1865. (Metropolitan Museum of Art, New York: Robert Charles Dudley collection.)

Figure 5: Great Eastern at the start of her cable laying voyage off Valentia, 23 July 1865 with HMS Terrible and HMS Sphinx as escorts. The cable being payed-out is just visible between the stern and the surface of the sea. The other vessels had brought out the shore cable, which was then spliced to the main cable, as well as guests including Sir Robert Peel, Chief Secretary for Ireland, to witness this momentous step in the creation of the Atlantic cable. (Metropolitan Museum of Art, New York: Robert Charles Dudley collection.)

4 The 1865 voyage

The wheels and brakes of the paying-out machinery on Great Eastern's afterdeck (Fig. 6) regulated the tension on the cable as indicated by a simple dynamometer based on the rising and falling of a heavy pulley resting on it. The tensile strength of the 1865 cable was 11 times its weight per mile in seawater, that is, it could support 11 miles hanging vertically in the ocean, and compared favourably with the 1858 cable's which was less than five times its weight per mile in seawater. This strength provided a safety factor of about four on the basis that the maximum depth encountered would be two and a half miles.

The 'electricians' of the Telegraph Construction and Maintenance Company had a continuous programme of electrical tests using Thomson's mirror galvanometer to be performed on the cable during laying. William Thomson was on-board on behalf of the Atlantic Telegraph Company as a consulting expert with the remit of reporting on the tests but without the power to interfere or control. Two days into the voyage, on 25 July 1865, contact was lost for the first time between Great Eastern and Valentia with the ship still only 66 miles from shore. The procedure for making a repair was to lash the cable to a wire rope and then cut the cable and allow it to fall into the sea; meanwhile, the ship was turned with the wire rope being transferred from the stern to the bow where it was then brought in through the picking-up apparatus, Fig. 7.

On this first occasion, the fault was identified when 10 miles of cable had been hauled back in and turned out to have been caused by a stray piece of wire that had penetrated the gutta-percha insulation. Once the damaged portion was cut out, a splice had to be made at this point, and again where the cable had originally been cut when being retrieved from the sea, Fig. 8. Communication was lost with Ireland, now 770 miles distant, for the second time on 31 July, Fig. 9.

When the cable failed for the third time on 2nd August and a piece of stray wire was once again found, on-board opinion was divided between those who suspected sabotage and those who

Figure 6: The cable paying-out machinery. (Metropolitan Museum of Art, New York: Robert Charles Dudley collection.)

Figure 7: The picking-up machinery on the foredeck of Great Eastern. (Metropolitan Museum of Art, New York: Robert Charles Dudley collection.)

Figure 8: The cable being spliced after the repair of a fault. (Metropolitan Museum of Art, Robert Charles Dudley collection: New York.)

Figure 9: The cable being examined for a fault on 31 July, the second time that communication was lost with Ireland, now 770 miles distant. Again the problem was caused by a stray piece of wire that had penetrated the gutta-percha insulation, but on this occasion it was decided (wrongly) that the cable had been deliberately damaged in the holding tank. From now on, crew members handling the cable in the tanks were supervised at all times. (Metropolitan Museum of Art, New York: Robert Charles Dudley collection.)

now thought that the problems were due to poor quality control during manufacture. The latter view became the accepted version when subsequent examination of cable recovered from the seabed showed numerous locations where the armour wire wound on the outside of the insulation had frayed. Although Thomson and Mr. C.F. Varley, Chief Electrician of the Atlantic Telegraph Company, thought that this fault was sufficiently incomplete that the cable could be used and generate a return on investment, the decision to retrieve it was made by Mr. Canning, Principal Engineer of the Telegraph Construction and Maintenance Company and person in overall charge of the operation. He was well aware that, despite what was being said by those around him, the Atlantic Telegraph Company would not accept a cable of marginal quality, even if fully laid and working. Unfortunately when the normal routine for lifting was being followed, the cable broke as Great Eastern was being turned and the end was lost overboard. The only option left was to attempt to grapple for the cable and bring it back to the surface.

Great Eastern was first moved to the windward side of the line of the cable and then, with a grappling hook resting on the seabed, was allowed to drift while the dynamometer monitored the tension in the wire rope. Although the cable was lost at a depth of 2500 fathoms or 15,000 feet, it was found the next day, 3rd August. The lift progressed well until about 1000 fathoms of the grappling rope had been wound in, but then the swivel coupling between two of the constituent 100 fathom lengths making up the wire rope fractured allowing the grappling iron and cable to sink back to the seabed.

After marking the line of the cable by a buoy, Figs 10 and 11, the second attempt to grapple started on the 7th August but with a similarly disastrous outcome; another grappling iron and over 1000 fathoms of wire rope were lost when the cable was about halfway to the surface. Canning immediately marked the position with yet another buoy and decided to make a final attempt with the remaining lengths of rope.

Figure 10: A large buoy being prepared for launch to mark the approximate line of the cable. (Metropolitan Museum of Art, Robert Charles Dudley collection: New York.)

Figure 11: The buoy being launched on the morning of 8th August to mark where the cable lay after the failure of the second attempt to lift it. (Metropolitan Museum of Art, Robert Charles Dudley collection: New York.)

This buoy was lost for over a day until the afternoon of 9th August due to a sudden gale that came up in the evening and caused Great Eastern to shift its position by 35 miles. By mid-morning of the following day, a grappling iron and rope had been prepared and at 10.30 a.m. it was thrown over the bow and allowed to sink for the next 48 minutes. Despite the tension in the grappling line increasing a couple of times, it was clear by mid-afternoon that the route of the cable had been crossed without it being found. The line and hook were brought back in overnight and prepared again for what turned out to be the final attempt on 11th August. With much of the wire rope in poor condition due to unravelling, the replacement included over 700 fathoms of various hemp ropes. Although this bid was being made out of a sense of duty, the dynamometer indicated an increased tension during the tracking across the route of the cable, but it came as no surprise to the assembled witnesses that the line parted during recovery. As a result, the 1865 cable expedition was over for the time being with only 600 miles still to be laid.

5 The 1866 voyage

On 12 October 1865, at an Extraordinary General Meeting, the Atlantic Telegraph Company resolved to seek additional investment to make another attempt at laying a cable using Great Eastern and to recover and complete the recently abandoned cable. In reaching this decision, the board was encouraged by the demonstrated ability to lay a cable in the deepest part of the Atlantic, to transmit signals through it despite the enormous hydrostatic pressure, and to successfully retrieve and repair a faulty or broken cable.

The new cable was ready by the early summer of the following year and was loaded again onto Great Eastern in the Thames. This time, the general public was allowed to inspect the ship at a cost of a shilling each, but William Thomson felt compelled to write to a nephew to make it clear that he was not to be asked for if visiting because he would be too busy. On 7 July 1866, Great Eastern sailed for the west of Ireland with the new cable and improved laying and picking-up machinery (Figs 12 and 13) together with the unlaid portion of the previous year's cable.

The new shore-end was picked up off Valentia on 13th July and after the new cable was spliced to it, the voyage began westwards; Thomson was on-board again having joined the previous day. In contrast with the events of August 1865, the operation proceeded according to plan and Heart's Content, Trinity Bay, Newfoundland, was reached on 27th July with the shore-end being successfully landed the next day, Fig. 14.

After being refuelled with 8000 tons of Welsh coal delivered from Cardiff in the preceding weeks, Great Eastern set sail again as a cable ship on 9th August. Three days later, it and an escort vessel joined two other ships that had left earlier to mark the line of the lost 1865 cable near the break but by the time they arrived one of the crews had already managed to find and lift the cable, although their grappling rope broke in the attempt. Although the cable was subsequently brought to the surface on 17th August, it was not until 2nd September that Great Eastern and two of the other ships working together to share the strain lifted it and brought it on-board. There now followed a few tense hours when the end was prepared and connected to

Figure 12: The 1866 paying out machinery on Great Eastern viewed from the stern. The cable is visible leaving the machinery and passing under the wheel of the dynamometer en route to the stern. (Photograph courtesy of the IET Archives.)

Figure 13: The grappling hook on Great Eastern used for finding and retrieving the abandoned 1865 cable in August 1866. (Photograph courtesy of the IET Archives.)

Figure 14: Great Eastern off Heart's Content, Trinity Bay, Newfoundland in 1866. (Photograph courtesy of the IET Archives.)

signalling apparatus. Confirmation that the cable had not suffered its year of immersion was obtained when contact was made with Valentia where telegraphists had been standing by since the 1866 cable had started transmitting at the end of July. Great Eastern then sailed west for the second time and completed the 1865 cable on 8th September.

References

[1] Russell, W.H., *The Atlantic Cable*, Day & Son: London, 1865.
[2] Thompson, S.P., *Life of Lord Kelvin, Baron Kelvin of Largs*, Macmillan & Co.: London, 1910.
[3] Field, H.M., *History of the Atlantic Cable*, 2nd edn, Scribner & Co.: New York, 1867.
[4] Thomson, W., *Atlantic Telegraph Cable*, Brown & Co.: London, 1866.

Appendix B

The Model Stirling Engine

C.S. König[1] & C.A. Greated[2]
[1]*College of Engineering, Design and Physical Sciences, Brunel University, United Kingdom.*
[2]*School of Physics and Astronomy, The University of Edinburgh, United Kingdom.*

This air engine is exhibited at the Hunterian Museum in Glasgow, Scotland. It was presented in 1827 to Glasgow University by its inventor the Reverend Dr Robert Stirling. The exhibit notes indicate that it is one of only two specimens with the other (a slightly different, earlier model) having been presented to Edinburgh University. Further, the exhibit notes read that 'it is probable that at least parts of these models were made by Stirling's brother James, a Dundee-based Iron Master or Founder'. Lord Kelvin claimed to have found this model in an 'Augean stable' in 1847. This term is generally used to refer to a messy, dirty and usually crowded or neglected place. The date is contemporary with Kelvin's adaptation of a cellar situated in the basement of the Old College, High Street, Glasgow, into the world's first student physical research laboratory. Kelvin subsequently repaired the model and used it in lecture demonstrations. The engine runs by using a recurrent cycle of the same trapped air, flowing repeatedly through a connected displacer cylinder (containing a 'displacer' plunger) and power piston, with momentum provided by the turn of a flywheel. Air in the large displacer cylinder is heated, rises up past the plunger and flows through a connecting pipe to the piston, causing it to rise and the flywheel to turn. The displacer (plunger) is moved downwards, forcing air into the top of its cylinder. This air cools, due to the presence of a water-filled cooling jacket, situated at the top of the cylinder. The continued momentum of the flywheel causes all the air to move to the top of the cylinder, where it quickly cools. The air is then forced downwards again into the heated part of the cylinder when the piston moves down, pulling the plunger upwards. The air, now in the heated bottom part of the cylinder, expands and increases pressure in the confined displacer cylinder space until it moves past the plunger, through the connector, and into the piston compartment causing the cycle to repeat. The engine is almost totally silent in operation and can potentially be fuelled with any heat source. Because the combustion of fuel is external to the machine, this process can be carried out in an efficient manner. Due to the abundance of coal in the 19th century and Stirling's obsession with the economic use of fuel, the air engine was used in the Scottish mining industry as a water pump and in churches as an air pump for organs. There are many modern applications for the air engine, including the powering of orbiting satellites and driving environmentally friendly power sources.' [1]

Figure 1: The Model Stirling Engine of the Hunterian Museum, Glasgow.

Reference

[1] Model Stirling Engine (External Combustion Engine), artefact index, Hunterian Museum & Art Gallery Collections GLAHM 105225, http://www.huntsearch.gla.ac.uk/cgi-bin/foxweb/huntsearch/DetailedResults.fwx?collection=all&SearchTerm=105225&mdaCode=GLAHM

Appendix C

Maxwell's Demon

M.W. Collins[1], R.C. Dougal[2] & C.S. König[1]
[1]*College of Engineering, Design and Physical Sciences, Brunel University London, United Kingdom.*
[2]*James Clerk Maxwell Foundation, United Kingdom.*

Maxwell's demon is an enduring iconic character. Born in 1867 as Maxwell's brainchild, the midwife was Kelvin himself and followed suggestions made 5 years previously by James Thomson. These contrasted the known processes of controlling energy in bulk with at least the possibility of 'mind or life' becoming able to combat, even reverse, the fundamental processes of nature at molecular level. A sequel was played out in the ensuing 10 years or so. It is included in detail in the exhaustive Kelvin biography *Energy & Empire*, by Smith & Wise, published in 1989 [1].

Early in these years, Maxwell acknowledged that there could be several sub-species of demons. Their talents had to match the tasks set for them. Thus, the original 1867 demon was a finite being dealing with finite numbers of finite molecules, and was observant, mathematically gifted and neat-fingered. This demon made decisions from its own observations of properties of the molecules – decisions on whether or not to let any molecule continue to move unhindered. If not, the demon – using only his intelligence – could control, without any input of energy, a continually changing, artificial distribution of molecules. One in particular corresponded to an ever increasing temperature difference between two adjoining volumes constituting a complete closed sample.

By 1870, there were analogue demons presented for clarification by Maxwell, usually inspired by some convenient technology of the *Victorian* era – such as the actions of the railway *pointsman*. At his fairly easily visualised place of work (Fig. 1), he could set friction-free 'points' to send fast (express) trains along one line and slow (goods) trains along the other. His intelligence, got from being able to recognize any one train from all the others, was derived from experience and commitment. Another, dispensing with intelligence altogether, was a demon acting as a *valve* providing unhindered passage only to particles moving in one direction.

However, in 1874 there arrived Kelvin's 'superior' types of demons. They had all the attributes of *Homo sapiens* but at molecular level, having free will and organisational abilities including exemplary teamwork. Such a demon could reverse the direction of individual molecules. Kelvin's Victorian example was of a *cricketer* (Fig. 2) resulting in an 'army' [in suitable formation] of demons … wielding 'molecular cricket bats'. Acting in concert, they could close ranks to all molecules except certain fast ones going in one direction and slow ones going in the other. Thus, they could readily meet Maxwell's summary outcome to Kelvin: when [a demon] sees a swift [molecule] let him turn it to the place of the swift and when he sees a slow one let him turn it to go among the slow.

Figure 1: Pointsman. (Illustration from [2].)

Figure 2: Cricketer. (Illustration from Akash Chopra, *The New Cricket Manual* [3].)

The back stories in a sequence of excerpts from recent publications (1998–2013) reveal the continuing influence of a physically unchanged Maxwell demon on developments in physics. Impasse-breaking led to the development of the science of *information* gathering and storing, with feedback asking for an even more talented (or, alternatively, burdened) demon. The final example – with input also now from nanoscience – refers to an actual experiment which showed that information can be converted into energy.

A satisfying starting summary, quoted and adapted here, was given by Smith in 1998 [4]. In Maxwell's thought experiment, the two vessels of equal volume but unequal temperatures (A being hotter than B) are separated by a fixed diaphragm CD (see Fig. 3). A 'finite being', subsequently known as Maxwell's demon, can open and close a hole in CD by means of a sliding door without mass. If the 'very observant and neat fingered being' knows the paths and velocities of all the gas molecules in A and in B, he may select molecules in A having less than the mean square velocity in B and allow such through the hole in CD. Similarly, he may select and allow to pass from B to A those molecules with more than the mean square velocity in B. Hence, the numbers of molecules in A and B are unchanged, but A's temperature increases and B's temperature decreases without work being done. This reverses the natural direction of dissipation of heat. Otherwise, the second law remained inviolable.

An enhanced version, here summarised, was given by Hunt [5] in 2010 (Fig. 4). In 1860, James Clerk Maxwell had derived a formula for the distribution of velocities of the molecules of a gas in equilibrium, providing a definitive statistical interpretation of their mean square velocity. Drawing on this, he introduced in 1867 an imaginary "demon" whose actions would illustrate the statistical nature of the second law of thermodynamics. The diagrams are a guide to the association of temperature with average velocity and velocity range, in line with Maxwell's own description. 'Imagine a box of air divided into two parts with … a small door between them … Over time the demon is thus able to gather all of the faster "hot" molecules on the left hand side of the box and all of the slower "cold" ones on the right … Such a separation of hot from cold without any net expenditure of work [with the hot becoming hotter at the expense of the cold, which becomes colder] would violate the second law. Moreover … we could connect the two sides through a heat engine and proceed to extract useful work. We would … in thermodynamic terms, have been able to get something for nothing.'

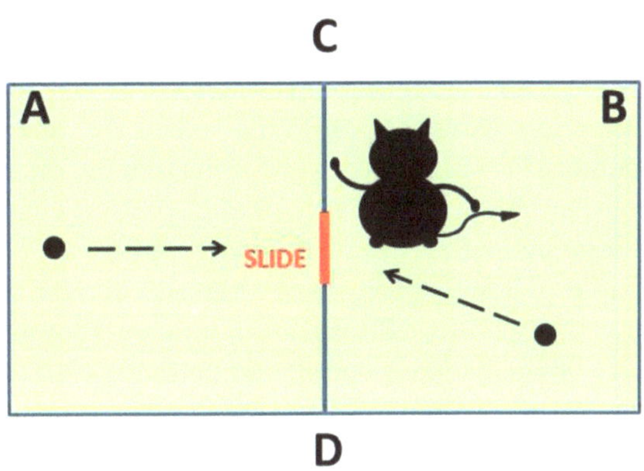

Figure 3: The basics after Smith [4].

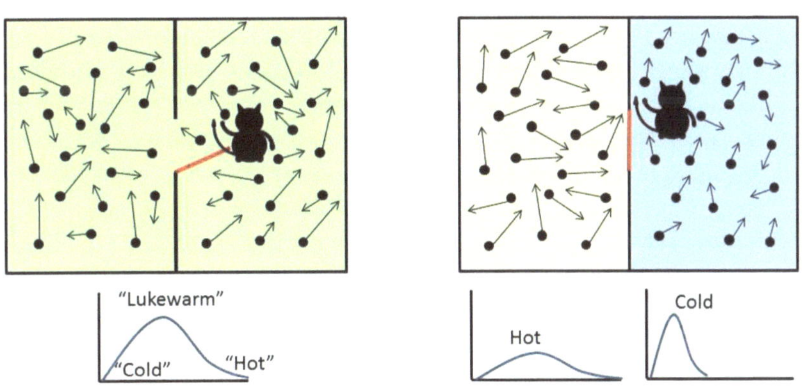

Figure 4: Enhanced version after Hunt [5].

The ordered selection of key points in an article by Al-Khalili [6] in 2013 introduces the beginning of the pivotal development of theory by Szilard in 1929 [7]. This neatly links changes in 'order', and thereby in entropy, to transfers of information. The second law is now expressed as a constraint on entropy: it must always increase – or, in the limit, be constant. Drawing on [6] 'imagine a closed box partitioned into two halves by a thick insulating wall. In the middle of this partition is a trap door that opens and closes very quickly … in 1867, James Clerk Maxwell described his famous thought experiment (Fig. 5) in which a demon sits on the box and can see all the individual molecules of air inside it and knows how fast they are moving … we let the demon control when it opens … It only allows faster moving molecules through from the left chamber to the right and only slow moving molecules from right to left … Using the demon's knowledge alone we seem to have created a temperature difference between the two halves [of the now more ordered system – with corresponding decrease in entropy], in violation of the second law of thermodynamics … In 1929, Leo Szilard proposed a version of Maxwell's demon that has since become known as Szilard's engine … it was the demon's intelligence and knowledge about the state of the molecules that made all the difference … [acquisition of such knowledge of random motion of molecules would be accompanied by an increase in entropy within the demon] … Szilard's resolution was a brilliant affirmation of the universality of the second law and the notion of increasing entropy … To an entropy tally of zero, any resulting from heat dissipation in the process could only make the tally positive … Thus, the total entropy of the box plus demon always goes up. The Second Law is saved.'

In 2000, in his book 'Investigations' [8], Kauffman addressed, among others, the question 'What is life?' In the Introduction, in seeking criteria to move forward, he chose to look back to the faculties of Maxwell's demon. He speculated on how the demon, on its own, might cope with a non-equilibrium system 'in one of the major places in physics where matter, energy, work, and information come together.' [By way of illustration here, a non-equilibrium system is exemplified by the hot and cold halves, produced by Maxell's demon, of a gas sample originally in equilibrium at a uniform temperature throughout.] William Thomson (Lord Kelvin) and his brother James and several others had also speculated on the science beyond that of inanimate matter in the few years before the arrival of Maxwell's demon.

More than a century later, Kauffman had the benefit of the much higher level of sophistication across the sciences, *particularly now including information*. Quoting from the main text directly, 'Maxwell's demon [is] flummoxed (Fig. 6). The demon is considering the gas system in which he finds himself and is trying to decide what feature of the system to measure in order to detect if the system is displaced from equilibrium in some way such that work can be [subsequently] extracted. Unfortunately, he decides to measure the positions of all gas particles at a single instant, a strenuous measurement from which he cannot detect a displacement from equilibrium from which work can, in principle, be extracted [in preference to other processes which preclude it] . How does the demon know what to measure?

Beginning with some of Maxwell's contemporaries, many have speculated on the possibility of a very close control of molecules in inanimate systems. Central to an article by Ball [9], also in 2013, is a discussion of how such speculation led to interest in a range of (in hindsight) peripheral concerns such as invisible and spiritual domains. Moving on from there, however, is his undistracted telling of the stepwise approach to the close control of matter at the microscopic

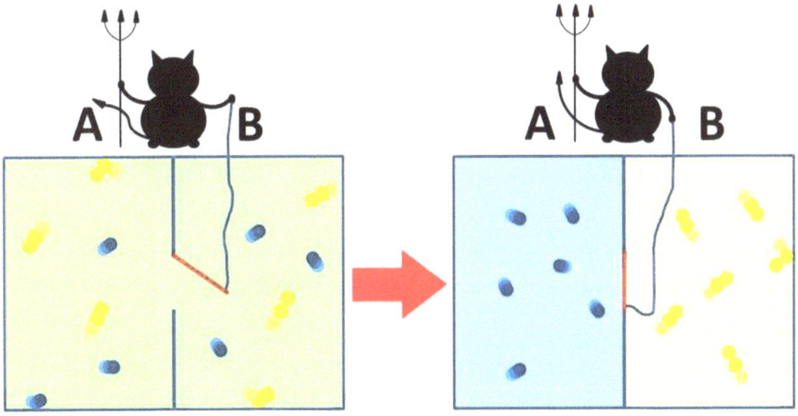

Figure 5: The basic system: sorting molecules according to speed.

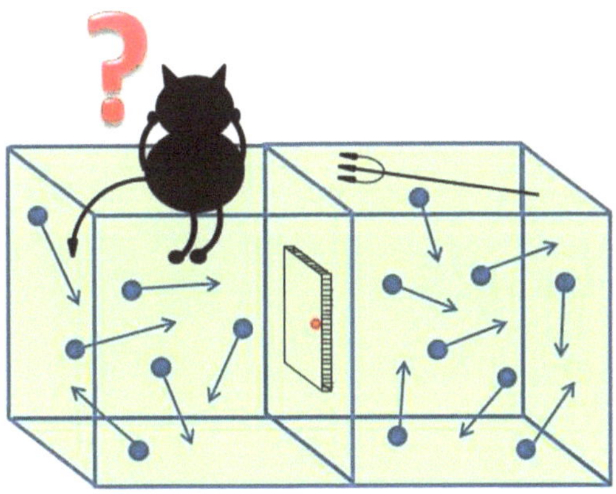
Figure 6: The demon struggles [8].

level. Again the concept of Maxwell's demon is fundamental – but at risk of being usurped. Ball's illustration, as retold here, was chosen from recent studies establishing that Maxwell's thought experiment is now accessible to direct experimental probing, and that such efforts are at the forefront of information science and technology. A microscopic (typically silica) bead, in a suitable environment, has a thermal motion during which it experiences tiny random changes in energy. A structured electromagnetic field can match the energy fluctuations with optical traps into which the bead can be coaxed. Quoting selectively from Ball's particular example [10] '… a group of physicists led by Shoichi Toyabe of Chuo University in Tokyo … put a nanoscale polystyrene bead in a system of traps [with energies represented symbolically by] a spiral staircase of sorts, on which the bead could hop up or down one step at a time between traps, using thermal energy. Left to its own devices, the ball would, on average, move down the staircase (left side of Fig. 7) … if a demon knew the position of the ball, it could place [an electromagnetic] barrier to prevent any downhill motion, so that (right side of Fig. 7) the ball only moves uphill [to a trap of ever higher potential energy]. In the experiment, the physicists [controlled

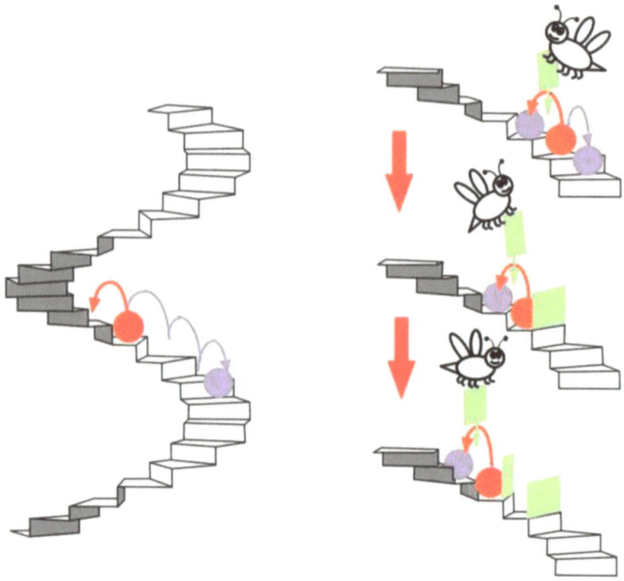

Figure 7: A replacement demon's role [9,10].

the barrier placing and so] took on the role of the demon … This demonstrated experimentally that – using no energy as such – information [in the form of the position of the bead] can be [sourced and hence] converted into energy.'

References

[1] Smith, C. & Wise, M.N., *Energy & Empire*, Cambridge University Press: Cambridge, 1989.

[2] Hammerton, Sir J.A., (ed). *Mr Punch's Railway Book*, The Educational Book Co. Ltd.: London, 1910.

[3] OPEN Magazine. The New Cricket Manual. http://www.openthemagazine.com/article/sports/the-new-cricket-manual

[4] Smith, C., *The Science of Energy*, The Athlone Press: London, 1998.

[5] Hunt, B.J., *Pursuing Power and Light*, Johns Hopkins University Press: Baltimore, 2010.

[6] Al-Khalili, J., Maxwell's Demon, *Newsletter of the James Clerk Maxwell Foundation*, Issue no. 3, Spring 2013.

[7] Szilard, L., On the decrease in entropy in a thermodynamic system by the intervention of intelligent beings. *Zeitschrift für Physik*, **53**, pp. 840–856, 1929.

[8] Kauffman, S., *Investigations*, Oxford University Press: Oxford, 2000.

[9] Ball, P., A demon-haunted theory. *Physics World*, **26**, pp. 36–39, 2013.

[10] Toyabe, S., Sagawa, T., Ueda, M., Muneyuki, E., & Sano, M., Experimental demonstration of information-to-energy conversion and validation of Jarzynski equality. *Nature Physics*, **6**, pp. 988–992, 2010.

Appendix D

Universities of the Heat Engine

C.S. König[1], M.W. Collins[1] & I.S. Ruddock[2]
[1]*College of Engineering, Design and Physical Sciences, Brunel University London, United Kingdom.*
[2]*Department of Physics, University of Strathclyde, United Kingdom.*

The higher education institutions that are strongly associated with the development of the heat engine were hotspots for thermodynamic research and were the places of study and work of the eminent scientists and engineers surrounding Kelvin's work and that have found special address in this book. In appreciation of their advancement of thermodynamic science, these institutions frequently featured on the reverse of banknotes as seen in Chapter 16. In this appendix section, we want to provide the reader with a visual impression of the universities related historically to the development of the heat engine as they were often architectural marvels of their time and indeed, the buildings still standing today, remain silent witnesses to extraordinary progress and enhancement of engineering science.

James Watt (1736–1819)

James Watt's association with the University of Glasgow, or Glasgow College as it was then known, is not as a scholar per se but as a mathematical instrument maker. In this role, he became interested in the technology of steam engines. He realised that contemporary engine designs wasted a great deal of energy by repeatedly cooling and re-heating the cylinder. As a result Watt introduced a design enhancement, the separate condenser, which avoided this waste of energy and radically improved the power, efficiency and cost-effectiveness of steam engines. Watt was initially schooled at home by his mother before attending an elementary school and then a grammar school. After a couple of years assisting in his father's carpentry shop, he spent a year as an apprentice instrument maker in London before returning to the west of Scotland and his first paid employment in the University. He had a room overlooking one of the quadrangles, Fig. 1 [1].

Nicolas Léonard Sadi Carnot (1796–1832)

The French military engineer and physicist Sadi Carnot entered the École Polytechnique in Paris as cadet at the age of 16. His classmates included Gaspard-Gustave Coriolis. Being principally an institution to train engineers for the military service, the school was renowned for its teaching of mathematics. The École Polytechnique was founded in 1794 and has a reputation of being the most selective and prestigious of the French Grandes Écoles. Whilst studying there Carnot befriended the scientist Nicolas Clément leading him to attend lectures on physics and chemistry. He subsequently became interested in understanding the limitation to improving the performance of steam engines. His investigations led to the publication of his *Reflections on the Motive Power of Fire* in 1824 [2,3].

Benoît Paul Émile Clapeyron (1799–1864)

Clapeyron was born in Paris and like Carnot studied there at the École Polytechnique (see Fig. 2), but also at the École des Mines (see Fig. 3), before leaving for Saint Petersburg in 1820 to teach at the *École des Travaux Publics* where he taught both pure and applied mathematics [5]. In 1844, Clapeyron was appointed to a chair at the École des Ponts et Chaussées and in 1848, he

Figure 1: Glasgow College in the seventeenth century, looking eastwards from the High Street frontage. The University moved from this east end location to its present site in the west of the city in 1870. (Illustration from J. Slezer, *Theatrum Scotiae*, 1693. Reproduced with the permission of the National Library of Scotland.)

Figure 2: The École Polytechnique in Paris around 1900. (Photograph from [4].)

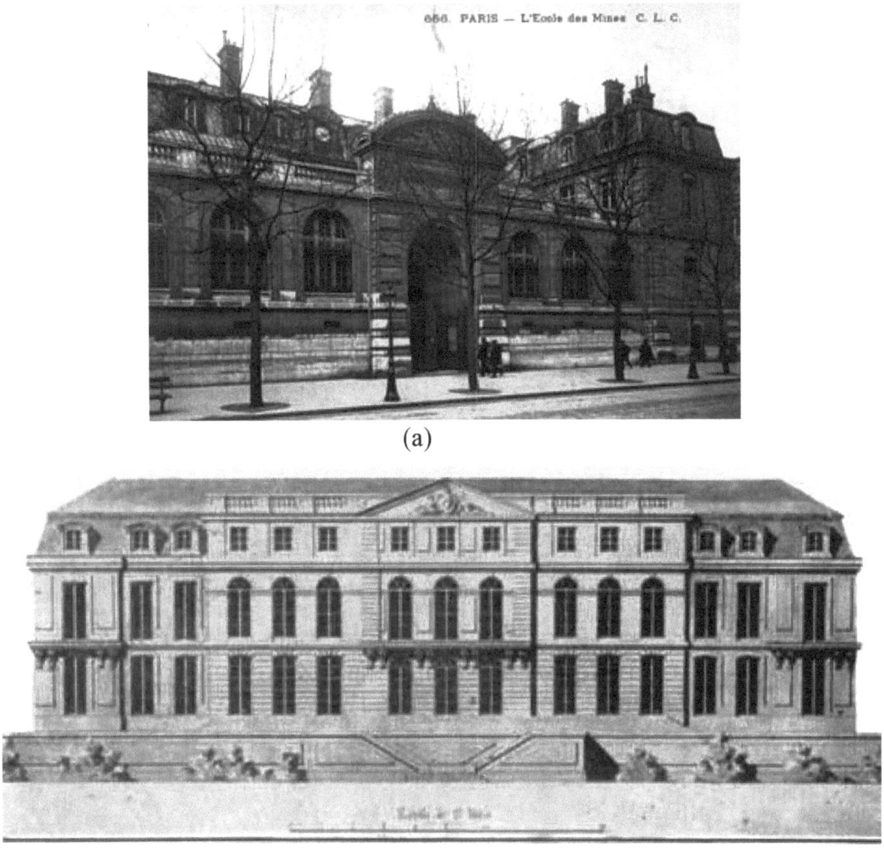

(a)

(b)

Figure 3: (a) Main entrance and (b) the Hôtel de Vendôme, central building of École des Mines. (Photograph and illustration from [7] and [8], respectively.)

was elected to the Paris Academy of Sciences. He served the Academy on many committees, in particular serving on the committee which awarded the mechanics prize. He also served on a committee investigating the construction of the Suez Canal and on a committee which considered how steam engines could be used in the navy [6].

Rudolf Clausius (1822–1888)

Clausius, who was born in Pomerania, Prussia, graduated from the University of Berlin (Humboldt University, see Fig. 4) in 1844. He was reading mathematics and physics with, among others, Peter Gustav Lejeune Dirichlet. There he also read history with Leopold von Ranke. In 1847, he obtained his doctorate from the University of Halle (see Fig. 5) on optical effects in the Earth's atmosphere. Subsequently, he became a professor of physics at the Royal Artillery and Engineering School in Berlin and held the position of 'Privatdozent' at the University of Berlin. In 1855, he became a professor at the ETH Zürich, the Swiss Federal Institute of Technology in Zürich, where he stayed until 1867. In that year he moved to Würzburg as a professor of physics, and 2 years later he was appointed to the same chair at Bonn, where he died on the 24 August 1888 [9–11].

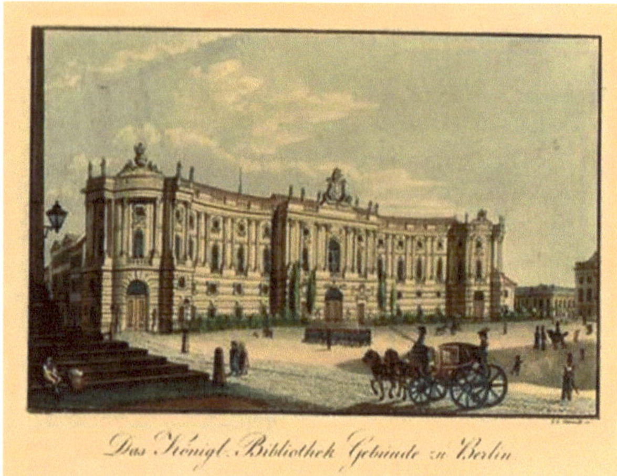

Figure 4: The Royal Library of the Humboldt University of Berlin, 1832. (Illustration from [12].)

Figure 5: Historic view of the Lion Building (Löwengebäude, 1832–1834) of the Martin Luther University in Halle-Wittenberg. (Image from [13].)

Josiah Willard Gibbs (1839–1903)

Gibbs was born on 11 February 1839, in New Haven, CT. He graduated from Yale College (see Fig. 6) in 1858. Subsequently, he continued his studies at Yale, receiving his doctorate of philosophy in 1863 which was the first American doctorate to be conferred in engineering. He was them appointed as tutor in the college, where he taught Latin for 2 years and natural philosophy for a third year. In 1871, he was appointed as a professor of mathematical physics in Yale College and he served in that capacity for the rest of his life [16–18]. Yale's Sloane Physical Laboratory is depicted in Fig. 7.

Figure 6: Yale University, the Old Brick Row at 1807. (Illustration from [14].)

Figure 7: Yale's Sloane Physical Laboratory, as it stood between 1882 and 1931. According to [15] Gibbs's office was on the second floor, to the right of the tower in the picture. (Photograph from [15].)

References

[1] Dickinson, H.W., *James Watt: Craftsman and Engineer*, Cambridge University Press: Cambridge, 1936.

[2] Carnot, S. & Thurston, R.H. (editor and translator), *Reflections on the Motive Power of Heat*, John Wiley & Sons: New York, 1890.

[3] Sadi Nicolas Léonard Carnot: http://www-groups.dcs.st-and.ac.uk/history/Printonly/Carnot_Sadi.html
[4] L'ancienne Ecole Polytechnique: http://paris1900.lartnouveau.com/paris05/ecoles/ecole_polytechnique.htm
[5] Benoit Paul Émile Clapeyron: http://www-groups.dcs.st-and.ac.uk/history/Printonly/Clapeyron.html
[6] Benoit Paul Émile Clapeyron: http://www-history.mcs.st-andrews.ac.uk/Biographies/Clapeyron.html
[7] Cartes postales 97 de Paris 6e Arrondissement: http://www.communes.com/ile-de-france/paris/paris-6e-arrondissement_75006/cartes-postales-anciennes, 97.html
[8] Histoire de l'Ecole des mines: http://www.annales.org/archives/x/ecole.html
[9] Cardwell, D.S.L, *From Watt to Clausius: The Rise of Thermodynamics in the Early Industrial Age*, Heinemann: London, 1971.
[10] Rudolf Julius Emmanuel Clausius: http://www-groups.dcs.st-and.ac.uk/history/Printonly/Clausius.html
[11] Rudolf Clausius: http://www.nndb.com/people/951/000100651/
[12] Humboldt Universität zu Berlin, Architectural history of the University Library: https://bauten.hu-berlin.de/en/grimm-en/baugeschichte_der_universitaetsbibliothek-en
[13] Martin Luther University of Halle-Wittenberg: http://en.wikipedia.org/wiki/Martin_Luther_University_of_Halle-Wittenberg
[14] Yale University: http://en.wikipedia.org/wiki/Yale_University
[15] Josiah Willard Gibbs: http://en.wikipedia.org/wiki/Josiah_Willard_Gibbs
[16] L.P. Wheeler, *Josiah Willard Gibbs, The History of a Great Mind*, Ox Bow Press: Woodbridge, CT, 1998 [1951].
[17] J. Willard Gibbs, American Physical Society Sites: http://www.aps.org/programs/outreach/history/historicsites/gibbs.cfm
[18] Josiah Willard Gibbs: http://www-history.mcs.st-and.ac.uk/Biographies/Gibbs.html

Appendix E

Honoured by monuments

I.S. Ruddock
Department of Physics, University of Strathclyde, United Kingdom.

Lord Kelvin will remain a significant historical figure on account of his distinguished career and achievements, but many physical monuments also honour his life, mainly in Belfast and Glasgow.

The first major one to be erected was the statue in Belfast's Botanic Gardens, dedicated in 1913 by fellow Ulster scientist, Sir Joseph Larmor MP [1], Figs 1 and 2 (left). Larmor was a most appropriate choice for guest of honour having also attended the Belfast Academical Institution in the mid-nineteenth century and been a strong opponent of Home Rule. This statue, near the rear of Queen's University's Department of Pure and Applied Physics, shows Kelvin standing beside a compass binnacle and holding a tablet, possibly representing his 'green book'. The epitaph on the rear of the plinth could serve as a one sentence biography – 'pre-eminent in elucidating the laws of nature and applying them to the service of man'.

Figure 1: Sir Joseph Larmor unveiling the statue of Lord Kelvin in Belfast's Botanic Gardens in 1913. (Photograph by A.R. Hogg.)

The Glasgow statue in Kelvingrove Park overlooked by Glasgow University, Fig. 2 (right), was also unveiled in 1913. This time, Kelvin is shown seated but again with a compass binnacle as part of the sculpture.

He was also one of the original 29 figures honoured by Glasgow University's Memorial Gates erected in 1952 to mark the University's first 500 years, Fig. 3. Interestingly as far as this volume is concerned, James Watt's name is immediately below his.

The site of Kelvin's childhood home in College Square East, Belfast is marked by a plaque placed there by the Ulster History Circle in 1996, Fig. 4. Coincidentally, a similar plaque is on the site of the neighbouring house to commemorate Robert Shipboy McAdam (1808–1895), a Presbyterian Irish scholar, founder of the Ulster Gaelic Society, inventor and owner of a Belfast foundry; James Thomson owned this house and rented it to the McAdam family.

Figure 2: Left: the statue of Lord Kelvin in the Botanic Gardens, Belfast. Right: the statue of Lord Kelvin in Kelvingrove Park, Glasgow. (Photographs by I.S. Ruddock.)

Figure 3: The University of Glasgow's Memorial Gates with 'Kelvin' included in the upper right panel of the left hand gate. (Photographs by I.S. Ruddock.)

Belfast is also home to the most unusual monument. Each of the 14 columns in St. Anne's (Church of Ireland) Cathedral has a theme – industry, agriculture, music and so on. Kelvin is included in the Column of Science's capital along with Archimedes, Bacon and Newton, Fig. 5.

The December 2007 centenary of Lord Kelvin's death was marked by events in Belfast, Glasgow and London. In the case of Glasgow, it was an opportunity for the dedication of a new monument sponsored by the Royal Philosophical Society of Glasgow, of which Kelvin was President in 1856–1858 and 1874–1877, and an Honorary Member from 1896. The Society was granted permission by Glasgow City Council to restore the Thomson family headstone and to add a stone to the burial plot in the Necropolis behind Glasgow Cathedral in the form of a gyrostat, Fig. 6.

Figure 4: The Ulster History Circle's plaque on the modern building that now occupies the site of the Thomson home at No. 17 College Square East, Belfast. (Photograph by I.S. Ruddock.)

Figure 5: The carving of Lord Kelvin in the capital of the Column of Science, St Anne's Cathedral, Belfast. (Photograph by I.S. Ruddock.)

Figure 6: The Thomson family burial plot in the Necropolis behind Glasgow Cathedral. The stone added by the Royal Philosophical Society of Glasgow on 17 December 2007 is on the left. The inscription reads 'This commemorative stone was erected by admirers from industry and science in 2007 to mark the centenary of the death of William Thomson, Lord Kelvin. The renovation was initiated by the Royal Philosophical Society of Glasgow, of which he was twice President; his father and brother, James, here buried, were also members'. (Photograph by I.S. Ruddock.)

References

[1] Hidden, A.E. & Latimer, C.J., *Science and Technology: Belfast and Its Region*, Queen's University of Belfast: Belfast, 1987.

Appendix F

Lord Kelvin and the Royal Society of Edinburgh

C.A. Greated
School of Physics and Astronomy, The University of Edinburgh, United Kingdom.

William Thomson was elected as a Fellow of the Royal Society of Edinburgh (Fig. 1) in 1847 and became Vice-President in 1868; a post which he held for 5 years. He is unique in the annals of the Society by going on to serve three separate terms as President, the first from 1873 to 1878, the second from 1886 to 1890 and the third from 1895 to 1907. To mark his momentous contribution to the Society, one of the prestige meeting rooms at their premises in George Street Edinburgh has been named the Kelvin Room, which is shown in Fig. 2. This houses both a bust of Kelvin and one of his Depth Sounding Machines, see Fig. 3.

Many of Kelvin's papers containing his most brilliant investigations and discoveries were published in the Society's Transactions and Proceedings. His ground-breaking series of articles on thermodynamics were published during the period 1851–1854, including his classic article on the reconciliation of the Joule's and Carnot's principles (*Transactions* 1851, 20, 261–289). Sir Alfred Ewing said of him 'No other contributor has done so much to give our publications a worldwide and lasting fame' (Neil Campbell and R Martin Smellie, *The Royal Society of Edinburgh 1783–1983*, Latimer Trend and Co: Plymouth, 1983).

In 2012, the society instigated the Lord Kelvin prize and medal to be awarded for to senior scientists who had made an outstanding contribution in the field of either physical, engineering or informatics sciences. The prize winners to date are:

Figure 1: The Royal Society of Edinburgh building.

Figure 2: The Kelvin Room at the Royal Society of Edinburgh.

Figure 3: One of Kelvin's Depth Sounding Machines at the Royal Society of Edinburgh.

2012:
Professor Colin McInnes FRSE, Professor of Engineering Science, University of Strathclyde, for his outstanding contribution to space systems engineering.

2013:
RSE/Lord Kelvin Prize (Senior Prize) to Professor Peter Davies FRSE, Professor of Fluid Dynamics, University of Dundee, for his outstanding contribution to the discipline of Fluid Mechanics, particularly Environmental Fluid Mechanics.

2014:
Professor Miles Padgett FRSE, Kelvin Chair of Natural Philosophy and Dean for Research, University of Glasgow, for his outstanding contribution to the field of optics through his pioneering work on orbital angular momentum, his promotion of a global community of researchers in this field and his commitment to the public engagement of science.

www.ingramcontent.com/pod-product-compliance
Lightning Source LLC
LaVergne TN
LVHW071522220926
848486LV00040B/305
* 9 7 8 1 8 4 5 6 4 1 4 9 8 *